NUCLEAR PHYSICS AND INTERACTION OF PARTICLES WITH MATTER

YADERNAYA FIZIKA I VZAIMODEISTVIE CHASTITS S VESHCHESTVOM

ЯДЕРНАЯ ФИЗИКА И ВЗАИМОДЕЙСТВИЕ ЧАСТИЦ С ВЕЩЕСТВОМ

The Lebedev Physics Institute Series

Editor: Academician D. V. Skobel'tsyn

Director, P. N. Lebedev Physics Institute, Academy of Sciences of the USSR

Proceedings (Trudy) of the P. N. Lebedev Physics Institute

Volume 44

NUCLEAR PHYSICS AND INTERACTION OF PARTICLES WITH MATTER

Edited by
Academician D. V. Skobel'tsyn
Director, P. N. Lebedev Physics Institute
Academy of Sciences of the USSR, Moscow

Translated from Russian

SPRINGER SCIENCE+BUSINESS MEDIA, LLC

1971

The Russian text was published by Nauka Press in Moscow in 1969 for the
Academy of Sciences of the USSR as Volume 44 of the Proceedings (Trudy)
of the P. N. Lebedev Physics Institute.   The present translation is published under an agreement with Mezhdunarodnaya Kniga, the Soviet book ex-
port agency.

Library of Congress Catalog Card Number 70-120025
ISBN 978-1-4757-6034-7   ISBN 978-1-4757-6032-3 (eBook)
DOI 10.1007/978-1-4757-6032-3

© 1971 Springer Science+Business Media New York
Originally published by Plenum Publishing Corporation, New York in 1971

CONTENTS

ANALYSIS OF THE RESONANCE REACTIONS
$H_3(d,n)He_4$ AND $He_3(d,p)He_4$ IN THE EFFECTIVE
INTERACTION RADIUS APPROXIMATION

I. Ya. Barit and V. A. Sergeev

Introduction

Rather comprehensive experimental information on the reactions $H^3(d,n)He^4$ and $He^3(d,p)$ · He^4 has become available in the low-energy region in which resonances are observed at $E_d =$ 0.064 MeV* in the (d,n) reaction and at $E_d = 0.255$ MeV in the (d,p) reaction. It could be established that the resonances result from the channel with $J^\pi = 3/2^+$. The energy dependence of the cross sections of these reactions near the resonances (0.006 MeV $< E_d <$ 0.250 MeV for the (d,n) reaction and 0.024 MeV $< E_d <$ 0.350 MeV for the (d,p) reaction) has been considered in numerous papers (see, e.g., [1] and [2]). The well-known formula for a single level in the Wigner-Aizenbad theory [3] was employed in those papers:

$$\frac{d\sigma_{dN}}{d\Omega} = q\,k_d^{-2}\,\frac{P_d\gamma_d^2 P_N\gamma_N^2}{(E_\lambda - S_N\gamma_N^2 - S_d\gamma_d^2 - E_d)^2 + (P_d\gamma_d^2 + P_N\gamma_N^2)^2}. \tag{1}$$

The notations are interpreted as follows: $q = (2J + 1)/(2i_1 + 1)(2i_2 + 1) = 2/3$ is a statistical factor; $P_d = P_0^+(k_d a, \eta)$, $S_d = S_0^+(k_d a, \eta)$, $P_N = P_2^+(k_N a, 0)$, $S_N = S_2^+(k_N a, 0)$; and N denotes the neutron or proton channel.

It is a characteristic feature of the research performed so far that the energy dependence of the resonance cross sections can be satisfactorily described by greatly different sets of resonance parameters a, E_λ, γ_d^2, γ_N^2. Even when one simultaneously analyzes data on the above reactions and data on the elastic resonance scattering of deuterons in this energy range [2], the indeterminacy in the parameter selection cannot be completely removed. On the other hand, it is a feature common to all sets of parameters that the reduced deuteron width is large: $\gamma_d^2 \gg$ $|E_\lambda'|$, $\gamma_d^2 \gg \Gamma_N/2$,† where $\gamma_d^2 \gtrsim \hbar^2/\mu_d a^2$ ($\hbar^2/\mu_d a^2$ denotes the characteristic reduced single-particle width). The quantity $E_\lambda' = E_\lambda - \gamma_N^2 S_N$ and $\Gamma_N/2 = \gamma_N^2 P_N$ can be considered constant because P_N and S_N are practically independent of the energy: $k_N = \sqrt{(Q_{dN} + E_d)2\mu_N/\hbar^2}$; the energy liberated in the reactions ($Q_{dn} = 17.58$ MeV and $Q_{dp} = 18.34$ MeV) exceed considerably the energy interval under inspection ($\Delta E_d \lesssim 0.35$ MeV).

---

*Throughout our discussion we use the center of mass system.

†More precisely, $\gamma_d^2/\bar\psi_0\varphi_0 \gg E_\lambda' - \gamma_N^2\bar\psi_0^*/\bar\psi_0$ and $\gamma_d^2/\bar\psi_0\varphi_0 \gg \Gamma_N/2$, where $\bar\psi_0\bar\psi_0^*$, φ_0 denotes the coefficients of the expansion in Coulomb functions [see below, Eqs. (14)].

However, we have for the minimum wavelength of the deuteron $\lambda_{min} = \hbar/\sqrt{2\mu_d \Delta E_d} > \bar{a}$, where $\bar{a}$ denotes the approximate interaction radius assumed to be the sum of the radii of the colliding nuclei.

As has been shown in [4], in such a situation it is convenient to use the effective interaction radius approximation for the analysis of resonances. Three parameters with a simple physical meaning are used in this approximation and the analytical forms of the energy dependence of both reaction cross section and scattering cross section are greatly simplified since the parameters in the effective radius theory are more directly related to the scattering amplitudes and the reaction cross sections than the resonance parameters of the R-matrix theory.

The effective radius approximation has been successfully employed in several cases in which resonance effects were observed in a particular energy interval, e.g., in the analysis of nucleon/nucleon scattering [5], in p/H^3 scattering below the threshold of the (p,n) reaction [6], and in He^4/He^4 scattering [7].

The present article shows that the energy dependence of the cross sections of the reactions H^3(d,n)He^4 and He^3(d,p)He^4 in the resonance region can be described by the effective radius approximation using three parameters. The various types of undeterminedness which occur in the analysis of the reaction cross sections are discussed. The results are compared with an analysis of experimental data interpreted with the R-matrix theory. The question is discussed to what extent the single level approximation in the Humblet — Rosenfeld theory [8] resembles the two methods mentioned above for the description of resonance effects.

§1. Basic Relations of the Effective Radius Approximation

The effective radius approximation, which is limited to the low-energy region, is based on extremely general properties of the scattering amplitudes. This approximation is therefore more convenient than the single level approximation of the R-matrix theory in which definite assumptions regarding the energy dependence of the logarithmic derivative of the internal wave function at the boundary of the nucleus are made.

In the multichannel effective radius theory [9], one considers the energy dependence of a symmetric real matrix M which is related to the scattering matrix S and the amplitude f by the following matrix relations:

$$M = ik^{l+1/2}(S-1)^{-1}(S+1)\,k^{l+1/2}, \tag{2}$$

$$f = k^l (M - ik^{2l+1})^{-1} k^l. \tag{3}$$

The notations are interpreted as follows: k^l is a diagonal matrix with $(k^l)_{ab} = k_a \delta_{ab}$; the amplitude f_{ab} defines the partial cross section of an $a \to b$ process [5]:

$$\frac{d\sigma_{ab}}{d\Omega} = |f_{ab}|^2 \frac{k_b}{k_a}. \tag{4}$$

The energy dependence of the M-matrix elements has a simple form at low energies, and when certain conditions are satisfied, the energy dependence can be simply related to the interaction radius of the systems considered [5, 9, 10].

When only one channel exists, we have $S = \exp(2i\delta_l)$, $M = k^{2l+1}\mathrm{ctg}\,\delta_l$, and $f = (k\,\mathrm{ctg}\,\delta_l - ik)^{-1}$. In the following discussion we will consider the case of two coupled channels: d + H^3(d + He^3) and n + He^4(p + He^4), where $l_d = 0$, $I_d = 3/2$, $l_N = 2$, $I_N = 1/2$, and $J = 3/2$ corresponding to the momentum and parity of the level involved. It follows from the matrix equation (3) that the partial amplitudes of the deuteron scattering $f_{dd} = (d, l_d, I_d | f^J | d, l_d, I_d)$ and of the (d, N) reaction, $f_{dN} = (d, l_d, I_d | f^J | N, l_N, I_N)$, can be represented in the form

$$f_{dd} = [g^c(k_d) - ik_d]^{-1}, \tag{5}$$

$$f_{dN} = -k_N^{-1/2} \exp\left[i\,\mathrm{arctg}\,(k_N^5/M_{NN}^c)\right]|\mathrm{Im}\,g^c(k_d)|^{1/2}[g^c(k_d)-ik_d]^{-1}, \tag{6}$$

wherein the superscript c indicates that it is necessary to take into acccount the long-range Coulomb interaction in the deuteron channel; we have

$$\mathrm{Re}\,g^c(k_d) = M_{dd}^c(k_d) - (M_{dN}^c(k_d))^2\,[(M_{NN}^c(k_d))^2 + k_N^{10}]^{-1}\,M_{NN}^c, \tag{7}$$

$$\mathrm{Im}\,g^c(k_d) = -k_N^5\,(M_{dN}^c(k_d))^2\,[(M_{NN}^c(k_d))^2 + k_N^{10}]^{-1}. \tag{7'}$$

We obtain from Eqs. (5), (6), and (4) the portions of the scattering cross section and reaction cross section which result from the initial particles in the quadruplet spin state ($I_d = 3/2$):

$$\frac{d\sigma_{dd}}{d\Omega} = q\left| F_c(k_d,\theta) + \frac{1}{\mathrm{Re}\,g^c(k_d) - i\,[k_d + |\mathrm{Im}\,g^c(k_d)|]} \right|^2, \tag{8}$$

$$\frac{d\sigma_{dN}}{d\Omega} = q\,\frac{1}{k_{il}}\,\frac{|\mathrm{Im}\,g^c(k_d)|}{[\mathrm{Re}\,g^c(k_d)]^2 + [k_d + |\mathrm{Im}\,g^c(\kappa_d)|]^2}, \tag{9}$$

where

$$F_c(k_d,\theta) = -\eta_d\,(2k_d\sin^2\theta/2)^{-1}\exp\left(-i\eta_d\ln\sin^2\theta/2\right)$$

denotes the amplitude of the pure Coulomb scattering; it was taken into account that always $\mathrm{Im}\,g^c(k_d) < 0$.

If no Coulomb interaction existed in the deuteron channel, the elements of the matrix M and the complex function g could be expanded in a series of k_d^2. In the effective radius approximation, we restrict ourselves to two terms of the expansion; in analogy to the single-channel case, the two terms are written in the form

$$g(k_d^2) = -(A_R + iA_J)^{-1} + (1/2)(r_R + ir_J)\,k_d^2. \tag{10}$$

The quantity $-(A_R + iA_J)$ corresponds to the scattering amplitude at $k_d = 0$. In the isolated level approximation of the R-matrix theory, $g(k_d^2)$ can be expressed through the resonance parameters:

$$\mathrm{Re}\,g(k_d^2) = -k_d\,\mathrm{ctg}\,k_d a\,[(E_\lambda' - E_d + \gamma_d^2 k_d a\,\mathrm{tg}\,k_d a)(E_\lambda' - E_d -$$
$$- \gamma_d^2 k_d a\,\mathrm{ctg}\,k_d a) + (\Gamma_N/2)^2]\,[(E_\lambda' - E_d - \gamma_d^2 k_d a\,\mathrm{ctg}\,k_d a)^2 + (\Gamma_N/2)^2]^{-1}, \tag{11}$$

$$\mathrm{Im}\,g(k_d^2) = -ak_d^3\,(\sin k_d a)^{-2}\,\gamma_d^2\,(\Gamma_N/2)\,[(E_\lambda' - E_d - \gamma_d^2 k_d a\,\mathrm{ctg}\,k_d a)^2 + (\Gamma_N/2)^2]^{-1}. \tag{11'}$$

When the interaction in the deuteron channel is close to a single-particle interaction ($\gamma_d^2 \gtrsim \hbar^2/\mu_d a^2$), the nucleon width is small ($\Gamma_N/2 \ll \gamma_d^2$) and the resonance occurs rather close to $E_d = 0$ ($E_\lambda' \ll \gamma_d^2$), then the quantity $r_R = 2\frac{d}{dk_d^2}\mathrm{Re}\,g(k_d^2)|_{k_d=0}$ has the meaning of an interaction radius ($r_R \sim a$), whereas $r_J = 2\frac{d}{dk_d^2}\mathrm{Im}\,g(k_d^2)|_{k_d=0} \ll a$ [4]. In view of the fact that the resonance parameters obtained from experimental data satisfy these conditions more or less, we will assume $r_J \approx 0$.

The Coulomb interaction in the deuteron channel can be included by matching at a certain distance $a \gtrsim \bar{a}$ a wave function of the form

$$\Psi_d = \begin{cases} M_{dd} \dfrac{\sin k_d r_d}{k_d r_d} + k_d \dfrac{\cos k_d r_d}{k_d r_d}, \\[2mm] \sqrt{\mu_N/\mu_d}\, M_{dN}\, k_N^{-2} j_2(k_N r_N) \end{cases}$$

with a wave function which has the asymptotic form required in our case:

$$\Psi_d^c = B \begin{cases} M_{dd}^c \dfrac{F(k_d r_d,\, \eta_d)}{k_d r_d} + k_d \dfrac{G(k_d r_d,\, \eta_d)}{k_d r_d}, \\[2mm] \sqrt{\dfrac{\mu_N}{\mu_d}}\, M_{dN}^c\, k_N^{-2} j_2(k_N r_N). \end{cases}$$

Thus we obtain*

$$M_{dd}^c = -k_d B^{-1}[M_{dd} k_d^{-1}(G' \sin k_d a - G \cos k_d a) + (G \sin k_d a + G' \cos k_d a)], \tag{12}$$

$$M_{dN}^c = B^{-1} M_{dN}, \tag{12'}$$

$$B = M_{dd} k_d^{-1}(F' \sin k_d a - F \cos k_d a) + (F \sin k_d a + F' \cos k_d a).$$

Similarly, we obtain an expression for M_{NN}^c:

$$M_{NN}^c = M_{NN} - k_d^{-1} B^{-1} M_{dN}^2 (F' \sin k_d a - F \cos k_d a). \tag{12''}$$

The quantities M_{dd}, M_{dN}, and M_{NN}, have the meaning of M-matrix elements which would characterize a particular process if for $r_d > a$ no Coulomb interaction existed. F, G, F', and G' denote the usual Coulomb functions with $l = 0$ and their derivatives with respect to $k_d r_d$ at $r_d = a$. In the following discussion we will omit the subscripts d in k_d, E_d, η_d, and μ_d. By substituting the expressions of Eqs. (12), (12'), and (12'') in Eqs. (7) and (7'), we obtain a relation between $g^c(k)$ and $g(k^2)$:

$$\begin{aligned} \mathrm{Re}\, g^c(k) = -kD^{-1}\{[(F\,\mathrm{tg}\,ka + F') + (F'\,\mathrm{tg}\,ka - F)k^{-1}\mathrm{Re}\,g(k^2)] \times \\ \times [(G\,\mathrm{tg}\,ka + G') + (G'\,\mathrm{tg}\,ka - G)k^{-1}\mathrm{Re}\,g(k^2)] + \\ + (F'\,\mathrm{tg}\,ka - F)(G'\,\mathrm{tg}\,ka - G)(k^{-1}\mathrm{Im}\,g(k^2))^2\}, \end{aligned} \tag{13}$$

$$\mathrm{Im}\, g^c(k) = (\cos ka\, D)^{-1} \mathrm{Im}\, g(k^2),$$

$$\begin{aligned} D = [(F\,\mathrm{tg}\,ka + F') + (F'\,\mathrm{tg}\,ka - F)k^{-1}\mathrm{Re}\,g(k^2)]^2 + \\ + (F'\,\mathrm{tg}\,ka - F)^2 (k^{-1}\mathrm{Im}\,g(k^2))^2. \end{aligned} \tag{13'}$$

It is generally accepted that the Coulomb functions and their derivatives can be written in the form

$$F = C\, ka\, \Phi, \quad F' = C\Phi^*,$$
$$G = C^{-1}[\overline{\Psi} + (2a/a_c) h(\eta^{-1})\Phi],$$
$$kaG' = C^{-1}[\overline{\Psi}^* + (2a/a_c) h(\eta^{-1})\Phi^*],$$

where $C^2 = 2\pi\eta(e^{2\pi\eta} - 1)^{-1}$,

$$h(\eta^{-1}) = \sum_{n=1}^{\infty} \frac{1}{n(n^2\eta^{-2} + 1)} - 0.5772 - \ln\eta, \quad a_c = \hbar^2/\mu Z e^2,$$

and power series in k^2 (or η^{-2}) are available for the functions Φ, Φ^*, $\overline{\Psi}$, and $\overline{\Psi}^*$ [11]. We write for the first two coefficients the expansions of these functions in a series in powers of η^{-2} and use the notation $x = \sqrt{8a/a_c}$:

---

*In similar expressions of [9], several errors were made.

$$\varphi_0 = (2/x)\, I_1(x), \qquad \varphi_1 = (^1/_{24})\, (x/2)^3\, [I_3(x) - I_1(x)],$$
$$\varphi_0^* = I_0(x), \qquad \varphi_1^* = (^1/_{24})\, (x/2)^3\, [I_3(x) - 3I_1(x)],$$
$$\overline{\psi}_0 = -x K_1(x), \qquad \overline{\psi}_1 = -(^1/_{12})\, (x/2)^5\, [K_3(x) - K_1(x)] - (x/_{24})\, I_1(x), \tag{14}$$
$$\overline{\psi}_0^* = -(x^2/2)\, K_0(x), \; \overline{\psi}_1^* = -(^1/_{12})\, (x/2)^5\, [K_3(x) - 3K_1(x)] - (x^2/48) I_0(x).$$

The function $h(\eta^{-1})$ has been plotted in [5]. For $\eta > 1$ we have $h(\eta^{-1}) \approx 1/12\eta^2$, and for $\eta < 0.4$ we obtain $h(\eta^{-1}) \approx -0.5772 + \ln 1/\eta + 1.2\,\eta^2$. The following approximation can be recommended for the intermediate region:

$$h(\eta^{-1}) \approx -0.02 + 0.1193\,\eta^{-2} - 0.00577\,\eta^{-4}, \tag{15}$$

which results when $h(\eta^{-1})$ is expanded in a power series near $\eta^{-2} = 2$:

$$h(\eta^{-1}) \approx h(\sqrt{2}) + \frac{dh(\eta^{-1})}{d(\eta^{-2})}\Big|_{\eta^{-2}=2} (\eta^{-2} - 2) + \frac{1}{2}\frac{d^2h(\eta^{-1})}{d(\eta^{-2})^2}\Big|_{\eta^{-2}=2} (\eta^{-2} - 2)^2.$$

When we solve Eqs. (13) and (13') for the universal functions C^2 and $h(\eta^{-1})$, we obtain

$$C^2 \operatorname{Re} g^c(k) + 2a_c^{-1} h(\eta^{-1}) = \mathscr{K}(k^2), \tag{16}$$

$$C^2 \operatorname{Im} g^c(k) = \mathscr{L}(k^2). \tag{16'}$$

When we expand $\mathscr{K}(k^2)$ and $\mathscr{L}(k^2)$ in powers of k^2, we obtain an approximation for the effective (complex) radius of charged particles:

$$C^2 \operatorname{Re} g^c(k) + \frac{2}{a_c} h(\eta^{-1}) = \operatorname{Re}[-1/(A_R^c + iA_J^c)] + r_R^c k^2/2, \tag{17}$$

$$C^2 \operatorname{Im} g^c(k) = \operatorname{Im}[-1/(A_R^c + iA_J^c)] + r_J^c k^2/2. \tag{17'}$$

As far as the form is concerned, the first equation coincides with the well-known expression for $k \operatorname{ctg} \delta$ derived by Landau and Smorodinskii in the proton scattering problem at low energies [12].

When we take into account that in the case considered $\operatorname{Im} g(k^2) \approx$ const holds and that the energy dependence of the functions Φ and Φ^* [according to Eq. (14)] does not substantially contribute to $\mathscr{L}(k^2)$, we can make the approximation $r_J^c = 0$.

Each of the parameters A_R^c, A_J^c, r_R^c [or $\operatorname{Re}(-A_R^c - iA_J^c)^{-1}$, $\operatorname{Im}(-A_R^c - iA_J^c)^{-1}$, and r_R^c] is a definite function of the parameters A_R, A_J, and r_R [or $\operatorname{Re} g(0)$, $\operatorname{Im} g(0)$, and r_R] and of the radius a of matching. In particular, we have

$$\mathscr{K}(0) = \operatorname{Re}(-A_R^c - iA_J^c)^{-1} = -a^{-1} n_0/d_0, \tag{18}$$

$$\mathscr{L}(0) = \operatorname{Im}(-A_R^c - iA_J^c)^{-1} = \operatorname{Im} g(0)/d_0, \tag{18'}$$

$$n_0 = [\varphi_0^* + (\varphi_0^* - \varphi_0)\, a\, \operatorname{Re} g(0)]\, [\overline{\psi}_0^* + (\overline{\psi}_0^* - \overline{\psi}_0)\, a\, \operatorname{Re} g(0)] +$$
$$+ (\varphi_0^* - \varphi_0)(\overline{\psi}_0^* - \overline{\psi}_0)\, [a\, \operatorname{Im} g(0)]^2, \tag{18''}$$

$$d_0 = [\varphi_0^* + (\varphi_0^* - \varphi_0)\, a\, \operatorname{Re} g(0)]^2 + (\varphi_0^* - \varphi_0)^2\, [a\, \operatorname{Im} g(0)]^2. \tag{18'''}$$

Finally, when we insert Eqs. (17) and (17') with $r_J^c = 0$ into Eqs. (8) and (9), we obtain

$$\frac{d\sigma_{dd}}{d\Omega} = q\left| F_c(k,\theta) + \left[\frac{C^2}{\left[\operatorname{Re}(-A_R^c - iA_J^c)^{-1} + \frac{r_R^c k^2}{2} - \frac{2}{a_c} h(\eta^{-1})\right] - i\left[C^2 k + |\operatorname{Im}(-A_R^c - iA_J^c)^{-1}|\right]} \right] \right|^2, \tag{19}$$

$$\frac{d\sigma_{dN}}{d\Omega} = q\,\frac{C^2}{k}\,\frac{|\operatorname{Im}(-A_R^c - iA_J^c)^{-1}|}{\left[\operatorname{Re}(-A_R^c - iA_J^c)^{-1} + \frac{r_R^c k^2}{2} - \frac{2}{a_c}h\,(\eta^{-1})\right]^2 + \left[C^2 k + |\operatorname{Im}(-A_R^c - iA_J^c)^{-1}|\right]^2}. \tag{20}$$

Thus, in the effective radius approximation used below, the differential (d,d)-scattering cross section and the (d,N)-reaction cross section are given by the three parameters A_R^c, A_J^c, and r_R^c. The quantities A_R^c and A_J^c characterize the amplitudes of the nucleon scattering and the reaction cross section at $\eta \gg 1$, i.e., we have

$$f_{dd}\big|_{\eta\gg1} = -C^2(A_R^c + iA_J^c), \quad \frac{d\sigma_{dN}}{d\Omega}\bigg|_{\eta\gg1} = q\,\frac{C^2}{k}\,|A_J^c|. \tag{21}$$

The quantity r_R^c is a measure for the size of the region in which the nuclear forces act. When Coulomb interaction takes place, r_R^c is reduced relative to r_R; then r_R^c decreases with decreasing magnitude of the parameter $2\bar{a}/a_c$.

The energy dependence of the cross sections defined by Eqs. (19) and (20) remains unchanged when one assumes that the functions $\mathcal{K}(k^2)$ and $\mathcal{L}(k^2)$ of Eqs. (16) and (16') were expanded near $k^2 = k_m^2 \neq 0$. In this case, the relation between the parameters A_R^c, A_J^c, and r_R^c on the one hand, and the scattering amplitude on the other assumes the form

$$\operatorname{Re}(-A_R^c - iA_J^c)^{-1} = \mathcal{K}(k_m^2) - \frac{d\mathcal{K}(k^2)}{dk^2}\bigg|_{k^2=k_m^2} k_m^2,$$

$$\operatorname{Im}(-A_R^c - iA_J^c)^{-1} = \mathcal{L}(k_m^2), \qquad r_R^c = 2\frac{d\mathcal{K}(k^2)}{dk^2}\bigg|_{k^2=k_m^2}. \tag{22}$$

We wish to note that for a full determination of the amplitudes of all processes, including the N$-$He^4 scattering, one must introduce a fourth parameter $\varphi_N = \operatorname{arctg}(k_N^5/M_{NN}^c)\big|_{k_N=\sqrt{\frac{2\mu_N Q}{\hbar^2}}}$

which is related in a simple fashion to the phase of the N$-$He^4 scattering at the threshold point:

$$\delta_N = \varphi_N + \operatorname{arctg}(A_J^c/A_R^c).$$

§2. Determination of the Parameters of the Effective Radius Approximation from Experimental Data for the Reactions H^3(d,n)He^4 and He^3(d,p)He^4

It is convenient to introduce in Eq. (20) dimensionless quantities for the reaction cross sections:

$$\operatorname{Re}[-a_c/(A_R^c + iA_J^c)] = g_R \quad \text{or} \quad A_R^c/a_c = -g_R(g_R^2 + g_J^2)^{-1},$$

$$\operatorname{Im}[-a_c/(A_R^c + iA_J^c)] = -g_J \quad \text{or} \quad A_J^c/a_c = -g_J(g_R^2 + g_J^2)^{-1},$$

$$r_R^c/2a_c = \rho, \quad C^2 k a_c = 2\pi(e^{2\pi\eta} - 1)^{-1} = \varepsilon(\eta^{-1}).$$

The Coulomb unit of the wavelength is $a_c = 24$ F for the channel d + H^3 and $a_c = 12$ F for the channel d + He^3. With the new notations, the reaction cross section assumes the form

$$\frac{d\sigma_{dN}}{d\Omega} = q\,\frac{C^2}{k}\,a_c\,\frac{g_J}{[g_R + \rho\eta^{-2} - 2h(\eta^{-1})]^2 + [\varepsilon(\eta^{-1}) + g_J]^2} = q\,\frac{C^2}{k}\,a_c\Sigma(E). \tag{23}$$

Instead of directly relating the measured cross section to this formula, we make use of the fact that the experimental data on the H^3(d,n)He^4 and He^3(d,p)He^4 reactions can be adequately described when one uses for $\Sigma(E)$ a simple Lorentz form of the energy dependence [13], i.e., when we set

$$\frac{d\sigma_{dN}}{d\Omega} = q\,\frac{C^2}{k}\,a_c\,\frac{\Sigma_m\,(\Gamma_m/2)^2}{(E-E_m)^2+(\Gamma_m/2)^2}.\tag{24}$$

Equation (24) is the approximation of a single level in the complex eigenvalue theory of Humblet-Rosenfeld [8]; $(E_m - i\Gamma_m/2)$ denotes the complex energy eigenvalue at which the amplitude of the convergent wave tends to zero in an adequately determined asymptotic wave function; Σ_m is the value of the function $\Sigma(E)$ at the maximum. In §4 we will dwell on a comparison of the analytical forms of Eqs. (1), (23), and (24).

In order to express the parameters of the effective radius approximation by the parameters used in Eq. (24), we have to expand $[\Sigma(E)]^{-1}$ from Eq. (23) into a series near $E = E_m$:

$$\Sigma^{-1}(E) = \Sigma^{-1}(E_m) + \frac{d(\Sigma^{-1})}{dE}\Big|_{E=E_m}(E-E_m) + \frac{1}{2}\frac{d^2(\Sigma^{-1})}{dE^2}\Big|_{E=E_m}(E-E_m)^2$$

and, in accordance with Eq. (24), we must take into account that

$$\Sigma(E_m) = \Sigma_m,\; \frac{d(\Sigma^{-1})}{dE}\Big|_{E=E_m} = 0,\; \Sigma_m^{-1}\Big/\frac{1}{2}\frac{d^2(\Sigma^{-1})}{dE^2}\Big|_{E=E_m} = (\Gamma_m/2)^2.$$

Thus, we obtain three equations:

$$g_R = 2h - \rho\eta_m^{-2} + \varepsilon'(\varepsilon + g_J)(2h'-\rho)^{-1},\tag{25}$$

$$\Sigma_m = g_J(\varepsilon + g_J)^{-2}\left[1 + \left(\frac{\varepsilon}{2h'-\rho}\right)^2\right]^{-1},\tag{25'}$$

$$\left(\Gamma_m\Big/\frac{\hbar^2}{\mu a_c^2}\right)^2 = \frac{g_J}{\Sigma_m}\left\{(2h'-\rho)^2 + (\varepsilon')^2 + (\varepsilon+g_J)\left(\varepsilon'' - \frac{2h''\varepsilon'}{2h'-\rho}\right)\right\}^{-1}.\tag{25''}$$

We have the following notations:

$$\varepsilon = \varepsilon(\eta_m^{-1}),\qquad \varepsilon' = \frac{d\varepsilon(\eta^{-1})}{d(\eta^{-2})}\Big|_{\eta=\eta_m},\qquad \varepsilon'' = \frac{d^2\varepsilon(\eta^{-1})}{d(\eta^{-2})^2}\Big|_{\eta=\eta_m},$$

$$h = h(\eta_m^{-1}),\qquad h' = \frac{dh(\eta^{-1})}{d(\eta^{-2})}\Big|_{\eta=\eta_m},\qquad h'' = \frac{d^2h(\eta^{-1})}{d(\eta^{-2})^2}\Big|_{\eta=\eta_m}.$$

Equations (25') and (25'') can easily be transformed to the following form:

$$\left(\Gamma_m\Big/\frac{\hbar^2}{\mu a_c^2}\right)^2 = \frac{g_J}{\Sigma_m}\left\{(\varepsilon')^2\left[1 - \frac{(\varepsilon+g_J)^2\Sigma_m}{g_J}\right]^{-1} + (\varepsilon+g_J)\left[\varepsilon'' - 2h''\left(\frac{g_J}{\Sigma_m(\varepsilon+g_J)^2} - 1\right)^{1/2}\right]\right\}^{-1},\tag{26}$$

$$\rho = 2h' \mp \varepsilon'\left[\frac{g_J}{(\varepsilon+g_J)^2\Sigma_m} - 1\right]^{-1/2}.\tag{26'}$$

We have evaluated experimental data on the $H^3(d,n)He^{4*}$ reaction, taking averages over various measurements; we found that $E_m = 0.05$ MeV and $\Gamma_m/2 = 0.037$ MeV. Almost the same values were stated in [13]: $E_m = 0.05$ MeV and $\Gamma_m/2 = 0.039$ MeV. When we assume that $(d\sigma_{dn}/d\Omega)_{E=E_m} = 32.8$ F^2/sr, we find $\Sigma_m = 2.9$.

When Eq. (26) is solved for g_J it turns out that at the g_J values at which the half-width $\Gamma_m/2$ is close to the "experimental" half-width, the right side of Eq.(26) has a maximum. Thus, a continuous series of g_J values exists (ranging from 0.12 to 0.15) and corresponding ρ and g_R values can be indicated, which equally well describe the reaction cross section. Table 1 lists

---

*A detailed list of the experimental work on the reactions $H^3(d,n)He^4$ and $He^3(d,p)He^4$ can be found in [2]; see also [14].

TABLE 1

g_R	g_J	ρ	$(\Gamma_m/2)_{calc}$, MeV	A_R^c, F	A_J^c, F	r_R^c. F
0.414	0.194	0	0.0323	—47.5	—22.3	0
0.321	0.15	0.0687	0.0373	—61.5	—28.7	3.30
0.308	0.14	0.078	0.0375	—64.5	—29.4	3.75
0.297	0.13	0.085	0.0375	—67.6	—29.6	4.08
0.287	0.12	0.0914	0.0373	—71.3	—29.7	4.40

several sets of parameters g_R, g_J, and ρ (A_R^c, A_J^c, r_R^c), along with the $\Gamma_m/2$ values calculated with these parameters for the reaction $H^3(d,n)He^4$.

The first line of Table 1 lists a set of parameters obtained in the limit $\rho \to 0$ from Eqs. (25) and (25'). It follows from the table that despite the continuous ambiguity, studies of the reaction cross section make it possible to deduct with good accuracy the parameters of the effective radius theory, among them the parameter r_R^c, which has the meaning of an average interaction radius. The approximation in which the interaction radius zero is assumed disagrees with the experimental results.

An experimental error of ±3% in the determination of Σ_m, Γ_m introduces a small error in the determination of A_R^c, A_J^c, and r_R^c.

The experimental data on the $He^3(d,p)He^4$ reaction are less accurate, i.e., the reaction cross section near the maximum is known with an accuracy of ± 10%. We found that the measured reaction cross section of the $He^3(d,p)He^4$ reaction is most adequately described by $E_m = 0.218$ MeV and $\Gamma_m/2 = 0.150$ MeV in the energy interval 0.048-0.321 MeV. In [13], the values $E_m = 0.208$ MeV and $\Gamma_m/2 = 0.129$ MeV were found. When we assume $\left(\frac{d\sigma_{dp}}{d\Omega}\right)_{E=E_m} = 5.54$ F^2/sr for the energy $E_m = 0.218$ MeV, we obtain $\Sigma_m = 1.74$. Table 2 lists several sets of parameters of the effective radius approximation for the reaction $He^3(d,p)He^4$, with $(\Gamma_m/2)_{calc}$ varying from 0.130 to 0.150 MeV ($E_m = 0.218$ MeV everywhere).

For given E_m, Γ_m, and Σ_m values, the right side of Eq. (26) depends almost linearly upon g_J and this means that regions of continuous indeterminacy do not exist. The errors in the determinations of g_R, g_J, and ρ are the result of experimental errors. In the approximation involving radius zero, the half-width $(\Gamma_m/2)_{calc}$ is obviously too large.

The overall results of this analysis, as well as direct calculations with Eq. (23) indicate that it is possible to describe the reaction cross section, within the limits of experimental error, by the effective radius approximation. A large value $|A_R^c| \gg r_R^c$ indicates that the resonance is close to the threshold. A small value $(r_R^c / \lambda_{min}) \approx 1/2$ permits one to disregard higher-order terms in the expansion in terms of (r_R^c / λ).

We calculated the cross section of $d-H^3$ and $d-He^3$ scattering in the same energy range with the parameter values A_R^c, A_J^c, and r_R^c which we had obtained. The strong agreement with

TABLE 2

g_R	g_J	ρ	$(\Gamma_m/2)_{calc}$, MeV	A_R^c, F	A_J^c, F	r_R^c. F
0.525	0.365	0	0.217	—15.4	—10.7	0
0.233	0.035	0.125	0.131	—50.4	—7.57	3.00
0.238	0.040	0.127	0.139	—49	—8.22	3.05
0.244	0.045	0.129	0.147	—47.6	—8.80	3.08

TABLE 3

a, F	E_λ, MeV	γ_d^2, MeV	γ_n^2, MeV	χ^2	Re $g(0)$, F^{-1}	Im $g(0)$, F^{-1}	r_R, F	(A_R^c)calc. F	(A_J^c)calc. F	(r_R^c)calc. F
3	−1.58	6.03	0.2	16.1	−0.0658	−0.0101	3.66	−59.9	−29.0	3.32
5	−0.794	3.00	0.067	20	−0.0419	−0.0095	6.6	−72.5	−29.3	4.31
7	−0.142	0.75	0.0184	26.8	−0.0239	−0.0130	7.17	−72.0	−30.3	4.40
10	+0.004	0.29	0.0081	17.6	−0.003	−0.0125	7.65	−72.7	−34.9	4.66

the experimental results of [2] indicates that one can use the effective radius approximation for describing scattering amplitudes. We wish to note that parameter sets corresponding to the second solution for ρ ($\rho > 2h'$, $g_R < 0$) disagree with the scattering data.

§3. Results of an Analysis of the $H^3(d,n)He^4$ and $He^3(d,p)He^4$ Reactions Using the Formulas of the R-Matrix Theory and the Effective Radius Approximation

In the previous chapters we have considered the effective radius approximation based on very simple and logical assumptions regarding the behavior of the functions Re $g^c(k)$ and Im $g^c(k)$, which determine the scattering amplitude and the reaction cross section defined by Eqs. (8) and (9). It is interesting to examine the energy dependence of these functions, which is obtained when the experimental results are evaluated with the formula for an isolated level of the R-matrix theory [see Eq. (1)], and to compare the resulting dependence with the results of the preceding section.

The resonance parameters a, E_λ, γ_d^2, and γ_N^2 were determined by fitting experimental cross sections of the reactions $H^3(d,n)He^4$ and $He^3(d,p)He^4$ to Eq. (1), using the least square method on a computer. Table 3 lists sets of parameters for the reaction $H^3(d,n)He^4$ at various fixed radii a of the channel and a quantity χ^2, which characterizes the deviation of the calculated curves from all experimental points in the energy interval 0.006-0.250 MeV. One must bear in mind that the method of fitting the measured cross sections to the effective radius approximation defined by Eq. (23), i.e., the method used in the preceding chapter, is less accurate insofar as we did not calculate the quantity χ^2 from all experimental points and did not attempt to minimize the deviations, as we did in the computer analysis.

We calculated the function $g(k^2)$, with which one can express the function $g^c(k)$ [see Eqs. (13) and (13')], with the parameters a, E_λ, γ_d^2, and γ_n^2 of Table 3 and with the relations defined by Eqs. (11) and (11'). Re $g(k^2)$ has an approximately linear energy dependence, whereas Im $g(k^2)$ remains practically constant. The table lists the quantities Re $g(0)$, Im $g(0)$, and $2\frac{d}{dk^2}\text{Re}\,g(k^2)|_{k=0}$ $= r_R$ Obviously, these quantities depend upon the radius a of the channel to a lesser extent than the parameters of the R-matric theory.

Furthermore, when we use Eqs. (18) and (18'), we can determine $\mathcal{K}(0)$ and $\mathcal{L}(0)$ from which we can calculate $(A_R^c)_\text{calc}$ and $(A_J^c)_\text{calc}$. An expression which is similar to that of Eqs. (18) and (18'), yet more cumbersome, can be used for determining $2\frac{d\mathcal{K}(k^2)}{dk^2}\big|_{k=0} = (r_R^c)_\text{calc}$.

The results of the calculations, which are listed in Table 3, indicate that despite their great differences, all sets of parameters of the R-matric theory lead to almost the same energy dependence of the scattering amplitude near $E = 0$.

A comparison of Tables 1 and 3 reveals that the calculated parameters A_R^c, A_J^c, and r_R^c are close to the parameters which were obtained from an analysis of the experimental data with

TABLE 4

a, F	E_λ, MeV	γ_d^2, MeV	γ_p^2, MeV	$x^\flat$	$(A_R^c)_{\text{calc}}$, F	$(A_J^c)_{\text{calc}}$, F	$(r_R^c)_{\text{calc}}$, F
3	−5.1	18.5	0.4	59	−45.9	−6.24	2.64
3	−20.8	74	1.6	48	−47.0	−6.55	3.3
5	−0.191	2.21	0.044	31.8	−50.5	−7.83	3.02
7	+0.205	0.463	0.0100	43.0	−38.7	−7.48	2.0

the effective radius approximation defined by Eq. (23) and with the procedure outlined in §2. In other words, when we analyze the measured reaction cross section with Eq. (1) of the R-matrix theory, which involves four parameters, we determine, in essence, three coefficients of the expansion of the functions $\mathcal{K}(k^2)$ and $\mathcal{L}(k^2)$, namely, $\mathcal{K}(0)$, $\mathcal{L}(0)$, and $2\frac{d\mathcal{K}(k^2)}{d(k^2)}\Big|_{k=0}$. This is the reason why there appears a "physical" ambiguity of the resonance parameters when the radius of the channel is considered; a continuous (nonphysical) ambiguity discussed in §2 is superimposed on the physical ambiguity. The sets of resonance parameters at a = 5 and 7 F on the one hand, and the set of parameters at a = 3 F on the other hand, mark the limits of the region of continuous ambiguity.

Table 4 lists the computer results of the resonance analysis of the cross section for the reaction $He^3(d,p)He^4$ in the energy interval 0.024–0.538 MeV. The E dependence of the functions $a_c C^2 \, \text{Re} g^c(k) + 2h(\eta^{-1}) = a_c \mathcal{K}(k^2)$ and $a_c C^2 \, \text{Im} \, g^c(k) = a_c \mathcal{L}(k^2)$ was determined for the parameter sets with a = 3 and 5 F. The corresponding curves are indicated by the solid lines in the figure. The dashed straight lines correspond to an effective radius approximation as defined by Eq. (23) with the set g_R = 0.238, g_J = 0.04, and ρ = 0.127 of Table 2. Table 4 includes numerical values for the quantities A_R^c, A_J^c, and r_R^c which were obtained with the isolated level approximation of the R-matrix theory for each set of parameters a, E_λ, γ_d^2, and γ_p^2.

When one compares Tables 2 and 4 and Fig. 1, one finds that the scattering amplitudes are actually determined by the three parameters $\mathcal{K}(0)$, $\mathcal{L}(0)$, and $2\frac{d\mathcal{K}(k^2)}{dk^2}\Big|_{k=0}$ in the entire energy interval considered. The ambiguity of these parameters results in noticeable errors of the measured reaction cross section ($\pm 10\%$).

We see that the possibility of using the effective radius approximation with the three parameters A_R^c, A_J^c, and r_R^c is corroborated by the data obtained from a resonance analysis based on R-matrix theory. The fact that it is possible to measure the resonance parameters a,

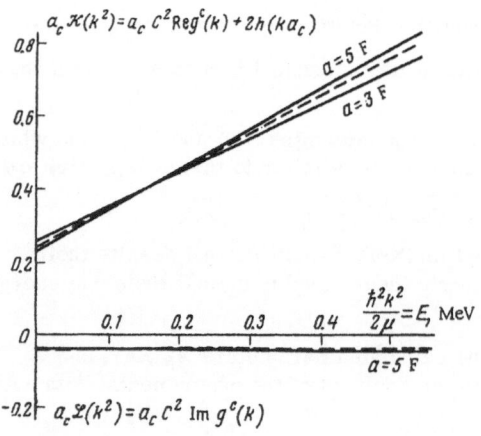

Fig. 1. Energy dependence of the functions $a_c C^2 \, \text{Re} \, g^c(k)+2h(ka_c)=a_c\mathcal{K}(k^2)$ and $a_c C^2 \, \text{Im} \, g^c(k) = a_c \mathcal{L}(k^2)$ for the reaction $He^3(d,p)He^4$. The solid lines result from the evaluation of the experimental cross section by using the isolated level approximation of the R-matrix theory with the set of resonance parameters E_λ, γ_d^2, and γ_p^2 of a = 3 F and 5 F (from Table 4). The dashed lines result from an analysis using the effective radius approximation with the parameter set A_R^c, A_J^c, and r_R^c for g_J = 0.04 (Table 2). At a = 3 F, the solid and the dashed lines are almost coincident for $\mathcal{L}(k^2)$.

E_λ, γ_d^2, and γ_N^2 within rather wide intervals* results, first, from the existence of a fourth parameter (the channel radius a), second, from the specific ambiguity in the H^3(d,n)He^4 reaction, and, third, from the inaccuracy of the measurements, particularly in the case of the He^3(d,p)He^4 reaction.

The question of whether it will be possible to state more precisely the parameters A_R^c, A_J^c, and r_R^c can be decided after measuring with greater precision the cross sections of the reactions H^3(d,n)He^4 and He^3(d,p)He^4, including the region $\eta > 1$. One must bear in mind that, strictly speaking, one has to separate from the reaction cross section the contribution of the channel with J$^\pi$ = ½$^+$, whose existence is confirmed by polarization measurements.

§4. Particular Features of Isolated Resonances

in the Various Theories

In §2, we used Eq. (24), which is a single-level approximation in the complex eigenvalue theory of Humblet-Rosenfeld [8], for the analysis of experimental data based on the effective radius approximation defined by Eq. (23). On the other hand, the energy dependence given by these formulas is obviously not the same in the entire region under inspection. We note that in the energy range close to zero ($\eta > 1$), the reaction cross sections calculated with Eq. (23) and certain parameter values A_R^c, A_J^c, and r_R^c are always smaller than the reaction cross sections calculated with Eq. (24) and the parameters Σ_m, $\Gamma_m/2$, and E_m, which were obtained from the conditions stated in Eqs. (25), (25'), and (25"). These conditions guarantee that the energy relations defined by Eqs. (23) and (24) are approximately identical in the energy interval ($E_m - \Gamma_m/2$, $E_m + \Gamma_m/2$).

It follows from Eq. (23) that for $\eta > 1$ the reaction cross section is given by $\Sigma(0) = g_J (g_R^2 + g_J^2)^{-1} = | A_J^c / a_c$. Equation (24), which is equivalent within the restrictions stated above, is obtained when $\Sigma^{-1}(E)$ is expanded around E_m. This means that in the particular approximation the values of the functions $\varepsilon(\eta^{-1})$ and $h(\eta^{-1})$ at $E = 0$ ($\eta^{-1} = 0$) are given by

$$\widetilde{h}(0) = h - h'\eta_m^{-2} + \frac{1}{2}h''\eta_m^{-4},$$

$$\widetilde{\varepsilon}(0) = \varepsilon - \varepsilon'\eta_m^{-2} + \frac{1}{2}\varepsilon''\eta_m^{-4},$$

$$\widetilde{h^2}(0) = (h - h'\eta_m^{-2})^2 + hh''\eta_m^{-4},$$

$$\widetilde{\varepsilon^2}(0) = (\varepsilon - \varepsilon'\eta_m^{-2})^2 + \varepsilon\varepsilon''\eta_m^{-4},$$

$$(h(\eta^{-1})\eta^{-2})_{\eta^{-1}=0} = \frac{1}{2}h''\eta_m^{-3}$$

rather than by $\varepsilon(0) = 0$ and $h(0) = 0$. When we insert the above values into Eq. (23), we find that in the expansion the quantity $\Sigma(0)$ is approximated by

$$\widetilde{\Sigma}(0) = g_J [g_R^2 + g_J^2 - 4(g_R + \rho\eta_m^{-2} - 2h)\widetilde{h}(0) - 4(h - h'\eta_m^{-2})(h + h'\eta_m^{-2} - \rho\eta_m^{-2}) + \widetilde{\varepsilon^2}(0) + 2g_J\widetilde{\varepsilon}(0)]^{-1}.$$

For an η_m value which is situated in a rather broad interval around the values which we considered ($\eta_m = 0.775$ for the (d,n) reaction and $\eta_m = 0.741$ for the (d,p) reaction), it is obvious by Eq. (15) that $h \approx h'\eta_m^{-2}$ and $h'' < 0$. Accordingly, we have $\widetilde{\varepsilon^2}(0) > 0$ and $\widetilde{\varepsilon}(0) < 0$, and with these values we obtain $g_J\widetilde{\varepsilon^2}(0) + 2g_J\widetilde{\varepsilon}(0) > 0$.

Since the maximum condition renders $g_R + \rho\eta_m^{-2} - 2h > 0$, we obtain

$$\widetilde{\Sigma}(0) \approx g_J [g_R^2 + g_J^2 + 4|\widetilde{h}(0)| | g_R + \rho\eta_m^{-2} - 2h | + |\widetilde{\varepsilon^2}(0) + 2g_J\widetilde{\varepsilon}(0)|]^{-1} < \Sigma(0).$$

_____

*Deuteron-resonance scattering studies based on R-matrix theory lead to a clear rejection of the parameter set for $a = 10$ F.

For the (d,n) reaction, numerical estimates render $\tilde{\Sigma}(0)/\Sigma(0) \approx 0.9$; accordingly, for the (d,p) reaction, we obtain $\tilde{\Sigma}(0)/\Sigma(0) \approx 0.8$.

Direct calculations of the function $\Sigma(E)$ according to Eq. (1) of the R-matrix theory always render $\Sigma(0) > \tilde{\Sigma}(0)$. This is quite logical, because Eqs. (1) and (23) are actually similar:

$$(P_d/a)_{a \to 0} = C^2 k, \quad (S_d/a)_{a \to 0} = \text{const} + 2h(\eta^{-1})/a_c.$$

Furthermore, we note that for $\eta > 1$ the cross section increases with increasing channel radius a.

Unfortunately, for a number of reasons it is not possible to select on the basis of experimental results one of the Eqs. (1), (23), or (24) for describing the resonance in a particular reaction.

A first reason is the low accuracy of the reaction cross section measurements at $\eta > 1$; a second reason is the unknown contribution of the channel with $J^\pi = \frac{1}{2}^+$, and finally, a third reason is the lack of precision in the measurements of the $He^3(d,p)He^4$ reaction.

However, one can safely state that a description of a resonance with Eqs. (1) or (23) is advantageous, because this description makes it possible to analyze with the same parameters the scattering cross section of deuterons and to take into account the effect which the Coulomb interaction has upon the scattering amplitude at low energies.

Conclusion

Let us summarize the basic conclusions. The cross sections of the resonance reactions $H^3(d,n)He^4$ at 0.006 MeV $< E < 0.250$ MeV and $He^3(d,p)He^4$ at 0.024 MeV $< E < 0.35$ MeV can be described with the effective radius approximation using the three parameters A_R^c, A_J^c, and r_R^c. There exists a simple relation between the quantities A_R^c and A_J^c and the amplitude of the nuclear scattering at $E = 0$ ($\eta \gg 1$); r_R^c is the effective interaction radius. When the experimental results are interpreted with Eq. (23) of the effective radius approximation, the ambiguity is removed from the resonance analysis based on R-matrix theory [Eq. (1)]. The ambiguity results from the fact that the channel radius can vary within a large interval. A small continuous ambiguity remains for statistical reasons in the parameters A_R^c, A_J^c, and r_R^c for the $H^3(d,n)$. He^4 reaction, an ambiguity which also appears when Eq. (1) is used for the analysis. Experimental errors in the cross section of the reaction $He^3(d,p)He^4$ lead to a certain degree of indeterminacy (about 20%) of the parameters A_R^c, A_J^c, and r_R^c.

For $\eta > 1$, Eqs. (1) and (23) give (by 10-20%) larger reaction cross sections than the single level approximation of the Humblet-Rosenfeld theory expressed by Eq. (24). However, in view of the lack of accuracy in the determination of the reaction cross section and the indeterminacy of the contribution provided by the channel with $J^\pi = \frac{1}{2}^+$ one can conclude that the three methods for describing the resonances are about equally well applicable to experimental data on a reaction.

The parameters A_R^c, A_J^c, and r_R^c, which were obtained in the analysis of reactions with the effective radius approximation, can be employed for describing the deuteron resonance scattering on H^3 and He^3 in the same energy interval.

The authors express their gratitude to Yu. G. Balashko for stimulating discussions and for his assistance in the resonance analysis of experimental data with a computer. The authors are indebted to É. N. Chaikovskaya for programming the resonance-parameter determination.

References

1. G. Breit, Theory of Nuclear Resonance Reactions, IL, Moscow (1961) [Russian translation].
2. Yu. G. Balashko, Trudy FIAN, 33:66 (1965); Yu. G. Balashko and I. Ya. Barit, Nuclear Forces and the Few-Nucleon Problem, Pergamon Press, New York (1960), p. 615.
3. A. Lane and R. Thomas, Theory of Nuclear Reactions at Low Energies, I L, Moscow (1960) [Russian translation].
4. V. A. Sergeev, FIAN Preprint No. 154 (1967); Izv. AN SSSR, ser. fiz., 32:312 (1968).
5. L. D. Landau and E. M. Lifshits, Quantum Mechanics, Fizmatgiz, Moscow (1963).
6. A. M. Baldin, Phys. Lett., 17:47 (1965).
7. T. A. Tombrello, Phys. Lett., 23:106 (1966).
8. J. Humblet and L. Rosenfeld, Nucl. Phys., 26:529 (1961); 50:1 (1964).
9. M. H. Ross and G. L. Shaw, Ann. Phys., 9:391 (1960); 13:147 (1961).
10. G. L. Shaw and M. H. Ross, Phys. Rev., 126:806 (1962).
11. G. Breit and W. G. Bouricius, Phys. Rev., 75:1029 (1949).
12. L. D. Landau and Ya. A. Smorodinskii, Zh. Éksp. Teor. Fiz., 14:269 (1944).
13. J. P. Jeukenne, Nucl. Phys., 58:1 (1964).
14. A. P. Kobzev, V. I. Salatskii, and S. A. Telezhnikov, OIYaI Preprint R-2386.

SINGLE-PARTICLE VIBRATIONAL INPUT STATES
IN INTERACTIONS BETWEEN NEUTRONS
AND NUCLEI

V. I. Popov

Introduction

The intermediate state or input state model, which was proposed by Block and Feshbach [1] has been extensively discussed in recent years [2-5]. This model is of particular interest because at the present time rather comprehensive data has been accumulated which points to the inadequacy of the optical model, the direct-interaction model, and the statistical model for providing a theoretical description of nuclear reactions. There is reason to hope that the concept of intermediate states, which is a very general approach and could qualitatively explain a number of experimental results, will make it possible to describe various nuclear processes from a unified viewpoint, and will become a useful tool for research on the structure of complex nuclei.

Input states must manifest themselves mainly in the form of a so-called intermediate structure in the energy dependence of the nuclear reaction cross sections. As a matter of fact, in some experiments, one can observe a structure in the energy dependence of the effective cross sections (among them processes involving medium and heavy nuclei) and this structure cannot be explained by random fluctuations [3]. But since the structure observed is rather complicated — the width of the maxima in the cross sections and the distances between the maxima have about the same order of magnitude (about 100 keV) — a detailed analysis of "intermediate resonances" similar to the analysis of isolated resonance peaks of a compound nucleus is usually not feasible. The only exceptions are resonances corresponding to the formation of analogous isobaric states which may be considered as input states.

One can expect that effects resulting from really existing input states will appear more clearly in experiments, when a very small number of reaction channels is considered. We think particularly of reactions induced by slow neutrons. In these reactions, only input channels with low orbital momentum of the incident neutron ($l = 0, 1$) are important. In Shakin's article [2], the input-state model was used for calculating the force function of s neutrons for nuclei with closed shells (Pb and Sn). Three-quasi-particle states were assumed as input states. Based on the three-quasi-particle input states, Block and Feshbach attempted a quantitative interpretation of the form of the mass-number dependence of the force function of s neutrons. However, their simplified approach to the evaluation of the deviations of the force function from the values calculated with the optical model is not correct (this was pointed out later

14

by Feshbach [3]). This is the reason why one must consider the agreement with the experimental data of [1] as pure coincidence and why, consequently, the existence of the three-quasi-particle input states postulated by Block and Feshbach could not yet be confirmed.

Interesting hints on the appearance of intermediate states in the interaction between slow neutrons and nuclei were obtained in the work Ikegamy and Emery [4] and Groshev and Demidov [5] did in comparing the probabilities of the excitation of low levels of nuclei involved in (n,γ) and (d,p) reactions. The fact that in some nuclei (Fe^{57}, Ni^{63}, and Zn^{65}) an increased intensity of γ transitions is observed on levels which are weakly excited in (d,p) reactions is considered as a hint that these γ transitions stem from input states of the three-quasi-particle type, which develop in interactions between neutrons and nuclei.

Though at the present time a rather large number of papers on the effects of input states have appeared, the discussion is usually qualitative. This is the reason why the justification of input states was sometimes doubted.

A certain nuclear model must be assumed for a quantitative analysis of concrete nuclear processes with the input-state theory. The selection of a model appropriate for the description of an input state is of great importance because the model decides over the applicability of the input-state concept. The shell model of the nucleus [1, 2] is used in discussions involving three-quasi-particle intermediate states. Recently, several articles were published which indicated that the collective model in conjunction with input states of the single-particle vibrational type [6-8] may be convenient for the analysis of nuclear processes. These two types of input states are not mutually exclusive, because a vibrational state of a nucleus can be represented as a coherent superposition of two quasi-particle excitations. One can assume that the three-quasi-particle input-state model will prove its merits in the case of nuclei with closed shells or with a small number of nucleons for which collective effects are immaterial. One can foresee for the majority of complicated nuclei that intermediate collective states will be of overriding importance. This conclusion is corroborated by the fact that in the direct inelastic scattering of nucleons, collective states are excited with a much greater probability than other, simpler states. The formation of an input state in the interaction between a nucleon and a nucleus resembles in many respects direct inelastic scattering; the difference is that in the first case the nucleon passes into a bound state, due to the interaction, rather than into a continuous spectrum.

As early as 1953, Bohr and Mottelson [9] pointed to the importance of collective interactions for the formation of a compound nucleus; these authors proved that a reaction between the incident particle and the surface vibrations of the nucleus can provide an important contribution to the imaginary part of the optical potential. The single-particle vibrational input-state model was used by Lane in his theory of the so-called collective capture of nucleons [10]. This theory assumes that the radiative capture of nucleons with energies of about 10 MeV proceeds in two stages: first, the incident nucleon excites dipole vibrations of the nucleus and passes into one of the lower single-particle states; after that, the excited oscillator passes into the ground state under emission of a γ quantum.

In the present article, a single-particle vibrational intermediate state model is employed for the analysis of interactions between low-energy neutrons and medium or heavy spherical nuclei. At neutron energies which are smaller than 1 MeV, and consequently, at excitation energies of the compound nucleus < 10 MeV, only quadrupole and octupole phonons will participate in the formation of the input states. The energies of these phonons are known from experimental data on the positions of the first levels 2^+ and 3^+ of even−even spherical nuclei. Neutron-force functions (described in Section 2) were calculated with a simple model using a single-phonon excitation of the quadrupole or octupole type as the input states. Section 3 describes the effects which single-particle vibrational input states have upon the radiative capture of low-energy neutrons.

§1. Intermediate States of the Phonon/Particle Type

for Low-Energy Neutrons

In the following discussion we consider the interaction of a neutron with a spherical even—even nucleus. According to the generalized model of [9], the states of such a system are characterized by the number of photons of the type λ N_λ, the moment R of surface motion and its projection R_z, and the quantum numbers l, j, m of the state of the neutron. The states can be denoted by the symbol $|j, NR; JM \rangle$, where J and M denote the total moment of the system and its projection, respectively. The first-order term in the expansion of the real part of the optical potential in spherical harmonics is usually employed as the interaction potential:

$$V_{\text{int}} = -r \frac{dV_0}{dr} \sum_{\lambda, \mu} \alpha_{\lambda\mu} Y_{\lambda\mu}(\vartheta, \varphi). \tag{1}$$

The states of a nucleus with an odd number of nucleons can be found in this model by solving a system of coupled wave equations, provided that one limits the considerations to a certain number of phonons which are present in these states. This particular approach was taken in [7] and [8], wherein single-particle vibrational states of the Ni^{59} nucleus were determined. A maximum of six quadrupole and three octupole phonons were taken into account. The authors consider the states determined in this fashion as intermediate states which appear in the $Ni^{58}(d,p)Ni^{59}$ reaction and in the interaction between low-energy neutrons with $l = 0$ and the Ni^{58} nucleus. Though there is no reason to believe that multiphonon excitations of a harmonic oscillator occur in nuclei, the question of whether this approach can be used can be decided from a comparison with experimental data. A comparison of the theoretical calculations of both the energy and the reduced neutron width of the Ni^{59} levels with the experimental data obtained from research on the (d,p) reaction indicates that the model renders a correct qualitative picture of the Ni^{59} levels, but disagrees with the experiment in many details. The correct order of magnitude was obtained in [8] for the force function of s neutrons.

Feshbach's input-state theory helps to include, in a somewhat different fashion, the relation between a nucleon and the surface vibrations of the nucleus in research on nuclear processes. The intermediate state of a nucleon—nucleus system can be considered as the combination of a single-phonon excitation of the nucleus and a single-particle nucleon state. The relation of this state to more complicated states of the system can be taken into account by introducing a width $\Gamma_d^\downarrow$, which characterizes the probability that the system undergoes a transition from a given intermediate state into more complicated states. The latter states may comprise single-particle vibrational states with a large number of phonons, or other types of states. The probability for the formation of an input state or for the reverse transition from a state to an open channel is characterized, according to Feshbach by the width $\Gamma_d^\uparrow$:

$$\Gamma_d^\uparrow = 2\pi |\langle \psi_0^{(-)} V_{\text{int}} \psi_d \rangle|^2. \tag{2}$$

In the general case, a system in an intermediate state can have several open channels. Then one has to include in the considerations the partial widths of the corresponding reactions which pass through a particular intermediate state without formation of a compound nucleus. For example, in the description of the radiative capture of slow neutron (see §3) the parameter $\Gamma_{\gamma d}$ is is introduced, which characterizes the probability of γ quantum emission from an intermediate state.

For calculating the force function with the intermediate state model, one must calculate the width $\Gamma_d^\uparrow$ of the states which are essential for the formation of a compound nucleus. When a transition occurs from an intermediate state with one phonon of the type λ, $\Psi_d = |j, 1\lambda; JM\rangle$ into an open channel without phonons ($\Psi_0^{(-)} = |j', 00; J = j', M\rangle$), one obtains the following expression for $\Gamma_d^\uparrow$ with the potential stated in Eq. (1):

$$\Gamma_d^{\uparrow} = \frac{2\pi}{2\lambda + 1} \, | \langle j' \, \| \, Y_\lambda \, \| \, j \rangle |^2 \, \beta_\lambda^2 \, | \langle \chi_{l'}(r) \, | \, k(r) \, | \, u_l(r) \rangle |^2. \tag{3}$$

The notations are interpreted as follows: $\langle j' \, \| \, Y_\lambda \, \| \, j \rangle$ is the reduced matrix element for $Y_{\lambda\mu}$; β_λ is the deformation parameter of the nucleus: $\beta_\lambda = [(2\lambda + 1) \, \hbar\omega_\lambda \, /2c_\lambda]^{1/2}$; $\chi_l(r)/k(r) \, | \, u_l(r)$ denotes the radial matrix element in which $k(r) = r(dV_0/dr)$; $\chi_l(r)$ and $u_l(r)$ are the radial wave functions of the neutron in the open channel and in the intermediate state, respectively.

The energy of the intermediate phonon–particle state can be taken as equal to the sum of the energies of the single-particle state and the corresponding vibrational state of an even-even target nucleus.

Obviously, the existence of input states whose widths are small relative to the distance between the states can lead to a structure which depends upon the relation between the neutron force function and the mass number. In the case of slow neutrons, the maxima of the force function must be observed in nuclei in which the binding energy of the neutrons is close to the energy required for exciting one of the intermediate states. Intermediate phonon–particle states which can be excited in nuclei with A = 40 to 200 neutrons having orbital momenta $l = 0$ and $l = 1$ are listed in the table. The energy of these states can be determined with an accuracy of about 1 MeV due to the undeterminedness in the position of the single-particle levels in the nuclei. Table 1 lists the mass numbers of the nuclei (A_d) for which the energy required for exciting a certain input state is equal to the binding energy of the neutron. The energies of the single-particle neutron energy levels, which are necessary for the determination of A_d, were taken from the work of Green [11]. Green's calculation for the Woods-Saxon potential is in rather strong agreement with experimental data. The inaccuracy in the energy of these states amounts to 1 MeV and corresponds to an undeterminedness in the A_d values which ranges from 4 at A ~ 50 to 15 at A ~ 190.

TABLE 1. Intermediate States of the
Phonon–Particle Type for Neutrons of Zero
Energy

Open channel with $l' = 0$			Open channel with $l' = 1$		
State of the neutron	Phonon type	A_d	State of the neutron	Phonon type	A_d
$1f_{5/2}$	$\lambda = 3$	53	$2d_{5/2}$	$\lambda = 3$	75
$2d_{5/2}$	$\lambda = 2$	65	$1g_{7/2}$	$\lambda = 2$	85
$2d_{3/2}$	$\lambda = 2$	81	$2d_{3/2}$	$\lambda = 3$	95
$2f_{7/2}$	$\lambda = 3$	131	$2f_{7/2}$	$\lambda = 2$	111
$2f_{5/2}$	$\lambda = 3$	145	$3p_{3/2}$	$\lambda = 2$	131
$3d_{5/2}$	$\lambda = 2$	191	$3p_{1/2}$	$\lambda = 2$	141
			$2f_{5/2}$	$\lambda = 2$	141

According to the table, for each orbital momentum of the incident neutrons there exists a small number of input states of the phonon–particle type and the distances between these states (expressed in energy units) amount usually to $\gtrsim 1$ MeV. Below, when we calculate the force function, we will assume that corresponding intermediate resonances can be isolated, which is fully correct only when the widths of the resonances, Γ_d, are small relative to the spacing of the resonances. It will be obviated by the following discussion that for small neutron energies the full width Γ_d is practically equal to $\Gamma_d^{\uparrow}$. In the present article, we will not calculate the width $\Gamma_d^{\uparrow}$, but consider it to be a parameter whose magnitude can be determined from experiments. The value $\Gamma_d^{\uparrow} = 1.2$ MeV is used in calculations of the force function in §2.

§2.   Neutron Force Functions

According to the input-state theory of [3], by averaging over the resonances of the compound nucleus in the energy interval $\Delta E < \Gamma_d$, one obtains the following expression for an isolated intermediate resonance:

$$\langle \Gamma \rangle / D = \Gamma_d^{\uparrow} \Gamma_d^{\downarrow} \Big/ 2\pi \left[(E - E_d)^2 + \left(\tfrac{1}{2} \Gamma_d \right)^2 \right]. \tag{4}$$

D denotes the average distance between the resonances of the compound nucleus under the assumption that $\langle \Gamma \rangle / D \ll 1$. Similarly, by averaging over a large interval of energies $\Delta E \gg D_d$ (where D_d denotes the average distance between intermediate resonances) the relation

$$\langle \Gamma \rangle / D = \langle \Gamma_d^{\uparrow} \rangle / D_d \tag{5}$$

must be satisfied.

Equations (4) and (5) can be used as initial equations for the calculation of neutron force functions when Γ is taken as the neutron width of the resonances of the compound nucleus. Since the goal of the present article is to clarify the structure of the force functions resulting from intermediate states of the phonon—particle type, Eq. (4) is used for calculating the contribution of each of the states to the force function. The calculated force function is simply a sum of terms stated by Eq. (4) for input states listed in the table. Thus, the contribution of input states of other types is completely disregarded. The assumption that other input states, e.g., three-quasi-particle states which do not lead to coherent excitations, are less important does not have a logical basis and can be checked only by experiments.

Since the number of possible three-quasi-particle states is much larger than the number of states of the phonon—particle type, the contribution of the first type of states to the force function could be described by a term in the form of Eq. (5), with the term having a smooth A dependence, provided that the averaging interval ΔE is sufficiently large. A quantitative estimate of the contribution of this term should serve in comparing the calculated force function with the experimental force function in regions A which are far from intermediate resonances of the phonon—particle type (see below).

In this article, the aim of the calculation of the force function was to describe qualitatively its form for a large interval of A values; it was not intended to obtain force functions for individual nuclei. Therefore, several simplifications were made for the calculation of the widths $\Gamma_d^{\uparrow}$. More particularly, in the calculation of the radial integral in Eq. (3), the approximation involving a surface interaction was used, i.e., it was assumed that $k(r) = RV_0 \delta(r - R)$. Then Eq. (3) can be written in the form

$$\Gamma_d^{\uparrow} = \frac{2\pi}{2\lambda + 1} \, \langle j' \| Y_\lambda \| j \rangle \, |^2 \beta_\lambda^2 R^6 V_0^2 \, | \chi_{l'}(R) |^2 \, | u_l(R) |^2. \tag{6}$$

One must bear in mind that one cannot obtain the correct matrix element of the interaction between the particle and the nucleus surface in the surface-interaction approximation when the constant V_0 of the binding force is assumed to be equal to the depth of the potential well. Therefore, V_0 was assumed to be an additional parameter and its value was chosen such that agreement with the experimental data on the force functions was obtained.

However, in view of the results obtained in [8], one may think that in a more systematic calculation (in which a diffuse boundary of the nucleus is introduced) one could obtain agreement with the experimental results without introducing this additional parameter.

We see from Eq. (6) that for calculating $\Gamma_d^{\uparrow}$ one must also know the wave function $\chi_{l'}(r)$ of the neutron in the open channel. It was assumed that the wave function resulting from a rectangular optical potential with a real part and a radius such that the correct position of the single-particle resonances is observed in low-energy neutron scattering will render a good approximation for $\chi_{l'}(r)$. The imaginary part of the potential was taken as a free parameter. The radial wave functions on the surface of the nuclei are of the order of $1/R^3$ for neutrons in bound states $u_l(R)$ and differ only slightly for the various neutron states.

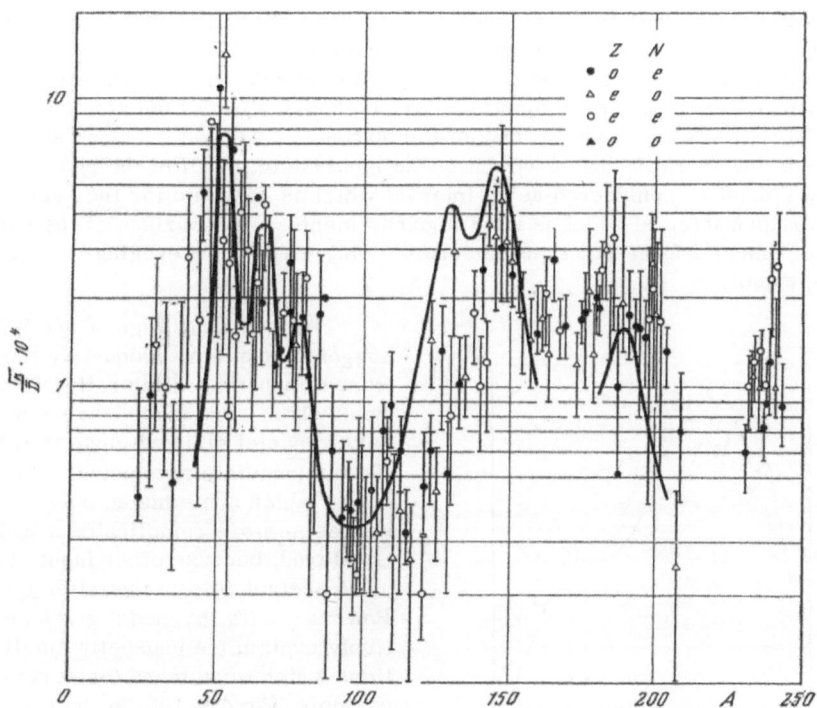

Fig. 1. Force function for s neutrons. The experimental values are indicated for target nuclei with certain Z and N values (o denotes odd and e even nuclei).

Figure 1 shows the curve which was obtained for the force function of s neutrons at the following parameter values entering into the calculation of $\Gamma_d^\uparrow$: radius of the nucleus $R = 1.35 \cdot 10^{-13} A^{1/3}$ cm; real part of the optical potential $V = 45$ MeV; imaginary part $W = 4.5$ MeV; and binding constant of the particle at the surface $V_0 = 90$ MeV. The deformation β_3 for octupole vibrations was assumed to be the same for all nuclei: $\beta_3 = 0.12$; this value is close to the known experimental values for spherical nuclei. β_2 values for nuclei in the maxima of the force functions were taken from the experimental values of B (E 2) for 2^+ levels. In order to obtain a smooth curve, in the regions A between the maxima intermediate β_2 values were used. The Γ_d value was assumed as 1.2 MeV for the entire range of nuclei considered. (In order to express Γ_d in units of A, the above-mentioned results of [11] were used for the energies of the single-particle levels.) The parameter Γ_d can be assumed as equal to the sum of the three widths $\Gamma_d = \Gamma_d^\downarrow + \Gamma_d^\uparrow + \Gamma_{\gamma d}$, where $\Gamma_{\gamma d}$ has the meaning of the radiative width of an intermediate state. According to an estimate given in §3, $\Gamma_{\gamma d} \sim 100$ eV and hence, the contribution to Γ_d can be ignored. The order of magnitude of $\Gamma_d^\uparrow$ can be estimated from the force function maxima: $S_0 \sim \Gamma_d^\uparrow / \Gamma_d^\downarrow \lesssim 5 \cdot 10^{-4}$. For neutron energies of 1 eV we obtain $\Gamma_d^\uparrow \lesssim 5 \cdot 10^{-4} \; \Gamma_d^\downarrow$, whereas $\Gamma_d^\uparrow \lesssim 0.2 \; \Gamma_d^\downarrow$ results for energies of about 200 keV. Thus, for low energy neutrons we have $\Gamma_d \approx \Gamma_d^\downarrow$. The Γ_d

value which we obtained is an estimate of the upper limit of this parameter, because the width of the maxima of the force function (averaged over A) can strongly depend upon fluctuations in the energy of the intermediate states in neighboring nuclei.

The parameter values which were used for calculating the curve of Fig. 1 were also used for calculating the force function of p neutrons (see Fig. 2). For the sake of simplicity the wave function of the open channel was taken for a potential without spin—orbit binding. In comparing the calculated curves with the experimental data shown in Figs. 1 and 2 one must bear in mind that the experimental force functions were usually obtained by averaging over resonances in a relatively narrow interval of energies with $\Delta E \lesssim 10$ keV. This is the reason why the large observed differences in the force functions for neighboring nuclei (particularly in the region of small A values) can be explained by random fluctuations as well as by a fine structure of intermediate resonances. From the viewpoint of the input-state model of the phonon—particle type, $\Delta E \sim 100$ keV must be considered as an interval which is adequate for the averaging. With the smaller averaging interval which is used in experiments, a comparison of the calculated curve with the experimental force functions is sensible only when the averaging is extended over neighboring nuclei.

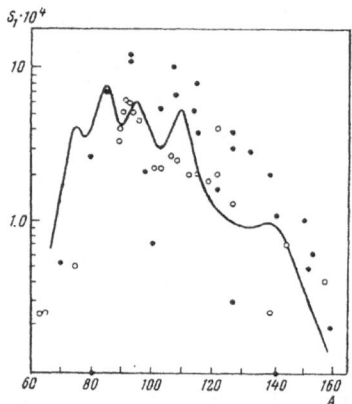

Fig. 2. Force function for p neutrons. The black circles indicate experimental values which were obtained from the capture cross sections; the open circles indicate values obtained from the total cross section.

We see from Figs. 1 and 2 that the model suggested explains adequately the force functions of neutrons. Naturally, with the assumptions which were made, the model is applicable only to nuclei situated close to the maxima of the intermediate resonances. In the case of nuclei which are situated between these resonances, a more complicated situation is encountered, because other input states may play an important role in the latter type of nuclei. However, with the model given one can qualitatively explain the unusually small force functions which were found for several nuclei, for example, for A ~ 100, in the case of s neutrons. The fact that these nuclei, which are far from the intermediate resonances considered, have abnormally small force functions confirms our assumption that other types of input states are less important.

The successful application of the single-particle vibrational intermediate state model in the simple form stated above for describing neutron force functions proves convincingly that the model is capable of correctly representing the interaction between neutrons and nuclei. It would be very helpful to establish the effect which the states discussed in this article have upon other nuclear processes involving neutrons. The radiative capture of neutrons is one of these processes.

§3. Radiative Neutron Capture by Intermediate States

A large number of experimental results concerning the radiative capture of neutrons with energies $\lesssim 1$ MeV can be adequately interpreted either with the optical or with the statistical model. These models help to obtain the capture cross section and its dependence upon the energy and the mass number. However, it is known that the statistical model cannot explain the spectra of the γ quanta which are emitted in neutron-capture processes. This fact led to the development of a model for the direct capture of neutrons in the resonance region (Lane and Lynn [12]). As mentioned by Lane and Lynn, the characteristic feature of γ spectra resulting from the capture of thermal neutrons by nuclei with A = 25-75 and A = 170-200 is an "abnormal" region corresponding to γ transitions into lower states of the nucleus formed in the reaction. In this region, the intensity of the transitions is much greater than predicted by the statistical model. The abnormal region has a width of about 1 MeV and its contribution to the total capture cross section is about 0.1 or, occasionally, higher.

The theory of a direct capture makes it possible to include radiative E1 transitions which occur from a single-particle state of the neutron in an open channel to bound single-particle states. More particularly, in the regions A = 25-75 and A = 170-200, such transitions can end in p states which are near the ground state, when s-neutron-capture processes take place. The model of Lane and Lynn can in several cases qualitatively explain the observed transition intensities in the regions A = 25-75. However, in the region A = 170-200, the observed intensities are usually several times greater than the theoretical values.

As has been mentioned above, in the γ spectra of (n,γ) reactions in several nuclei, another property was observed which cannot be explained by the direct capture model. Increased γ transitions occur to levels which are weakly excited in (d,p) reactions [4, 5]. Moreover, it was noted in [4] that in the case of the Fe^{57} nucleus a correlation of the excitation probabilities of certain levels exists between (n,γ) and (p,p') reactions. It is generally accepted that in the latter reaction collective levels are excited with a high probability.

When an intermediate state with a sufficiently long lifetime is formed during the interaction between a neutron and a nucleus, the emission of a γ quantum can occur directly from this intermediate state. According to the collective capture model of [10], in the capture of fast neutrons a γ quantum is emitted due to a collective E1 transition. In the case of single-particle vibrational states in which quadrupole and octupole phonons play a role, single-particle E1 transitions have the highest probabilities. A capture of neutrons through such intermediate states must increase the probability for the excitation of lower single-particle vibrational levels. For example, in the region A = 50-80 or A = 180-200, a capture of s neutrons by intermediate states with one quadrupole phonon and a neutron in the d state can increase the probability of γ transitions to levels which are a mixture of a single-particle state of a neutron with $l = 1$ and a quadrupole vibration of the core. If this is the case, a numerical estimate of the effect is possible.

The probability for a direct radiative transition from an intermediate d state to a finite state can be characterized by the width $\Gamma_{\gamma d}$. When a radiative capture is observed for a neutron resonance λ, the partial radiative width of this resonance is $\Gamma_{\lambda \gamma d} = |a_{\lambda d}|^2 \Gamma_{\gamma d}$, where $|a_{\lambda d}|^2$ denotes the contribution of the d state to the actual state λ of the nucleus. A similar expression can be formulated for the neutron width of the resonance λ: $\Gamma_{\lambda n} = |a_{\lambda d}|^2 \Gamma^{\uparrow}$. By averaging the latter expression over the resonances and comparing it with Eq. (4), we can express $\overline{|a_{\lambda d}|^2}$ with the parameters of the intermediate state; this gives us the following expression for $\overline{\Gamma}_{\lambda \gamma d}$:

$$\overline{\Gamma}_{\lambda \gamma d} = \frac{D}{2\pi} \frac{\Gamma_d^{\downarrow} \Gamma_{\gamma d}}{(E - E_d)^2 + \frac{1}{4} \Gamma_d^2}. \tag{7}$$

For estimating $\overline{\Gamma}_{\lambda \gamma d}$ we can assume that

$$\overline{\Gamma}_{\lambda \gamma d} \approx D \frac{\Gamma_{\gamma d}}{\Gamma_d^{\downarrow}} \tag{8}$$

holds in the maximum of an intermediate resonance.

We can set $\Gamma_d^\frac{1}{2} \lesssim 1$ MeV according to the above calculations of the force functions in which the parameter $\Gamma_d^\frac{1}{2}$ enters. For a calculation of $\Gamma_{\gamma d}$ it suffices to use the probabilities of single-particle transitions in a rectangular, infinitely deep well (those calculations which were performed in [13]). In the case of E1 transitions between levels with the same principal quantum number, these probabilities are 2-3 times smaller than those resulting from the calculations of Weisskopf. When we assume for the effective charge of the neutron e = $\frac{1}{2}$e, for the nuclear radius R = $1.35 \cdot A^{1/3} \cdot 10^{-13}$ cm, and for the statistical factor S = 1, we obtain the following values for the width of radiative E1 transitions at the γ-quantum energy $E_\gamma = 7$ MeV: for the 2d → 2p transition (A = 50-80) we find $\Gamma_{\gamma d} \approx 40$ eV, and for the 3d → 3p transition (A = 180-200) we obtain $\Gamma_{\gamma d} \approx 70$ eV. In other words, the order of magnitude of $\overline{\Gamma}_{\lambda \gamma d} \approx 10^{-4} \cdot D$, where D denotes the average distance between the resonances.

Let us compare this estimate of $\overline{\Gamma}_{\lambda \gamma d}$ with the experimentally obtained average radiative width $\overline{\Gamma}_{\lambda \gamma}$. In the region A = 50-80, we have $\overline{\Gamma}_{\lambda \gamma} \approx 0.5$ eV and D ~ 500 eV; consequently, we $\overline{\Gamma}_{\lambda \gamma d} \sim 0.1 \ \overline{\Gamma}_{\lambda \gamma}$. In the region A = 180-200, we find $\overline{\Gamma}_{\lambda \gamma} \approx 0.1$ eV and the average distance between resonances (for even-even target nuclei) is D ~ 50-100 eV, which gives us $\overline{\Gamma}_{\lambda \gamma d} \sim (0.05-0.1) \ \overline{\Gamma}_{\lambda \gamma}$. In other words, the contribution of radiative transitions from intermediate phonon/particle states to the cross section of the (n, γ) reaction may amount to about 10%, which corresponds to the observed abnormal region of the γ spectrum, as far as the order of magnitude of the effect is concerned.

It is worthwhile to compare the estimate of $\overline{\Gamma}_{\lambda \gamma d} / D$ given by Eq. (8) with the corresponding quantity obtained in the direct capture model of [12]. Lane and Lynn used the optical model with the Woods-Saxon potential and calculated the quantity $\Gamma_{\lambda \gamma f} / D \varepsilon_\gamma^3$, where $\Gamma_{\lambda \gamma f}$ denotes the average radiative width for the direct transition from an open channel to a single-particle state with $l = 1$ (ε_γ denotes the energy of the γ quantum in MeV). This quantity amounts to $(3 - 0.7) \cdot 10^{-7}$ for the region A = 60-70, and to $(2 - 0.7) \cdot 10^{-7}$ for A = 170-200. Our estimate of $\overline{\Gamma}_{\lambda \gamma d} / D \varepsilon_\gamma^3$ is ~ $3 \cdot 10^{-7}$. On the other hand, according to Cameron's calculations of the particle radiative widths, the statistical model renders for $\overline{\Gamma}_{\lambda \gamma f} / D \varepsilon_\gamma^3$ values of about $0.7 \cdot 10^{-8}$ for A = 60-70 and of about 10^{-8} for A = 170-200 [12]. Both the direct capture and the capture through an intermediate state of the phonon/particle type result in radiative transition probabilities which exceed by at least an order of magnitude the probabilities predicted by the statistical model.

So far we have considered radiative neutron capture in which intermediate states with one phonon occur. One must keep in mind that in (n, γ) reactions, γ transitions resulting from admixtures of other, relatively simple states to the initial state of the system can strongly contribute to the abnormal region in the γ spectrum. With the model under consideration, one would expect the following initial state components with two-phonon excitations of the core and a single-particle state neutron (the initial state components are listed in the order of increasing complexity).

For example, when we start from the assumption of single-particle neutron levels, we can expect in the region A = 50-80 that at excitation energies close to the binding energy of the neutron, states with two quadrupole phonons and a neutron in the 1 $g_{7/2}$ state, as well as states with one quadrupole and one octupole phonon and a neutron in the 1 $h_{11/2}$ state, will significantly contribute to the wave function of the nucleus. These states result from E1 transitions involving states which are a superposition of the two-phonon excitation of the core and a 1 $f_{5/2}$ and a 1 $g_{9/2}$ state of the neutron (these latter states may be important for levels close to the ground state). In the Fe^{56}(n, γ)Fe^{57} reaction (in which in addition to the Fe^{57} ground state and levels close to the ground state — 14 keV, levels with the energies 1625 and 1723 keV as well as a group of levels between 3 and 4 MeV are excited with a high probability), some of the boosted transitions result mainly from admixtures of two-phonon excitations in both the initial and final states.

We see that the excitation probability of the various levels in (n,γ) reactions can be determined by several relatively simple single-particle components or more complicated components in both the initial and the final state. Obviously, the relative importance of radiative transitions resulting from these components can vary from case to case.

In order to clarify the capture mechanism involving single-particle vibrational intermediate states, it is suitable to consider not only the γ-emission spectra resulting from (n,γ) reactions, but also the energy dependence of the average excitation cross section of the various lines. In the case of an isolated intermediate state, the average cross section for a neutron capture without formation of a compound nucleus can be written in the following form (in analogy to the collective capture cross section of [10]):

$$\bar{\sigma}_{n\gamma d} = \pi\lambda^2 \frac{\Gamma_d^\uparrow \Gamma_{\gamma d}}{(E - E_d)^2 + \frac{1}{4}\Gamma_d^2} \, ,$$

or, when Eq. (4) is taken into account:

$$\bar{\sigma}_{n\gamma d} = 2\pi^2\lambda^2 \frac{\overline{\Gamma}_{\lambda n}}{D} \frac{\Gamma_{\gamma d}}{\Gamma_d^\downarrow} \, .$$

These expressions reveal that in the energy dependencies of $\bar{\sigma}_{n\gamma d}$, the same structure as in the neutron force functions must be observed. Interestingly enough, $\bar{\sigma}_{n\gamma d}$ is independent of the average distance between the levels, whereas the average cross sections of the (n, γ) reactions which involve a compound nucleus are inversely proportional to D at rather high neutron energies. Therefore, one can anticipate a particularly strong appearance of intermediate states in the case of nuclei with a low level density.

Conclusion

The above examples for the application of the intermediate state model indicate that the model can provide a logical explanation of several effects which are observed in interactions between low-energy neutrons and complicated nuclei. This is another proof of the validity of the intermediate state concept for nuclear systems and of the value of this concept.

The question of how important the various intermediate states are will be answered when the model is developed further and used. From the results of our work and from the other publications on single-particle vibrational intermediate states one can conclude that in a number of cases intermediate states which correspond to coherent excitation of several particles are very important, and can be described with the collective model.

The principle of intermediate states in nuclear reactions is closely related to the problem of residual interactions between the nuclei in a nucleus. The dominating role of intermediate collective states in reactions involving neutrons attests to the importance of the quadrupole and octupole terms in the multipole expansion of the residual potential.

Section 3 proves that the intermediate state concept can be successfully employed in research on the radiative capture of neutrons. One can foresee that a deeper understanding of the radiative capture mechanism will make radiative capture an important means for analyzing nuclear states, among them low levels (including the ground state) and nonstationary states with energies exceeding the binding energy of the neutron.

References

1. B. Block and H. Feshbach, Ann. Phys., 23:49 (1963).
2. C. M. Shakin, Ann. Phys., 22:373 (1963).

3. H. Feshbach, Nuclear Structure Study with Neutrons, Proc. Intern. Conf., Antwerpen, July 19–23, 1965.

4. H. Ikegamy and G. Emery, Phys. Rev. Lett., 13:26 (1964).

5. L. V. Groshev and A. M. Demidov, Yadern. Fiz. 4:785 (1966).

6. V. I. Popov, Report on the 16th Annual Conference on Nuclear Spectroscopy, Moscow, January 1966.

7. A. Lande and G. E. Brown, Nucl. Phys., 75:344 (1966).

8. N. Azziz, Phys. Lett., 23:337 (1966).

9. A. Bohr and B. Mottelson, Dan. Mat. Fys. Medd., Vol. 27, No. 16 (1953).

10. A. M. Lane, Nuclear Structure Study with Neutrons, Proc. Intern. Conf., Antwerpen, July 19–23, 1965.

11. A. E. Green, Proc. Intern. Conf. Nucl. Opt. Model, Florida State Univ., 1959.

12. A. M. Lane and J. E. Lynn, Nucl. Phys., 17:563 (1960).

13. D. Wilkinson, Physica, 22:1039 (1956).

RADIATION OF A CHARGED PARTICLE IN THE PRESENCE OF A SEPARATING BOUNDARY

V. E. Pafomov

Introduction

Charged particles emit electromagnetic waves even in the case of uniform, linear motion. This effect was predicted by Ginzburg and Frank [1]. The transition radiation which results when a charged particle passes through a plane boundary between two media, as discussed in [1], is a classic example of radiation in the presence of inhomogeneities. Later on, other researchers considered particular features of radiation by relativistic particles, the radiation from a diffuse boundary, from a boundary with a ferrodielectric substance or with a dielectric crystal, the transition radiation in the case of spatial dispersion and under gyrotropic conditions, in the case of inclined incidence of the particle on the boundary, etc.*

Several years ago there appeared the first experimental papers on transition radiation. At the present time, ways for practical applications[†] of transition radiation have been suggested, in addition to the continuation of the experimental work. For example, it was suggested that resonance-radiation effects in a layered medium (which develop when the transition radiation from several boundaries adds up) be used for the detection of high-energy particles. Transition radiation can also be used for measurements of the optical properties of metals and for detecting and examining thin surface films. In order to become familiar with the present state of transition-radiation research, the reader is referred to the articles [2] and [3]. The potential practical value of the effect has considerably stimulated additional research work.

The radiation of a charged particle passing through a separating boundary is usually considered under the assumption that the particle motion is uniform and linear [2]. The theoretical results obtained under this assumption agree strongly with the experimental results, provided that the changes in the particle velocity are small along the path over which the waves are absorbed. If this is not the case, substantial deviations from the theory are observed and then the theory must take into account the real conditions of the particle motion in the medium. To do this, one must first of all calculate the radiation field of an arbitrarily moving charged particle when boundaries separating various media are present. The so-called image-representation technique is used throughout this article for this purpose. This technique is based on the possibility of representing the field of a charged particle by a sum of dipole fields, with the dipoles

---

*See review article by F. G. Bass and V. M. Yakovenko [2] and the literature cited there.
†A review of the experimental work has been given in the article by I. M. Frank [3].

situated along the particle trajectory [4]; the images of these dipoles can be introduced and, making use of the reciprocity theorem, the field of an arbitrarily moving charged particle can be expressed directly by the law of motion of the particle. Below, we will consider solutions obtained under certain particular assumptions regarding the particle motion.

The article comprises six chapters. In Chapter 1, the field of a moving charged particle is determined for the case that one boundary between media with different dielectric constants and magnetic susceptibilities is present. This chapter discusses recent articles [5, 6] by the author. Chapter 2 considers the radiation of a particle moving in a ferrodielectric medium ($\varepsilon \neq 1$, $\mu \neq 1$) perpendicular to the vacuum−ferrodielectric-medium boundary. Particular attention is paid to the radiation in the frequency range involving negative group velocities, a case which can occur at $\varepsilon \neq 1$ and $\mu \neq 1$. Radiation in the presence of one boundary is considered under various specific assumptions on the particle motion. Chapter 3 describes the field of a moving charged particle in the presence of two boundaries and considers the emission of an oscillator in a thin surface film as well as the radiation of a charged particle moving uniformly and linearly through two boundaries. Chapter 4 concerns the radiation field of a charged particle which in some way passes through a layered medium; the radiation in the optical frequency range is described for the simplest case of uniform and linear motion. Chapter 5 concerns the radiation of a particle which is once, or several times, scattered in the medium. Chapter 6 describes the excitation of surface waves.

All specific considerations assume a radiation field of an arbitrarily moving charged particle in the presence of boundaries, a field which can be directly expressed by the law of the particle motion. Several of the results developed herein were obtained by the author in previous work [7–18], and in collaboration with I. M. Frank [19, 20] and E. P. Fetisov [21].

CHAPTER 1

The Image-Representation Technique

§1.   Introduction

When boundaries separating media are present, the solution of any problem of electrodynamics must satisfy the continuity condition for the tangential components of the electric and magnetic fields. Usually, the solution is found by matching the fields at the boundaries. However, it is possible to solve these problems without direct "joining" of the fields. One can use the well-known result that the Fourier component of the field of a moving charge is equivalent to the field created by a set of dipoles fixed along the particle trajectory [4] and employ the reciprocity theorem. Twenty years ago, Ginzburg and Frank [1] used this method (side by side with the direct field matching), which, to date, has not been further developed and applied. Ginsburg and Frank calculated the radiation field of a charged particle moving uniformly and linearly along the normal to the boundary separating two media. In [1], a test dipole was introduced and the radiation field was calculated at large distances from the boundary with the reciprocity theorem. The point of observation at which the test dipole was placed had to be far from the boundary in order to facilitate the use of the reciprocity theorem, since under this condition the radiation field of the test dipole could be considered as a plane wave along the path from the boundary to the particle trajectory.

This restriction (that the point of observation be far away) is removed (and a solution valid at any distance can be found) when dipole images are introduced in place of the test dipole and the reciprocity theorem is employed for determining a plane wave expansion of the field. Thus, in order to determine the field of an arbitrarily moving charged particle in the presence of plane boundaries separating media, we suggest another method instead of the test-dipole method (which is generalized to dipoles of arbitrary orientation). Dipole images are introduced instead of the boundaries, and the parameters of these dipoles can be determined from the condition that the fields of the following two cases be identical: 1) a moving charge (equivalent to a set of fixed dipoles) and a boundary are present, and 2) the same charge and its image representation (equivalent to a set of dipole images) are present, but no boundaries.

In developing the image method we will use the well-known result that the Fourier component of the field of a moving charge is equivalent to the field generated by a set of fixed dipoles [4].

The current density of an arbitrarily moving charged particle has the form

$$\mathbf{j} = e\mathbf{v}(t)\,\delta\,[\mathbf{r} - \mathbf{r}_0(t)]. \tag{1.1}$$

In the following discussion, all quantities will be expanded in Fourier integrals:

$$\mathbf{E} = \int_{-\infty}^{+\infty} \mathbf{E}_\omega e^{-i\omega t}\,d\omega, \quad \mathbf{H} = \int_{-\infty}^{+\infty} \mathbf{H}_\omega e^{-i\omega t}\,d\omega, \ldots, \text{etc.,} \tag{1.2}$$

and we will consider the Fourier components. The component of the current density is

$$\mathbf{j}_\omega = \int \mathbf{J}_\omega \delta\,[\mathbf{r} - \mathbf{r}_0(t)]\,dt, \tag{1.3}$$

where $\mathbf{J}_\omega$ denotes the Fourier component of the current per unit time:

$$\mathbf{J}_\omega = \frac{e\mathbf{v}}{2\pi}\,e^{i\omega t}. \tag{1.4}$$

A current distribution of this type is created by dipoles which are situated along the particle trajectory; these dipoles have the moment

$$\mathbf{p}_\omega = \frac{ie\mathbf{v}}{2\pi\omega v}\,e^{i\omega t} \tag{1.5}$$

per unit of the path. Thus, the radiation field of a moving charged particle can be considered as the result of the interference of waves emitted by sources (the dipoles) having the current distribution of Eq. (1.4) or, equivalently, having the dipole-moment distribution of Eq. (1.5) along the particle trajectory.

§2. Dipole Images

The problem of radiation of a moving charged particle in the presence of boundaries is reduced to determining dipole-image representations. Electric as well as magnetic dipoles can be used for the representation. To be more specific, let us assume that the dipole-image representation consists of electric dipoles. We will find their moments by using the reciprocity theorem which, for electric dipoles, states that the scalar product of the electric vector $\mathbf{E}_A^B$ at at point B, of the wave emitted by a dipole at point A times the moment of another dipole situated at point B is equal to the scalar product of the electric vector $\mathbf{E}_B^A$ at point A of the wave emitted by a similar dipole situated at point B times the moment of another dipole situated at point A. Since any field can be expanded in plane waves, the following result, which will be used below, follows from the reciprocity theorem. When two dipoles interact in the same fashion with a plane wave propagating in a certain direction, the scalar product of the electric field of the wave times the moments of these dipoles is constant in time. The plane waves emitted by the dipoles in a given direction are identical.

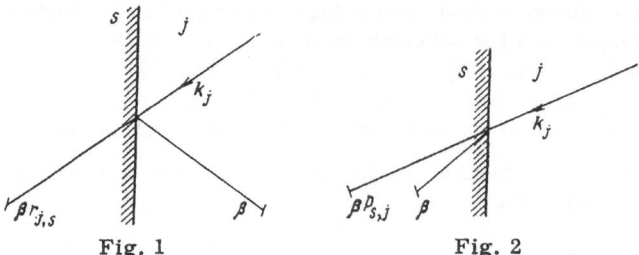

Fig. 1   Fig. 2

When we determine the moments of the dipole–image representations, then, according to the reciprocity theorem, we must postulate only that the scalar products of the electric field vector times the moment of the real dipole image, and of the electric field vector times the moment of the unknown dipole image are the same and, consequently, the orientation of the unknown dipole image can be arbitrary. In order to be more specific in the ensuing discussion, we will assume that the moments are collinear.

First, let us determine the moment of the dipole image describing the reflected wave field. We assume that a dipole with the moment β is in a half-space filled with a medium j at a certain distance from the plane boundary separating medium j from medium s (Fig. 1). When the interaction of the plane wave (in Fig. 1, its wave vector is denoted by $\mathbf{k}_j$) with a dipole βr_{js} is the same in the absence of a boundary as the interaction of the reflected wave with a real dipole, then the dipole βr_{js} is the image of dipole β and describes the reflected radiation field. Thus, the coefficient of the dipole image $r_{j,s}$ is equal to $r_{j,s} = \beta E_j^0/\beta E_j$, where E_j^0 and E_j denote the electric field of the reflected and incident plane waves at the point of the dipole and its image, respectively. In the following calculations, the coordinates of the image are immaterial and can be arbitrarily assumed.

The image determining the refracted wave field in medium j is found similarly. Let us consider the interaction between a plane wave propagating in medium j with the dipole vector β (Fig. 2) and its image $\beta \mathbf{p}_{s,j}$ (the subscripts of the coefficients $r_{j,s}$ and $p_{s,j}$ are written in the sequence which corresponds to the direction of the motion of the refracted or reflected waves emitted by the real dipole). Since the real dipole is behind the separating boundary in medium s, a plane wave is refracted on its path towards the real dipole. When we consider the image, we assume that the boundary is not present, i.e., we assume that the entire space is filled with medium j. The interaction between the wave and the dipole is the same in the two cases, provided that $p_{s,j} = \beta E_s^{\pi}/\beta E_j$, where E_s^{π} denotes the electric field vector of a plane refracted wave at the point at which the real dipole is located. According to the reciprocity theorem, the radiation of the dipoles in a particular direction is the same in these two cases.

Depending upon the wave polarizations, both the reflection and refraction coefficients are determined with the various Fresnel formulas. We must distinguish between the different polarizations in the wave images. The following calculations are made on the assumption that we consider the optical properties of the media. In the simplest case of isotropic media (which we will consider), waves polarized in the plane of incidence (the corresponding quantities will be denoted by the symbol ‖) must be distinguished from waves polarized in a plane perpendicular to the plane of incidence (symbol ⊥). Then we find

$$r_{j,s;\parallel} = \beta E_{j;\parallel}^0/\beta E_{j;\parallel}, \qquad p_{s,j;\parallel} = \beta E_{s;\parallel}^{\pi}/\beta E_{j;\parallel}, \qquad r_{j,s;\perp} = \beta E_{j;\perp}^0/\beta E_{j;\perp}, \qquad p_{s,j;\perp} = \beta E_{s;\perp}^{\pi}/\beta E_{j;\perp}. \quad (2.1)$$

Let us express these coefficients by the components of the wave vector. In a Cartesian x, y, z coordinate system with the xy plane coinciding with the boundary and a z axis pointing into medium j, the electric field of plane waves of various polarizations has the following form:

$$\mathbf{E}_{j;\parallel} \sim [(i\mathbf{k}_x + \mathbf{j}k_y) k_{jz} - \mathbf{k}\varkappa^2] \exp(-ik_x x - ik_y y - ik_{jz} z),$$

$$\mathbf{E}_{j;\parallel}^0 \sim -r'_{j,\,s;\,\parallel} [(i\mathbf{k}_x + \mathbf{j}k_y) k_{jz} + \mathbf{k}\varkappa^2] \exp(-ik_x x - ik_y y + ik_{jz} z),$$

$$\mathbf{E}_{s;\parallel}^{\Pi} \sim \frac{\varepsilon_j}{\varepsilon_s} p'_{j,\,s;\,\parallel} [(i\mathbf{k}_x + \mathbf{j}k_y) k_{sz} - \mathbf{k}\varkappa^2] \exp(-ik_x x - ik_y y - ik_{sz} z); \qquad (2.2)$$

$$\mathbf{E}_{j;\perp} \sim (i\mathbf{k}_y - \mathbf{j}k_x) \exp(-ik_x x - ik_y y - ik_{jz} z),$$

$$\mathbf{E}_{j;\perp}^0 \sim r'_{j,\,s;\,\perp} (i\mathbf{k}_y - \mathbf{j}k_x) \exp(-ik_x x - ik_y y + ik_{jz} z),$$

$$\mathbf{E}_{s;\perp}^{\Pi} \sim p'_{j,\,s;\,\perp} (i\mathbf{k}_y - \mathbf{j}k_x) \exp(-ik_x x - ik_y y - ik_{sz} z), \qquad (2.3)$$

where

$$k_x = k_{jx} = k_{sx}, \qquad k_y = k_{jy} = k_{sy}, \qquad \varkappa^2 = k_x^2 + k_y^2,$$

$$k_{jz} = \sqrt{k_j^2 - \varkappa^2}, \qquad k_{sz} = \sqrt{k_s^2 - \varkappa^2}, \qquad k_j = \omega\sqrt{\varepsilon_j\mu_j}/c, \qquad k_s = \omega\sqrt{\varepsilon_s\mu_s}/c; \qquad (2.4)$$

i, j, and k denote the unit vectors in the directions of the coordinate axes x, y, and z, respectively; $r'_{j,s;\parallel}$, $p'_{j,s;\parallel}$, $r'_{j,s;\perp}$, and $p'_{j,s;\perp}$ are the Fresnel coefficients:

$$r'_{j,\,s;\,\parallel} = (\varepsilon_s k_{jz} - \varepsilon_j k_{sz})/(\varepsilon_s k_{jz} + \varepsilon_j k_{sz}); \qquad p'_{j,\,s;\,\parallel} = 2\varepsilon_s k_{jz}/(\varepsilon_s k_{jz} + \varepsilon_j k_{sz}); \qquad (2.5)$$

$$r'_{j,\,s;\,\perp} = (\mu_s k_{jz} - \mu_j k_{sz})/(\mu_s k_{jz} + \mu_j k_{sz}); \qquad p'_{j,\,s;\,\perp} = 2\mu_s k_{jz}/(\mu_s k_{jz} + \mu_j k_{sz}). \qquad (2.6)$$

In Eq. (2.1), the electric field of the reflected and refracted waves must be taken at the point of the real dipole. The coordinates of these points will be denoted by x_ζ, y_ζ, and z_ζ. When the components of the dipole-image field expansion in plane waves are determined, both the time lag and the attenuation of the waves on the path from the dipole images to the point of observation must be taken into account. The results become independent of the coordinates of the dipole images. This is the reason why, in the intermediate calculations, these coordinates can be chosen arbitrarily. To be more specific, let us assume that the coordinates of the dipole images coincide with the point at which a straight line drawn from the real dipole's location perpendicular to the boundary intersects the boundary. When we assume that the real dipole is very close to the boundary so that the time lag and the attenuation of the waves on the path from the boundary to the dipole can be ignored, the unknown coefficients are given by the amplitudes of the electric field at the boundary and are independent of the dipole coordinates. The coefficients are

$$r_{j,\,s;\,\parallel} = \frac{\varepsilon_s k_{jz} - \varepsilon_j k_{sz}}{\varepsilon_s k_{jz} + \varepsilon_j k_{sz}} \frac{\beta_z \varkappa^2 + (\beta_x k_x + \beta_y k_y) k_{jz}}{\beta_z \varkappa^2 - (\beta_x k_x + \beta_y k_y) k_{jz}},$$

$$p_{s,\,j;\,\parallel} = \frac{2\varepsilon_j k_{jz}}{\varepsilon_s k_{jz} + \varepsilon_j k_{sz}} \frac{\beta_z \varkappa^2 - (\beta_x k_x + \beta_y k_y) k_{sz}}{\beta_z \varkappa^2 - (\beta_x k_x + \beta_y k_y) k_{jz}}; \qquad (2.7)$$

$$r_{j,\,s;\,\perp} = \frac{\mu_s k_{jz} - \mu_j k_{sz}}{\mu_s k_{jz} + \mu_j k_{sz}}, \qquad p_{s,\,j;\,\perp} = \frac{2\mu_s k_{jz}}{\mu_s k_{jz} + \mu_j k_{sz}}. \qquad (2.8)$$

In calculating the coefficients of the dipole images which determine the field in region s, we must include the interaction between dipoles and waves propagating under an acute angle relative to the z axis. This means that the subscripts s and j must be exchanged in Eqs. (2.7) and (2.8) and, in addition, that the signs before the z components of the wave vector must be changed. We obtain the following result:

$$r_{s,\,j;\,\parallel} = \frac{\varepsilon_j k_{sz} - \varepsilon_s k_{jz}}{\varepsilon_j k_{sz} + \varepsilon_s k_{jz}} \frac{\beta_z \varkappa^2 - (\beta_x k_x + \beta_y k_y) k_{sz}}{\beta_z \varkappa^2 + (\beta_x k_x + \beta_y k_y) k_{sz}},$$

$$p_{j,\,s;\,\parallel} = \frac{2\varepsilon_s k_{sz}}{\varepsilon_j k_{sz} + \varepsilon_s k_{jz}} \frac{\beta_z \varkappa^2 + (\beta_x k_x + \beta_y k_y)\,k_{jz}}{\beta_z \varkappa^2 + (\beta_x k_x + \beta_y k_y)\,k_{sz}}\,; \tag{2.9}$$

$$r_{s,\,j;\,\perp} = \frac{\mu_j k_{sz} - \mu_s k_{jz}}{\mu_j k_{sz} + \mu_s k_{jz}} \qquad p_{j,\,s;\,\perp} = \frac{2\mu_j k_{sz}}{\mu_j k_{sz} + \mu_s k_{jz}}\,. \tag{2.10}$$

When the real dipoles are not close to the boundary, the time lag and the attenuation of the waves are taken into account through exponential factors. In the case of dipoles in the media s ($z_\zeta < 0$) and j ($z_\zeta > 0$), these factors are $\exp\,(-ik_{sz}z_\zeta)$ and $\exp\,(ik_{jz}z_\zeta)$, respectively.

In other media, the moments of dipole images can be found similarly and the image representation method can be generalized and applied to crystal media, gyrotropic media, media with spatial dispersion, and moving media.

The coefficients $r_{s,j}$ and $p_{s,j}$ satisfy the following equations:

$$r_{s+1,\,s} + \frac{p_{s+1,\,s}\,p_{s,\,s-1}\,r_{s,\,s-1}}{1 - r_{s,\,s-1}\,r_{s,\,s+1}} = r_{s+1,\,s-1}, \qquad \frac{p_{s-1,\,s}\,p_{s,\,s+1}}{1 - r_{s,\,s-1}\,r_{s,\,s+1}} = p_{s-1,\,s+1}. \tag{2.11}$$

The meaning of these equations will be examined below when we consider two boundaries.

§3. Dipole Field and A Moving Charged Particle in the Presence of a Boundary

We will use the results of the preceding paragraph to determine the field generated by an arbitrarily oriented dipole when a boundary is present.

It is convenient to introduce the Hertz vector Π_ω for the following considerations:

$$\mathbf{E}_\omega = \frac{1}{\varepsilon}\,\mathrm{grad\,div}\,\Pi_\omega + \frac{\omega^2}{c^2}\,\mu\Pi_\omega,$$
$$\mathbf{H}_\omega = -\,(i\omega/c)\,\mathrm{rot}\,\Pi_\omega, \tag{3.1}$$

which gives us for the plane waves the following equations:

$$\mathbf{E}_\omega = \mu\left[\frac{\omega^2}{c^2}\,\Pi_\omega - \frac{1}{\varepsilon}\,\mathbf{k}\,(\mathbf{k}\Pi_\omega)\right],$$
$$\mathbf{H}_\omega = \frac{\omega}{c}\,[\mathbf{k}\Pi_\omega]. \tag{3.2}$$

The field of a dipole having the moment $\mathbf{p}_\omega$ satisfies the following wave equation for Π:

$$\Delta\Pi_\omega + k^2\Pi_\omega = -4\pi\mathbf{p}_\omega. \tag{3.3}$$

The solution of this equation has the form

$$\Pi_\omega = \mathbf{p}_\omega \exp\,(ikR)/R, \tag{3.4}$$

where R denotes the distance from the dipole to the point of observation. We rewrite this formula and resolve the spherical wave e^{ikR}/R into nonuniform plane waves:

$$\Pi_\omega = \frac{i}{2\pi}\iint \frac{\mathbf{p}_\omega}{k_z}\exp\,[ik_x(x-x_\zeta) + ik_y(y-y_\zeta) + ik_z\,|\,z-z_\zeta\,|]\,dk_x\,dk_y, \tag{3.5}$$

where x_ζ, y_ζ, z_ζ denotes the dipole coordinates and x, y, z the coordinates of the point of observation.

In order to use this formula along with the results of the preceding paragraph, we must write the Hertz vector as the sum of two vectors. One of these vectors corresponds to the waves polarized in the plane of their propagation ($\Pi_{\omega\,\parallel}$) and the other to the waves polarized in a perpendicular plane $(\Pi_{\omega\perp})$. In an x, y, z coordinate system with the z axis perpendicular to the boundary, the components of a plane wave expansion are related by the following expressions:

$$\Pi_{\omega\parallel}^{k} = \mathbf{k} \left[\Pi_{\omega z}^{k} - (\Pi_{\omega x}^{k} k_{x} + \Pi_{\omega y}^{k} k_{y}) \frac{k_{z}}{\varkappa^{2}} \right],$$

$$\Pi_{\omega\perp}^{k} = \mathbf{i}\Pi_{\omega x}^{k} + \mathbf{j}\Pi_{\omega y}^{k} + \mathbf{k} (\Pi_{\omega x}^{k} k_{x} + \Pi_{\omega y}^{k} k_{y}) \frac{k_{z}}{\varkappa^{2}}, \tag{3.6}$$

where $\mathbf{i}$, $\mathbf{j}$, and $\mathbf{k}$ denote the unit vectors in the direction of the coordinate axes. With the direction of the Hertz vector coinciding with the orientation of the dipole β, we obtain from Eq. (3.6)

$$\Pi_{\omega\parallel}^{k} \sim \mathbf{k} \left[\beta_{z} - (\beta_{x} k_{x} + \beta_{y} k_{y}) \frac{k_{z}}{\varkappa^{2}} \right],$$

$$\Pi_{\omega\perp}^{k} \sim \mathbf{i}\beta_{x} + \mathbf{j}\beta_{y} + \mathbf{k} (\beta_{x} k_{x} + \beta_{y} k_{y}) \frac{k_{z}}{\varkappa^{2}}. \tag{3.7}$$

When we employ Eqs. (3.6) and (3.7) and the coefficients of the dipole-image representations, we can calculate in the region j (z > 0) the dipoles having the moment given by Eq. (1.5) in the region s $(z_{\zeta} < 0)$:

$$\Pi_{\omega s\parallel}^{\zeta} = - \frac{e\mathbf{k}e^{i\omega t}}{4\pi^{2}\omega\beta} \iint \frac{P_{s,\,j;\,\parallel}}{k_{jz}} \left[\beta_{z} - (\beta_{x} k_{x} + \beta_{y} k_{y}) \frac{k_{jz}}{\varkappa^{2}} \right] \exp{(if_{1})}\, dk_{x}\, dk_{y}, \tag{3.8}$$

$$\Pi_{\omega s\perp}^{\zeta} = - \frac{e \cdot e^{i\omega t}}{4\pi^{2}\omega\beta} \iint \frac{P_{s,\,j;\,\perp}}{k_{jz}} \left[\mathbf{i}\beta_{x} + \mathbf{j}\beta_{y} + \mathbf{k} (\beta_{x} k_{x} + \beta_{y} k_{y}) \frac{k_{jz}}{\varkappa^{2}} \right] \exp{(if_{1})}\, dk_{x}\, dk_{y} \tag{3.9}$$

where

$$f_{1} = k_{x} (x - x_{\zeta}) + k_{y} (y - y_{\zeta}) + k_{jz} z - k_{sz} z_{\zeta}. \tag{3.10}$$

In the region j (z > 0), the field of a dipole located in region j $(z_{\zeta} > 0)$ is equal to the sum of the following Hertz vectors:

$$\Pi_{\omega j\parallel}^{\zeta} = - \frac{e\mathbf{k}e^{i\omega t}}{4\pi^{2}\omega\beta} \iint \frac{1}{k_{jz}} \left[\beta_{z} - (\beta_{x} k_{x} + \beta_{y} k_{y}) \frac{k_{jz}}{\varkappa^{2}} \right] \exp{(if_{1}')}\, dk_{x}\, dk_{y}, \tag{3.11}$$

$$\Pi_{\omega j\perp}^{\zeta} = - \frac{e \cdot e^{i\omega t}}{4\pi^{2}\omega\beta} \iint \frac{1}{k_{jz}} \left[\mathbf{i}\beta_{x} + \mathbf{j}\beta_{y} + \mathbf{k} (\beta_{x} k_{x} + \beta_{y} k_{y}) \frac{k_{jz}}{\varkappa^{2}} \right] \exp{(if_{1}')}\, dk_{x}\, dk_{y}, \tag{3.12}$$

$$\Pi_{\omega j\parallel}^{\prime\prime\zeta} = - \frac{e\mathbf{k}e^{i\omega t}}{4\pi^{2}\omega\beta} \iint \frac{r_{j,\,s;\,\parallel}}{k_{jz}} \left[\beta_{z} - (\beta_{x} k_{x} + \beta_{y} k_{y}) \frac{k_{jz}}{\varkappa^{2}} \right] \exp{(if_{1}')}\, dk_{x}\, dk_{y}, \tag{3.13}$$

$$\Pi_{\omega s\perp}^{\prime\prime\zeta} = - \frac{e \cdot e^{i\omega t}}{4\pi^{2}\omega\beta} \iint \frac{r_{j,\,s;\,\perp}}{k_{jz}} \left[\mathbf{i}\beta_{x} + \mathbf{j}\beta_{y} + \mathbf{k} (\beta_{x} k_{x} + \beta_{y} k_{y}) \frac{k_{jz}}{\varkappa^{2}} \right] \exp{(if_{1}')}\, dk_{x}\, dk_{y}, \tag{3.14}$$

where

$$f_{1}' = k_{x} (x - x_{\zeta}) + k_{y} (y - y_{\zeta}) + k_{jz} (z - z_{\zeta}), \tag{3.15}$$

$$f_{1}' = k_{x} (x - x_{\zeta}) + k_{y} (y - y_{\zeta}) + k_{jz} (z + z_{\zeta}). \tag{3.16}$$

Since we have the condition that the field be finite at infinity, $\mathrm{Im}\,k_{jz} \geq 0$ and $\mathrm{Im}\,k_{sz} \geq 0$. These inequalities are assumed to hold everywhere, provided that $z > z_{\zeta}$. In order to obtain the field in the region $z < z_{\zeta}$ the signs before the integrals and before k_{jz} must be changed in Eqs. (3.11) and (3.12). Then, as has been indicated above, we have everywhere $\mathrm{Im}\,k_{jz} \geq 0$ and $\mathrm{Im}\,k_{sz} \geq 0$. The Hertz vectors for the region s are found from those calculated for the region j by transposing the subscripts j and s and changing the signs before the integrals and the components k_{jz} and k_{sz}.

These results are a rigorous solution to the problem of the field generated by an arbitrarily oriented dipole situated at some distance from the boundary separating media with different dielectric constants and magnetic permeabilities. Therefore, this solution is more general than that of the well-known Sommerfeld problem, in which the field of a vertically or horizontally oriented dipole situated at the boundary separating nonmagnetic media is considered (see, e.g., [22]).

The field of an arbitrarily moving charged particle in the presence of a boundary can be calculated from integrals over the dipole fields along the particle trajectory.

In the wave zone, the relative small region of variables around the extrema of the functions appearing in the exponent of the exponential functions contributes to the integrals. The integration is performed according to the stationary phase technique. We begin this integration by switching to new variables φ_1 and $\varkappa$, which are related to k_x and k_y by the expressions

$$k_x = \varkappa \cos \varphi_1, \qquad k_y = \varkappa \sin \varphi_1 \qquad (\varkappa^2 = k_x^2 + k_y^2). \tag{3.17}$$

The functions of Eqs. (3.10), (3.15), and (3.16) assume the following form:

$$f_1 = \varkappa \rho_\zeta + \sqrt{k_j^2 - \varkappa^2}\, z - \sqrt{k_s^2 - \varkappa^2}\, z_\zeta, \tag{3.18}$$

$$f_1' = \varkappa \rho_\zeta + \sqrt{k_j^2 - \varkappa^2}\,(z - z_\zeta), \tag{3.19}$$

$$f_1'' = \varkappa \rho_\zeta + \sqrt{k_j^2 - \varkappa^2}\,(z + z_\zeta), \tag{3.20}$$

where

$$\varkappa = i\varkappa \cos \varphi_1 + j\varkappa \sin \varphi_1, \quad \rho_\zeta = i\rho_\zeta \cos \varphi + j\rho_\zeta \sin \varphi. \tag{3.21}$$

Expanding the functions f_1, f_1', and f_2'' in power series of $\Delta\varphi_1$ near the extrema of these functions and integrating over φ_1, we obtain results which reduce to a replacement of k_x by $\varkappa \cos \varphi$ and of k_y by $\varkappa \sin \varphi$ and a multiplication by

$$\sqrt{-2i\pi/\varkappa\rho_\zeta}. \tag{3.22}$$

After integration over φ, the functions in the exponents of the exponential functions assume the following form:

$$f = \varkappa \rho_\zeta + \sqrt{k_j^2 - \varkappa^2}\, z - \sqrt{k_s^2 - \varkappa^2}\, z_\zeta, \tag{3.23}$$

$$f' = \varkappa \rho_\zeta + \sqrt{k_j^2 - \varkappa^2}\,(z - z_\zeta), \tag{3.24}$$

$$f'' = \varkappa \rho_\zeta + \sqrt{k_j^2 - \varkappa^2}\,(z + z_\zeta), \tag{3.25}$$

where

$$\rho_\zeta = \sqrt{(x - x_\zeta)^2 + (y - y_\zeta)^2}, \tag{3.26}$$

and the components k_x and k_y of the wave vector are related by the following equation

$$k_x/k_y = (x - x_\zeta)/(y - y_\zeta). \tag{3.27}$$

We will use once more the stationary phase method and integrate over $\varkappa$. By equating the functions f, f', and f'' to zero, we obtain

$$\rho_\zeta - \frac{\varkappa z}{\sqrt{k_j^2 - \varkappa^2}} + \frac{\varkappa z_\zeta}{\sqrt{k_s^2 - \varkappa^2}} = 0, \tag{3.28}$$

$$\rho_\zeta - \frac{\varkappa (z - z_\zeta)}{\sqrt{k_j^2 - \varkappa^2}} = 0, \tag{3.29}$$

$$\rho_\zeta - \frac{\varkappa (z + z_\zeta)}{\sqrt{k_j^2 - \varkappa^2}} = 0. \tag{3.30}$$

The second derivatives of the functions f, f', and f'' are equal to

$$\frac{d^2 f}{dx^2} = - \frac{k_j^2 z}{(k_j^2 - \varkappa^2)^{3/2}} + \frac{k_s^2 z_\zeta}{(k_s^2 - \varkappa^2)^{3/2}},$$

$$\frac{d^2 f'}{dx^2} = - \frac{k_j^2 (z - z_\zeta)}{(k_j^2 - \varkappa^2)^{3/2}}, \quad \frac{d^2 f''}{dx^2} = - \frac{k_j^2 (z + z_\zeta)}{(k_j^2 - \varkappa^2)^{3/2}}. \tag{3.31}$$

The regions of integration which give substantial contributions to the integrals are of the order of the square root of the inverse of the second derivatives in the extrema. We introduce the notations $\Delta \varkappa$, $\Delta \varkappa'$, and $\Delta \varkappa''$:

$$\Delta \varkappa = \sqrt{2} \left[\frac{k_j^2 z}{(k_j^2 - \varkappa^2)^{3/2}} - \frac{k_s^2 z_\zeta}{(k_s^2 - \varkappa^2)^{3/2}} \right]^{-1/2},$$

$$\Delta \varkappa' = \frac{\sqrt{2} (k_j^2 - \varkappa^2)^{3/4}}{k_j \sqrt{z - z_\zeta}}, \quad \Delta \varkappa'' = \frac{\sqrt{2} (k_j^2 - \varkappa^2)^{3/4}}{k_j \sqrt{z + z_\zeta}}, \tag{3.32}$$

where $\varkappa$ satisfies each of the equations (3.28), (3.29), and (3.30).

The integration over $\varkappa$ reduces to multiplying the expressions in the integral by

$$\varkappa \sqrt{- i\pi} \Delta \varkappa, \quad \varkappa \sqrt{- i\pi} \Delta \varkappa', \quad \varkappa \sqrt{- i\pi} \Delta \varkappa'' \tag{3.33}$$

and to replacing $\varkappa$ by values satisfying Eqs. (3.28), (3.29), and (3.30).

In other words, the result of the integration over k_x and k_y in Eqs. (3.8) and (3.9) reduces to multiplying the expressions in the integral by

$$- 2\pi i \sqrt{\frac{\varkappa}{\rho_\zeta}} \left[\frac{k_j^2 z}{(k_j^2 - \varkappa^2)^{3/2}} - \frac{k_s^2 z_\zeta}{(k_s^2 - \varkappa^2)^{3/2}} \right]^{-1/2} \tag{3.34}$$

and to replacing k_x, k_y, and $\varkappa$ by values satisfying Eqs. (3.27) and (3.28). The results of the integration in Eqs. (3.11) and (3.12) reduce to multiplying by

$$- 2\pi i \frac{k_j^2 - \varkappa^2}{k_j (z - z_\zeta)} \tag{3.35}$$

and replacing k_x, k_y, and $\varkappa$ by values satisfying Eqs. (3.27) and (3.29); accordingly, in Eqs. (3.13) and (3.14) we must multiply with

$$- 2\pi i \frac{k_j^2 - \varkappa^2}{k_j (z + z_\zeta)} \tag{3.36}$$

and replace k_x, k_y, and $\varkappa$ by values satisfying Eqs. (3.27) and (3.30).

Let us write the results of the integration in Eqs. (3.8) and (3.9) in the form

$$\Pi_{\omega s \parallel}^\zeta = \frac{iek}{2\pi\omega} \frac{p_{s,\,j;\,\parallel}}{\beta R_{s,\,j}} \left[\beta_z - (\beta_x^! k_x + \beta_y k_v) \frac{k_{jz}}{\varkappa^2} \right] \exp \left[i \left(\omega t + k_s R_s + k_j R_j \right) \right], \tag{3.37}$$

$$\Pi^{\zeta}_{\omega s \perp} = \frac{ie}{2\pi\omega} \frac{P_{s,\,j;\,\perp}}{\beta R_{s,\,j}} \left[\mathbf{i}\beta_x + \mathbf{j}\beta_y + \mathbf{k}\,(\beta_x k_x + \beta_y k_y)\frac{k_{jz}}{\varkappa^2} \right] \exp\left[i\,(\omega t + k_s R_s + k_j R_j)\right], \tag{3.38}$$

where

$$R_s = -z_\zeta k_{sz}/k_s, \qquad R_j = \frac{\varkappa}{k_j}\rho_\zeta + \frac{k_{jz}}{k_j}z, \tag{3.39}$$

$$R_{s,\,j} = k_{jz}\sqrt{\frac{\rho_\zeta}{\varkappa}\left(\frac{z}{k_{jz}^3} - \frac{z_\zeta}{k_{sz}^3}\right)}; \tag{3.40}$$

k_x, k_y, and $\varkappa$ satisfy Eqs. (3.27) and (3.28).

The results of the integration in Eqs. (3.11) and (3.12) have the form

$$\Pi^{'\zeta}_{\omega j \|} = \frac{iek}{2\pi\omega} \frac{1}{\beta R_j'} \left[\beta_z - (\beta_x k_x + \beta_y k_y)\frac{k_{jz}}{\varkappa^2} \right] \exp\left[i\,(\omega t + k_j R_j')\right], \tag{3.41}$$

$$\Pi^{'\zeta}_{\omega j \perp} = \frac{ie}{2\pi\omega} \frac{1}{\beta R_j'} \left[\mathbf{i}\beta_x + \mathbf{j}\beta_y + \mathbf{k}\,(\beta_x k_x + \beta_y k_y)\frac{k_{jz}}{\varkappa^2} \right] \exp\left[i\,(\omega t + k_j R_j')\right], \tag{3.42}$$

where

$$R_j' = \frac{k_j}{k_{jz}}(z - z_\zeta); \tag{3.43}$$

in this case, the components k_j of the wave vector satisfy Eqs. (3.27) and (3.29)

Finally, integration of the integrals in Eqs. (3.13) and (3.14) leads to the following results in the wave zone:

$$\Pi^{''\zeta}_{\omega j \|} = \frac{iek}{2\pi\omega} \frac{r_{j,\,s;\,\|}}{\beta R_{s,\,j}''} \left[\beta_z - (\beta_x k_x + \beta_y k_y)\frac{k_{jz}}{\varkappa^2} \right] \exp\left[i\,(\omega t + k_j R_{s,\,j}'')\right], \tag{3.44}$$

$$\Pi^{''\zeta}_{\omega j \perp} = \frac{ie}{2\pi\omega} \frac{r_{j,\,s;\,\perp}}{\beta R_{s,\,j}''} \left[\mathbf{i}\beta_x + \mathbf{j}\beta_y + \mathbf{k}\,(\beta_x k_x + \beta_y k_y)\frac{k_{jz}}{\varkappa^2} \right] \exp\left[i\,(\omega t + k_j R_{s,\,j}'')\right], \tag{3.45}$$

where

$$R_{s,\,j}'' = \frac{k_j}{k_{jz}}(z + z_\zeta); \tag{3.46}$$

the wave-vector components are solutions of Eqs. (3.27) and (3.30).

Equations (3.37) and (3.38) hold under the condition

$$|k_s R_s + k_j R_j| \gg 1. \tag{3.47}$$

The range in which the results stated in Eqs. (3.41) and (3.42) can be employed is defined by the inequality

$$|k_j|\,R_j' \gg 1, \tag{3.48}$$

while Eqs. (3.44) and (3.45) are correct for

$$|k_j|\,R_{s,\,j}'' \gg 1. \tag{3.49}$$

When neither absorption nor total reflection occur in medium s, R_s and R_j are real and equal to the distance (in the direction of the wave vector of the refracted wave) between dipole A (Fig. 3) and the boundary (R_s = AB) and to the distance between the boundary and point C of observation (R_j = BC). When absorption or total reflection of the wave occurs, R_s and R_j are complex and no longer have that simple meaning in terms of physics. R_j' denotes the distance

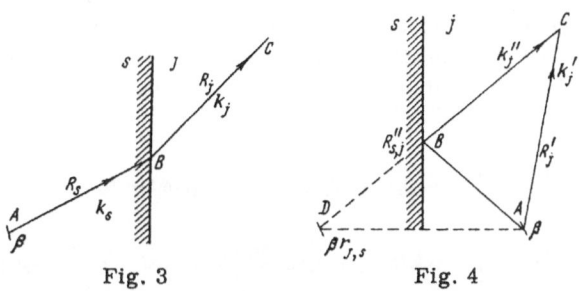

Fig. 3   Fig. 4

between dipole A in region j and the point of observation (Fig. 4); $R_{s,j}''$ is the distance (along the wave vector of the reflected wave) between the dipole and the point of observation (from the dipole image D to the point of observation, we have $R_{s,j}'' = AB + BC = DB + BC$). In Fig. 3, $\mathbf{k}_j$ denotes the wave vector whose components satisfy Eqs. (3.27) and (3.28). In Fig. 4, $\mathbf{k}_j'$ is the wave vector whose components satisfy Eqs. (3.27) and (3.29). The components of the wave vector $\mathbf{k}_j''$ are defined by Eqs. (3.27) and (3.30).

The interference of the radiation fields generated by all dipoles is obtained by integrating Eqs. (3.37), (3.38), (3.41), (3.42), (3.44), and (3.45) over the trajectory of the charged particle.

In region j, the field generated by a particle moving in region s is given by the equations

$$\Pi_{\omega s \parallel} = \frac{iek}{2\pi\omega} \int \frac{P_{s,j;\parallel}}{\beta R_{s,j}} \left[\beta_z - (\beta_x k_x + \beta_y k_y) \frac{k_{jz}}{\varkappa^2} \right] \exp\left[i\left(\omega t + k_s R_s + k_j R_j\right) \right] d\zeta, \tag{3.50}$$

$$\Pi_{\omega s \perp} = \frac{ie}{2\pi\omega} \int \frac{P_{s,j;\perp}}{\beta R_{s,j}} \left[\mathbf{i}\beta_x + \mathbf{j}\beta_y + \mathbf{k}(\beta_x k_x + \beta_y k_y) \frac{k_{jz}}{\varkappa^2} \right] \exp\left[i\left(\omega t + k_s R_s + k_j R_j\right) \right] d\zeta. \tag{3.51}$$

The radiation field created by a particle moving in region j is given by the sum of the following Hertz vectors:

$$\Pi'_{\omega j \parallel} = \frac{iek}{2\pi\omega} \int \frac{1}{\beta R_j'} \left[\beta_z - (\beta_x k_x + \beta_y k_y) \frac{k_{jz}}{\varkappa^2} \right] \exp\left[i\left(\omega t + k_j R_j'\right) \right] d\zeta, \tag{3.52}$$

$$\Pi'_{\omega j \perp} = \frac{ie}{2\pi\omega} \int \frac{1}{\beta R_j'} \left[\mathbf{i}\beta_x + \mathbf{j}\beta_y + \mathbf{k}(\beta_x k_x + \beta_y k_y) \frac{k_{jz}}{\varkappa^2} \right] \exp\left[i\left(\omega t + k_j R_j'\right) \right] d\zeta, \tag{3.53}$$

$$\Pi''_{\omega j \parallel} = \frac{iek}{2\pi\omega} \int \frac{r_{j,s;\parallel}}{\beta R_{s,j}''} \left[\beta_z - (\beta_x k_x + \beta_y k_y) \frac{k_{jz}}{\varkappa^2} \right] \exp\left[i\left(\omega t + k_j R_{s,j}''\right) \right] d\zeta, \tag{3.54}$$

$$\Pi''_{\omega j \perp} = \frac{ie}{2\pi\omega} \int \frac{r_{j,s;\perp}}{\beta R_{s,j}''} \left[\mathbf{i}\beta_x + \mathbf{j}\beta_y + \mathbf{k}(\beta_x k_x + \beta_y k_y) \frac{k_{jz}}{\varkappa^2} \right] \exp\left[i\left(\omega t + k_j R_{s,j}''\right) \right] d\zeta. \tag{3.55}$$

The quantities under the integral signs of Eqs. (3.50)–(3.55) are functions of both the particle coordinates and particle-velocity components. These expressions describe the radiation field of a moving particle independently of the law governing the particle's motion.

§ 4. Spherical Waves

When the point of observation is moved farther out, the angles between the wave vectors $\mathbf{k}_j$, $\mathbf{k}_j'$, and $\mathbf{k}_j''$ decrease. At sufficiently large distances, the dependence of the wave vectors on the particle coordinates can be ignored and the wave vectors can be assumed as parallel. At the same time, the dependence of the distance to the point of observation upon the particle coordinates becomes less pronounced. When $R_{s,j}$, R_j', and $R_{s,j}''$ are far enough away, they differ only slightly from R, the distance from some fixed point near the trajectory. Then the quantities $R_{s,j}$, R'_j, and $R''_{s,j}$ in the denominators of the expressions under the integrals of Eqs. (3.50)-(3.55) can be replaced by the constant R and drawn before the integral sign. Spherical waves are the result. We switch from contour integrals to integrals over the time and write them in the following form:

$$\Pi_{\omega s\parallel} = \frac{ieck}{2\pi\omega R}\exp(ik_j R)\int P_{s,j;\parallel}\left[\beta_z - (\beta_x k_x + \beta_y k_y)\frac{k_{jz}}{\varkappa^2}\right]\exp\left[i(\omega t - k_x x - k_y y - \sqrt{k_s^2 - \varkappa^2}\,z)\right]dt, \quad (4.1)$$

$$\Pi_{\omega s\perp} = \frac{iec p_{s,j;\perp}}{2\pi\omega R}\exp(ik_j R)\int\left[i\beta_x + j\beta_y + k(\beta_x k_x + \beta_y k_y)\frac{k_{jz}}{\varkappa^2}\right]\exp\left[i(\omega t - k_x x - k_y y - \sqrt{k_s^2 - \varkappa^2}\,z)\right]dt, \quad (4.2)$$

$$\Pi'_{\omega j\parallel} = \frac{ieck}{2\pi\omega R}\exp(ik_j R)\int\left[\beta_z - (\beta_x k_x + \beta_y k_y)\frac{k_{jz}}{\varkappa^2}\right]\exp\left[i(\omega t - k_x x - k_y y - k_{jz} z)\right]dt, \quad (4.3)$$

$$\Pi'_{\omega j\perp} = \frac{iec}{2\pi\omega R}\exp(ik_j R)\int\left[i\beta_x + j\beta_y + k(\beta_x k_x + \beta_y k_y)\frac{k_{jz}}{\varkappa^2}\right]\exp\left[i(\omega t - k_x x - k_y y - k_{jz} z)\right]dt, \quad (4.4)$$

$$\Pi''_{\omega j\parallel} = \frac{ieck}{2\pi\omega R}\exp(ik_j R)\int r_{j,s;\parallel}\left[\beta_z - (\beta_x k_x + \beta_y k_y)\frac{k_{jz}}{\varkappa^2}\right]\exp\left[i(\omega t - k_x x - k_y y + k_{jz} z)\right]dt, \quad (4.5)$$

$$\Pi''_{\omega j\perp} = \frac{iec r_{j,s;\perp}}{2\pi\omega R}\exp(ik_j R)\int\left[i\beta_x + j\beta_y + k(\beta_x k_x + \beta_y k_y)\frac{k_{jz}}{\varkappa^2}\right]\exp\left[i(\omega t - k_x x - k_y y + k_{jz} z)\right]dt, \quad (4.6)$$

where x, y, z denote the particle coordinates at the time t.

In the following, we will consider radiation in vacuum ($\varepsilon_j = 1$, $\mu_j = 1$). Both the electric and magnetic fields of spherical waves of different polarization are determined by the direction cosines of the wave vector in vacuum and the Hertz vector according to the equations

$$E_{\omega\parallel} = -\omega^2\Pi_{\omega\parallel}(i\cos\vartheta_x\cos\vartheta_z + j\cos\vartheta_y\cos\vartheta_z - k\sin^2\vartheta_z)/c^2,$$

$$E_{\omega\perp} = \omega^2(\Pi_{\omega\perp x}\cos\vartheta_y - \Pi_{\omega\perp y}\cos\vartheta_x)(i\cos\vartheta_y - j\cos\vartheta_x)/c^2\sin^2\vartheta_z,$$

$$H_{\omega\parallel} = \omega^2\Pi_{\omega\parallel}(i\cos\vartheta_y - j\cos\vartheta_x)/c^2, \quad (4.7)$$

$$H_{\omega\perp} = \omega^2(\Pi_{\omega\perp x}\cos\vartheta_y - \Pi_{\omega\perp y}\cos\vartheta_x)[(i\cos\vartheta_x \ j\cos\vartheta_y)\cos\vartheta_z - k\sin^2\vartheta_z]/c^2\sin^2\vartheta_z.$$

When we insert the values of the coefficients given by Eqs. (2.7) and (2.8) in the particular case $\varepsilon_s = \varepsilon$, $\varepsilon_j = 1$, $\mu_j = \mu_s = 1$, we obtain from Eqs. (4.1)-(4.6) the following results, which are expressed through the direction cosines of the wave vector in vacuum (the subscripts s and j are replaced by 1 and 2, respectively):

$$\Pi_{\omega 1\parallel} = \frac{ieck\exp\left(i\frac{\omega}{c}R\right)\cos\vartheta_z}{\pi\omega R(\varepsilon\cos\vartheta_z + \sqrt{\varepsilon - \sin^2\vartheta_z})}\int\Big[\beta_z - (\beta_x\cos\vartheta_x +$$

$$+ \beta_y\cos\vartheta_y)\frac{\sqrt{\varepsilon - \sin^2\vartheta_z}}{\sin^2\vartheta_z}\Big]\exp\left[i\omega t - i\frac{\omega}{c}(x\cos\vartheta_x + y\cos\vartheta_y + z\sqrt{\varepsilon - \sin^2\vartheta_z})\right]dt, \quad (4.8)$$

$$\Pi_{\omega 1\perp} = \frac{iec\exp\left(i\frac{\omega}{c}R\right)\cos\vartheta_z}{\pi\omega R(\cos\vartheta_z + \sqrt{\varepsilon - \sin^2\vartheta_z})}\int\Big[i\beta_x + j\beta_y + k(\beta_x\cos\vartheta_x +$$

$$+ \beta_y\cos\vartheta_y)\frac{\cos\vartheta_z}{\sin^2\vartheta_z}\Big]\exp\left[i\omega t - i\frac{\omega}{c}(x\cos\vartheta_x + y\cos\vartheta_y + z\sqrt{\varepsilon - \sin^2\vartheta_z})\right]dt, \quad (4.9)$$

$$\Pi'_{\omega 2\parallel} = \frac{ieck\exp\left(i\frac{\omega}{c}R\right)}{2\pi\omega R}\int\Big[\beta_z - (\beta_x\cos\vartheta_x + \beta_y\cos\vartheta_y)\frac{\cos\vartheta_z}{\sin^2\vartheta_z}\Big]\times$$

$$\times\exp\left[i\omega t - i\frac{\omega}{c}(x\cos\vartheta_x + y\cos\vartheta_y + z\cos\vartheta_z)\right]dt, \quad (4.10)$$

$$\mathbf{\Pi}'_{\omega 2 \perp} = \frac{iec \exp\left(i\frac{\omega}{c}R\right)}{2\pi\omega R} \int \left[i\beta_x + j\beta_y + \mathbf{k}\left(\beta_x \cos\vartheta_x + \right.\right.$$

$$\left.\left. + \beta_y \cos\vartheta_y\right) \frac{\cos\vartheta_z}{\sin^2\vartheta_z}\right] \exp\left[i\omega t - i\frac{\omega}{c}(x\cos\vartheta_x + y\cos\vartheta_y + z\cos\vartheta_z)\right] dt, \qquad (4.11)$$

$$\mathbf{\Pi}'_{\omega 2 \parallel} = \frac{iec\mathbf{k}\exp\left(i\frac{\omega}{c}R\right)}{2\pi\omega R} \frac{\varepsilon\cos\vartheta_z - \sqrt{\varepsilon - \sin^2\vartheta_z}}{\varepsilon\cos\vartheta_z + \sqrt{\varepsilon - \sin^2\vartheta_z}} \int \left[\beta_z + (\beta_x \cos\vartheta_x + \right.$$

$$\left. + \beta_y \cos\vartheta_y) \frac{\cos\vartheta_z}{\sin^2\vartheta_z}\right] \exp\left[i\omega t - i\frac{\omega}{c}(x\cos\vartheta_x + y\cos\vartheta_y - z\cos\vartheta_z)\right] dt, \qquad (4.12)$$

$$\mathbf{\Pi}''_{\omega 2 \perp} = \frac{iec \exp\left(i\frac{\omega}{c}R\right)}{2\pi\omega R} \frac{\cos\vartheta_z - \sqrt{\varepsilon - \sin^2\vartheta_z}}{\cos\vartheta_z + \sqrt{\varepsilon - \sin^2\vartheta_z}} \int \left[i\beta_x + j\beta_y + \right.$$

$$\left. + \mathbf{k}(\beta_x \cos\vartheta_x + \beta_y \cos\vartheta_y) \frac{\cos\vartheta_z}{\sin^2\vartheta_z}\right] \exp\left[i\omega t - i\frac{\omega}{c}(x\cos\vartheta_x + y\cos\vartheta_y - z\cos\vartheta_z)\right] dt. \qquad (4.13)$$

The solution can be simplified when we add the vectors $\mathbf{\Pi}_{\omega 1 \parallel}$, $\mathbf{\Pi}'_{\omega 2 \parallel}$, and $\mathbf{\Pi}''_{\omega 2 \parallel}$ to the vectors $\mathbf{\Pi}_{\omega 1 \perp}$, $\mathbf{\Pi}'_{\omega 2 \perp}$, and $\mathbf{\Pi}''_{\omega 2 \perp}$. The result is

$$\mathbf{\Pi}_{\omega 1} = \frac{iec \exp\left(i\frac{\omega}{c}R\right)\cos\vartheta_z}{\pi\omega R(\varepsilon\cos\vartheta_z + \sqrt{\varepsilon - \sin^2\vartheta_z})} \int \left\{ (i\beta_x + j\beta_x)(\sin^2\vartheta_z + \right.$$

$$+ \cos\vartheta_z \sqrt{\varepsilon - \sin^2\vartheta_z}) + \mathbf{k}[\beta_z + (\beta_x \cos\vartheta_x + \beta_y \cos\vartheta_y)(\cos\vartheta_z -$$

$$\left. - \sqrt{\varepsilon - \sin^2\vartheta_z})]\right\} \exp\left[i\omega t - i\frac{\omega}{c}(x\cos\vartheta_x + y\cos\vartheta_y + z\sqrt{\varepsilon - \sin^2\vartheta_z})\right] dt, \qquad (4.14)$$

$$\mathbf{\Pi}'_{\omega 2} = \frac{iec \exp\left(i\frac{\omega}{c}R\right)}{2\pi\omega R} \int (i\beta_x + j\beta_y + \mathbf{k}\beta_z) \exp\left[i\omega t - i\frac{\omega}{c}(x\cos\vartheta_x + y\cos\vartheta_y + z\cos\vartheta_z)\right] dt, \qquad (4.15)$$

$$\mathbf{\Pi}''_{\omega 2} = \frac{iec \exp\left(i\frac{\omega}{c}R\right)}{2\pi\omega R(\varepsilon\cos\vartheta_z + \sqrt{\varepsilon - \sin^2\vartheta_z})} \int \left\{ (i\beta_x + j\beta_y)[2\cos\vartheta_z(\sin^2\vartheta_z + \cos\vartheta_z \sqrt{\varepsilon - \sin^2\vartheta_z}) - \right.$$

$$- (\varepsilon\cos\vartheta_z + \sqrt{\varepsilon - \sin^2\vartheta_z})] + \mathbf{k}[2\cos\vartheta_z(\beta_x \cos\vartheta_x + \beta_y \cos\vartheta_y)(\cos\vartheta_z -$$

$$\left. - \sqrt{\varepsilon - \sin^2\vartheta_z}) + \beta_z(\varepsilon\cos\vartheta_z - \sqrt{\varepsilon - \sin^2\vartheta_z})]\right\} \exp\left[i\omega t - i\frac{\omega}{c}(x\cos\vartheta_x + y\cos\vartheta_y - z\cos\vartheta_z)\right] dt. \qquad (4.16)$$

The spectral density of the radiation energy per unit solid angle is given by any one of the following five formulas:

$$W_{\mathbf{n}\omega} = cR^2(\mathbf{E}_\omega \mathbf{H}^*_\omega + \mathbf{E}^*_\omega \mathbf{H}_\omega)/2, \qquad (4.17)$$

$$W_{\mathbf{n}\omega} = cR^2|\mathbf{E}_\omega|^2, \qquad (4.18)$$

$$W_{\mathbf{n}\omega} = cR^2|\mathbf{H}_\omega|^2, \qquad (4.19)$$

$$W_{\mathbf{n}\omega} = \omega^4 R^2 [|\mathbf{\Pi}_\omega|^2 - |(\mathbf{n}\mathbf{\Pi}_\omega)|^2]/c^3, \qquad (4.20)$$

$$W_{\mathbf{n}\omega} = \omega^4 R^2 |[\mathbf{n}\mathbf{\Pi}_\omega]|^2/c^3. \qquad (4.21)$$

In calculations of the angular distribution of the radiation intensity of waves having one or the other polarization, the vectors of the spherical waves of the corresponding polarization must be inserted in these equations.

These results are the basis for radiation calculation which will be performed under various assumptions regarding the motion of a charged particle in the presence of a boundary.

CHAPTER 2

Radiation in the Presence of a Separating Boundary

§ 5. Time and Path of Coherence

The path transgressed by a particle during a time within which the phase of the wave changes by about unity at the point of the particle plays an important role in the theory of electromagnetic wave radiation. This path characterizes the interaction between the particle and the electromagnetic wave and determines the radiation intensity. The change of the wave's phase at the point of the particle (i.e., the phase as a function of time) is determined by the exponents in the exponential functions of Eqs. (4.1)-4.6). According to these formulas, the rate of phase change differs for the primary, reflected, and refracted radiation fields. In the following, this characteristic section of the particle trajectory will be termed "path of coherence" or "path of coherent interaction between the particle and the wave." Similarly, the time in which the particle is on the crest of the wave will be termed coherence time. In the simplest case (uniform, linear motion with $\mathbf{v} = \mathbf{r}t$) the phase of the wave propagating under an angle ϑ relative to the particle's velocity vector has the time dependence $\varphi = \omega t(1 - n\beta\cos\vartheta)$ at the point of the particle. When we assume that φ is of the order of unity, we obtain for the coherence path and the coherence time

$$\tau \sim (\omega\,|\,1 - n\beta\cos\vartheta\,|)^{-1}, \qquad s \sim v\,(\omega\,|\,1 - n\beta\cos\vartheta\,|)^{-1}. \tag{5.1}$$

This means that for $n\beta > 1$ the phase of a wave propagating under an angle $\vartheta = \arccos 1/n\beta$ relative to the velocity vector of the particle does not change in time at the point of the particle: τ and s tend to infinity. We know that the Cerenkov radiation is generated in this direction, a radiation which can be explained by the coherent interaction effect between the field and the particle.

A particular result of Eq. (5.1) is that the time and the path of coherence in vacuum are equal to

$$\tau \sim [\omega\,(1 - \beta\cos\vartheta)]^{-1}, \qquad s \sim v\,[\omega\,(1 - \beta\cos\vartheta)]^{-1}. \tag{5.2}$$

When, during a time which is smaller than the coherence time in the case of uniform and linear motion, a particle is scattered (for example, twice), the radiation emitted during these changes is characterized by interference effects. Accordingly, when a particle passes through an inhomogeneity of the medium (e.g., through a foil) the radiation generated during the passage through the first and second boundaries creates an interference pattern, provided that the time between the two boundary passes is shorter than the coherence time. When two processes involving changes of the particle velocity and the phase velocity of the wave are separated by a time interval which is much greater than the coherence time, the interference can be ignored in averaging over a small frequency range, and the two radiation processes can be considered as independent. In this case, the coherence path plays the role of the effective diameter of the radiation "source," and this diameter is related to the particular change of the velocity.

When a separating boundary exists, reflected and refracted wave fields occur along with the primary radiation field. When a particle moves uniformly and linearly and perpendicular to the boundary separating two media, the time of coherent interaction between the particle and the refracted wave is determined by the condition that the absolute value of the exponents of the exponential function in the integral of Eq. (4.1) must be unity. In the case of a wave propagating in vacuum after refraction at an angle ϑ relative to the velocity vector, the time of coherent interaction is

$$\tau \sim [\omega\,|\,1 - \beta\,\sqrt{\varepsilon\mu - \sin^2\vartheta}\,|]^{-1}. \tag{5.3}$$

In the event that at the point of the particle the phase of the wave changes insignificantly, but the wave amplitude changes substantially, τ is given by the attentuation. Then τ denotes the full interaction time. The corresponding path of coherent interaction is equal to

$$s \sim v \left[\omega \left| 1 - \beta \sqrt{\varepsilon\mu - \sin^2\vartheta} \right| \right]^{-1}. \tag{5.4}$$

The time and the path of coherent interaction with reflected waves in vacuum (the wave vectors of these waves are assumed to include an acute angle ϑ with the particle-velocity vector after reflection) are

$$\tau \sim [\omega(1 + \beta\cos\vartheta)]^{-1}, \qquad s \sim v[\omega(1 + \beta\cos\vartheta)]^{-1}. \tag{5.5}$$

Equations (5.1)-(5.4) and Eq. (5.5) have been stated for particle motion from a medium into vacuum. When the direction of motion is reversed, β must be replaced by $-\beta$.

We wish to note that the coherence path is of the same order of magnitude as the diameter of the "source" of radiation generated in a particular interaction. The coherence path determines the extension of the region in which the radiation field is transformed into a spherical wave. Equations (3.50)-(3.55) are the basis for the treatment of this problem.

§6. The Radiation Field Generated over a Finite Path of a Uniformly Moving Charge in a Ferrodielectric Medium

In the following we will discuss the radiation from a particle which passes through a vacuum-ferrodielectric ($\varepsilon \neq 1$, $\mu \neq 1$) boundary normal to the direction of particle motion [8]. Since the field of the transition radiation is equivalent to the radiation field of a stopped or an ejected charge and its images (one of which is ejected while another one is stopped at the same point and at the same time), it is logical to consider first the radiation of the charge and then its images.

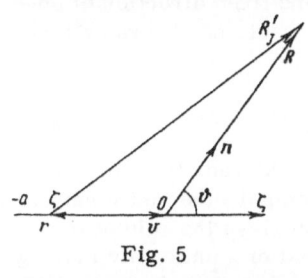

Fig. 5

Let us consider the radiation which is generated when a uniformly moving charge runs through a finite length of path in a medium without a boundary; we assume that the medium has a dielectric constant different from unity and a magnetic permeability different from unity. We start from Eq. (3.52) and assume $\beta_x = \beta_y = 0$, $\beta_z = \beta = \text{const}$ for $-a < \zeta < 0$ and $\beta_x = \beta_y = \beta_z = 0$ for $\zeta < -a, \zeta > 0$. We place the coordinate origin at the point at which the particle is suddenly stopped and expand the absolute value of the radius vector R_j' (extending from the dipole ζ to the point of observation) in powers of the absolute value of the radius vector of the dipole. When we restrict ourselves to the quadratic term in the expansion, we obtain

$$R_j' = R - \mathbf{rn} + \frac{\mathbf{r}^2}{2R} - \frac{(\mathbf{rn})^2}{2R}, \tag{6.1}$$

where $\mathbf{R}$ and $\mathbf{r}$ denote the radius vectors of the point of observation and the radiating dipole, respectively, and $\mathbf{n}$ the unit vector in the direction of $\mathbf{R}$ (see Fig. 5).

When we use this expansion, which is valid for $r \ll R$, we obtain from Eq. (3.52)

$$\Pi_\omega = \frac{ie v}{2\pi\omega R v} \int\limits_{-a}^{0} \exp\left[i\frac{\omega}{v}\left(\zeta - \zeta n\beta\cos\vartheta + \frac{n\beta\zeta^2\sin^2\vartheta}{2R}\right) + i\frac{\omega n}{c}R\right]d\zeta \quad (n = \sqrt{\varepsilon\mu}). \tag{6.2}$$

The linear term in ζ of the expansion has the same value as the ratio ζ/s in the exponent of the exponential function in the integral expression. When the particle travels over a certain path length a (with $a \gg s$), the phase difference of the waves emitted at the ends of the path segment a is much greater than unity. Therefore, we can neglect the interference radiation resulting from the instantaneous stopping of the particle at the point $\zeta = 0$ and from the ejection of the particle at the point $\zeta = -a$ when we average over a relatively small frequency interval. The two radiation processes can be considered as independent.

Let us consider the radiation resulting from an instantaneous arrest of the particle. According to what has been outlined above, along a path of length s the particle motion is uniform and linear up to the point at which the particle is stopped. At sufficiently large distances from the trajectory, a substantial range of the ζ integration in Eq. (6.2) has a size of the order of the path length s of coherent interaction. In the case in which the distance R is so large that the term proportional to the square of ζ in the expansion of the exponent of the exponential function in the integral of Eq. (6.2) can be neglected along the path s, i.e., if

$$R \gg \frac{\omega}{c} n s^2 \sin^2 \vartheta \approx \frac{nv^2 \sin^2 \vartheta}{\omega c \,|\, 1 - n\beta \cos \vartheta \,|^{\,2}} \,, \qquad (6.3)$$

we obtain, after integration over ζ, a spherical wave as the radiation field:

$$\Pi_\omega = \frac{e\mathbf{v}}{2\pi\omega^2 R} \frac{\exp\left(i\,\frac{\omega n}{c}\,R\right)}{1 - n\beta \cos \vartheta} \,. \qquad (6.4)$$

The inequality (6.3) determines the size of the region in which the spherical wave field exists.

We wish to note that the amplitude of the field of a spherical wave generated when a uniformly moving particle is suddenly stopped, or when a particle is suddenly ejected and moves on uniformly and linearly, is proportional to the path of coherent interaction.

It is well-known that the entire radiation field generated by some finite source transforms into a spherical wave at distances R beyond which for $R \to \infty$ (and fixed direction of observation) the phase difference of waves arriving from various source sections differs by less than unity. This condition is fulfilled, if

$$R \gg (\omega/c)\, n s_m^2, \qquad (6.5)$$

where s_m^2 denotes the maximum linear dimensions of the projections of the radiation source onto a plane perpendicular to the direction of observation. When we compare the last formula with that of inequality (6.3), we see that the path of coherent interaction plays the role of the diameter of the radiation source resulting from the instantaneous arrest of a uniformly moving particle (or the ejection and subsequent linear and uniform motion of a particle). The quantity $s \sin \vartheta$ is equal to the projection of the radiation source onto the direction of observation.

When a particle moves in a medium with a velocity exceeding the phase velocity of light ($n\beta > 1$), the radiation field near the trajectory is that a cylindrical wave (Cerenkov radiation). Let us calculate the field of the cylindrical wave and determine the region in which this wave exists when the particle travels over a finite path segment. In order to avoid the introduction of new notations, we will assume that the particle does not stop at the point $\zeta = 0$, but travels on with the initial velocity.

We denote with ϑ_r' the angle between the radius vector of the point of observation and the velocity vector of the particle; this angle satisfies the equation $n\beta \cos \vartheta = 1$ ($\vartheta_r' = \text{arc cos } 1/n\beta$). In the direction $\vartheta = \vartheta_r'$, the linear term in ζ in the exponent of the exponential function appearing in Eq. (6.2) tends to zero, and the point $\zeta = 0$ is the center of the region of integration over ζ. The size of this region is equal to

$$\Delta\zeta \sim \sqrt{cR/\omega n}/\sin\vartheta_r', \tag{6.6}$$

within which the quadratic term of the expansion in terms of ζ of the exponent of the exponential function is of the order of unity or smaller than unity.

The expansion stated in Eq. (6.1) can be used in Eq. (6.2) when the condition $\Delta\zeta \sin\vartheta_r' \ll R$ holds, i.e., if

$$R \gg c/\omega n. \tag{6.7}$$

In this case, it is correct to replace ϑ (which depends upon ζ) with the constant ϑ_r' as outlined above, because the $\Delta\vartheta_r'$ interval corresponding to $\Delta\zeta$ of Eq. (6.6) is given by

$$\Delta\vartheta_r' \sim \sqrt{c/\omega R n}, \tag{6.8}$$

i.e., the $\Delta\vartheta_r'$ interval is much greater than unity.

After substituting the integration variable

$$\zeta = u \sqrt{2cR/\omega n}/\sin\vartheta_r', \tag{6.9}$$

Eq. (6.2) assumes the form

$$\Pi_\omega = \frac{iev\,\sqrt{c}\exp\left(i\,\dfrac{\omega n}{c}\,R\right)}{\omega v\,\sqrt{2R\omega n}\,\pi\sin\vartheta_r'}\int e^{iu^2}\,du. \tag{6.10}$$

The integral appearing in this formula is equal to $\sqrt{i\pi}$. Thus, our result is

$$\Pi_\omega = -\frac{ev\,\sqrt{-i\omega c}\exp\left(i\,\dfrac{\omega n}{c}\,R\right)}{\omega^2 v\,\sqrt{2\pi n R}\,\sin\vartheta_r'}, \tag{6.11}$$

where R denotes the distance between the coordinate origin and the point of observation. The point of observation is situated on a straight line which includes the angle $\vartheta_r' = \arccos 1/n\beta$ with the velocity vector of the particle and passes through the coordinate origin.

Obviously, a similar result is obtained for other points in the space which are not situated on this straight line:

$$\Pi_\omega = -\frac{ev\,\sqrt{-i\omega c}\exp\left[i\,\dfrac{\omega n}{c}\,R\cos(\vartheta-\vartheta_r')\right]}{\omega^2 v\,\sqrt{2\pi R n \sin\vartheta\,\sin\vartheta_r'}}, \tag{6.12}$$

where R denotes the distance from the coordinate origin to the point of observation and ϑ the angle between **R** and **v**. Equation (6.12) describes the radiation field in those points of space from which straight lines drawn at an angle ϑ_r' relative to the particle trajectory intersect the trajectory, while the particle velocity remains constant over a length of the order of $\Delta\zeta$ (given by Eq. (6.6)), in the neighborhood of the point of intersection.

The consequence is that when the particle travels a finite length of path, the cylindrical wave can fill only a fraction of the space between the circular cones having their apeces at the coordinate origin and at the trajectory point under consideration, respectively (see Fig. 6, in which this region has been shaded). The nonshaded region has angular dimensions of the order of $\Delta\vartheta_r'$ [see Eq. (6.8)]. Within this region, the field differs from that stated in Eq. (6.12) because a substantial portion of the region of ζ integra-

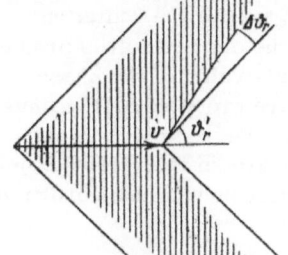

Fig. 6

tion in Eq. (6.2) comprises points of observation, among them the point of particle ejection or sudden particle arrest. On the resulting cones, the point at which the particle is ejected or suddenly stopped is located in the center of the region of integration and, consequently, there the amplitude of the wave is half the amplitude of the cylindrical wave described by Eq. (6.12).

When the distance R increases, from a certain R value on, the region $\Delta\zeta$ [defined by Eq. (6.6)] becomes of the order of magnitude of the path length a traveled by the particle. Since $\Delta\zeta \le a$, the region of ζ integration includes the beginning as well as the end of the particle path at increasing distances to the point of observation. Then, $\Delta\vartheta_r'$ exceeds the angular distance between the resulting cones. The corresponding distances satisfy the inequality

$$R > a^2\omega n \sin^2 \vartheta_r'/c. \tag{6.13}$$

Obviously, no cylindrical wave occurs at these distances. This result agrees with the condition that the entire radiation field transforms into a spherical wave. As a matter of fact, the projection of the radiation source onto the direction of the Cerenkov radiation is equal to $a \sin \vartheta_r'$. When we assume in Eq. (6.5) that $s_m = a \sin \vartheta_r'$, the transformation of the radiation field into a spherical wave is equivalent to the disappearance of the cylindrical wave. In the direction of the Cerenkov radiation, the Hertz vector calculated from Eq. (6.2) is equal to

$$\Pi_\omega = \frac{iev a}{2\pi\omega R v} \exp\left(i \frac{\omega n}{c} R\right). \tag{6.14}$$

In the case of a finite particle path, the Cerenkov radiation resembles a wave passing through a slit. The transformation of the cylindrical wave into the spherical wave is equivalent to wave diffraction at slits.

The cylindrical wave of the Cerenkov radiation was discussed for the first time in [23]. The radiation field at large distances from the finite path traveled by a uniformly moving charged particle was described in [24] for a homogeneous medium. Later, the same questions were discussed in [25] and [26]. It is easy to see from a comparison of the calculation of the Hertz vector for a radiation field introduced by Eq. (3.52) for $\mu \neq 1$ with the calculations of [23-26] (in which $\mu = 1$ was assumed), that the results are different insofar as in [23-26] the refractive index of the medium is $n = \sqrt{\varepsilon}$, whereas in the present case with $\mu \neq 1$ we have $n = \sqrt{\varepsilon\mu}$. In principle, the generalization to the case $\mu \neq 1$ makes it possible to consider a radiation in the frequency range involving negative group velocities.

§7. Cerenkov Radiation in the Frequency Range Involving

Negative Group Velocities

When the group-velocity vector is directed to the side of the wave vector, the group velocity is termed positive. The group velocity in a medium can be negative, i.e., a frequency range can exist in which the phase velocity of light and the velocity of energy propagation are antiparallel. In this frequency range, the phase of the waves emitted runs toward the radiation source rather than away from it. The direction in which the phase of the emitted waves propagates is important in all radiation processes depending upon the absolute value of the phase velocity of light. The radiation in the frequency range involving negative group velocities must be considered only in media in which such a frequency range can exist. Ferrodielectric substances are such media ($\varepsilon \neq 1$, $\mu \neq 1$) [8]. Ferrodielectric substances are an interesting object of radiation studies because, as has been shown in [27], in ferrodielectric substances conditions can be created under which the group velocity becomes negative.

The sign of the group velocity can be uniquely determined when the propagation of a monochromatic wave is considered. Whenever the Poynting vector of the monochromatic wave is parallel to the wave vector, the group velocity in all real media is positive; the group velocity is negative when the Poynting vector and the wave vector are antiparallel. In the case of strong attenuation, the wave packet is rapidly dissipated and a velocity of energy propagation (group velocity) per se does not exist. But when the attentuation is small, a direction of energy propagation can be indicated for the wave packet as well as for the monochromatic wave. It is therefore sensible to introduce in addition to the concept of a "negative group velocity" another concept, which is not affected by the degree of attenuation, for example, the so-called "negative direction of energy propagation." Whenever it is immaterial whether the wave attentuation in a medium is strong or weak, we will speak of a "negative direction of energy propagation."

The scalar product of the wave vector of waves emitted by a uniformly moving particle times the particle velocity is equal to the frequency of the emitted waves: $\mathbf{kv} = \omega$, i.e., this scalar product is a positive quantity. Hence, in the general case, the angle between the vectors $\mathbf{k}$ and $\mathbf{v}$ is an acute angle. In the frequency range involving negative group velocities, the wave vector $\mathbf{k}$ and the Poynting vector $\mathbf{S}$ are antiparallel, and therefore, the angle between $\mathbf{S}$ and $\mathbf{v}$ is greater than a right angle. This means that in the frequency range imvolving a negative group velocity, the radiation energy of the Cerenkov waves is emitted under an obtuse angle relative to the particle's velocity vector [8]. A similar situation arises in an anisotropic medium in the frequency range in which the signs of the wave vector projection and of the Poynting vector projection onto the direction of particle motion are different [28, 11].

The radiation from a moving oscillator is characterized by some interesting features in the frequency range involving negative group velocities. We introduce the Doppler formula

$$|\mathbf{kv} - \omega| = \omega_0, \tag{7.1}$$

where $\mathbf{v}$ and ω_0 denote the velocity and the eigenfrequency of the oscillator, respectively; $\mathbf{k}$ is the wave vector of the waves emitted ($k = \omega n(\omega, \vartheta)/c$); and $n(\omega, \vartheta)$ is the refractive index of the medium. It follows from Eq. (7.1) that waves of high frequencies are emitted in the forward direction, and waves of low frequencies in the backward direction. When the projections of the vectors $\mathbf{k}$ and $\mathbf{S}$ on the direction of motion of the oscillator have different signs, then, contrary to the case of positive group velocities, the high-frequency waves carry away energy into a direction which includes an obtuse angle with the particle's velocity vector (whereas the low-frequency waves imply an acute angle). This is the so-called inverse Doppler effect [8, 29–31].

In principle, the group velocity can be negative, for example, in ferrodielectric substance.* To show this, we insert into the Maxwell equation the electromagnetic field of a plane wave $(\mathbf{E}, \mathbf{H} \sim \exp[i(\mathbf{kr} - i\omega t)])$. The result is

$$kE^2 = \omega\mu\,[\mathbf{EH}]/c = \omega\mu\,S/4\pi, \qquad k = \omega\sqrt{\varepsilon\mu}/c. \tag{7.2}$$

When no absorption occurs, the unattenuated electromagnetic waves can propagate in the ferrodielectric substance as in the case $\varepsilon > 0$, $\mu > 0$, or in the case $\varepsilon < 0$, $\mu < 0$, because the refractive index is a real quantity in these two cases. According to Eq. (7.2), in the first case the direction of the Poynting vector coincides with the direction of the wave vector, whereas in the second case $\mathbf{S}$ and $\mathbf{k}$ are antiparallel, and hence, the group velocity is negative.

In the frequency range involving negative group velocities, the advancing potentials carry away the energy from the particle trajectory, but the phase of the waves related to these potentials does not run away from the emitting source but rather converges to the same. Thus, the phase of the emitted wave at the point of observation lags behind the phase at the point at which

---

*The group velocity can be negative even in media characterized by spatial dispersion [32]. The conditions which we are considering do not depend on the mechanism which leads to a frequency range involving negative group velocities.

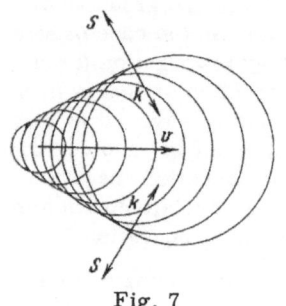

Fig. 7

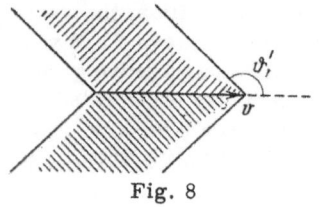

Fig. 8

the particle actually is. Since this corresponds to an outflow of energy, the disturbance arriving at some point of space is retarded. Figure 7 shows schematically the formation of the waves of the Cerenkov radiation in the frequency range involving negative group velocities. When the condition $|n|\beta > 1$ holds, all elementary waves emitted from various path sections (in Fig. 7, they are indicated by circles) propagate with the same phase in the direction $\vartheta'_r = \arccos 1/n\beta$. The waves of the Cerenkov radiation, which result from interference effects, have a conical enveloping surface on which all waves have the same phase; the apex of this cone points backward. The phase of these waves runs toward the particle trajectory under an acute angle ϑ relative to the particle's velocity vector, whereas the energy is emitted in reverse direction, under an obtuse angle.

With the advancing potentials carrying away the energy from the radiating system, the emitted waves are attenuated as they move away from the particle trajectory, provided that absorption takes place. Dissipation of electromagnetic energy means that Im $\varepsilon > 0$ and Im $\mu > 0$. In the frequency range involving negative group velocities (Re $\varepsilon < 0$, Re $\mu < 0$), the imaginary part of the product $\varepsilon\mu$ is negative. Since the square root of a complex quantity whose imaginary part is negative is equal to a complex number having real and imaginary parts with different signs, advancing potentials correspond to a solution with attenuation (for Re $ik < 0$ we have Re $k < 0$):

$$\text{Im } n^2 < 0, \quad \text{Re } n < 0, \quad \text{Im } n > 0. \tag{7.3}$$

If the medium is completely transparent and $|n|\beta > 1$ holds, then at $\vartheta = \vartheta'_r = \arccos 1/n\beta$ (where $n < 0$) in the frequency region involving negative group velocities the linear term of the ζ expansion in the exponent of the exponential function in the integral of Eq. (6.2) tends to zero. Hence, the angle ϑ'_r is greater than a right angle. This means that when the particle's path in the medium has a finite length, the cylindrical wave of the Cerenkov radiation fills the region of the space which is behind the particle (Fig. 8). The cylindrical wave of the Cerenkov radiation is computed with Eq. (6.12), in which we have Im $n > 0$ (Im $n \ll |\text{Re } n|$) in the frequency range involving negative group velocities. We change the sign before n and write the cylindrical wave of the Cerenkov radiation in the frequency range involving negative group velocities in the form

$$\Pi_\omega = \frac{ev\sqrt{i\omega c}\exp\left[-i\dfrac{\omega n}{c}R\cos(\vartheta - \vartheta'_r)\right]}{\omega^2 v\sqrt{2\pi Rn\sin\vartheta\sin\vartheta'_r}}, \tag{7.4}$$

where

$$\vartheta'_r = \pi - \arccos\frac{1}{n\beta}, \qquad \text{Re } n > 0, \quad \text{Im } n < 0. \tag{7.5}$$

When the path traveled by the particle is finite, the cylindrical wave (as in the case of positive group velocities) fills only the space region between the cones and the apeces at the ends of the particle's path. This region is shown in Fig. 8. The angular dimensions of the space regions at the cones (the wave defined by Eq. (7.4) does not exist in these regions) can be calculated in the same fashion as in the frequency range involving positive group velocities, i.e., with Eq. (6.8), where $n > 0$.

§ 8. Radiation from a Charged Particle Moving Perpendicular

to the Boundary

Let us consider the consequences of Eqs. (3.50)–(3.55) in the case of a charged particle moving perpendicular to the boundary separating a medium from vacuum. We will assume that in the medium the particle is ejected from some point situated at a distance a from the boundary and that the particle is directed toward the boundary and moves uniformly along a straight line. We will consider the raidation field in vacuum.

When the particle moves normal to the boundary, the radiation field is polarized in the plane of incidence. This radiation field is given by Eqs. (3.50), (3.52), and (3.54), wherein we must set $\beta_x = \beta_y = 0$.

Let us consider the radiation field generated along the path in the medium. According to Eq. (3.50), the field is given by the following Hertz vector:

$$\Pi_{\omega s} = \frac{iek}{2\pi\omega} \int_{-a}^{0} \frac{p_{s,j}}{R_{s,j}} \exp(if) \, dz_s, \tag{8.1}$$

where

$$f = \frac{\omega}{v} z_s - \sqrt{k_s^2 - \varkappa^2} \, z_s + \varkappa\rho + \sqrt{k_j^2 - \varkappa^2} \, z; \tag{8.2}$$

the components of the wave vector satisfy Eqs. (3.27) and (3.29). We omit the symbol $\parallel$. The z axis of the Cartesian coordinate system points toward the vacuum side and is perpendicular to the boundary; the coordinate origin is located on the boundary.

The quantities $p_{s,j}$, $R_{s,j}$, and f in Eq. (8.1) are functions of z_s. When the point of observation moves away from the particle trajectory, the derivatives $dp_{s,j}/dz_s$ and $dR_{s,j}/dz_s$ decrease, and when the distance between the point of observation and the particle trajectory is large enough, the functions $p_{s,j}$ and $R_{s,j}$ can be assumed as constants and drawn before the integral sign, which simplifies the integration considerably. In order to establish the conditions under which this can be done, and for the ensuing integration, we expand the function f in the exponent of the exponential function (this function determines the essential region of integration) in a power series of $(z_s - z_0)$ around some point z_0 on the trajectory. We restrict ourselves to the quadratic term of the expansion and obtain

$$f = \frac{\omega}{v} z_0 - \sqrt{k_s^2 - \varkappa^2} \, z_0 + \varkappa\rho + \sqrt{k_j^2 - \varkappa^2} \, z + \left(\frac{\omega}{v} - \sqrt{k_s^2 - \varkappa^2}\right)(z_s - z_0) +$$

$$+ \left[\frac{(k_s^2 - k_j^2) z}{(k_j^2 - \varkappa^2)^{3/2}} + \frac{k_s^2 \rho}{\varkappa^3}\right]^{-1} (z_s - z_0)^2, \tag{8.3}$$

where the wave-vector component $\varkappa$ is fixed and satisfies Eq. (3.29) at $z_s = z_0$.

Let us assume $z_0 = 0$. Then, we obtain from Eq. (3.32) that $\varkappa = k_j \sin\vartheta$, where ϑ denotes the angle between the radius vector $\mathbf{R}$ of the point of observation and the z axis. In this case, the function f assumes the following form:

$$f = k_j R + \left(\frac{\omega}{v} - \sqrt{k_s^2 - \varkappa^2}\right) z_s + \frac{\varkappa^2 (k_j^2 - \varkappa^2)}{k_j R (k_s^2 - \varkappa^2)} z_s^2. \tag{8.4}$$

Under the condition that in the entire region of integration the quadratic term of the expansion of the function f is smaller than unity, i.e., for

$$\frac{\varkappa^2 (k_j^2 - \varkappa^2)}{k_j R (k_s^2 - \varkappa^2)} a^2 \ll 1, \tag{8.5}$$

the quantity $R_{s,j}$ is almost constant and equal to R in the wave zone; we can assume $\varkappa = k_j \sin \vartheta$ in the expression for $p_{s,j}$. In this case, the result of the integration in Eq. (8.1) is a spherical wave. For $k_j = \omega/c$, and $k_s = \omega \sqrt{\varepsilon\mu}/c$, this spherical wave assumes the following form:

$$\Pi_{\omega s} = \frac{ev}{\pi\omega^2 R} \frac{\cos\vartheta \exp\left(i\frac{\omega}{c}R\right)}{(\varepsilon\cos\vartheta + \sqrt{\varepsilon\mu - \sin^2\vartheta})(1 - \beta\sqrt{\varepsilon\mu - \sin^2\vartheta})} [1 - e^{-i\frac{\omega a}{v}(1-\beta\sqrt{\varepsilon\mu - \sin^2\vartheta})}]. \tag{8.6}$$

When the particle path section is much greater than the coherence path, i.e., when

$$a \gg \frac{v}{\omega\,|\,1 - \beta\sqrt{\varepsilon\mu - \sin^2\vartheta}\,|} \tag{8.7}$$

holds, we can neglect interference radiation in averaging over some relatively small frequency interval (in both cases, i.e., when the particle is ejected from the point $z = -a$ or when the particle is suddenly stopped). Then the two radiation processes can be treated separately. When absorption occurs and when the waves undergo total reflection in the medium, i.e., when the path of interaction is determined by the imaginary part of $\varepsilon\mu - \sin^2\vartheta$, the radiation resulting from particle ejection is absorbed in the medium and does not enter into the vacuum [see Eq. (8.7)]. In this case, the exponent of the term in sqaure brackets in Eq.(8.6) has an absolute value much smaller than unity and can be neglected.

When the particle is stopped, the radiation generated along the path in the medium determines the transition radiation. The corresponding spherical wave has the following form:

$$\Pi_{\omega s} = \frac{ev}{\pi\omega^2 R} \frac{\cos\vartheta \exp\left(i\frac{\omega}{c}R\right)}{(\varepsilon\cos\vartheta + \sqrt{\varepsilon\mu - \sin^2\vartheta})(1 - \beta\sqrt{\varepsilon\mu - \sin^2\vartheta})}. \tag{8.8}$$

Since the region of integration has the size of the coherence path, the region of integration is given by the inequality (8.5), where we have to insert the coherence path

$$R \gg \frac{v^2 \sin^2\vartheta \cos^2\vartheta}{\omega c\,|\,1 - \beta\sqrt{\varepsilon\mu - \sin^2\vartheta}\,|^2\,|\,\varepsilon\mu - \sin^2\vartheta\,|} \tag{8.9}$$

or

$$R \gg \frac{s_s^2 \omega \sin^2\vartheta \cos^2\vartheta}{c\,|\,\varepsilon\mu - \sin^2\vartheta\,|} \tag{8.10}$$

instead of a. We note that when a field of diffracted radiation transforms into the spherical wave of Eq. (8.8), the coherence path appears in a fashion different from that in a boundary-free medium. This difference results from the fact that in the case under consideration, the apparent path of coherent interaction (due to wave refraction) $s_s^! = s_s \cos\vartheta/|\varepsilon\mu - \sin^2\vartheta|$ plays the role of the size of the "source."

The spherical wave emitted by the real dipoles was considered in §6. It remains to consider the radiation field from the dipole-images whose moments are proportional to $r_{j,s}$. The initial formula is

$$\Pi_{\omega j}^{"} = \frac{iev r_{j,s}}{2\pi\omega v} \int_0^\infty \frac{1}{R_\zeta} \exp\left[i\omega\left(\frac{z}{v} - \frac{R_\zeta}{c}\right)\right] dz, \tag{8.11}$$

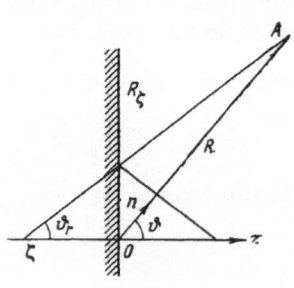

Fig. 9

where R_ζ denotes the distance between the dipole image with the coordinate $z = \zeta$ ($\zeta < 0$) and the point of observation A (Fig. 9).

The field at large distances is calculated as in §6. One must take into account that the charge image under consideration moves in the direction of the negative z axis ($v_z < 0$). The corresponding path of coherent interaction and the existence range of the spherical wave are determined with formulas which are obtained from Eqs. (5.2) and (6.4) (which were calculated for $v_z > 0$) by replacing $\mathbf{v}$ by $-\mathbf{v}$. The spherical wave which was calculated from Eq. (8.11) differs from that of Eq. (6.4) at n = 1 not only by the sign of $\mathbf{v}$, but also by the factor $r_{2,1}$.

Summarizing, the Hertz vectors of spherical waves generated by a charge and its images during the uniform and linear motion of the particle from the medium into vacuum (a motion perpendicular to the boundary) have the following form for $\mu = 1$:

$$\Pi_{\omega s} = \frac{ev}{2\pi\omega^2 R} \frac{2\cos\vartheta}{\varepsilon\cos\vartheta + \sqrt{\varepsilon - \sin^2\vartheta}} \frac{\exp\left(i\frac{\omega}{c}R\right)}{1 - \beta\sqrt{\varepsilon - \sin^2\vartheta}}, \tag{8.12}$$

$$\Pi'_{\omega j} = -\frac{ev}{2\pi\omega^2 R} \frac{\exp\left(i\frac{\omega}{c}R\right)}{1 - \beta\cos\vartheta}, \tag{8.13}$$

$$\Pi''_{\omega j} = -\frac{ev}{2\pi\omega^2 R} \frac{\varepsilon\cos\vartheta - \sqrt{\varepsilon - \sin^2\vartheta}}{\varepsilon\cos\vartheta + \sqrt{\varepsilon - \sin^2\vartheta}} \frac{\exp\left(i\frac{\omega}{c}R\right)}{1 + \beta\cos\vartheta}. \tag{8.14}$$

The corresponding paths of coherent interaction are

$$s'_s \sim \frac{v\cos\vartheta}{\omega\,|1 - \beta\sqrt{\varepsilon - \sin^2\vartheta}\,|\,|\sqrt{\varepsilon - \sin^2\vartheta}\,|}, \tag{8.15}$$

$$s'_j \sim \frac{v}{\omega\,(1 - \beta\cos\vartheta)}, \tag{8.16}$$

$$s''_j \sim \frac{v}{\omega\,(1 + \beta\cos\vartheta)}; \tag{8.17}$$

the regions in which the spherical waves defined by Eqs. (8.12), (8.13), and (8.14) exist are given by the formulas

$$R \gg \frac{v^2\sin^2\vartheta\cos^2\vartheta}{\omega c\,|1 - \beta\sqrt{\varepsilon - \sin^2\vartheta}\,|^2\,|\varepsilon - \sin^2\vartheta|}, \tag{8.18}$$

$$R \gg \frac{v^2\sin^2\vartheta}{\omega c\,(1 - \beta\cos\vartheta)^2}, \tag{8.19}$$

$$R \gg \frac{v^2\sin^2\vartheta}{\omega c\,(1 + \beta\cos\vartheta)^2}. \tag{8.20}$$

The radiation field is given by the sum of the vectors defined by Eqs. (8.12), (8.13), and (8.14).

In Eqs. (8.12)–(8.20) we have taken into account the actual form of the coefficient $p_{s,j}$ and $r_{j,s}$ and expressed the angles between the wave vectors in the medium and the normal to the boundary by ϑ.

We recall that the Fresnel formulas are also correct when attenuation takes place and this is the reason why the wave absorption in the medium does not limit the applicability range of Eqs. (8.12)–(8.20). We will make use of this fact when we consider the region in which the radiation field transforms into a spherical wave propagating in the direction of the Cerenkov radiation entering into vacuum.

In the case of a completely transparent medium, the path of coherent interaction between the particle and the wave in the medium (a wave, which after refraction at the boundary propagates in the direction $\vartheta_r = \arcsin(\sqrt{\varepsilon\beta^2 - 1}/\beta)$ tends to infinity. The waves which pass into the vacuum in this direction correspond to the Cerenkov radiation resulting from a strong coherent interaction with the uniformly moving particle along its entire path in the medium, a path which is assumed now as infinitely long. When absorption occurs, the coherence path defined by Eq. (8.15) does not tend toward infinity. In the event that the absorption is weak and that the inequality $\varepsilon''\beta^2 \ll (\varepsilon'\beta^2 - 1)$ holds, where ε' and ε'' denote the real and imaginary parts of the dielectric constant, respectively ($\varepsilon = \varepsilon' + i\varepsilon''$, $\varepsilon'' > 0$), the path s_s is large in the direction $\vartheta_r = \arcsin(\sqrt{\varepsilon\beta^2 - 1}/\beta)$ and the radiation field transforms into a spherical wave at relatively large distances. We obtain from Eq. (8.18) in the direction ϑ_r:

$$\varepsilon'' \sqrt{R\omega/c} \gg \sin 2\vartheta_r. \tag{8.21}$$

The cylindrical wave of the Cerenkov radiation passing into the vacuum is completely transformed into a spherical wave at distances satisfying the above inequality. This result could have been obtained from simple qualitative considerations. As a matter of fact, when absorption takes place, the Cerenkov radiation reaches the boundary only from a finite section of the particle path. We recall that the wave attenuation along the path EC (Fig. 10) is given by the factor $\exp(-s\varepsilon''\omega/2\sqrt{\varepsilon'}c)$, and therefore, the path along a ray on which the wave is almost completely absorbed is equal to $EC \sim \sqrt{\varepsilon'}c/\varepsilon''\omega$. Obviously, the wave must be a spherical wave at a distance R' from the point at which the particle leaves the medium for the phase difference between

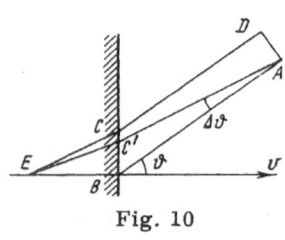

Fig. 10

waves emitted from the ends of section EB to remain practically constant when R' is tending toward infinity. This condition is satisfied when the optical path difference along the rays ECD and EC'A is less than λ. When the distance to the point of observation is much greater than the linear dimensions of the region from which the radiation is emitted into the vacuum (R' $\gg$ BC), the optical path difference results mainly from the sections CD and C'A. Hence, we must have C'A — CD $\ll$ c/ω. Since $\Delta\vartheta \ll 1$, we must have $C'A = \frac{CD}{\cos\Delta\vartheta} = CD\left[1 + \frac{(\Delta\vartheta)^2}{2}\right]$, where $\Delta\vartheta = \frac{EC}{CD}\sin\vartheta'\cos\vartheta = \frac{4\sqrt{\varepsilon'}c}{\varepsilon''\omega CD}\sin\vartheta'\cos\vartheta = \frac{4\lambda}{\varepsilon''CD}\sin\vartheta\cos\vartheta$. Denoting CD by R, we can set: C'A — CD = $R(\Delta\vartheta)/2$. Using the expression for $\Delta\vartheta$, we find the condition for a small optical path difference in the following form:

$$\varepsilon''^2 R\omega \gg c \sin^2\vartheta \cos^2\vartheta. \tag{8.22}$$

As had to be expected, for an emission under the angle ϑ_r, the above condition coincides with that stated in (8.21).

If Eq. (8.22) does not hold, the quadratic term in the expansion of the function f is not much smaller than unity. In this case, a large contribution to the integral stems from the relatively small integration region around the point at which the first derivative of the function in the exponent of the exponential tends to zero, i.e., when we have

$$\beta \sqrt{k_s^2 - \varkappa^2} = 1. \tag{8.23}$$

The important portion Δz_r of the integration interval is determined by the inverse square root of the second derivative of the function in the exponent of the exponential function:

$$\Delta z_r \sim \left|\frac{(k_s^2 - \varkappa^2)z}{(k_j^2 - \varkappa^2)^{3/2}} + \frac{k_s^2\rho}{\varkappa^3}\right|^{1/2}. \tag{8.24}$$

In this interval, the functions $p_{s,j}$ and $R_{s,j}$ vary but slightly in the wave zone and they can be taken before the integral; at the same time, the wave vector component $\varkappa$ is replaced by the quantity which satisfies Eq. (8.23). In the event that the distance between z_0 and the ends of the integration path is greater than Δz_r, we obtain for $\mu = 1$*

$$\Pi_{\omega s} = \frac{-2ev\sqrt{-i\omega}\exp\left[i\frac{\omega}{v}R\left(\cos\vartheta\sqrt{1+\beta^2-\varepsilon\beta^2}+\sin\vartheta\sqrt{\varepsilon\beta^2-1}\right)\right]}{\omega^2\sqrt{2\pi vR\sin\vartheta}\,(\varepsilon\beta^2-1)^{1/4}\,(\varepsilon\sqrt{1+\beta^2-\varepsilon\beta^2}+1)}, \qquad (8.25)$$

i.e., the cylindrical wave of the Cerenkov radiation passing into the vacuum. The space region in which the cylindrical wave exists is given by the condition $z_0 < \Delta z_r$ and, using Eqs. (3.26) and (8.23), we obtain from this condition

$$\vartheta > \vartheta_r + \Delta\vartheta_r, \ \Delta\vartheta_r \approx \sqrt{c/\omega R}, \qquad (8.26)$$

where

$$\vartheta_r = \arcsin\left(\sqrt{\varepsilon\beta^2-1}\,/\beta\right). \qquad (8.27)$$

$\Delta\vartheta_r$ denotes the angle of diffraction.

When absorption takes place, the function f in the exponent of the exponential function in Eq. (8.1) is complex and its first derivative does not tend to zero. The first derivative of the function f is imaginary for the refraction angle of the Cerenkov radiation. When the linear term of the expansion of the function f is small relative to unity in the interval Δz, the result of Eq. (8.25) is also valid in the case of attenuation. This condition coincides with the inequality which is the inverse of inequality (8.22) for the angle of refraction of the Cerenkov radiation. Besides that, the attenuation must be small.

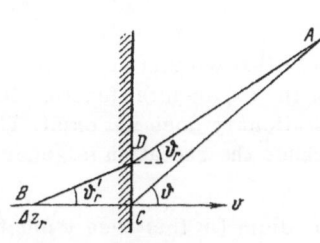

Fig. 11

Let us write Eq. (8.25) for the attenuated case in the form

$$\Pi_{\omega s} = -\frac{2ev\sqrt{-i\omega}\exp\left[i\frac{\omega}{c}R\cos(\vartheta-\vartheta_r)-\frac{R\omega\varepsilon''}{c}\frac{\sin(\vartheta-\vartheta_r)}{\sin 2\vartheta_r}\right]}{\omega^2\sqrt{2\pi vR\sin\vartheta}\,(\varepsilon\beta^2-1)^{1/4}\,(\varepsilon\sqrt{1+\beta^2-\varepsilon\beta^2}+1)}. \qquad (8.28)$$

Actually, a portion of the exponent of the exponential function corresponds to wave absorption in the medium. When the angle ϑ is fixed, Cerenkov radiation generated on particle-path sections far from the boundary arrives at a remote point of observation. Radiation generated near some point B at a distance BC from the boundary arrives at the point A with the coordinates R and ϑ (Fig. 11). The absorption reduces the wave amplitude by the exponential factor $\exp\left(-DB\varepsilon''/2\sqrt{\varepsilon'}\lambda\right)$, where BD denotes the path along the ray over which the radiation passes to the boundary. We infer from Fig. 11 that $BD = CD/\sin\vartheta_r' = CD\sqrt{\varepsilon'}/\sin\vartheta_r$, where $CD = R\frac{\sin(\vartheta-\vartheta_r)}{\cos\vartheta_r}$. With these relations we obtain a factor which characterizes the wave absorption in the medium: $\exp\left[-\frac{R\omega\varepsilon''}{c}\frac{\sin(\vartheta-\vartheta_r)}{\sin 2\vartheta_r}\right]$. As had to be expected, the exponent of this exponential function coincides with the exponent portion related to wave absorption in the exponential function of Eq. (8.28).

When the distance R satisfies the inequality

$$\varepsilon''\sqrt{R\omega/c} \ll \sin 2\vartheta_r, \qquad (8.29)$$

---

*This result was previously obtained in [33].

waves generated on a particle-trajectory section of the order of Δz_r near the point B (these waves form a cylindrical wave at the point A) are attenuated almost in the same way along the path to the boundary. This condition defines the existence region of the cylindrical wave. We found above that at greater distances, i.e., when the inequality (8.21) holds, the entire field of the Cerenkov radiation transforms into the spherical wave defined by Eq. (8.12).

Of particular interest is the exit (into the vacuum) of the Cerenkov radiation generated in the frequency range involving negative group velocities. We know from the discussion of §6 that the phase of the waves resulting from a radiation field which is attenuated at infinity (absorption case) runs toward the radiation source. Let us consider how this affects the exit of the Cerenkov radiation into the vacuum. We use Eq. (8.1) as the starting point. In the frequency range involving negative group velocities, we have $\operatorname{Im} k_s^2 < 0$ and

$$\operatorname{Im} \sqrt{k_s^2 - \varkappa^2} > 0 \tag{8.30}$$

corresponds to the solution in the absorption case. It follows from Eq. (8.1) that in this case a wave generated in the medium at the point z_s ($z_s < 0$) is attenuated and completely absorbed at $z_s \to -\infty$, i.e., this wave does not pass into the vacuum. In the frequency region involving negative group velocities, the inequality

$$\operatorname{Re} \sqrt{k_s^2 - \varkappa^2} < 0 \tag{8.31}$$

holds along with the inequality (8.30). Then, it follows from Eq. (8.3) that without absorption, the linear term of the expansion of the function f in the exponent of the exponential function in Eq. (8.1) does not tend to zero, i.e., a point at which the phase is stationary does not exist. The Cerenkov radiation does not pass into the vacuum in this case, because the radiation is generated under an obtuse angle relative to the velocity vector.

When the charged particle moves from the vacuum into the medium [in this case v must be replaced by $-$v in Eqs. (8.1) and (8.3)], the first derivative of the function f tends to zero, provided that $\frac{c^2}{\omega^2} \beta^2 (k_s^2 - \varkappa^2) = 1$, where $\varkappa^2$ satisfies Eq. (3.26) and where $-a < z_s < 0$ holds. The phase is stationary at the corresponding point and integration according to the stationary phase method leads to a diverging cylindrical wave for $\varkappa > 0$:

$$\Pi_{\omega s} = -\frac{2ev\sqrt{i\omega}\,\mu \exp\left[i\,\frac{\omega}{c}\,R\cos(\vartheta - \vartheta_r)\right]}{\omega^2\sqrt{2\pi vR \sin\vartheta}\,(\varepsilon\mu\beta^2 - 1)^{1/4}\,(\varepsilon\sqrt{1 + \beta^2 - \varepsilon\mu\beta^2} + \mu)}. \tag{8.32}$$

According to Eqs. (3.26) and (8.3), in the frequency range involving negative group velocities the phase can be stationary, provided that $\vartheta < \vartheta_r$ holds. However, as in the case of positive group velocities, a stationary phase is not a sufficient condition for the integration to result in the cylindrical wave of Eq. (8.32). The integration leads to a cylindrical wave defined by Eq. (8.32), if the principal region of integration does not include the coordinates of the beginning and end of the particle's path. This condition is fulfilled at points of space whose radius vector includes an angle larger than $\Delta\vartheta_r > \sim \sqrt{c/\omega R}$ with the angle of refraction of the Cerenkov radiation. Thus, the cylindrical wave of Eq. (8.32) fills the space region inside a circular cone having generatrices including the angle ϑ_r with the z axis and having the apex at the point at which the particle passes into the medium. We note that the Eq. (3.26), which defines a relation between the particle coordinates and the point of observation on the one hand, and the wave-vector component $\varkappa$ on the other, also holds for $\varkappa < 0$. In this case, integration leads to a converging cylindrical wave:

$$\Pi_{\omega s} = - \frac{2ev \sqrt{i\omega}\, \mu \exp\left[i\,\frac{\omega}{c}\,R\cos(\vartheta+\vartheta_r)\right]}{\omega^2 \sqrt{2\pi vR\sin\vartheta}\,(\varepsilon\mu\beta^2-1)^{1/4}\,(\varepsilon\sqrt{1+\beta^2-\varepsilon\mu\beta^2}+\mu)}. \tag{8.33}$$

Figure 12 depicts the space regions in which cylindrical waves of the Cerenkov radiation generated in a medium with a negative group velocity exist at finite particle paths in the medium.

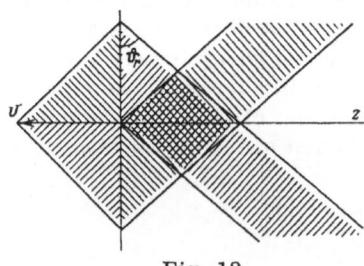

Fig. 12

In the immediate neighborhood of the boundary, the cylindrical wave of the Cerenkov radiation is a converging wave, but transforms into a diverging wave after intersecting the z axis.

At sufficiently large distances from the boundary, the quadratic terms in the expansion of the functions appearing in the exponent of the exponential functions in Eqs. (4.1), (4.3), and (4.5) can be ignored. Then, when a charged particle passes in uniform motion from a ferromagnetic substance into vacuum, the spherical wave generated along the path in the medium has the following form:

$$\Pi_{\omega s} = \frac{ev}{2\pi\omega^2 R}\,\frac{2\cos\vartheta}{\varepsilon\cos\vartheta+\sqrt{\varepsilon\mu-\sin^2\vartheta}}\,\frac{\exp\left(i\,\frac{\omega}{c}\,R\right)}{1-\beta\sqrt{\varepsilon\mu-\sin^2\vartheta}}, \tag{8.34}$$

where

$$R \gg \frac{v^2\sin^2\vartheta\cos^2\vartheta}{\omega c\,|1-\beta\cdot\sqrt{\varepsilon\mu-\sin^2\vartheta}\,|^2\,|\varepsilon\mu-\sin^2\vartheta|}. \tag{8.35}$$

The spherical wave generated along the path in the vacuum has the form

$$\Pi_{\omega j} = -\frac{ev\cos\vartheta}{\pi\omega^2 R}\,\frac{\varepsilon+\beta\sqrt{\varepsilon\mu-\sin^2\vartheta}}{\varepsilon\cos\vartheta+\sqrt{\varepsilon\mu-\sin^2\vartheta}}\,\frac{\exp\left(i\,\frac{\omega}{c}\,R\right)}{1-\beta^2\cos^2\vartheta}, \tag{8.36}$$

wherein it was assumed that

$$R \gg \frac{v^2\sin^2\vartheta}{\omega c\,(1-\beta^2\cos^2\vartheta)^2}. \tag{8.37}$$

The spherical wave resulting from the total radiation field is determined by the sum of the Hertz vectors defined by Eqs. (8.34) and (8.36) and is equal to

$$\Pi_{\omega} = \frac{ev\cos\vartheta}{\pi\omega^2 R}\,\frac{[(\varepsilon-1)(\beta\sqrt{\varepsilon\mu-\sin^2\vartheta}-1)+\beta^2(\varepsilon\mu-1)]\exp\left(i\,\frac{\omega}{c}\,R\right)}{(1-\beta^2\cos^2\vartheta)(1-\beta\sqrt{\varepsilon\mu-\sin^2\vartheta})(\varepsilon\cos\vartheta+\sqrt{\varepsilon\mu-\sin^2\vartheta})}. \tag{8.38}$$

The corresponding spectral energy density of the radiation per unit solid angle is (according to [8])

$$W_{n\omega} = \frac{e^2\beta^2}{\pi^2 c}\left|\frac{(\varepsilon-1)(1-\beta\sqrt{\varepsilon\mu-\sin^2\vartheta})-\beta^2(\varepsilon\mu-1)}{(1-\beta\sqrt{\varepsilon\mu-\sin^2\vartheta})(\varepsilon\cos\vartheta+\sqrt{\varepsilon\mu-\sin^2\vartheta})}\right|^2\frac{\sin^2\vartheta\cos^2\vartheta}{(1-\beta^2\cos^2\vartheta)^2}. \tag{8.39}$$

When the direction of particle motion is reversed, i.e., when the particle passes from the vacuum into the medium, β must be replaced by $-\beta$ in Eqs. (8.34)-(8.39). Evidently, when the attenuation is small, then in the frequency range involving negative group velocities, the radiation intensity near the refracting Cerenkov angle reaches relatively large values. This sharp intensity peak results from the Cerenkov radiation emitted under an obtuse angle relative to the particle's velocity vector. The cylindrical wave described above tranforms completely into a spherical wave at correspondingly large distances.

§9. Charge Images

A set of fixed dipole images corresponds to moving charge images. The parameters of these charges, namely their magnitude, velocities, and coordinates at some particular instance of time can be found by comparing the results of the preceding three sections. First, let us compare the spherical wave defined by Eq. (8.12) which results when a charge image is abruptly stopped, with the spherical wave emitted from a real charge [Eq. (6.4)]. Since the wave defined by Eq. (8.12) propagates in vacuum, n must be set equal to unity in Eq. (6.4). Then, a comparison reveals that the field of Eq. (8.12) can be considered as the result of the abrupt stopping of a charge e_1 moving with the velocity $\mathbf{v}_1$:

$$e_1 = e p_{1,2} \frac{v}{v_1} = \frac{2e \cos \vartheta}{\varepsilon \cos \vartheta + \sqrt{\varepsilon - \sin^2 \vartheta}} \frac{v}{v_1} ,$$

$$v_1 = v \sqrt{\varepsilon - \sin^2 \vartheta} / \cos \vartheta, \qquad z_1 = z \sqrt{\varepsilon - \sin^2 \vartheta} / \cos \vartheta. \tag{9.1}$$

We have also stated the relation between the coordinate z_1 of the image and the coordinate z of the real charge.

The spherical wave defined by Eq. (8.14) is generated by another image which is suddenly ejected from the point $z = 0$ at the instant at which the particle passes through the boundary. The charge, velocity, and coordinate of this image are given by

$$e_2 = e r_{2,1} = e \frac{\varepsilon \cos \vartheta - \sqrt{\varepsilon - \sin^2 \vartheta}}{\varepsilon \cos \vartheta + \sqrt{\varepsilon - \sin^2 \vartheta}} , \quad v_2 = - v, \, z_2 = - z. \tag{9.2}$$

If wave attenuation occurs, the applicability range of the results forming the basis for the present considerations is not affected. In other words, the absorption is immaterial.

We infer from Eqs. (9.1) and (9.2) that the quantities e_1, e_2, $\mathbf{v}_1$, and z_1 are generally complex, and their imaginary parts may be relatively large. However, this is quite sensible, because charge images were introduced only for the purpose of explaining the results. In the case under consideration, the explanation no longer has the obvious meaning in terms of physics which it had before. When we speak of charge images, we will assume in the following discussion that we are concerned with the frequency and angle regions in which the parameters of the images are real.

It is interesting to note that the velocity v_1 of the charge e_1 can exceed the velocity of light in vacuum. This corresponds to Cerenkov radiation leaving the vacuum and does not imply any inconsistencies, because the charge e_1 (and e_2 as well) determines the field in the vacuum, i.e., in that part of space into which this charge does not penetrate. It determines the field in the region $z > 0$, whereas the charge e_1 moves in the region $z < 0$ and stops suddenly at the point $z = 0$. This is the formal difference between the image and the real charge which determines the field in the region of space in which the charge really is.

In the case of weak absorption, the imaginary part of v_1 is relatively small, provided that $\varepsilon' > \sin^2 \vartheta$, but the real part is greater than the velocity of light in vacuum when we have $\beta^2 (\varepsilon' - \sin^2 \vartheta) > 1$. The charge image of e_1 generates Cerenkov radiation in this case, and the radiation is that passing into the vacuum after refraction at the boundary. Let us write the parameters of the corresponding charge image:

$$e_1' = \frac{2e}{\varepsilon \sqrt{1 + \beta^2 - \varepsilon\beta^2} + 1}, \qquad v_1' = \frac{v}{\sqrt{1 + \beta^2 - \varepsilon\beta^2}}, \qquad z_1' = z\sqrt{1 + \beta^2 - \varepsilon\beta^2}. \qquad (9.3)$$

When we set n = 1 in Eq. (6.12), which describes the cylindrical wave resulting from the Cerenkov radiation in an infinite medium, and replace e and **v** by e_1' and v_1' from Eq. (9.3), we obtain the refracted wave of the Cerenkov radiation in vacuum as in Eq. (8.28). This agreement between the equations proves that our image parameters are correct.

We note that the assumption $\vartheta = \vartheta_r$, transforms v_1 into v_1', whereas e_1 and z_1 remain different from e_1' and z_1', respectively. We have

$$e_1 (1 + \beta^2 - \varepsilon\beta^2) = e_1', \quad z_1' (1 + \beta^2 - \varepsilon\beta^2) = z_1. \qquad (9.4)$$

This difference can be explained with simple qualitative considerations.

The important point in the case of a cylindrical wave is that the radiation arriving at a particular point of space develops in a limited particle-trajectory region situated at a certain distance from the point at which the particle leaves the vacuum. The direction which is opposite the direction of propagation of the corresponding light ray in vacuum determines the spot on the charge-image trajectory from which radiation arrives at a certain point of space. Taking into account the refraction of the cylindrical wave of the Cerenkov radiation, one finds immediately that the neighborhood of a certain point z on the trajectory of a real charge (which causes the observed radiation at a given point in space) corresponds to the neighborhood of the point z_1' on the trajectory of the image charge e_1'; the equation $z_1' = z\sqrt{1 + \beta^2 - \varepsilon\beta^2}$ holds.

We note that the velocity v_1' of the charge image e_1' is equal to the velocity v_1 of the charge image e_1 when observations are made under the angle $\vartheta = \vartheta_r$, whereas the distance to the boundary is smaller than the distance to the charge e_1 by the factor $1 + \beta^2 - \varepsilon\beta^2$. When the flight time T of the real charge is finite, the time T_1' during which the image charge e_1' moves is smaller than the time T_1 of motion of the image charge e_1 by the factor $1 + \beta^2 - \varepsilon\beta^2$. Hence, with Eqs. (9.1) and (9.3) we find that the equality $T_1'e_1' = T_1e_1$ holds in the direction $\vartheta = \vartheta_r$. The meaning of this equality is that the energy of the Cerenkov radiation in the cylindrical and spherical waves is the same.

The parameters of the charge images, which were obtained in the particular case under consideration from a comparison of the results, agree with the parameters found in [34] and [35] by direct "joining" of the fields.

§10. Transition Radiation of a Relativistic Charged
Particle in the Optical Frequency Range*

The radiation of a charged particle which passes through a boundary separating media (transition radiation) was studied in various experiments [36]-[41]. A detailed comparison between theory and experiment was made for electrons with energies of up to 100 keV, i.e., for nonrelativistic particles. It could be shown that good agreement exists between experimental results and the predictions of theory, as far as the outgoing radiation energy in dependence of the particle energy, and the angular distribution of the radiation are concerned. The observed radiation spectrum agrees with the spectrum calculated according to theoretical formulas, provided that one inserts the real and imaginary parts of the refractive index at various wavelengths.

It was discovered that other processes also give a substantial contribution to the radiation. In the case of metals, the luminescence from a surface film on the metal and the optical

---

*This section outlines the work of [19], which was done in collaboration with I. M. Frank.

portion of the bremsstrahlung spectrum are involved. As far as luminescence is concerned, direct experiments, which were made at various temperatures of the target [38], have shown that the luminescence contribution is negligibly small when the metal surface has been carefully cleaned.

The scattering of the particle per unit path and the intensity of the bremsstrahlung increase with increasing particle energy. The intensity of the bremsstrahlung of nonrelativistic electrons is inversely proportional to their energy, whereas the intensity of the transition radiation increases with β^2. At electron energies of the order of 100 keV, the bremsstrahlung in metals can be neglected compared to the transition radiation in the spectral regions in which the metal is relatively transparent. This means that, as far as an increased output of the transition radiation and improved observation conditions are concerned, it is convenient to employ high-energy electrons, in particular electrons of relativistic velocities. Since the predictions of the theory of transition radiation have been well confirmed, practical applications should be considered, e.g., the use of transition radiation as a method for determining the optical properties of substances with strong absorption of light (e.g., metals). In order to establish optimum conditions for practical applications, we have to discuss the qualitative results of the theory.

We write the spectral density of the radiation energy per unit solid angle in the following form:

$$W_{n\omega} = \frac{e^2 v^2}{\pi^2 c^3} \frac{|(1-n^2)(1-\beta^2+\beta\sqrt{n^2-\sin^2\vartheta})|^2 \sin^2\vartheta \cos^2\vartheta}{|(\sqrt{n^2-\sin^2\vartheta}+n^2\cos\vartheta)(1+\beta\sqrt{n^2-\sin^2\vartheta})|^2(1-\beta^2\cos^2\vartheta)^2}, \tag{10.1}$$

where ϑ denotes the angle between the normal to the boundary and the direction of observation; $\beta = v/c$; and n is the complex refractive index of the medium. Equation (10.1) describes the transition radiation in vacuum when a particle passes from the vacuum into a medium in the direction perpendicular to the boundary. When the direction of particle motion is reversed, the particle velocity β in Eq. (10.1) must be replaced by $-\beta$.

Let us assume that the particle has a relativistic velocity. The transition radiation generated by a relativistic particle has been discussed in various papers, with the attention focused mainly on the radiation emitted under very small angles relative to the particle velocity vector: $\vartheta \sim \sqrt{1-\beta^2}$. This case is of particular interest insofar as the emission spectrum extends into the region of high frequencies beyond the optical frequency range.

Less attention has been paid to the radiation in the angular interval beyond the relativistic maximum. This radiation, as in the case of low-energy particles, is determined by the optical properties of the medium. No experimental work has been done in this radiation region.

Let us assume $\beta \approx 1$ in Eq. (10.1). Then, we obtain for $\vartheta \gg \sqrt{1-\beta^2}$:

$$W_{n\omega H} = \frac{e^2}{\pi^2 c} \frac{|(1-n^2)\sqrt{n^2-\sin^2\vartheta}|^2 \operatorname{ctg}^2\vartheta}{|(\sqrt{n^2-\sin^2\vartheta}+n^2\cos\vartheta)(1+\sqrt{n^2-\sin^2\vartheta})|^2}, \tag{10.2}$$

$$W_{n\omega B} = \frac{e^2}{\pi^2 c} \frac{|(1-n^2)\sqrt{n^2-\sin^2\vartheta}|^2 \operatorname{ctg}^2\vartheta}{|(\sqrt{n^2-\sin^2\vartheta}+n^2\cos\vartheta)(1-\sqrt{n^2-\sin^2\vartheta})|^2}. \tag{10.3}$$

Equation (10.2) describes the backward radiation (i.e., the radiation in the backward direction) from a particle passing into the medium; Eq. (10.3) describes the forward radiation (i.e., the radiation in the forward direction) for a particle leaving the medium and entering into vacuum (as before, ϑ denotes the angle relative to the normal to the surface of the medium and, consequently, in the case of backward radiation, the angle relative to the particle's velocity vector is equal to $\pi - \vartheta$). These formulas indicate that the energy of the radiation is independent of the energy of the relativistic particle. Both the angular distribution and the radiation spectrum (as in the

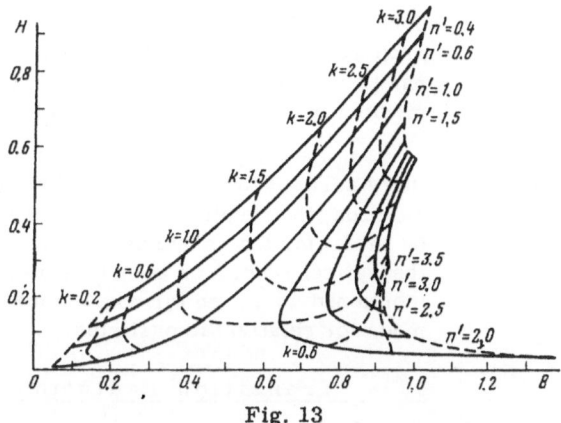

Fig. 13

nonrelativistic case) are exclusively determined by n, but the results in the forward and backward direction are different.

For comparisons with experiments, the functions $W_{n\omega H}$ and $W_{n\omega B}$ were calculated in [19] for various values of the real and imaginary parts of the refractive index (n' and k, with n = n' + ik). These values fall into the interval 0.2 to 3.5, in which a large portion of the n' and k values of real metals can be expected in the range of optical frequencies. The results of the calculations are displayed in 23 figures. We bring only some of these figures as examples. Figure 13 shows the results of the calculations for the angle $\vartheta = 55°$. The function of Eq. (10.2) has been plotted to the ordinate axis, which is defined by Eq. (1.3) to the abscissa axis in units of $e^2/\hbar^2 c$. The solid lines join the points which correspond to n' = const, whereas the dashed lines are characterized by k = const. We see that the curves of each set do not intersect among themselves. Similar results are obtained for the radiation observed under other angles ϑ. This means that when one measures the radiation intensity in the backward and forward direction along a certain line ϑ, one can uniquely determine n' and k from the curves. The absolute values of $W_{n\omega H}$ and $W_{n\omega B}$ can be found, for example, by comparing the brightness of the transition radiation with the Cerenkov radiation, or with the transition radiation generated by a standard substance with known n' and k values.

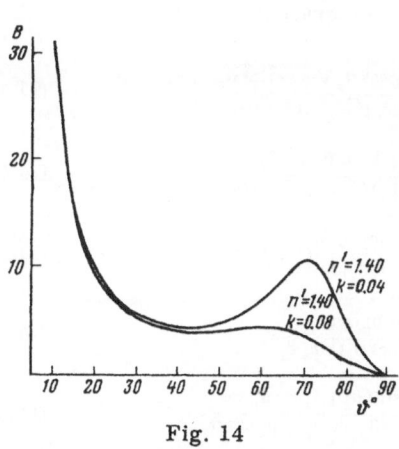

Fig. 14

At small k values, the angular distribution of the radiation intensity generated by a particle leaving a medium is characterized by a typical maximum which stems from the exit of the Cerenkov radiation into the vacuum. This maximum is visible in Figs. 14 and 15. Both the magnitude and the position of the maximum depend upon the real and imaginary parts of the refractive index. When a relativistic particle passes perpendicular to the boundary from the medium into the vacuum, the real part of the refractive index is greater than unity and less than about 1.4. At n' > 1.4, the Cerenkov radiation generated in the medium experiences total reflection and does not pass into the vacuum. In order to eliminate total reflection, the target must be inclined relative to the charged particle beam. Naturally, the angular distribution then follows a law different from that stated in Eq. (10.3). Since both the form and the position of the

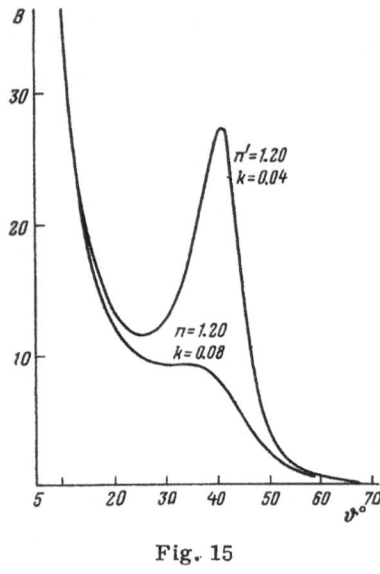

Fig. 15

intensity maximum resulting from the Cerenkov radiation passing into the vacuum are highly dependent on the optical properties of the medium, observations of the maximum can be used for determining the optical properties of the medium in cases in which k is not very large.

In the frequency intervals in which the real and imaginary parts of the refractive index were measured by other methods, the results of the calculations given in [19] can be useful for the comparison of experimental results with theoretical results on transition radiation. As has been mentioned above, so far no such measurements have been made for relativistic particles.

§11. Transition Radiation at Inclined Incidence

The transition radiation generated by a charged particle incident at a certain angle relative to the boundary was discussed in [42]-[45]. In those papers, either the calculations were not conducted to the end (i.e., the angular distribution of the radiation intensity was not obtained), or incorrect results were derived (in [42]). The author obtained the first correct final results on the transition-radiation problem in the case of inclined particle incidence in [12], which includes a critical review of previous work; Korkhmazyan [46] derived correct final results in independent work. In [15], the author considered polarization of the transition radiation in the case of inclined incidence; generally speaking, the polarization is elliptical. Furthermore, in [15] the radiation generated by a relativistic particle at grazing incidence was considered at frequencies which are large relative to the optical frequencies. This paragraph outlines the results of the author's work [12, 15] based on the image method.

Let us consider the radiation generated by a particle which moves uniformly and along a straight line including an angle with the boundary. The time at which the particle intersects the boundary is taken as the origin of the time count. Equations (4.8)-(4.13) are the starting point for our discussion. In Eqs. (4.8) and (4.9), the time integration must be performed from minus infinity to zero, and in Eqs. (4.10)-(4.13), from zero to infinity. When we assume that the yz plane coincides with a plane defined by the normal to the boundary and the velocity vector of the particle, and when we set $y = v_y t$ and $z = v_z t$, we obtain after integration

$$\Pi_{\omega 1\parallel} = \frac{eck \exp\left(i\frac{\omega}{c}R\right)\cos\vartheta_z}{\pi\omega^2 R\left(\varepsilon\cos\vartheta_z + \sqrt{\varepsilon - \sin^2\vartheta_z}\right)}\frac{\beta_z\sin^2\vartheta_z - \beta_y\cos\vartheta_y\sqrt{\varepsilon - \sin^2\vartheta_z}}{(1 - \beta_y\cos\vartheta_y - \beta_z\sqrt{\varepsilon - \sin^2\vartheta_z})\sin^2\vartheta_z}, \tag{11.1}$$

$$\Pi_{\omega 1\perp} = \frac{ev_y \exp\left(i\frac{\omega}{c}R\right)\cos\vartheta_z}{\pi\omega^2 R\left(\cos\vartheta_z + \sqrt{\varepsilon - \sin^2\vartheta_z}\right)}\frac{j\sin^2\vartheta_z + k\cos\vartheta_y\cos\vartheta_z}{(1 - \beta_y\cos\vartheta_y - \beta_z\sqrt{\varepsilon - \sin^2\vartheta_z})\sin^2\vartheta_z}, \tag{11.2}$$

$$\Pi'_{\omega 2\parallel} = -\frac{eck \exp\left(i\frac{\omega}{c}R\right)}{2\pi\omega^2 R}\frac{\beta_z\sin^2\vartheta_z - \beta_y\cos\vartheta_y\cos\vartheta_z}{(1 - \beta_y\cos\vartheta_y - \beta_z\cos\vartheta_z)\sin^2\vartheta_z}, \tag{11.3}$$

$$\Pi_{\omega 2\perp} = -\frac{ev_y \exp\left(i\frac{\omega}{c}R\right)}{2\pi\omega^2 R}\frac{j\sin^2\vartheta_z + k\cos\vartheta_y\cos\vartheta_z}{(1 - \beta_y\cos\vartheta_y - \beta_z\cos\vartheta_z)\sin^2\vartheta_z}, \tag{11.4}$$

$$\Pi''_{\omega 2\parallel} = -\frac{eck \exp\left(i\frac{\omega}{c}R\right)}{2\pi\omega^2 R\sin^2\vartheta_z}\frac{\varepsilon\cos\vartheta_z - \sqrt{\varepsilon - \sin^2\vartheta_z}}{\varepsilon\cos\vartheta_z + \sqrt{\varepsilon - \sin^2\vartheta_z}}\frac{\beta_z\sin^2\vartheta_z + \beta_y\cos\vartheta_y\cos\vartheta_z}{1 - \beta_y\cos\vartheta_y + \beta_z\cos\vartheta_z}, \tag{11.5}$$

$$\Pi_{\omega 2 \perp}'' = - \frac{e v_y \exp\left(t\,\frac{\omega}{c}\,R\right)}{2\pi\omega^2 R \sin^2 \vartheta_z} \frac{\cos\vartheta_z - \sqrt{\varepsilon - \sin^2\vartheta_z}}{\cos\vartheta_z + \sqrt{\varepsilon - \sin^2\vartheta_z}} \frac{\mathbf{j}\sin^2\vartheta_z + \mathbf{k}\cos\vartheta_y\cos\vartheta_z}{1 - \beta_y\cos\vartheta_y + \beta_z\cos\vartheta_z}. \tag{11.6}$$

The sum of the vectors (11.1), (11.3), and (11.5) describes the radiation of the waves polarized in the plane of wave propagation, whereas the sum of Eqs. (11.2), (11.4), and (11.6) refers to the radiation of the waves polarized perpendicular to the plane of propagation:

$$\Pi_{\omega\|} = \frac{e v_z \mathbf{k}(1 - \varepsilon)\cos\vartheta_z \exp\left(i\,\frac{\omega}{c}\,R\right)}{\pi\omega^2 R\,(\varepsilon\cos\vartheta_z + \sqrt{\varepsilon - \sin^2\vartheta_z})\sin^2\vartheta_z} [(1 - \beta_z\sqrt{\varepsilon - \sin^2\vartheta_z} - \beta_z^2 -$$

$$- \beta_y\cos\vartheta_y)\sin^2\vartheta_z + \beta_y\beta_z\cos\vartheta_y\sqrt{\varepsilon - \sin^2\vartheta_z}]\,[(1 - \beta_y\cos\vartheta_y)^2 -$$

$$- \beta_z^2\cos^2\vartheta_z]^{-1}\,(1 - \beta_y\cos\vartheta_y - \beta_z\sqrt{\varepsilon - \sin^2\vartheta_z})^{-1}, \tag{11.7}$$

$$\Pi_{\omega\perp} = \frac{e v_y \beta_z^2(1 - \varepsilon)\cos\vartheta_z \exp\left(i\,\frac{\omega}{c}\,R\right)}{\pi\omega^2 R\,(\cos\vartheta_z + \sqrt{\varepsilon - \sin^2\vartheta_z})\sin^2\vartheta_z}\,(\mathbf{j}\sin^2\vartheta_z + \mathbf{k}\cos\vartheta_y\cos\vartheta_z) \times$$

$$\times\,[(1 - \beta_y\cos\vartheta_y)^2 - \beta_z^2\cos^2\vartheta_z]^{-1}\,(1 - \beta_y\cos\vartheta_y - \beta_z\sqrt{\varepsilon - \sin^2\vartheta_z})^{-1}. \tag{11.8}$$

The spherical wave of the radiation field of the waves with both types of polarization is given by the sum of the Hertz vectors defined in Eqs. (11.7) and (11.8). This sum is

$$\Pi_\omega = \frac{ec\exp\left(i\,\frac{\omega}{c}\,R\right)\beta_z(1 - \varepsilon)\cos\vartheta_z}{\pi\omega^3 R\,(\varepsilon\cos\vartheta_z + \sqrt{\varepsilon - \sin^2\vartheta_z})}\,\{-\mathbf{j}\beta_y\beta_z\,(\sin^2\vartheta_z + \cos\vartheta_z\sqrt{\varepsilon - \sin^2\vartheta_z}) +$$

$$+ \mathbf{k}\,[(1 - \beta_y\cos\vartheta_y)(1 - \beta_z\sqrt{\varepsilon - \sin^2\vartheta_z}) - \beta_z^2 - \beta_y\beta_z\cos\vartheta_y\cos\vartheta_z]\} \times$$

$$\times\,[(1 - \beta_y\cos\vartheta_y)^2 - \beta_z^2\cos^2\vartheta_z]^{-1}\,(1 - \beta_y\cos\vartheta_y - \beta_z\sqrt{\varepsilon - \sin^2\vartheta_z})^{-1}. \tag{11.9}$$

This result agrees with the result which the author had found earlier in [12], when he used the other method.

With the results stated in Eqs. (11.7) and (11.8) and one of the Eqs. (4.17)-(4.21), we can find the spectral density of the radiation energy per unit solid angle for waves of various polarizations in the form

$$W_{n\omega\|} = \frac{e^2\beta_z^2\cos^2\vartheta_z\,|1 - \varepsilon|^2}{\pi^2 c\,[(1 - \beta_y\cos\vartheta_y)^2 - \beta_z^2\cos^2\vartheta_z]^2\sin^2\vartheta_z} \times$$

$$\times \left|\frac{(1 - \beta_z\sqrt{\varepsilon - \sin^2\vartheta_z} - \beta_z^2 - \beta_y\cos\vartheta_y)\sin^2\vartheta_z + \beta_y\beta_z\cos\vartheta_y\sqrt{\varepsilon - \sin^2\vartheta_z}}{(1 - \beta_y\cos\vartheta_y - \beta_z\sqrt{\varepsilon - \sin^2\vartheta_z})\,(\varepsilon\cos\vartheta_z + \sqrt{\varepsilon - \sin^2\vartheta_z})}\right|^2, \tag{11.10}$$

$$W_{n\omega\perp} = \frac{e^2\beta_y^2\beta_z^4\cos^2\vartheta_x\cos^2\vartheta_z\,|1 - \varepsilon|^2}{\pi^2 c\,[(1 - \beta_y\cos\vartheta_y)^2 - \beta_z^2\cos^2\vartheta_z]^2\sin^2\vartheta_z} \times$$

$$\times\,|(1 - \beta_y\cos\vartheta_y - \beta_z\sqrt{\varepsilon - \sin^2\vartheta_z})\,(\cos\vartheta_z + \sqrt{\varepsilon - \sin^2\vartheta_z})|^{-2}. \tag{11.11}$$

The sum of these formulas gives us the spectral density of the radiation energy of waves of both polarizations per unit solid angle, as determined previously in [12].

I. M. Frank directed the author's attention to the fact that, similar to the reflection of plane waves from an absorbing medium, the polarization of the transition radiation incident at a certain angle with respect to the boundary must be elliptic, depending upon the real and imaginary parts of the dielectric constant. I. M. Frank also emphasized the importance of this dependence for the study of optical properties of absorbing media.

Let us consider the polarization of transition radiation. According to Eq. (3.2), the electric vector of the spherical wave lies on a plane which is defined by the wave vector and the Hertz vector, whereas the magnetic vector of the spherical wave is perpendicular to that plane. When the medium is transparent, and when $\sin^2 \vartheta_z < \varepsilon$ holds, the vector $\Pi_\omega e^{-i\omega t}$ does not change its direction in space during the course of time, and the waves are linearly polarized. If the above conditions are not applicable, the values $\Pi_{\omega y} e^{-i\omega}$ and $\Pi_{\omega z} e^{-i\omega t}$ are nonvanishing and the corresponding relative phase shift is π. Then the vector $\Pi_\omega e^{-i\omega t}$, and, consequently, also the vectors $E_\omega e^{-i\omega t}$ and $H_\omega e^{-i\omega t}$ change in time (magnitude and direction). The change in the direction corresponds to elliptic polarization of the waves.

For discussing the polarization of the radiation, the vectors H_ω and E_ω of the spherical wave are conveniently represented as the sum of two vectors:

$$H_\omega = H_{\omega 1} + H_{\omega 2}, \quad E_\omega = E_{\omega 1} + E_{\omega 2}, \tag{11.12}$$

where

$$H_{\omega 1} = \frac{\omega^2}{c^2} \Pi_{\omega z} (i \cos \vartheta_y - j \cos \vartheta_x), \tag{11.13}$$

$$H_{\omega 2} = \frac{\omega^2}{c^2} \Pi_{\omega y} (- i \cos \vartheta_z + k \cos \vartheta_x), \tag{11.14}$$

$$E_{\omega 1} = \frac{\omega^2}{c^2} \Pi_{\omega z} (- i \cos \vartheta_x \cos \vartheta_z - j \cos \vartheta_y \cos \vartheta_z + k \sin^2 \vartheta_z), \tag{11.15}$$

$$E_{\omega 2} = \frac{\omega^2}{c^2} \Pi_{\omega y} (- i \cos \vartheta_x \cos \vartheta_y + j \sin^2 \vartheta_y - k \cos \vartheta_y \cos \vartheta_z). \tag{11.16}$$

The positions of the vectors n, $H_{\omega 1}$, $H_{\omega 2}$, $E_{\omega 1}$, $E_{\omega 2}$ are such that the relations

$$
\begin{aligned}
(H_{\omega 1} E_{\omega 1}) = 0, \qquad (H_{\omega 2} E_{\omega 2}) = 0, \qquad (n H_{\omega 1}) = 0, \\
(n H_{\omega 2}) = 0, \qquad (n E_{\omega 1}) = 0, \qquad (n E_{\omega 2}) = 0
\end{aligned}
\tag{11.17}
$$

hold. We introduce the following notation:

$$
\cos(\widehat{H_{\omega 1} H_{\omega 2}}) = \cos(\widehat{E_{\omega 1} E_{\omega 2}}) = \cos \psi,
$$
$$
\cos \psi = - \operatorname{ctg} \vartheta_y \operatorname{ctg} \vartheta_z, \quad \sin \psi = \cos \vartheta_x \sin^{-1} \vartheta_y \sin^{-1} \vartheta_z.
\tag{11.18}
$$

The vector $H_{\omega 1}$ rotates by an angle ψ counterclockwise relative to the vector $H_{\omega 2}$, provided that $\psi > 0$; the rotation is clockwise, if $\psi < 0$ (the same is true for the vectors $E_{\omega 1}$ and $E_{\omega 2}$).

The polarization of the waves is determined by the absolute value Π and the angle φ of the complex ratio of the Hertz vector components $\Pi_{\omega y}/\Pi_{\omega z}$:

$$
\Pi = \left| \frac{\Pi_{\omega y}}{\Pi_{\omega z}} \right|, \qquad \varphi = \operatorname{arctg} \frac{\operatorname{Im}(\Pi_{\omega y}/\Pi_{\omega z})}{\operatorname{Re}(\Pi_{\omega y}/\Pi_{\omega z})},
$$
$$
\frac{\Pi_{\omega y}}{\Pi_{\omega z}} = \frac{\beta_y \beta_z (\sin^2 \vartheta_z + \cos \vartheta_z \sqrt{\varepsilon - \sin^2 \vartheta_z})}{\beta_z^2 + \beta_y \beta_z \cos \vartheta_y \cos \vartheta_z - (1 - \beta_y \cos \vartheta_y)(1 - \beta_z \sqrt{\varepsilon - \sin^2 \vartheta_z})}.
\tag{11.19}
$$

The parameter representation of the polarization ellipse has the form

$$
\begin{aligned}
H_1 &= (i \cos \vartheta_y - j \cos \vartheta_x) \cos \omega t, \\
H_2 &= \Pi (- i \cos \vartheta_z + k \cos \vartheta_x) \cos (\omega t - \varphi).
\end{aligned}
\tag{11.20}
$$

The above introduced vectors H_1 and H_2 agree in their direction with the vectors $H_{\omega 1}$ and $H_{\omega 2}$, respectively; they differ from these vectors by the same factor $\omega^2 \Pi_{\omega z}/c^2$ (at $t = 0$; this factor is immaterial when only polarization is considered). It is convenient to switch to χ, η coordinates in the following discussion. Assume that the χ axis is parallel to H_1 at $t = 0$, and the η

axis is rotated clockwise around **n** by 90°. The equations which are equivalent to Eq. (11.20) have the following form in this coordinate system:

$$\chi = \sin \vartheta_z \cos \omega t + \Pi \sin \vartheta_y \cos \psi \cos (\omega t - \varphi),$$

$$\eta = \Pi \sin \vartheta_y \sin \psi \cos (\omega t - \varphi).$$

(11.21)

We infer from these equations that the polarization is symmetric relative to the plane of incidence, the yz plane. At $\cos \vartheta_x > 0$, the polarization is counterclockwise, provided that we have $0 < \varphi < \pi$, and clockwise in the case $\pi < \varphi < 0$.

When we eliminate the parameter ωt from Eq. (11.21), we obtain the equation of the ellipse in the following form:

$$\chi = \left(\frac{\eta \cos \varphi}{\Pi \sin \vartheta_y \sin \psi} - \sin \varphi \sqrt{1 - \frac{\eta^2}{\Pi^2 \sin^2 \vartheta_y \sin^2 \psi}} \right) \sin \vartheta_z + \frac{\eta \cos \psi}{\sin \psi}$$

(11.22)

or, when we take into account Eq. (11.18), we obtain

$$\chi = \left(\frac{\eta \cos \varphi \sin \vartheta_z}{\Pi \cos \vartheta_x} - \sin \varphi \sqrt{1 - \frac{\eta^2 \sin^2 \vartheta_z}{\Pi^2 \cos^2 \vartheta_x}} \right) \sin \vartheta_z - \frac{\eta \cos \vartheta_y \cos \vartheta_z}{\cos \vartheta_x}.$$

(11.23)

The principal axes of the ellipse include the angles α_1 and α_2 with the vector $\mathbf{H}_{\omega 1}$; these angles are given by the following equations:

$$\operatorname{tg} \alpha_{1,2} = (-1 \pm \sqrt{1 + k^2})/k \qquad (\operatorname{tg} \alpha_1 \operatorname{tg} \alpha_2 = -1),$$

where

$$k = \frac{2\Pi \cos \vartheta_x (\sin^2 \vartheta_z \cos \varphi - \Pi \cos \vartheta_y \cos \vartheta_z)}{\Pi^2 \cos^2 \vartheta_x - \sin^2 \vartheta_y \sin^4 \vartheta_z \sin^2 \varphi - (\sin^2 \vartheta_z \cos \varphi - \Pi \cos \vartheta_y \cos \vartheta_z)^2}.$$

(11.24)

A positive angle α_1 indicates that the corresponding axis of the ellipse is rotated counterclockwise relative to $\mathbf{H}_{\omega 1}$. The ratio of the squares of the semiaxes is

$$\frac{a_1^2}{a_2^2} = \frac{\cos^2 \alpha_1 (\Pi \operatorname{tg} \alpha_1 \cos \vartheta_x - \sin^2 \vartheta_z \cos \varphi + \Pi \cos \vartheta_y \cos \vartheta_z)^2 + \sin^4 \vartheta_z \sin^2 \varphi}{\cos^2 \alpha_2 (\Pi \operatorname{tg} \alpha_2 \cos \vartheta_x - \sin^2 \vartheta_z \cos \varphi + \Pi \cos \vartheta_y \cos \vartheta_z)^2 + \sin^4 \vartheta_z \sin^2 \varphi}.$$

(11.25)

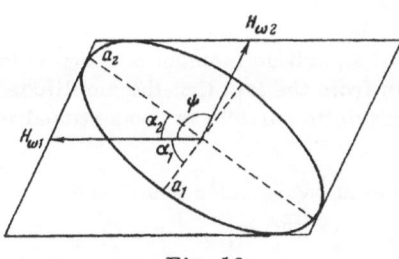

Fig. 16

The polarization ellipse for the magnetic (or electric) field of the waves emitted can be inscribed in a parallelogram whose sides are parallel to the vectors $\mathbf{H}_{\omega 1}$ and $\mathbf{H}_{\omega 2}$ (or $\mathbf{E}_{\omega 1}$ and $\mathbf{E}_{\omega 2}$) and equal to twice the absolute values of these vectors. Figure 16 shows the "magnetic" ellipse. The vectors $\mathbf{H}_{\omega 1}$ and $\mathbf{H}_{\omega 2}$ lie on the plane of the drawing, with the wave vector perpendicular to them, pointing toward the reader ($\psi > 0$, $\alpha_1 > 0$, $\alpha_2 < 0$, $-\pi/2 < \varphi < \pi/2$). We see from Eq. (11.24) that at $k = 0$ one of the angles α_i (i = 1 or 2) is equal to $\pi/2$, while the other angle is equal to zero. Let us assume $\alpha_1 = \pi/2$, $\alpha_2 = 0$. k vanishes when the particle incidence is perpendicular to the boundary ($\Pi = 0$) or, in the case of inclined particle incidence, for waves propagating in the plane of incidence ($\cos \vartheta_x = 0$). In the nonrelativistic approximation, k is close to zero. In any event, in rigorous calculations as well as in approximations, we obtain $a_1 = 0$ and $a_2 \neq 0$ from Eq. (11.25). The fact that one of the semiaxes of the ellipse vanishes indicates linear polarization. Since in our case we have $a_2 \neq 0$ ($\alpha_2 = 0$), the electric vector of the waves emitted is in the plane of wave propagation, and the magnetic vector in the plane parallel to the boundary. k also vanishes in the case $\Pi \cos \vartheta_y \cos \vartheta_z = \sin^2\vartheta_z \cos \varphi$. Then we have

$$\frac{a_1^2}{a_2^2} = \frac{\Pi^2 \cos^2 \vartheta_x}{\sin^4 \vartheta_z - \Pi^2 \cos^2 \vartheta_y \cos^2 \vartheta_z}. \tag{11.26}$$

We have circular polarization if $\Pi^2 \cos^2\vartheta_z = \sin^4\vartheta_z - \Pi^2\cos^2\vartheta_y\cos^2\vartheta_z$. If $\varphi = 0$, then

$$\text{tg } \alpha_1 = \frac{\sin^2 \vartheta_z - \Pi \cos \vartheta_y \cos \vartheta_z}{\Pi \cos \vartheta_x} \quad (a_1 = 0), \tag{11.27}$$

$$\text{tg } \alpha_2 = -\frac{\Pi \cos \vartheta_x}{\sin^2 \vartheta_z - \Pi \cos \vartheta_y \cos \vartheta_z} \quad (a_2 \neq 0). \tag{11.28}$$

For $\varphi = \pi$ we obtain

$$\text{tg } \alpha_1 = -\frac{\sin^2 \vartheta_z + \Pi \cos \vartheta_y \cos \vartheta_z}{\Pi \cos \vartheta_x} \quad (a_1 = 0), \tag{11.29}$$

$$\text{tg } \alpha_2 = \frac{\Pi \cos \vartheta_x}{\sin^2 \vartheta_z + \Pi \cos \vartheta_y \cos \vartheta_z} \quad (a_2 \neq 0). \tag{11.30}$$

Simple results are obtained from Eqs. (11.24) and (11.25) for $\cos \vartheta_y = 0$ (radiation in a plane which passes through the normal to the boundary and is perpendicular to the plane of incidence), provided that $\varphi = \pm \pi/2$:

$$a_1^2/a_2^2 = \Pi^2/\sin^2\vartheta_z, \ \alpha_1 = \pi/2, \ \alpha_2 = 0. \tag{11.31}$$

Circular polarization results for $\Pi = \sin \vartheta_z$.

Let us consider the transition radiation resulting from a nonrelativistic particle. When we neglect in Eq. (11.15) all terms containing the factor β, we see that the vector Π_ω is parallel to the normal to the boundary, i.e., we have the same situation as in the case of a particle incident perpendicular to the boundary. This means that the polarization of the waves emitted is the same as in the case of particle incidence perpendicular to the boundary. The angular distribution of the spectral density of the radiation energy is given by the formula

$$W_{n\omega} = \frac{e^2 \beta_z^2}{\pi^2 c} \frac{|1 - \varepsilon|^2 \sin^2 \vartheta_z \cos^2 \vartheta_z}{|\varepsilon \cos \vartheta_z + \sqrt{\varepsilon - \sin^2 \vartheta_z}|^2}. \tag{11.32}$$

Obviously, this result differs from that obtained in the case of a particle incident perpendicular to the boundary by the factor $(\beta_z/\beta)^2$. This difference results from the fact that the amplitude of the wave of the transition radiation generated by a nonrelativistic particle is proportional to the normal component of the velocity vector.

Let us consider the radiation from a relativistic particle at frequencies which are much larger than the frequencies of the optical interval ($\omega \gg \omega_0$, $\omega_0 = \sqrt{\frac{4\pi e^2 nZ}{m}}$, $\varepsilon = 1 - \frac{\omega_0^2}{\omega^2}$). We know that relativistic particles radiate electromagnetic waves under small angles (of the order of $\sqrt{1 - \beta^2}$) relative to their direction of motion. In the small-angle approximation, we can use the following equations:

$$\beta_y = \beta - \frac{\zeta^2}{2}, \ \beta_z = \zeta, \ \cos \vartheta_x = \theta_\perp, \tag{11.33}$$

$$\cos\vartheta_y = 1 - \frac{\theta_\parallel^2}{2} - \frac{\theta_\perp^2}{2} - \theta_\parallel\zeta - \frac{\zeta^2}{2}, \quad \cos\vartheta_z = \theta_\parallel + \zeta. \tag{11.33}$$

The notations are interpreted as follows: ζ denotes the angle between the direction of motion of the particle and the boundary ($\zeta \ll 1$); $\theta_\parallel$ and $\theta_\perp$ are the projections of the angle between the velocity vector of the particle and the wave vector upon the plane of incidence and the boundary, respectively ($-\zeta < \theta_\parallel \ll 1, -1 \ll \theta_\perp \ll 1$).

When we insert Eq. (11.33) into Eqs. (11.9) and (4.21), we obtain the angular distribution of the radiation intensity and the Hertz vector of the spherical wave at frequencies $\omega \gg \omega_0$ in the following form:

$$W_{n\omega} = \frac{16e^2}{\pi^2 c}\left(\frac{\omega_0}{\omega}\right)^4 \zeta^2(\theta_\parallel + \zeta)^2\{(1 - \beta^2 + \theta_\parallel^2 + \theta_\perp^2)(1 - \beta^2 + \theta_\parallel^2 + \theta_\perp^2 + 4\theta_\parallel\zeta) +$$

$$+ 4(\theta_\parallel^2 + \theta_\perp^2)\zeta^2\}\left|1 - \beta^2 + \theta_\parallel^2 + \theta_\perp^2 + 2\theta_\parallel\zeta + 2\zeta^2 - 2\zeta\sqrt{(\theta_\parallel + \zeta)^2 - \frac{\omega_0^2}{\omega^2}}\right|^{-2} \times$$

$$\times \left|\theta_\parallel + \zeta + \sqrt{(\theta_\parallel + \zeta)^2 - \frac{\omega_0^2}{\omega^2}}\right|^{-2}(1 - \beta^2 + \theta_\parallel^2 + \theta_\perp^2)^{-2}(1 - \beta^2 + \theta_\parallel^2 + \theta_\perp^2 + 4\theta_\parallel\zeta + 4\zeta^2)^{-2}, \tag{11.34}$$

$$\Pi_\omega = \frac{4ec\zeta(\theta_\parallel + \zeta)}{\pi\omega^2 R}\frac{\omega_0^2}{\omega^2}\left\{-\mathbf{j}\zeta\left[2 - 3\zeta^2 - 4\theta_\parallel\zeta - 2\theta_\parallel^2 + 2(\theta_\parallel + \zeta)\times\right.\right.$$

$$\left.\times\sqrt{(\theta_\parallel + \zeta)^2 - \frac{\omega_0^2}{\omega^2}}\right] + \mathbf{k}(1 - \beta^2 + \theta_\parallel^2 + \theta_\perp^2 - 2\zeta^2)\right\} \times$$

$$\times \left(1 - \beta^2 + \theta_\parallel^2 + \theta_\perp^2 + 2\theta_\parallel\zeta + 2\zeta^2 - 2\zeta\sqrt{(\theta_\parallel + \zeta)^2 - \frac{\omega_0^2}{\omega^2}}\right)^{-1} \times$$

$$\times (1 - \beta^2 + \theta_\parallel^2 + \theta_\perp^2)^{-1}(1 - \beta^2 + \theta_\parallel^2 + \theta_\perp^2 + 4\theta_\parallel\zeta + \zeta^2)^{-1}\left(\theta_\parallel + \zeta + \sqrt{(\theta_\parallel + \zeta)^2 - \frac{\omega_0^2}{\omega^2}}\right)^{-1}\exp\left(i\frac{\omega}{c}R\right) \tag{11.35}$$

In this approximation, the polarization of the waves is close to linear polarization and it follows from Eqs. (11.33), (11.35), and (11.24) that the electric vector of the wave includes the angle

$$\alpha_2 = \text{arc tg}\frac{2\theta_\perp\zeta}{1 - \beta^2 + \theta_\parallel^2 + \theta_\perp^2 + 2\theta_\parallel\zeta} \tag{11.36}$$

with the yz plane. Since all the angles are small, we can neglect the deviation of the electric and magnetic field vectors from the plane which passes through the normal to the boundary and is perpendicular to the plane of incidence (xz plane).

In the case

$$(1 - \beta^2, \theta_\parallel^2, \theta_\perp^2) \ll \zeta^2, \tag{11.37}$$

Eq. (11.34) assumes the form

$$W_{n\omega} = \frac{e^2}{\pi^2 c}\left(\frac{\omega_0}{\omega}\right)^4\frac{\theta_\parallel^2 + \theta_\perp^2}{(1 - \beta^2 + \theta_\parallel^2 + \theta_\perp^2)^2\left(1 - \beta^2 + \theta_\parallel^2 + \theta_\perp^2 + \frac{\omega_0^2}{\omega^2}\right)^2}, \tag{11.38}$$

except for the very small angular interval around the direction of motion of the particle ($\theta_\parallel = 0$, $\theta_\perp = 0$) wherein the density of the radiation per unit solid angle is relatively small. This region results from the refraction of the radiation generated along the path in the medium. In the region in which Eq. (11.38) holds, the expression given in Eq. (11.36) can be approximated in the following form:

$$\alpha_2 \approx \operatorname{arc} \operatorname{tg} \frac{\theta_\perp}{\theta_\parallel}. \tag{11.39}$$

Let us recall that $\theta_\parallel$ and $\theta_\perp$ can assume positive as well as negative values. The condition $\theta_\parallel > 0$ indicates that the angle included by the wave vector and the boundary is greater than the angle between the particle's velocity vector and the boundary (and vice versa, if $\theta_\parallel < 0$). Positive (negative) $\theta_\perp$ values correspond to a positive (negative) x-axis component of the wave vector. The coordinate system which we use is a right-handed coordinate system and therefore, according to Eq. (11.39), the electric vector of the wave propagating in the positive direction along the y axis lies on a plane which is defined by the wave vector and the direction of motion of the particle. More precisely, the absolute value of α_2 is slightly smaller than the approximate value given by Eq. (11.39): the plane of polarization seems to be somewhat shifted from the boundary. Disregarding these small deviations, we can conclude that the angular distribution and the polarization of the radiation are, with a high degree of accuracy, symmetric relative to the direction of motion of the particle. In the case $1 - \beta^2 + \theta_\parallel^2 + \theta_\perp^2 < \omega_0^2/\omega^2$, this result is related to the fact that the radiation is generated along the path in the vacuum. In the case $1 - \beta^2 + \theta_\parallel^2 + \theta_\perp^2 > \omega_0^2/\omega^2$, the relatively small wave refraction at the boundary is responsible for this particular result.

It follows from the symmetry of the radiation relative to the direction of motion of the particle and from Eq. (11.37) that in the case of a grazing incidence ($\zeta \ll 1$) the intensity ratio of the waves polarized in the plane of incidence to the waves polarized perpendicular to that plane is

$$W_{n\omega\parallel} / W_{n\omega\perp} \approx \theta_\parallel^2 / \theta_\perp^2. \tag{11.40}$$

This means that the spectral density of the radiation energy of the waves polarized in the plane of incidence is equal to the spectral density of the radiation energy of the waves polarized perpendicular to the plane of incidence*

$$W_{\omega\parallel} \approx W_{\omega\perp}. \tag{11.41}$$

The relation between the spectral density of the transition radiation and the frequency deserves particular attention. In order to reach qualitative conclusions on this relation we consider the derivative of the angular distribution of the spectral density of the radiation energy [defined by Eq. (11.34)] with respect to the frequency. The result is

$$\frac{dW_{n\omega}}{d\omega} = - \frac{\eta\eta_3}{\sqrt{\eta_2^2 - \eta_3}} \frac{[2\eta_2 \sqrt{\eta_2^2 - \eta_3} + 2(\eta_2^2 - \eta_3) + \eta_3]}{(\eta_1 + \eta_2 - \sqrt{\eta_2^2 - \eta_3})^3 (\eta_2 + \sqrt{\eta_2^2 - \eta_3})^3} \quad (\eta_2^2 > \eta_3), \tag{11.42}$$

$$\frac{dW_{n\omega}}{d\omega} = - \frac{\eta(\eta_1 + 2\eta_2)}{(\eta_1^2 + 2\eta_1\eta_2 + \eta_3)^3} \quad (\eta_3 < \eta_3), \tag{11.43}$$

where the following abbreviations were introduced for the sake of brevity:

$$\eta = \frac{2e^2}{\pi^2 c} \frac{\eta_2^2 \eta_3}{\omega\zeta^2 \eta_1} \frac{\eta_1(\eta_1 + 2\theta_\parallel) + \theta_\parallel^2 + \theta_\perp^2}{(\eta_1 + 2\eta_2)},$$

$$\eta_1 = (1 - \beta^2 + \theta_\parallel^2 + \theta_\perp^2)/2\zeta, \quad \eta_2 = \theta_\parallel + \zeta, \quad \eta_3 = \omega_0^2/\omega^2. \tag{11.44}$$

---

*This conclusion contradicts the result of [46], which stated that in the case of grazing incidence the transition radiation is almost completely polarized in the longitudinal direction. The result of [46] is a consequence of an insufficient analysis of the calculations.

Thus, for any small fixed angle ζ the derivative of the angular distribution of the spectral density of the radiation energy with respect to the frequency is negative in any direction $\theta_\parallel$, $\theta_\perp$. This means that the spectral density of the radiation energy decreases with increasing frequency. In a similar fashion we can arrive at the conclusion that the spectral density of the radiation energy decreases when the angle ζ between the particle trajectory and the boundary decreases. Since the derivative of the angular distribution with respect to the frequency is negative, the spectral density of the radiation energy at a smaller frequency is always greater than that at a higher frequency.*

§ 12. Radiation of an Oscillator Situated Near the Boundary

Equations (4.14)–(4.16) permit one to calculate the radiation of an oscillator which is situated near the boundary. When a charged particle moves periodically over a closed trajectory, the radiation field must be expanded in a Fourier series. The expansion coefficients of the field in the Fourier integral can be obtained from our results. To this end, we have to integrate over a period of particle motion and multiply the result by $n\omega$, where n denotes the order number of the terms (number of the harmonic) and ω the frequency of the oscillator.

We have to use Eqs. (4.15) and (4.16) for the calculation of the radiation field of an oscillator situated on the boundary. When the maximum particle velocity is much smaller than the velocity c of light, the time dependence of the particle coordinates in the exponents of Eqs. (4.15) and (4.16) can be ignored. In the event that the oscillator is nonrelativistic and harmonic, only one term in the expansion of the field in a Fourier series is nonvanishing in a first approximation, and this term corresponds to the principal frequency. When we assume $\mathbf{v} = \mathbf{v}_0 \cos \omega t$, integrate Eqs. (4.15) and (4.16) over a period of motion, multiply the results by $\omega e^{-i\omega t}$, and add them up, we obtain the Hertz vector of the oscillator-radiation field:

$$\Pi = \frac{ie}{2\omega R}\,(iv_{0x} + jv_{0v} + kv_{0z}) \exp\left(i\,\frac{\omega}{c}\,R - i\omega t - i\,\frac{\omega}{c}\,s\cos\vartheta_z\right) +$$
$$+ \frac{ie}{2\omega R}\,\frac{1}{(\varepsilon\cos\vartheta_z + \sqrt{\varepsilon - \sin^2\vartheta_z})}\,\{(iv_{0x} + jv_{0v})\,[2\cos\vartheta_z\,(\sin^2\vartheta_z +$$
$$+ \cos\vartheta_z\,\sqrt{\varepsilon - \sin^2\vartheta_z}) - (\varepsilon\cos\vartheta_z + \sqrt{\varepsilon - \sin^2\vartheta_z})] + k\,[2\cos\vartheta_z \times$$
$$\times\,(v_{0x}\cos\vartheta_x + v_{0v}\cos\vartheta_y)(\cos\vartheta_z - \sqrt{\varepsilon - \sin^2\vartheta_z}) +$$
$$+ v_{0z}(\varepsilon\cos\vartheta_z - \sqrt{\varepsilon - \sin^2\vartheta_z})]\}\exp\left(i\,\frac{\omega}{c}\,R - i\omega t + i\,\frac{\omega}{c}\,s\cos\vartheta_z\right), \qquad (12.1)$$

where it was assumed that the oscillator coordinates are x = 0, y = 0, and z = s (s denotes the distance between the oscillator and the boundary).

When s is much smaller than the wavelength ($\omega s/c \ll 1$), Eq. (12.1) assumes the following form:

$$\Pi = \frac{ie\cos\vartheta_z}{\omega R\,(\varepsilon\cos\vartheta_z + \sqrt{\varepsilon - \sin^2\vartheta_z})}\,\{(iv_{0x} + jv_{0v})(\sin^2\vartheta_z + \cos\vartheta_z\,\sqrt{\varepsilon - \sin^2\vartheta_z}) +$$
$$+ k\,[\varepsilon v_{0x} + (\vartheta_{0x}\cos\vartheta_x + v_{0v}\cos\vartheta_y)(\cos\vartheta_z - \sqrt{\varepsilon - \sin^2\vartheta_z})]\}\exp\left(i\,\frac{\omega}{c}\,R - i\omega t\right). \qquad (12.2)$$

Another consequence of Eq. (12.1) is that the Hertz vector of an oscillator oscillating parallel to the normal to the boundary ($v_{0x} = v_{0y} = 0$) is

---

*When the author of [45] considered a similar problem he came to a contradicting result, namely, that the radiation spectrum can be limited by fixed quanta when the angle of incidence of the particle trajectory relative to the boundary is increased.

$$\Pi = \frac{iekv_0}{2\omega R}\left[\exp\left(-i\frac{\omega}{c}s\cos\vartheta_z\right)+\frac{\varepsilon\cos\vartheta_z-\sqrt{\varepsilon-\sin^2\vartheta_z}}{\varepsilon\cos\vartheta_z+\sqrt{\varepsilon-\sin^2\vartheta_z}}\exp\left(i\frac{\omega}{c}s\cos\vartheta_z\right)\right]\exp\left(i\frac{\omega}{c}R-i\omega t\right). \quad (12.3)$$

If the oscillator is oscillating parallel to the boundary, along the y axis ($v_{0x} = v_{yz} = 0$), the Hertz vector of the corresponding radiation field is

$$\Pi = \frac{iev_0}{2\omega R}\left[\mathbf{j}\exp\left(-i\frac{\omega}{c}s\cos\vartheta_z\right)+\mathbf{j}\frac{\cos\vartheta_z-\sqrt{\varepsilon-\sin^2\vartheta_z}}{\cos\vartheta_z+\sqrt{\varepsilon-\sin^2\vartheta_z}}\exp\left(i\frac{\omega}{c}s\cos\vartheta_z\right)+\right.$$

$$\left.+\mathbf{k}\frac{2\cos\vartheta_y\cos\vartheta_z(\cos\vartheta_z-\sqrt{\varepsilon-\sin^2\vartheta_z})}{\varepsilon\cos\vartheta_z+\sqrt{\varepsilon-\sin^2\vartheta_z}}\exp\left(i\frac{\omega}{c}s\cos\vartheta_z\right)\right]\exp\left(i\frac{\omega}{c}R-i\omega t\right). \quad (12.4)$$

When the distance between the oscillator and the boundary is much smaller than the wavelength, the following well-known results (see [22]) are obtained from Eqs. (12.3) and (12.4):

$$\Pi = \frac{iekv_0\varepsilon\cos\vartheta_z\exp\left(i\frac{\omega}{c}R-i\omega t\right)}{\omega R(\varepsilon\cos\vartheta_z+\sqrt{\varepsilon-\sin^2\vartheta_z})}, \quad (12.5)$$

$$\Pi = \frac{iev_0}{\omega R}\cos\vartheta_z\exp\left(i\frac{\omega}{c}R-i\omega t\right)\left[\frac{\mathbf{j}}{\cos\vartheta_z+\sqrt{\varepsilon-\sin^2\vartheta_z}}+\mathbf{k}\frac{\cos\vartheta_y(\cos\vartheta_z-\sqrt{\varepsilon-\sin^2\vartheta_z})}{\varepsilon\cos\vartheta_z+\sqrt{\varepsilon-\sin^2\vartheta_z}}\right]. \quad (12.6)$$

These results describe the radiation when the vibrational motion of the particle takes place along a straight line. The radiation field generated by a particle moving along an elliptical path can be found in a similar way. The motion along the ellipse can be represented as a superposition of periodic motions along two mutually perpendicualr directions, with a phase shift between the two motions. Let us denote the phase shifts of the particle vibrations along the x and y axes with respect to the vibration along the z axis by φ_x and φ_y, respectively. The velocity components of the particle depend upon the time in the following fashion:

$$v_z = v_{0z}\cos\omega t,$$
$$v_x = v_{0x}\cos\omega t\cos\varphi_x + v_{0x}\sin\omega t\sin\varphi_x, \quad (12.7)$$
$$v_y = v_{0y}\cos\omega t\cos\varphi_y + v_{0y}\sin\omega t\sin\varphi_y.$$

The quantities v_{0x}, v_{0y}, v_{0z}, φ_x, and φ_y determine the position of the ellipse in space relative to the boundary ($z = 0$), the parameters of the ellipse, and the direction of motion of the particle. After inserting these expressions into Eqs. (4.15) and (4.16) and integrating over a period of the motion, summation renders the Hertz vector of the spherical wave of the radiation field of an elliptic oscillator:

$$\Pi = \frac{ie}{2\omega R}(\mathbf{i}v_{0x}e^{i\varphi_x}+\mathbf{j}v_{0y}e^{i\varphi_y}+\mathbf{k}v_{0z})\exp\left(i\frac{\omega}{c}R-i\omega t-i\frac{\omega}{c}s\cos\vartheta_z\right)+$$

$$+\frac{ie}{2\omega R}\frac{1}{\varepsilon\cos\vartheta_z+\sqrt{\varepsilon-\sin^2\vartheta_z}}\{(\mathbf{i}v_{0x}e^{i\varphi_x}+\mathbf{j}v_{0y}e^{i\varphi_y})[2\cos\vartheta_z(\sin^2\vartheta_z+$$

$$+\cos\vartheta_z\sqrt{\varepsilon-\sin^2\vartheta_z})-(\varepsilon\cos\vartheta_z+\sqrt{\varepsilon-\sin^2\vartheta_z})]+$$

$$+\mathbf{k}[2\cos\vartheta_z(v_{0x}e^{i\varphi_x}\cos\vartheta_x+v_{0y}e^{i\varphi_y}\cos\vartheta_y)(\cos\vartheta_z-\sqrt{\varepsilon-\sin^2\vartheta_z})+$$

$$+v_{0z}(\varepsilon\cos\vartheta_z-\sqrt{\varepsilon-\sin^2\vartheta_z})]\}\exp\left(i\frac{\omega}{c}R-i\omega t+i\frac{\omega}{c}s\cos\vartheta_z\right). \quad (12.8)$$

The waves emitted by this oscillator have elliptic polarization.

§13. Vacuum Radiation of an Oscillator Situated in a Medium

Separated by a Plane Boundary from the Vacuum

In a fashion similar to the calculation method applied to the oscillator on the boundary in vacuum (see preceding paragraph), we can calculate the radiation field of an oscillator in the medium. We have to use Eq. (4.14). The result has the form

$$\Pi = \frac{ie\cos\vartheta_z}{\omega R(\varepsilon\cos\vartheta_z + \sqrt{\varepsilon - \sin^2\vartheta_z})}\{(i v_{0x} + j v_{0y})(\sin^2\vartheta_z + \cos\vartheta_z\sqrt{\varepsilon - \sin^2\vartheta_z}) +$$
$$+ k[v_{0z} + (v_{0x}\cos\vartheta_x + v_{0y}\cos\vartheta_y)(\cos\vartheta_z - \sqrt{\varepsilon - \sin^2\vartheta_z})]\} \times$$
$$\times \exp\left(i\frac{\omega}{c}R - i\omega t + i\frac{\omega}{c}s\sqrt{\varepsilon - \sin^2\vartheta_z}\right), \tag{13.1}$$

where s denotes the distance between the oscillator and the boundary.

The particular result of Eq. (13.1) is that the radiation field of an oscillator oscillating in a direction normal to the boundary is defined by the following Hertz vector:

$$\Pi = \frac{iekv_0\cos\vartheta_z\exp\left(i\frac{\omega}{c}R - i\omega t\right)}{\omega R(\varepsilon\cos\vartheta_z + \sqrt{\varepsilon - \sin^2\vartheta_z})}\exp\left(i\frac{\omega}{c}s\sqrt{\varepsilon - \sin^2\vartheta_z}\right). \tag{13.2}$$

If the moment of the oscillator is parallel to the boundary and parallel to, say, the y axis, the radiation field is given by the equation

$$\Pi = \frac{iev_0}{\omega R}\cos\vartheta_z\left[j(\cos\vartheta_z + \sqrt{\varepsilon - \sin^2\vartheta_z})^{-1} +\right.$$
$$\left.+ k\cos\vartheta_y\frac{\cos\vartheta_z - \sqrt{\varepsilon - \sin^2\vartheta_z}}{\varepsilon\cos\vartheta_z + \sqrt{\varepsilon - \sin^2\vartheta_z}}\right]\exp\left(i\frac{\omega}{c}R - i\omega t + i\frac{\omega}{c}s\sqrt{\varepsilon - \sin^2\vartheta_z}\right). \tag{13.3}$$

If the projection of the velocity vector of a charged particle, vibrating around some fixed point, onto the normal to the boundary is nonvanishing, the particle can periodically pass through the boundary, provided that the distance between the center of the oscillator and the boundary is small enough. In this case the particle motion in the two media is not harmonic and the radiation spectrum must include higher harmonics in addition to the principal frequency. In other words, the oscillator generates transition radiation (see §14). There exists another possibility for an extension of the radiation spectrum of a harmonic oscillator. This situation arises when the particle moves in an absorbing medium. In the event that the attenuation of the waves is rather strong over a path of the order of magnitude of the projection of the amplitude of particle motion upon the normal to the boundary, the radiation generated by the particle over various sections of its path proceeds toward the boundary and is absorbed more or less. From a qualitative viewpoint, this is equivalent to the radiation of a particle with a periodically changing charge.

In deriving the results stated in Eq. (13.1), we neglected deviations from the harmonic form of the apparent dipole moment of the oscillator which result from wave attenuation in the medium. This is admissible when the condition

$$\Delta z_0\frac{\omega}{c}\operatorname{Im}\sqrt{\varepsilon - \sin^2\vartheta_z} \ll 1 \tag{13.4}$$

is fulfilled, where Δz_0 denotes the projection of the amplitude of a particle in harmonic motion upon the normal to the boundary. If this inequality is satisfied, the radiation intensity of the higher harmonic waves is evidently much smaller than the intensity of the principal wave.

When the particle does not move along a straight-line path but along an ellipse, and the velocity components depend upon time as defined by Eq. (12.7), the Hertz vector of the radiation field is equal to

$$\Pi = \frac{ie\cos\vartheta_z}{\omega R(\varepsilon\cos\vartheta_z + \sqrt{\varepsilon - \sin^2\vartheta_z})}\{(\mathbf{i}v_{0x}e^{i\varphi_x} + \mathbf{j}v_{0y}e^{i\varphi_y})(\sin^2\vartheta_z +$$

$$+ \cos\vartheta_z\sqrt{\varepsilon - \sin^2\vartheta_z}) + \mathbf{k}[v_{0z} + (v_{0x}e^{i\varphi_x}\cos\vartheta_x + v_{0y}e^{i\varphi_y}\cos\vartheta_y) \times$$

$$\times (\cos\vartheta_z - \sqrt{\varepsilon - \sin^2\vartheta_z})]\}\exp\left(i\frac{\omega}{c}R - i\omega t + i\frac{\omega}{c}s\sqrt{\varepsilon - \sin^2\vartheta_z}\right), \tag{13.5}$$

as can be derived by direct calculation. This result differs from Eq. (13.1), which was obtained for a linear oscillator [like Eq. (12.8) differs from Eq. (12.1)] insofar as complex quantities $v_{0x}e^{i\varphi_x}$ and $v_{0y}e^{i\varphi_y}$ appear instead of the real quantities v_{0x} and v_{0y}.

§14. Transition Radiation of a Fixed Oscillator

Let us consider the radiation from an oscillator which is situated at the boundary. We assume that a charged particle moves in harmonic motion along the normal to the boundary (along the z axis) around some fixed point (center of the oscillator) and that the particle intersects the boundary.

The starting points for calculating the radiation field are Eqs. (4.14), (4.15), and (4.16), wherein we have to assume $v_x = v_y = 0$ in the case under consideration. When we assume that the x and y coordinates of the particle are zero at any instant of time, we obtain a formula which is free of the direction cosines of the wave vector relative to the x and y axes. We omit the subscript z at the angle of the direction cosine referring to the z axis and write the results in the form:

$$\Pi_{\omega 1} = \frac{iek\cos\vartheta\exp\left(i\frac{\omega}{c}R\right)}{\pi\omega R(\varepsilon\cos\vartheta + \sqrt{\varepsilon - \sin^2\vartheta})}\int v\exp\left(i\omega t + i\frac{\omega}{c}z\sqrt{\varepsilon - \sin^2\vartheta}\right)dt, \tag{14.1}$$

$$\Pi'_{\omega 2} = \frac{iek\exp\left(i\frac{\omega}{c}R\right)}{2\pi\omega R}\int v\exp\left(i\omega t - i\frac{\omega}{c}z\cos\vartheta\right)dt, \tag{14.2}$$

$$\Pi''_{\omega 2} = \frac{iek(\varepsilon\cos\vartheta - \sqrt{\varepsilon - \sin^2\vartheta})}{2\pi\omega R(\varepsilon\cos\vartheta + \sqrt{\varepsilon - \sin^2\vartheta})}\exp\left(i\frac{\omega}{c}R\right)\int v\exp\left(i\omega t + i\frac{\omega}{c}z\cos\vartheta\right)dt. \tag{14.3}$$

In the case of a nonrelativistic oscillator, the time dependence of the particle coordinates can be neglected. Thus, we obtain

$$\Pi_{\omega 1} = \frac{iek\cos\vartheta\exp\left(i\frac{\omega}{c}R\right)}{\pi\omega R(\varepsilon\cos\vartheta + \sqrt{\varepsilon - \sin^2\vartheta})}\int v\exp(i\omega t)dt, \tag{14.4}$$

$$\Pi_{\omega 2} = \frac{iek\varepsilon\cos\vartheta\exp\left(i\frac{\omega}{c}R\right)}{\pi\omega R(\varepsilon\cos\vartheta + \sqrt{\varepsilon - \sin^2\vartheta})}\int v\exp(i\omega t)dt. \tag{14.5}$$

The sum of the Hertz vectors defined by Eqs. (14.4) and (14.5) is the solution [Eqs. (14.2) and (14.3) were combined into a single equation — Eq. (14.5)].

We assume the law of motion of the charged particle to be as follows:

$$z = z_0 + \Delta z_0\sin\omega t, \tag{14.6}$$

where z_0 denotes the coordinate of the center of the oscillator and Δz_0 the amplitude of the vibrations. Since the coordinate origin is on the boundary and the z axis points toward the vacuum, for $z_0 > 0$ the center of the oscillator is in the vacuum, and for $z_0 < 0$ in the medium. We assume that the particle periodically intersects the boundary, i.e., that

$$| z_0 | < \Delta z_0 \qquad (\Delta z_0 > 0). \tag{14.7}$$

It follows from Eq. (14.6) that

$$v = v_0 \cos \omega t, \qquad v_0 = \Delta z_0 \omega. \tag{14.8}$$

We denote the phase of the particle motion at the time at which the particle intersects the boundary by

$$\varphi = \text{arc sin} \; (| z_0 |/\Delta z_0). \tag{14.9}$$

For $z_0 < 0$ (the center of the oscillator is in the medium), φ denotes the phase at the time at which the particle passes into the vacuum. In this case, the phase at the moment at which the particle intersects the boundary in the opposite direction is symmetric relative to $\pi/2$. In other words, the phase of the particle during its motion in the vacuum changes from φ to $\pi - \varphi$. If $z_0 > 0$, φ determines the phase at which the particle intersects the boundary, reckoned from the end of the period of motion. The corresponding phase, reckoned from the beginning of the period ($t = 0$) is $2\pi - \varphi$. At that moment, the particle passes into the vacuum. The phase at the moment of re-entry of the particle into the medium is symmetric relative to $3\pi/2$. In other words, the phase at the moment of entry is equal to $\pi + \varphi$. For $z_0 < 0$, the phase assumes values of the interval $\varphi < \omega t < \pi - \varphi$ and corresponds to the motion of the particle in the vacuum; all other phase values (during the period $0 < t < T$) correspond to the motion in the medium; for $z_0 > 0$, phase values of the interval $\pi + \varphi < \omega t < 2\pi - \varphi$ correspond to motion in the medium and all other phase values to motion in the vacuum. These intervals define the time integration interval in Eqs. (14.4) and (14.5).

According to the theorem on the transition from a Fourier integral expansion to a Fourier series expansion (which must be used in the case of a periodic motion), for $z_0 < 0$ the terms of the latter series are given, in our case, by

$$\Pi_{1,\,n} = \frac{i e v_0 k \cos \vartheta \exp \left(i \dfrac{n\omega}{c} R - i n \omega t \right)}{\pi n \omega R \left(\varepsilon \cos \vartheta + \sqrt{\varepsilon - \sin^2 \vartheta} \right)} \left[\int_0^{2\pi} (\cos nx + i \sin nx) \cos x \, dx - \int_\varphi^{\pi - \varphi} (\cos nx + i \sin nx) \cos x \, dx \right], \tag{14.10}$$

$$\Pi_{2,\,n} = \frac{i e v_0 k e \cos \vartheta \exp \left(i \dfrac{n\omega}{c} R - i n \omega t \right)}{\pi n \omega R \left(\varepsilon \cos \vartheta + \sqrt{\varepsilon - \sin^2 \vartheta} \right)} \int_\varphi^{\pi - \varphi} (\cos nx + i \sin nx) \cos x \, dx. \tag{14.11}$$

For $z_0 > 0$, we obtain

$$\Pi_{1,\,n} = \frac{i e v_0 k \cos \vartheta \exp \left(i \dfrac{n\omega}{c} R - i n \omega t \right)}{\pi n \omega R \left(\varepsilon \cos \vartheta + \sqrt{\varepsilon - \sin^2 \vartheta} \right)} \int_{\pi + \varphi}^{2\pi - \varphi} (\cos nx + i \sin nx) \cos x \, dx, \tag{14.12}$$

$$\Pi_{2,\,n} = \frac{i e v_0 k e \cos \vartheta \exp \left(i \dfrac{n\omega}{c} R - i n \omega t \right)}{\pi n \omega R \left(\varepsilon \cos \vartheta + \sqrt{\varepsilon - \sin^2 \vartheta} \right)} \left[\int_0^{2\pi} (\cos nx + i \sin nx) \cos x \, dx - \right.$$
$$\left. - \int_{\pi + \varphi}^{2\pi - \varphi} (\cos nx + i \sin nx) \cos x \, dx \right] \; (n = \pm 1, \pm 2, \pm 3, \ldots). \tag{14.13}$$

The solution is given by the sum of the vectors defined by Eqs. (14.10) and (14.11) and by Eqs. (14.12) and (14.13). Summing up, we obtain for $z_0 < 0$

$$\Pi_n = \frac{iev_0 \mathbf{k} \cos\vartheta \exp\left(i\,\frac{n\omega}{c}\,R - in\omega t\right)}{\pi n\omega R\,(\varepsilon\cos\vartheta + \sqrt{\varepsilon - \sin^2\vartheta})} \left[\int_0^{2\pi}(\cos nx + i\sin nx)\cos x\,dx + (\varepsilon - 1)\int_\varphi^{\pi-\varphi}(\cos nx + i\sin nx)\cos x\,dx\right],$$

$$(14.14)$$

and for $z_0 > 0$

$$\Pi_n = \frac{iev_0 \mathbf{k} \cos\vartheta \exp\left(i\,\frac{n\omega}{c}\,R - in\omega t\right)}{\pi n\omega R\,(\varepsilon\cos\vartheta + \sqrt{\varepsilon - \sin^2\vartheta})} \left[\varepsilon\int_0^{2\pi}(\cos nx + i\sin nx)\cos x\,dx + (1-\varepsilon)\int_{\pi+\varphi}^{2\pi-\varphi}(\cos nx + i\sin nx)\cos x\,dx\right].$$

$$(14.15)$$

It follows from Eqs. (14.14) and (14.15) that for $\varepsilon = 1$ (radiation in vacuum without a boundary) only the first harmonic in the Fourier series expansion is nonvanishing, as had to be expected for the nonrelativistic oscillator under consideration. If $\varphi = \pi/2$, i.e., for $|z_0| = \Delta z_0$, the oscillator does not pass through the boundary. Also, in this case only the first harmonic is nonvanishing.

When we integrate in Eqs. (14.4) and (14.15), we obtain for $z_0 < 0$

$$\Pi_1 = \frac{iev_0 \mathbf{k} \cos\vartheta \exp\left(i\cdot\frac{\omega}{c}\,R - i\omega t\right)}{\pi\omega R\,(\varepsilon\cos\vartheta + \sqrt{\varepsilon - \sin^2\vartheta})} \left[\pi + (\varepsilon - 1)\left(\frac{\pi}{2} - \varphi - \frac{1}{2}\sin 2\varphi\right)\right],$$

$$\Pi_n = \frac{iev_0 \mathbf{k} \cos\vartheta \exp\left(i\,\frac{n\omega}{c}\,R - in\omega t\right)}{2\pi n\omega R\,(\varepsilon\cos\vartheta + \sqrt{\varepsilon - \sin^2\vartheta})}(\varepsilon - 1)\left\{\left[\frac{\sin(n-1)(\pi-\varphi)}{n-1} + \right.\right.$$
$$\left. + \frac{\sin(n+1)(\pi-\varphi)}{n+1} - \frac{\sin(n-1)\varphi}{n-1} - \frac{\sin(n+1)\varphi}{n+1}\right] + i\left[\frac{\cos(n-1)\varphi}{n-1} + \right.$$
$$\left.\left. + \frac{\cos(n+1)\varphi}{n+1} - \frac{\cos(n-1)(\pi-\varphi)}{n-1} - \frac{\cos(n+1)(\pi-\varphi)}{n+1}\right]\right\}$$

$$(14.16)$$

$$(n = \pm 2, \pm 3, \ldots),$$

and for $z_0 > 0$

$$\Pi_1 = \frac{iev_0 \mathbf{k} \cos\vartheta \exp\left(i\,\frac{\omega}{c}\,R - i\omega t\right)}{\pi\omega R\,(\varepsilon\cos\vartheta + \sqrt{\varepsilon - \sin^2\vartheta})} \left[\varepsilon\pi + (1-\varepsilon)\left(\frac{\pi}{2} - \varphi - \frac{1}{2}\sin 2\varphi\right)\right],$$

$$\Pi_n = \frac{iev_0 \mathbf{k} \cos\vartheta \exp\left(i\,\frac{n\omega}{c}\,R - in\omega t\right)}{2\pi n\omega R\,(\varepsilon\cos\vartheta + \sqrt{\varepsilon - \sin^2\vartheta})}(1-\varepsilon)\left\{-\left[\frac{\sin(n-1)\varphi}{n-1} + \right.\right.$$
$$\left. + \frac{\sin(n+1)\varphi}{n+1} + \frac{\sin(n-1)(\pi+\varphi)}{n-1} + \frac{\sin(n+1)(\pi+\varphi)}{n+1}\right] +$$
$$\left. + i\left[\frac{\cos(n-1)(\pi+\varphi)}{n-1} + \frac{\cos(n+1)(\pi+\varphi)}{n+1} - \frac{\cos(n-1)\varphi}{n-1} - \frac{\cos(n+1)\varphi}{n+1}\right]\right\}$$

$$(14.17)$$

$$(n = \pm 2, \pm 3, \ldots).$$

When we assume $\varphi = \pi/2$ in Eq. (14.16) (which corresponds to $z_0 = -\Delta z_0$), we obtain the radiation of the oscillator, which is situated in the medium, in the vacuum:

$$\Pi = \frac{iev_0 \mathbf{k} \cos\vartheta \exp\left(i\,\frac{\omega}{c}\,R - i\omega t\right)}{\omega R\,(\varepsilon\cos\vartheta + \sqrt{\varepsilon - \sin^2\vartheta})}.$$

$$(14.18)$$

When we assume $\varphi = \pi/2$ in Eq. (14.15) (which corresponds to $z_0 = \Delta z_0$), we obtain the radiation of an oscillator, which is situated in the vacuum, in the immediate vicinity of the boundary. This radiation has the form

$$\Pi = \frac{iev_0 \mathbf{k}\varepsilon \cos\vartheta \exp\left(i\,\frac{\omega}{c}\,R - i\omega t\right)}{\omega R\,(\varepsilon\cos\vartheta + \sqrt{\varepsilon - \sin^2\vartheta})}.$$

$$(14.19)$$

In the case $\varphi = 0$, i.e., when the center of the oscillator is situated on the boundary ($z_0 = 0$), Eqs. (14.16) and (14.17) give the same result, as we could expect:

$$\Pi_1 = \frac{iev_0 \mathbf{k}\,(1 + \varepsilon)\cos\vartheta \exp\left(i\,\frac{\omega}{c}\,R - i\omega t\right)}{2\omega R\,(\varepsilon\cos\vartheta + \sqrt{\varepsilon - \sin^2\vartheta})},$$

$$\Pi_n = \frac{2ev_0 \mathbf{k}\,(1 - \varepsilon)\cos\vartheta \exp\left(i\,\frac{n\omega}{c}\,R - in\omega t\right)}{\pi\omega R\,(\varepsilon\cos\vartheta + \sqrt{\varepsilon - \sin^2\vartheta})\,(n^2 - 1)},$$

(14.20)

where n denotes an even number. The odd-number harmonics vanish.

We infer from Eqs. (14.16) and (14.17) that the angular distribution of the radiation intensity contributed by the various harmonics is independent of the position of the oscillator center relative to the boundary. The oscillator position affects only the distribution of the radiation energy over the various harmonics. When the oscillating particle periodically intersects the boundary and its vibration center is displaced relative to the boundary, waves of all frequencies which are multiples of the principal vibration frequency of the particle are emitted. When the particle does not intersect the boundary, only waves of the principal frequency are emitted during the harmonic motion of the particle (in the nonrelativistic approximation considered). Finally, as we see from Eq. (14.20), when the oscillator center is situated on the boundary, only waves of even harmonics are emitted.

The transition radiation of a relativistic oscillator deserves attention, too. We assume the law of motion in the same form as above, i.e., in the form stated in Eq. (14.6). In order to simplify the considerations, we will assume that the oscillator center is situated on the boundary ($z_0 = 0$). Equations (14.1), (14.2), and (14.3) are the basis for calculating the radiation field of the vertical relativistic oscillator (i.e., of a particle moving along the normal to the boundary). According to the theorem on the transition from the Fourier integral expansion to the Fourier series expansion (a transition which is necessary in the case of a periodic motion), we obtain from Eqs. (14.1), (14.2), and (14.3) (when we make an expansion in the interval $-\pi < \omega t < \pi$):

$$\Pi_{1,n} = \frac{iev_0 \mathbf{k}\cos\vartheta \exp\left(i\,\frac{n\omega}{c}\,R - in\omega t\right)}{\pi n\omega R\,(\varepsilon\cos\vartheta + \sqrt{\varepsilon - \sin^2\vartheta})} \int_{-\pi}^{0} [\cos(nx - z_n'\sin x) + i\sin(nx - z_n'\sin x)]\cos x\,dx, \qquad (14.21)$$

$$\Pi_{2,n}' = \frac{iev_0 \mathbf{k}\exp\left(i\,\frac{n\omega}{c}\,R - in\omega t\right)}{2\pi n\omega R} \int_{0}^{\pi} [\cos(nx - z_n\sin x) + i\sin(nx - z_n\sin x)]\cos x\,dx, \qquad (14.22)$$

$$\Pi_{2,n}'' = \frac{iev_0 \mathbf{k}\,(\varepsilon\cos\vartheta - \sqrt{\varepsilon - \sin^2\vartheta})\exp\left(i\,\frac{n\omega}{c}\,R - in\omega t\right)}{2\pi n\omega R\,(\varepsilon\cos\vartheta - \sqrt{\varepsilon - \sin^2\vartheta})} \int_{0}^{\pi} [\cos(nx + z_n\sin x) +$$

$$+\,i\sin(nx + z_n\sin x)]\cos x\,dx, \qquad (14.23)$$

where

$$z_n' = n\beta_0\sqrt{\varepsilon - \sin^2\vartheta}, \qquad z_n = n\beta_0\cos\vartheta \qquad (\beta_0 = v_0/c). \qquad (14.24)$$

The sum of these vectors is the solution. We use the following transformation for the calculation of the integral which appears, for example, in Eq. (14.21):

$$\int_{0}^{\pi} [\cos(nx - z_n'\sin x) + i\sin(nx - z_n'\sin x)]\cos x\,dx = \int_{0}^{\pi} [\cos(nx - z_n'\sin x) - i\sin(nx - z_n'\sin x)]\cos x\,dx =$$

$$= \int\limits_0^\pi [\cos nx \cos (z'_n \sin x) + \sin nx \sin (z'_n \sin x) - i \sin nx \cos (z'_n \sin x) + i \cos nx \sin (z'_n \sin x)] \cos x \, dx =$$

$$= \frac{1}{2} \int\limits_0^\pi [\cos (n+1) x + \cos (n-1) x] [\cos (z'_n \sin x) + i \sin (z'_n \sin x)] \, dx +$$

$$+ \frac{1}{2} \int\limits_0^\pi [\sin (n+1) x + \sin (n-1) x][\sin (z'_n \sin x) - i \cos (z'_n \sin x)] \, dx. \tag{14.25}$$

The following relations hold:

$$\int\limits_0^\pi \cos (z \sin x) \cos mx \, dx = \frac{\pi}{2} [1 + (-1)^m] J_m (z),$$

$$\int\limits_0^\pi \cos (z \sin x) \sin mx \, dx = -m [1 - (-1)^m] S_{-1, m} (z),$$

$$\int\limits_0^\pi \sin (z \sin x) \cos mx \, dx = [1 + (-1)^m] S_{0, m} (z),$$

$$\int\limits_0^\pi \sin (z \sin x) \sin mx \, dx = \frac{\pi}{2} [1 - (-1)^m] J_m (z),$$

$$\tag{14.26}$$

where J_m denotes Bessel functions and $S_{-1, m}$ Lommel functions:

$$S_{-1, m} (z) = -\frac{1}{m^2} + \sum_{k=1}^\infty \frac{(-1)^{l-1} z^{2k}}{m^2 (2^2 - m^2) (4^2 - m^2) \ldots [(2k)^2 - m^2]},$$

$$S_{0, m} (z) = \sum_{l=1}^\infty \frac{(-1)^{l-1} z^{2k-1}}{(1^2 - m^2) (3^2 - m^2) (5^2 - m^2) \ldots [(2k-1)^2 - m^2]}.$$

$$\tag{14.27}$$

With the formulas of (14.26), we obtain

$$\int\limits_{-\pi}^0 [\cos (nx - z'_n \sin x) + i \sin (nx - z'_n \sin x)] \cos x \, dx =$$

$$= \frac{1}{2} \Big\{ \frac{\pi}{2} [1 + (-1)^{n+1}] J_{n+1} (z'_n) + \frac{\pi}{2} [1 + (-1)^{n-1}] J_{n-1} (z'_n) + i [1 + (-1)^{n+1}] S_{0, n+1} (z'_n) +$$

$$+ i [1 + (-1)^{n-1}] S_{0, n-1} (z'_n) + \frac{\pi}{2} [1 - (-1)^{n+1}] J_{n+1} (z'_n) + \frac{\pi}{2} [1 - (-1)^{n-1}] J_{n-1} (z'_n) +$$

$$+ i (n+1) [1 - (-1)^{n+1}] S_{-1, n+1} (z'_n) + i (n-1) [1 - (-1)^{n-1}] S_{-1, n-1} (z'_n) \Big\} =$$

$$= \frac{1}{2} \{ \pi [J_{n+1} (z'_n) + J_{n-1} (z'_n)] + i [1 + (-1)^{n+1}] [S_{0, n+1} (z'_n) + S_{0, n-1} (z'_n)] +$$

$$+ i [1 - (-1)^{n+1}] [(n+1) S_{-1, n+1} (z'_n) + (n-1) S_{-1, n-1} (z'_n)] \}. \tag{14.28}$$

Similarly, we obtain

$$\int\limits_0^\pi [\cos (nx - z_n \sin x) + i \sin (nx - z_n \sin x)] \cos x \, dx =$$

$$= \frac{1}{2} \{ \pi [J_{n+1} (z_n) + J_{n-1} (z_n)] - i [1 + (-1)^{n+1}] \times [S_{0, n+1} (z_n) + S_{0, n-1} (z_n)] -$$

$$- i [1 - (-1)^{n+1}] [(n+1) S_{-1, n+1} (z_n) + (n-1) S_{-1, n-1} (z_n)] \}, \tag{14.29}$$

$$\int_0^\pi [\cos(nx + z_n \sin x) + i \sin(nx + z_n \sin x)] \cos x \, dx =$$

$$= \frac{1}{2} \{\pi (-1)^{n+1} [J_{n+1}(z_n) + J_{n-1}(z_n)] + i [1 + (-1)^{n+1}] [S_{0, n+1}(z_n) + S_{0, n-1}(z_n)] -$$

$$- i [1 - (-1)^{n+1}] [(n+1) S_{-1, n+1}(z_n) + (n-1) S_{-1, n-1}(z_n)]\}. \tag{14.30}$$

Without a boundary (for $\varepsilon = 1$), we obtain

$$\Pi_n = i e v_0 \mathbf{k} \exp\left(i \frac{n\omega}{c} R - i n\omega t\right) [J_{n+1}(n\beta_0 \cos \vartheta) + J_{n-1}(n\beta_0 \cos \vartheta)] / 2n\omega R. \tag{14.31}$$

In the particular case $\varepsilon \to \infty$, the Hertz vector defined by Eq. (14.21) vanishes and the quantities before the integral signs in Eqs. (14.22) and (14.23) become identical. Thus, we obtain the radiation field of the n-th harmonic of the relativistic oscillator situated on the boundary between the vacuum and an ideal conductor in the form

$$\Pi_n = \frac{i e v_0 \mathbf{k} \exp\left(i \frac{n\omega}{c} R - i n\omega t\right)}{4\pi n\omega R} \{\pi [1 + (-1)^{n+1}] [J_{n+1}(n\beta_0 \cos \vartheta) +$$

$$+ J_{n-1}(n\beta_0 \cos \vartheta)] - 2i [1 - (-1)^{n+1}] [(n+1) S_{-1, n+1}(n\beta_0 \cos \vartheta) + (n-1) S_{-1, n-1}(n\beta_0 \cos \vartheta)]\}. \tag{14.32}$$

We see from this formula that the angular distribution of the radiation intensity generated by odd-number harmonic waves is given by Bessel functions and that of even-number harmonic waves by Lommel functions. A comparison of Eqs. (14.32) and (14.31) reveals that the radiation intensity of the odd harmonics is identical with the radiation intensity of the corresponding odd-number harmonics in the case of an oscillator in vacuum when no boundary is present. It follows from Eq. (14.32) that in the case of a nonrelativistic oscillator ($\beta_0 \ll 1$) the radiation intensity generated by the odd-number harmonics, which satisfy the inequality $n\beta_0 \ll 1$, is negligibly small, whereas in the case $n\beta_0 \gg 1$, the radiation energy of the odd-number harmonics is comparable to that of the even-number harmonics.

Let us write the Hertz vectors of the radiation field of the first few harmonics:

$$\Pi_1 = \frac{i e v_0 \mathbf{k} \exp\left(i \frac{\omega}{c} R - i\omega t\right)}{2\omega R} [J_2(\beta_0 \cos \vartheta) + J_0(\beta_0 \cos \vartheta)], \tag{14.33}$$

$$\Pi_2 = \frac{3 e v_0 \mathbf{k} \exp\left(i \frac{2\omega}{c} R - i2\omega t\right)}{2\pi\omega R} S_{-1, 3}(2\beta_0 \cos \vartheta), \tag{14.34}$$

$$\Pi_3 = \frac{i e v_0 \mathbf{k} \exp\left(i \frac{3\omega}{c} R - i3\omega t\right)}{6\omega R} [J_4(3\beta_0 \cos \vartheta) + J_2(3\beta_0 \cos \vartheta)], \tag{14.35}$$

$$\Pi_4 = \frac{e v_0 \mathbf{k} \exp\left(i \frac{4\omega}{c} R - i4\omega t\right)}{4\pi\omega R} [5 S_{-1, 5}(4\beta_0 \cos \vartheta) + 3 S_{-1, 3}(4\beta_0 \cos \vartheta)], \tag{14.36}$$

$$\Pi_5 = \frac{i e v_0 \mathbf{k} \exp\left(i \frac{5\omega}{c} R - i5\omega t\right)}{10\omega R} [J_6(5\beta_0 \cos \vartheta) + J_4(5\beta_0 \cos \vartheta)], \tag{14.37}$$

$$\Pi_6 = \frac{e v_0 \mathbf{k} \exp\left(i \frac{6\omega}{c} R - i6\omega t\right)}{6\pi\omega R} [7 S_{-1, 7}(6\beta_0 \cos \vartheta) + 5 S_{-1, 5}(6\beta_0 \cos \vartheta)], \tag{14.38}$$

etc.

The electromagnetic field of the n-th harmonic can be expressed by the Hertz vector with the formulas

$$\mathbf{E}_n = n^2\omega^2 [\Pi_n - \mathbf{n}(\mathbf{n}\Pi_n)] / c^2, \quad \mathbf{H}_n = n^2\omega^2 [\mathbf{n}\Pi_n] / c^2, \tag{14.39}$$

when $\mathbf{n}$ denotes the unit vector in the direction of observation.

The radiation energy provided by the waves of the n-th harmonic per unit time and unit solid angle is given by the corresponding Hertz vector according to the formula

$$W_{n,\,n} = \frac{n^4\omega^4}{2c^3\pi}\,R^2\,|\,\Pi_n\,|^2\sin^2\vartheta.$$ (14.40)

§15. Luminescence Radiation

The angular distribution and the degree of polarization of the luminescence radiation generated near the surface is obtained by averaging over the radiation intensity of the oscillator "immersed" in the medium, provided that orientations of the oscillator in all directions are equally likely.

We will state the equation for the radiation field (in vacuum) of an oscillator immersed in the medium, when a plane boundary is present [see Eq. (13.1)]:

$$\Pi = \frac{ie\cos\vartheta\exp\left(i\,\frac{\omega}{c}\,R - i\omega t\right)}{\omega R\,(\varepsilon\cos\vartheta_z + \sqrt{\varepsilon - \sin^2\vartheta_z})}\,\{(iv_{0x} + jv_{0y})\,(\sin^2\vartheta_z + \cos\vartheta_z\,\sqrt{\varepsilon - \sin^2\vartheta_z}) +$$

$$+ \,k\,[v_{0z} + (v_{cx}\cos\vartheta_x + v_{0y}\cos\vartheta_y)\,(\cos\vartheta_z - \sqrt{\varepsilon - \sin^2\vartheta_z})]\}\exp\left(i\,\frac{\omega}{c}\,s\sqrt{\varepsilon - \sin^2\vartheta_z}\right),$$ (15.1)

and for the Hertz vectors of waves which are polarized in the plane of wave propagation and perpendicular to that plane:

$$\Pi_{\parallel} = k\left[\Pi_z - \frac{\cos\vartheta_z}{\sin^2\vartheta_z}\,(\Pi_x\cos\vartheta_x + \Pi_y\cos\vartheta_y)\right],$$

$$\Pi_{\perp} = i\Pi_x + j\Pi_y + k\frac{\cos\vartheta_z}{\sin^2\vartheta_z}\,(\Pi_x\cos\vartheta_x + \Pi_y\cos\vartheta_y).$$ (15.2)

The radiation energies of the waves which are polarized in the plane of propagation and perpendicular to that plane per unit time and per unit solid angle are given by the formulas

$$W_{n\parallel} = \frac{\omega^4 R^2}{2\pi c^3}\,[|\,\Pi_{\parallel}\,|^2 - |\,(n\Pi_{\parallel})\,|^2].$$ (15.3)

$$W_{n\perp} = \frac{\omega^4 R^2}{2\pi c^3}\,[|\,\Pi_{\perp}\,|^2 - |\,(n\Pi_{\perp})\,|^2].$$ (15.4)

With Eqs. (15.1) and (15.2), we obtain the angular distribution of the radiation intensity generated by waves of different polarization emitted by a randomly oriented oscillator in a medium behind a plane boundary:

$$W_{n\parallel} = \frac{\omega^2 e^2\cos^2\vartheta_z\exp\left(-2\,\frac{\omega}{c}\,s\,\mathrm{Im}\sqrt{\varepsilon - \sin^2\vartheta_z}\right)}{2\pi c^3\,|\,\varepsilon\cos\vartheta_z + \sqrt{\varepsilon - \sin^2\vartheta_z}\,|^2\sin^2\vartheta_z}\,|\,v_{0z}\sin^2\vartheta_z - \sqrt{\varepsilon - \sin^2\vartheta_z}\,(v_{0x}\cos\vartheta_x + v_{0y}\cos\vartheta_y)\,|^2,$$ (15.5)

$$W_{n\perp} = \frac{\omega^2 e^2\cos^2\vartheta_z\exp\left(-2\,\frac{\omega}{c}\,s\,\mathrm{Im}\sqrt{\varepsilon - \sin^2\vartheta_z}\right)}{2\pi c^3\,|\,\varepsilon\cos\vartheta_z + \sqrt{\varepsilon - \sin^2\vartheta_z}\,|^2\sin^2\vartheta_z}\,|\sin^2\vartheta_z + \cos\vartheta_z\,\sqrt{\varepsilon - \sin^2\vartheta_z}\,|^2\,(v_{0x}\cos\vartheta_x + v_{0y}\cos\vartheta_y)^2.$$ (15.6)

When we assume that the oscillator orientations in all directions are equally probable, averaging of the above results over all possible oscillator orientations yields the equations

$$\langle W_{n\parallel}\rangle = \frac{\omega^2 e^2 v_0^2\cos^2\vartheta}{4\pi c^3\,|\,\varepsilon\cos\vartheta + \sqrt{\varepsilon - \sin^2\vartheta}\,|^2}\,(\sin^2\vartheta + |\,\varepsilon - \sin^2\vartheta\,|)\exp\left(-2\,\frac{\omega}{c}\,s\,\mathrm{Im}\sqrt{\varepsilon - \sin^2\vartheta}\right),$$ (15.7)

$$\langle W_{n\perp} \rangle = \frac{\omega^2 e^2 v_0^2 \cos^2 \vartheta}{4\pi c^3 |\varepsilon \cos \vartheta + \sqrt{\varepsilon - \sin^2 \vartheta}|^2} |\sin^2 \vartheta + \cos \vartheta \sqrt{\varepsilon - \sin^2 \vartheta}|^2 \exp\left(-2 \frac{\omega}{c} s \operatorname{Im} \sqrt{\varepsilon - \sin^2 \vartheta}\right), \quad (15.8)$$

where ϑ denotes the angle included by the wave vector and the normal to the boundary ($\vartheta \equiv \vartheta_z$). We can derive the angular distribution and the degree of polarization from these equations. We obtain

$$\langle W_n \rangle = \frac{\omega^2 e^2 v_0^2 \cos^2 \vartheta \exp\left(-2 \frac{\omega}{c} s \operatorname{Im} \sqrt{\varepsilon - \sin^3 \vartheta}\right)}{4\pi c^3 |\varepsilon \cos \vartheta + \sqrt{\varepsilon - \sin^2 \vartheta}|^2} [\sin^2 \vartheta + |\varepsilon - \sin^2 \vartheta| + |\sin^2 \vartheta + \cos \vartheta \sqrt{\varepsilon - \sin^2 \vartheta}|^2], (15.9)$$

$$P = \frac{\langle W_{n\parallel} \rangle - \langle W_{n\perp} \rangle}{\langle W_{n\parallel} \rangle + \langle W_{n\perp} \rangle} = \frac{\sin^2 \vartheta + |\varepsilon - \sin^2 \vartheta| - |\sin^2 \vartheta + \cos \vartheta \sqrt{\varepsilon - \sin^2 \vartheta}|^2}{\sin^2 \vartheta + |\varepsilon - \sin^2 \vartheta| + |\sin^2 \vartheta + \cos \vartheta \sqrt{\varepsilon - \sin^3 \vartheta}|^2}. \quad (15.10)$$

In the absorption case, the angular distribution and the intensity of the radiation which passes into the vacuum depend upon the distance between the radiation center and the boundary.

When the number of arbitrarily oriented oscillators per unit volume is the same at various depths, the angular distribution of the radiation, which is given by the integration of Eq. (15.9) over s, in our case is equal to

$$\langle W_n \rangle \sim \frac{\cos^2 \vartheta}{|\varepsilon \cos \vartheta + \sqrt{\varepsilon - \sin^2 \vartheta}|^2 \operatorname{Im} \sqrt{\varepsilon - \sin^2 \vartheta}} [\sin^2 \vartheta + |\varepsilon - \sin^2 \vartheta| + |\sin^2 \vartheta + \cos \vartheta \sqrt{\varepsilon - \sin^2 \vartheta}|^2]. \quad (15.11)$$

The angular distribution of the luminescence radiation of a compact absorbing sample (which has luminescence centers uniformly distributed over its entire volume) has the same form.

CHAPTER 3

Radiation in the Presence of Two Boundaries

§16. Spherical Waves Generated by an Arbitrarily Moving Charged Particle

The image method makes it relatively easy to calculate the radiation field generated by a particle in the case of several boundaries [5, 6]. For example, let us assume that three media separated by plane, parallel boundaries are considered. The media will be denoted by the subscripts 1, 2, and 3; the distance between the boundaries is a. The z axis is assumed perpendicular to the boundaries and the coordinate origin is assumed to be on the boundary separating medium 2 from medium 3 (Fig. 17). Let us calculate the spherical wave of the radiation field in region 3. This wave results from the interference of waves generated on particle trajectory sections situated in various media.

The waves which are polarized in the plane of propagation and perpendicular to this plane and generated while the particle moves through region 1 ($z < -a$) are given by

$$\Pi_{\omega_1 \parallel} = \frac{i e c k}{2\pi \omega R} \int P_{1,2;\parallel} \alpha_\parallel P_{2,3;\parallel} \left[\beta_z - (\beta_x \cos \vartheta_{3x} + \beta_y \cos \vartheta_{3y}) \frac{\cos \vartheta_{3z}}{\sin^2 \vartheta_{3z}}\right] \times$$
$$\times \exp\left[i\omega t - i\mathbf{k}_1 \mathbf{r} + i \frac{a}{z}(\mathbf{k}_2 \mathbf{r} - \mathbf{k}_1 \mathbf{r}) + i k_3 R\right] dt, \quad (16.1)$$

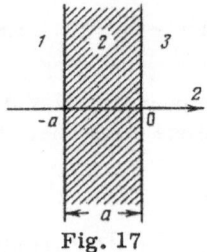

Fig. 17

$$\Pi_{\omega 1\perp} = \frac{iec}{2\pi\omega R}\, P_{1,2;\,\perp}\,\alpha_\perp P_{2,3;\,\perp} \int \Big[i\beta_x + j\beta_y + k\,(\beta_x\cos\vartheta_{3x} +$$

$$+ \beta_y\cos\vartheta_{3y})\frac{\cos\vartheta_{3z}}{\sin^2\vartheta_{3z}}\Big] \exp\Big[i\omega t - i\mathbf{k}_1\mathbf{r} + i\,\frac{a}{z}\,(\mathbf{k}_2\mathbf{r} - \mathbf{k}_1\mathbf{r}) + ik_3R\Big]\,dt, \qquad (16.2)$$

where R denotes the distance between the coordinate origin and the point of observation; $\alpha_\parallel$ and $\alpha_\perp$ are factors taking into account multiple wave reflections in medium 2. The exponential term in the integrals of Eqs. (16.1) and (16.2) determine the delay of the waves and, in the case of absorption, the wave attenuation.

In calculations of the field of waves generated during the particle motion in medium 2, we must take into account the contribution of the primary radiation in the forward direction, i.e., in the direction toward the boundary between medium 2 and medium 3; we obtain

$$\Pi'_{\omega 2\parallel} = \frac{ieck}{2\pi\omega R}\int \alpha_\parallel P_{2,3;\,\parallel}\Big[\beta_z - (\beta_x\cos\vartheta_{3x} + \beta_y\cos\vartheta_{3y})\frac{\cos\vartheta_{3z}}{\sin^2\vartheta_{3z}}\Big]\exp(i\omega t - i\mathbf{k}_2\mathbf{r} + ik_3R)\,dt, \qquad (16.3)$$

$$\Pi'_{\omega 2\perp} = \frac{iec}{2\pi\omega R}\,\alpha_\perp P_{2,3;\,\perp}\int\Big[i\beta_x + j\beta_y + k\,(\beta_x\cos\vartheta_{3x} + \beta_y\cos\vartheta_{3y})\frac{\cos\vartheta_{3z}}{\sin^2\vartheta_{3z}}\Big]\exp(i\omega t - i\mathbf{k}_2\mathbf{r} + ik_3R)\,dt; \qquad (16.4)$$

furthermore, we have to include in our considerations the backward radiation resulting from the reflection at the boundary separating medium 1 from medium 2:

$$\Pi''_{\omega 2\parallel} = \frac{ieck}{2\pi\omega R}\int r_{2,1;\,\parallel}\alpha_\parallel P_{2,3;\,\parallel}\Big[\beta_z - (\beta_x\cos\vartheta_{3x} + \beta_y\cos\vartheta_{3y})\frac{\cos\vartheta_{3z}}{\sin^2\vartheta_{3z}}\Big]\times$$

$$\times\,\exp(i\omega t + ik_{2z}z - ik_{2x}x - ik_{2y}y + i2ak_{2z} + ik_3R)\,dt, \qquad (16.5)$$

$$\Pi''_{\omega 2\perp} = \frac{iec}{2\pi\omega R}\,r_{2,1;\,\perp}\alpha_\perp P_{2,3;\,\perp}\int[i\beta_x + j\beta_y + k\,(\beta_x\cos\vartheta_{3x} + \beta_y\cos\vartheta_{3y})\frac{\cos\vartheta_{3z}}{\sin^2\vartheta_{3z}}]\times$$

$$\times\,\exp(i\omega t + ik_{2z}z - ik_{2x}x - ik_{2y}y + i2ak_{2z} + ik_3R)\,dt. \qquad (16.6)$$

Finally, the waves generated in region 3 result from the interference of the waves emitted in the forward direction, away from the boundary:

$$\Pi'_{\omega 3\parallel} = \frac{ieck}{2\pi\omega R}\int\Big[\beta_z - (\beta_x\cos\vartheta_{3x} + \beta_y\cos\vartheta_{3y})\frac{\cos\vartheta_{3z}}{\sin^2\vartheta_{3z}}\Big]\exp(i\omega t - i\mathbf{k}_3\mathbf{r} + ik_3R)\,dt, \qquad (16.7)$$

$$\Pi'_{\omega 3\perp} = \frac{iec}{2\pi\omega R}\int\Big[i\beta_x + j\beta_y + k\,(\beta_x\cos\vartheta_{3x} + \beta_y\cos\vartheta_{3y})\frac{\cos\vartheta_{3z}}{\sin^2\vartheta_{3z}}\Big]\exp(i\omega t - i\mathbf{k}_3\mathbf{r} + ik_3R)\,dt \qquad (16.8)$$

with the waves emitted in the backward direction and reflected at the boundary separating medium 2 from medium 3:

$$\Pi''_{\omega 3\parallel} = \frac{ieck}{2\pi\omega R}\int r_{3,2;\,\parallel}\Big[\beta_z - (\beta_x\cos\vartheta_{3x} + \beta_y\cos\vartheta_{3y})\frac{\cos\vartheta_{3z}}{\sin^2\vartheta_{3z}}\Big]\times$$

$$\times\exp(i\omega t - ik_{3x}x - ik_{3y}y + ik_{3z}z + ik_3R)\,dt, \qquad (16.9)$$

$$\Pi''_{\omega 3\perp} = \frac{iec}{2\pi\omega R}\int r_{3,2;\,\perp}\Big[i\beta_x + j\beta_y + k\,(\beta_x\cos\vartheta_{3x} + \beta_y\cos\vartheta_{3y})\frac{\cos\vartheta_{3z}}{\sin^3\vartheta_{3z}}\Big]\times$$

$$\times\exp(i\omega t - ik_{3x}x - ik_{3y}y + ik_{3z}z + ik_3R)\,dt; \qquad (16.10)$$

in this interference we have to include the waves which passed into medium 2 and underwent multiple reflections:

$$\mathbf{\Pi}''_{\omega 3\,\|} = \frac{ieck}{2\pi\omega R} \int P_{3,2;\,\|}\, r_{2,1;\,\|}\, \alpha_{\|} P_{2,3;\,\|} \left[\beta_z - (\beta_x \cos\vartheta_{3x} + \beta_y \cos\vartheta_{3y}) \frac{\cos\vartheta_{3z}}{\sin^2\vartheta_{3z}} \right] \times$$

$$\times \exp\left(i\omega t - ik_{3x}x - ik_{3y}y + ik_{3z}z + i2ak_{2z} + ik_3 R \right) dt, \tag{16.11}$$

$$\mathbf{\Pi}''_{\omega 3\,\perp} = \frac{iec}{2\pi\omega R}\, P_{3,2;\,\perp}\, r_{2,1;\,\perp}\, \alpha_{\perp} P_{2,3;\,\perp} \int \left[i\beta_x + j\beta_y + \mathbf{k}\,(\beta_x \cos\vartheta_{3x} + \beta_y \cos\vartheta_{3y}) \times \right.$$

$$\left. \times \frac{\cos\vartheta_{3z}}{\sin^2\vartheta_{3z}} \right] \exp\left(i\omega t - ik_{3x}x - ik_{3y}y + ik_{3z}z + i2ak_{2z} + ik_3 R \right) dt. \tag{16.12}$$

The coefficients $\alpha_{\|}$ and $\alpha_{\perp}$ in these formulas account for multiple reflections of waves of the corresponding polarization states. These coefficients are

$$\alpha_{\|} = [1 - r_{2,3;\,\|}\, r_{2,1;\,\|} \exp{(i2ak_{2z})}]^{-1},$$
$$\alpha_{\perp} = [1 - r_{2,3;\,\perp}\, r_{2,1;\,\perp} \exp{(i2ak_{2z})}]^{-1}. \tag{16.13}$$

We note that when the thickness a of layer 2 decreases (in the limit $a \to 0$), the dependence on the optical properties of layer 2 disappears, as one could expect. As a matter of fact, for $a \to 0$ the time during which the particles moves in layer 2 tends to zero and therefore, Eqs. (16.3)-(16.6) tend to zero. In view of the second Eq. (2.11), the dependence upon the optical properties of the medium disappears in Eqs. (16.1) and (16.2), which describe the radiation field generated in the first region. The first of the equations in (2.11) eliminates the dependence on the optical properties of the layer for $a \to 0$ (in the sum of the vectors defined by Eqs. (16.9) and (16.11) and, accordingly, in the sum of Eqs. (16.10) and (16.12), which describe reflected waves generated in the medium). The final result agrees with Eqs. (4.1)-(4.6). It is evident that when the optical properties of adjacent media are the same (which is equivalent to the assumption that a single boundary exists), the sum of Eqs. (16.1)-(16.12) becomes identical with the sum of Eqs. (4.1)-(4.6), i.e., we obtain the result which we obtained in the case of a single boundary.

It follows from the meaning of Eqs. (16.1)-(16.12) that when the particle moves only in the first medium, the particle radiation induced in medium 3 is described by the sum of the vectors defined by Eqs. (16.1) and (16.2). When the particle motion takes place only in the second medium, the radiation is described by the sum of Eqs. (16.3)-(16.6). Finally, when the particle moves only in the third medium, the radiation is determined by the sum of all the vectors defined by Eqs. (16.7)-(16.12). When the charged particle passes through all three media, the radiation is given by the sum of all the vectors defined in Eqs. (16.1)-(16.12).

We use the relations stated in Eq. (2.4) and write the coefficients r and p (defined by Eqs. (2.7)-(2.9) and Eq. (2.10), with i, k = 1, 2, 3) for nonmagnetic media 1 and 2 ($\mu_1 = \mu_2 = 1$, $\vartheta_{3x} \equiv \vartheta_x$, $\vartheta_{3y} \equiv \vartheta_y$, $\vartheta_{3z} \equiv \vartheta_z$ for $\varepsilon_3 = 1$, $\mu_3 = 1$):

$$r_{1,2;\,\perp} = \frac{\sqrt{\varepsilon_1 - \sin^2\vartheta_z} - \sqrt{\varepsilon_2 - \sin^2\vartheta_z}}{\sqrt{\varepsilon_1 - \sin^2\vartheta_z} + \sqrt{\varepsilon_2 - \sin^2\vartheta_z}}, \quad r_{2,1;\,\perp} = -\frac{\sqrt{\varepsilon_1 - \sin^2\vartheta_z} - \sqrt{\varepsilon_2 - \sin^2\vartheta_z}}{\sqrt{\varepsilon_1 - \sin^2\vartheta_z} + \sqrt{\varepsilon_2 - \sin^2\vartheta_z}},$$

$$r_{2,3;\,\perp} = \frac{\sqrt{\varepsilon_2 - \sin^2\vartheta_z} - \cos\vartheta_z}{\sqrt{\varepsilon_2 - \sin^2\vartheta_z} + \cos\vartheta_z}, \quad r_{3,2;\,\perp} = -\frac{\sqrt{\varepsilon_2 - \sin^2\vartheta_z} - \cos\vartheta_z}{\sqrt{\varepsilon_2 - \sin^2\vartheta_z} + \cos\vartheta_z},$$

$$P_{1,2;\,\perp} = \frac{2\sqrt{\varepsilon_2 - \sin^2\vartheta_z}}{\sqrt{\varepsilon_1 - \sin^2\vartheta_z} + \sqrt{\varepsilon_2 - \sin^2\vartheta_z}}, \quad P_{2,1;\,\perp} = \frac{2\sqrt{\varepsilon_1 - \sin^2\vartheta_z}}{\sqrt{\varepsilon_1 - \sin^2\vartheta_z} + \sqrt{\varepsilon_2 - \sin^2\vartheta_z}},$$

$$P_{2,3;\,\perp} = \frac{2\cos\vartheta_z}{\sqrt{\varepsilon_2 - \sin^2\vartheta_z} + \cos\vartheta_z}, \quad P_{3,2;\,\perp} = \frac{2\sqrt{\varepsilon_2 - \sin^2\vartheta_z}}{\sqrt{\varepsilon_2 - \sin^2\vartheta_z} + \cos\vartheta_z},$$

$$r_{1,2;\,\|} = -\frac{\varepsilon_1\sqrt{\varepsilon_2-\sin^2\vartheta_z}-\varepsilon_2\sqrt{\varepsilon_1-\sin^2\vartheta_z}}{\varepsilon_1\sqrt{\varepsilon_2-\sin^2\vartheta_z}+\varepsilon_2\sqrt{\varepsilon_1-\sin^2\vartheta_z}}\;\frac{\beta_z\sin^2\vartheta_z-(\beta_x\cos\vartheta_x+\beta_y\cos\vartheta_y)\sqrt{\varepsilon_1-\sin^2\vartheta_z}}{\beta_z\sin^2\vartheta_z+(\beta_x\cos\vartheta_x+\beta_y\cos\vartheta_y)\sqrt{\varepsilon_1-\sin^2\vartheta_z}},$$

$$r_{2,1;\,\|} = \frac{\varepsilon_1\sqrt{\varepsilon_2-\sin^2\vartheta_z}-\varepsilon_2\sqrt{\varepsilon_1-\sin^2\vartheta_z}}{\varepsilon_1\sqrt{\varepsilon_2-\sin^2\vartheta_z}+\varepsilon_2\sqrt{\varepsilon_1-\sin^2\vartheta_z}}\;\frac{\beta_z\sin^2\vartheta_z+(\beta_x\cos\vartheta_x+\beta_y\cos\vartheta_y)\sqrt{\varepsilon_2-\sin^2\vartheta_z}}{\beta_z\sin^2\vartheta_z-(\beta_x\cos\vartheta_x+\beta_y\cos\vartheta_y)\sqrt{\varepsilon_2-\sin^2\vartheta_z}},$$

$$r_{2,3;\,\|} = \frac{\sqrt{\varepsilon_2-\sin^2\vartheta_z}-\varepsilon_2\cos\vartheta_z}{\sqrt{\varepsilon_2-\sin^2\vartheta_z}+\varepsilon_2\cos\vartheta_z}\;\frac{\beta_z\sin^2\vartheta_z-(\beta_x\cos\vartheta_x+\beta_y\cos\vartheta_y)\sqrt{\varepsilon_2-\sin^2\vartheta_z}}{\beta_z\sin^2\vartheta_z+(\beta_x\cos\vartheta_x+\beta_y\cos\vartheta_y)\sqrt{\varepsilon_2-\sin^2\vartheta_z}},$$

$$r_{3,2;\,\|} = -\frac{\sqrt{\varepsilon_2-\sin^2\vartheta_z}-\varepsilon_2\cos\vartheta_z}{\sqrt{\varepsilon_2-\sin^2\vartheta_z}+\varepsilon_2\cos\vartheta_z}\;\frac{\beta_z\sin^2\vartheta_z+(\beta_x\cos\vartheta_x+\beta_y\cos\vartheta_y)\cos\vartheta_z}{\beta_z\sin^2\vartheta_z-(\beta_x\cos\vartheta_x+\beta_y\cos\vartheta_y)\cos\vartheta_z},$$

$$p_{1,2;\,\|} = \frac{2\varepsilon_2\sqrt{\varepsilon_2-\sin^2\vartheta_z}}{\varepsilon_1\sqrt{\varepsilon_2-\sin^2\vartheta_z}+\varepsilon_2\sqrt{\varepsilon_1-\sin^2\vartheta_z}}\;\frac{\beta_z\sin^2\vartheta_z-(\beta_x\cos\vartheta_x+\beta_y\cos\vartheta_y)\sqrt{\varepsilon_1-\sin^2\vartheta_z}}{\beta_z\sin^2\vartheta_z-(\beta_x\cos\vartheta_x+\beta_y\cos\vartheta_y)\sqrt{\varepsilon_2-\sin^2\vartheta_z}},$$

$$p_{2,1;\,\|} = \frac{2\varepsilon_1\sqrt{\varepsilon_1-\sin^2\vartheta_z}}{\varepsilon_1\sqrt{\varepsilon_2-\sin^2\vartheta_z}+\varepsilon_2\sqrt{\varepsilon_1-\sin^2\vartheta_z}}\;\frac{\beta_z\sin^2\vartheta_z+(\beta_x\cos\vartheta_x+\beta_y\cos\vartheta_y)\sqrt{\varepsilon_2-\sin^2\vartheta_z}}{\beta_z\sin^2\vartheta_z+(\beta_x\cos\vartheta_x+\beta_y\cos\vartheta_y)\sqrt{\varepsilon_1-\sin^2\vartheta_z}},$$

$$p_{2,3;\,\|} = \frac{2\cos\vartheta_z}{\sqrt{\varepsilon_2-\sin^2\vartheta_z}+\varepsilon_2\cos\vartheta_z}\;\frac{\beta_z\sin^2\vartheta_z-(\beta_x\cos\vartheta_x+\beta_y\cos\vartheta_y)\sqrt{\varepsilon_2-\sin^2\vartheta_z}}{\beta_z\sin^2\vartheta_z-(\beta_x\cos\vartheta_x+\beta_y\cos\vartheta_y)\cos\vartheta_z},$$

$$p_{3,2;\,\|} = \frac{2\varepsilon_2\sqrt{\varepsilon_2-\sin^2\vartheta_z}}{\sqrt{\varepsilon_2-\sin^2\vartheta_z}+\varepsilon_2\cos\vartheta_z}\;\frac{\beta_z\sin^2\vartheta_z+(\beta_x\cos\vartheta_x+\beta_y\cos\vartheta_y)\cos\vartheta_z}{\beta_z\sin^2\vartheta_z+(\beta_x\cos\vartheta_x+\beta_y\cos\vartheta_y)\sqrt{\varepsilon_2-\sin^2\vartheta_z}}. \tag{16.14}$$

Let us also state the coefficients $\alpha_\|$ and $\alpha_\perp$ and the wave vectors $\mathbf{k}_1$, $\mathbf{k}_2$, and $\mathbf{k}_3$ in the case $\varepsilon_3 = 1$, $\mu_1 = \mu_2 = \mu_3 = 1$:

$$\alpha_\perp = \left[1 - \left(\frac{\sqrt{\varepsilon_2-\sin^2\vartheta_z}-\sqrt{\varepsilon_1-\sin^2\vartheta_z}}{\sqrt{\varepsilon_2-\sin^2\vartheta_z}+\sqrt{\varepsilon_1-\sin^2\vartheta_z}}\right)\left(\frac{\sqrt{\varepsilon_2-\sin^2\vartheta_z}-\cos\vartheta_z}{\sqrt{\varepsilon_2-\sin^2\vartheta_z}+\cos\vartheta_z}\right)\exp\left(i\,\frac{2a\omega}{c}\sqrt{\varepsilon_2-\sin^2\vartheta_z}\right)\right]^{-1},$$

$$\alpha_\| = \left[1 - \left(\frac{\varepsilon_1\sqrt{\varepsilon_2-\sin^2\vartheta_z}-\varepsilon_2\sqrt{\varepsilon_1-\sin^2\vartheta_z}}{\varepsilon_1\sqrt{\varepsilon_2-\sin^2\vartheta_z}+\varepsilon_2\sqrt{\varepsilon_1-\sin^2\vartheta_z}}\right)\left(\frac{\sqrt{\varepsilon_2-\sin^2\vartheta_z}-\varepsilon_2\cos\vartheta_z}{\sqrt{\varepsilon_2-\sin^2\vartheta_z}+\varepsilon_2\cos\vartheta_z}\right)\exp\left(i\,\frac{2a\omega}{c}\sqrt{\varepsilon_2-\sin^2\vartheta_z}\right)\right]^{-1}, \tag{16.15}$$

$$\mathbf{k}_1 = \frac{\omega}{c}\left(\mathbf{i}\cos\vartheta_x + \mathbf{j}\cos\vartheta_y + \mathbf{k}\sqrt{\varepsilon_1-\sin^2\vartheta_z}\right),$$

$$\mathbf{k}_2 = \frac{\omega}{c}\left(\mathbf{i}\cos\vartheta_x + \mathbf{j}\cos\vartheta_y + \mathbf{k}\sqrt{\varepsilon_2-\sin^2\vartheta_z}\right), \tag{16.16}$$

$$\mathbf{k}_3 = \frac{\omega}{c}\left(\mathbf{i}\cos\vartheta_x + \mathbf{j}\cos\vartheta_y + \mathbf{k}\cos\vartheta_z\right).$$

After inserting Eqs. (16.14)-(16.16) into Eqs. (16.1)-(16.12), we obtain the spherical waves of the radiation in vacuum in the case of nonmagnetic media 1 and 2:

$$\mathbf{\Pi}_{\omega1\|} = \frac{2ieck}{\pi\omega R}\,A_\|\varepsilon_2\sqrt{\varepsilon_2-\sin^2\vartheta_z}\,\cos\vartheta_z\int\Big[\beta_z - (\beta_x\cos\vartheta_x +$$
$$+\,\beta_y\cos\vartheta_y)\frac{\sqrt{\varepsilon_1-\sin^2\vartheta_z}}{\sin^2\vartheta_z}\Big]\exp\Big[i\omega t - i\frac{\omega}{c}\,(x\cos\vartheta_x + y\cos\vartheta_y +$$
$$+\,z\sqrt{\varepsilon_1-\sin^2\vartheta_z} + a\sqrt{\varepsilon_1-\sin^2\vartheta_z}) + i\frac{\omega}{c}\,R\Big]dt, \tag{16.17}$$

$$\mathbf{\Pi}_{\omega1\perp} = \frac{2iec}{\pi\omega R}\,A_\perp\sqrt{\varepsilon_2-\sin^2\vartheta_z}\,\cos\vartheta_z\int\Big[i\beta_x + j\beta_y + \mathbf{k}\,(\beta_x\cos\vartheta_x +$$
$$+\,\beta_y\cos\vartheta_y)\frac{\cos\vartheta_z}{\sin^2\vartheta_z}\Big]\exp\Big[i\omega t - i\frac{\omega}{c}\,(x\cos\vartheta_x + y\cos\vartheta_y +$$
$$+\,z\sqrt{\varepsilon_1-\sin^2\vartheta_z} + a\sqrt{\varepsilon_1-\sin^2\vartheta_z}) + i\frac{\omega}{c}\,R\Big]dt, \tag{16.18}$$

$$\Pi'_{\omega 2 \parallel} = \frac{ieck}{\pi \omega R} A_{\parallel} (\varepsilon_1 \sqrt{\varepsilon_2 - \sin^2 \vartheta_z} + \varepsilon_2 \sqrt{\varepsilon_1 - \sin^2 \vartheta_z}) \cos \vartheta_z \times$$

$$\times \int \left[\beta_z - (\beta_x \cos \vartheta_x + \beta_y \cos \vartheta_y) \frac{\sqrt{\varepsilon_2 - \sin^2 \vartheta_z}}{\sin^2 \vartheta_z} \right] \exp \left[i \omega t - i \frac{\omega}{c} (x \cos \vartheta_x + y \cos \vartheta_y + z \sqrt{\varepsilon_2 - \sin^2 \vartheta_z} + a \sqrt{\varepsilon_2 - \sin^2 \vartheta_z}) + i \frac{\omega}{c} R \right] dt, \qquad (16.19)$$

$$\Pi'_{\omega 2 \perp} = \frac{iec}{\pi \omega R} A_{\perp} (\sqrt{\varepsilon_2 - \sin^2 \vartheta_z} + \sqrt{\varepsilon_1 - \sin^2 \vartheta_z}) \cos \vartheta_z \int \left[i \beta_x + j \beta_y + k (\beta_x \cos \vartheta_x + \beta_y \cos \vartheta_y) \frac{\cos \vartheta_z}{\sin^2 \vartheta_z} \right] \exp \left[i \omega t - i \frac{\omega}{c} (x \cos \vartheta_x + y \cos \vartheta_y + z \sqrt{\varepsilon_2 - \sin^2 \vartheta_z} + a \sqrt{\varepsilon_2 - \sin^2 \vartheta_z}) + i \frac{\omega}{c} R \right] dt, \qquad (16.20)$$

$$\Pi''_{\omega 2 \parallel} = \frac{ieck}{\pi \omega R} A_{\parallel} (\varepsilon_1 \sqrt{\varepsilon_2 - \sin^2 \vartheta_z} - \varepsilon_2 \sqrt{\varepsilon_1 - \sin^2 \vartheta_z}) \cos \vartheta_z \times$$

$$\times \int \left[\beta_z + (\beta_x \cos \vartheta_x + \beta_y \cos \vartheta_y) \frac{\sqrt{\varepsilon_2 - \sin^2 \vartheta_z}}{\sin^2 \vartheta_z} \right] \exp \left[i \omega t - i \frac{\omega}{c} (x \cos \vartheta_x + y \cos \vartheta_y - z \sqrt{\varepsilon_2 - \sin^2 \vartheta_z} - a \sqrt{\varepsilon_2 - \sin^2 \vartheta_z}) + i \frac{\omega}{c} R \right] dt, \qquad (16.21)$$

$$\Pi''_{\omega 2 \perp} = \frac{iec}{\pi \omega R} A_{\perp} (\sqrt{\varepsilon_2 - \sin^2 \vartheta_z} - \sqrt{\varepsilon_1 - \sin^2 \vartheta_z}) \cos \vartheta_z \int \left[i \beta_x + j \beta_y + k (\beta_x \cos \vartheta_x + \beta_y \cos \vartheta_y) \frac{\cos \vartheta_z}{\sin^2 \vartheta_z} \right] \exp \left[i \omega t - i \frac{\omega}{c} (x \cos \vartheta_x + y \cos \vartheta_y - z \sqrt{\varepsilon_2 - \sin^2 \vartheta_z} - a \sqrt{\varepsilon_2 - \sin^2 \vartheta_z}) + i \frac{\omega}{c} R \right] dt, \qquad (16.22)$$

$$\Pi'_{\omega 3 \parallel} = \frac{ieck}{2 \pi \omega R} \int \left[\beta_z - (\beta_x \cos \vartheta_x + \beta_y \cos \vartheta_y) \frac{\cos \vartheta_z}{\sin^2 \vartheta_z} \right] \times$$

$$\times \exp \left[i \omega t - i \frac{\omega}{c} (x \cos \vartheta_x + y \cos \vartheta_y + z \cos \vartheta_z) + i \frac{\omega}{c} R \right] dt, \qquad (16.23)$$

$$\Pi'_{\omega 3 \perp} = \frac{iec}{2 \pi \omega R} \int \left[i \beta_x + j \beta_y + k (\beta_x \cos \vartheta_x + \beta_y \cos \vartheta_y) \frac{\cos \vartheta_z}{\sin^2 \vartheta_z} \right] \times$$

$$\times \exp \left[i \omega t - i \frac{\omega}{c} (x \cos \vartheta_x + y \cos \vartheta_y + z \cos \vartheta_z) + i \frac{\omega}{c} R \right] dt, \qquad (16.24)$$

$$\Pi'_{\omega 3 \parallel} = \frac{ieck}{2 \pi \omega R} \frac{\varepsilon_2 \cos \vartheta_z - \sqrt{\varepsilon_2 - \sin^2 \vartheta_z}}{\varepsilon_2 \cos \vartheta_z + \sqrt{\varepsilon_2 - \sin^2 \vartheta_z}} \int \left[\beta_z + (\beta_x \cos \vartheta_x + \beta_y \cos \vartheta_y) \frac{\cos \vartheta_z}{\sin^2 \vartheta_z} \right] \times$$

$$\times \exp \left[i \omega t - i \frac{\omega}{c} (x \cos \vartheta_x + y \cos \vartheta_y - z \cos \vartheta_z) + i \frac{\omega}{c} R \right] dt, \qquad (16.25)$$

$$\Pi''_{\omega 3 \perp} = \frac{iec}{2 \pi \omega R} \frac{\cos \vartheta_z - \sqrt{\varepsilon_2 - \sin^2 \vartheta_z}}{\cos \vartheta_z + \sqrt{\varepsilon_2 - \sin^2 \vartheta_z}} \int \left[i \beta_x + j \beta_y + k (\beta_x \cos \vartheta_x + \beta_y \cos \vartheta_y) \times \frac{\cos \vartheta_z}{\sin^2 \vartheta_z} \right] \exp \left[i \omega t - i \frac{\omega}{c} (x \cos \vartheta_x + y \cos \vartheta_y - z \cos \vartheta_z) + i \frac{\omega}{c} R \right] dt, \qquad (16.26)$$

$$\Pi''_{\omega 3 \parallel} = \frac{2 ieck}{\pi \omega R} A_{\parallel} \frac{\varepsilon_2 \sqrt{\varepsilon_2 - \sin^2 \vartheta_z} \cos \vartheta_z}{\sqrt{\varepsilon_2 - \sin^2 \vartheta_z} + \varepsilon_2 \cos \vartheta_z} (\varepsilon_1 \sqrt{\varepsilon_2 - \sin^2 \vartheta_z} - \varepsilon_2 \sqrt{\varepsilon_1 - \sin^2 \vartheta_z}) \times$$

$$\times \int \left[\beta_z + (\beta_x \cos \vartheta_x + \beta_y \cos \vartheta_y) \frac{\cos \vartheta_z}{\sin^2 \vartheta_z} \right] \exp \left[i \omega t - i \frac{\omega}{c} (x \cos \vartheta_x + y \cos \vartheta_y - z \cos \vartheta_z - a \sqrt{\varepsilon_2 - \sin^2 \vartheta_z}) + i \frac{\omega}{c} R \right] dt, \qquad (16.27)$$

$$\Pi''_{\omega 3\perp} = \frac{2iec}{\pi\omega R}\, A_\perp \frac{\sqrt{\varepsilon_2 - \sin^2\vartheta_z}\,\cos\vartheta_z}{\sqrt{\varepsilon_2 - \sin^2\vartheta_z} + \cos\vartheta_z}\,(\sqrt{\varepsilon_2 - \sin^2\vartheta_z} - \sqrt{\varepsilon_1 - \sin^2\vartheta_z})\times$$

$$\times \int \Big[i\beta_x + j\beta_y + k(\beta_x\cos\vartheta_x + \beta_y\cos\vartheta_y)\frac{\cos\vartheta_z}{\sin^2\vartheta_z} \Big] \exp\Big[i\omega t - i\frac{\omega}{c}\times$$

$$\times(x\cos\vartheta_x + y\cos\vartheta_y - z\cos\vartheta_z - a\sqrt{\varepsilon_2 - \sin^2\vartheta_z}) + i\frac{\omega}{c}R\Big]dt \qquad (16.28)$$

where

$$A_\parallel = [(\varepsilon_1\sqrt{\varepsilon_2 - \sin^2\vartheta_z} + \varepsilon_2\sqrt{\varepsilon_1 - \sin^2\vartheta_z})(\sqrt{\varepsilon_2 - \sin^2\vartheta_z} + \varepsilon_2\cos\vartheta_z)\times$$

$$\times \exp\Big(- i\frac{a\omega}{c}\sqrt{\varepsilon_2 - \sin^2\vartheta_z}\Big) - (\varepsilon_1\sqrt{\varepsilon_2 - \sin^2\vartheta_z} - \varepsilon_2\sqrt{\varepsilon_1 - \sin^2\vartheta_z})\times$$

$$\times(\sqrt{\varepsilon_2 - \sin^2\vartheta_z} - \varepsilon_2\cos\vartheta_z)\exp\Big(i\frac{a\omega}{c}\sqrt{\varepsilon_2 - \sin^2\vartheta_z}\Big)]^{-1}, \qquad (16.29)$$

$$A_\perp = \Big[(\sqrt{\varepsilon_2 - \sin^2\vartheta_z} + \sqrt{\varepsilon_1 - \sin^2\vartheta_z})(\sqrt{\varepsilon_2 - \sin^2\vartheta_z} + \cos\vartheta_z)\times$$

$$\times \exp\Big(- i\frac{a\omega}{c}\sqrt{\varepsilon_2 - \sin^2\vartheta_z}\Big) - (\sqrt{\varepsilon_2 - \sin^2\vartheta_z} - \sqrt{\varepsilon_1 - \sin^2\vartheta_z})\times$$

$$\times(\sqrt{\varepsilon_2 - \sin^2\vartheta_z} - \cos\vartheta_z)\exp\Big(i\frac{a\omega}{c}\sqrt{\varepsilon_2 - \sin^2\vartheta_z}\Big)\Big]^{-1}. \qquad (16.30)$$

This solution is valid for an arbitrarily moving charged particle. The solution can be used when the effect of surface layers upon the transition radiation is considered, or when the radiation of a particle passing through a layer under a certain acute angle relative to the normal is calculated, or when the bremsstrahlung in a film is discussed, or when the luminescence radiation (its angular distribution and degree of polarization) in thin layers is calculated, etc.

§ 17. Radiation from an Oscillator Situated

in a Thin Film

Let us assume that a charged particle moves in medium 2 without intersecting the boundaries to media 1 and 3. In this case, the radiation field is determined by Eqs. (16.19)–(16.22). Let us calculate the radiation field under the assumption that the particle performs harmonic oscillations around a certain equilibrium position: $\mathbf{r} = \mathbf{r}_0 + \Delta\mathbf{r}_0 \sin\omega t$. The particle will not intersect the boundary if the following relation holds all the time:

$$-a < z_0 + \Delta z_0 \sin\,\omega t < 0. \qquad (17.1)$$

When the maximum velocity of the particle is much smaller than the velocity of light in vacuum, and when also

$$v_{0z}|\sqrt{\varepsilon_2 - \sin^2\vartheta_z}|/c \ll 1, \qquad (17.2)$$

holds, we can neglect the time dependence of the particle coordinates in the integration over the time of motion. Under these assumptions, and if layer 2 is very thin, i.e., when the condition

$$a\omega\,|\sqrt{\varepsilon_2 - \sin^2\vartheta_z}|/c \ll 1, \qquad (17.3)$$

is fulfilled, we obtain from Eqs. (16.19)–(16.22)

$$\Pi_{\omega\parallel} = \frac{ieck}{\pi\omega R}\frac{\cos\vartheta_z}{\sqrt{\varepsilon_1 - \sin^2\vartheta_z} + \varepsilon_1\cos\vartheta_z}\int\Big[\frac{\varepsilon_1}{\varepsilon_2}\beta_z - (\beta_x\cos\vartheta_x +$$

$$+ \beta_y\cos\vartheta_y)\frac{\sqrt{\varepsilon_1 - \sin^2\vartheta_z}}{\sin^2\vartheta_z}\Big]\exp\Big(i\omega t + i\frac{\omega}{c}R\Big)dt, \qquad (17.4)$$

$$\Pi_{\omega\perp} = \frac{iec}{\pi\omega R}\frac{\cos\vartheta_z}{\sqrt{\varepsilon_1 - \sin^2\vartheta_z} + \cos\vartheta_z}\int\Big[i\beta_x + j\beta_y +$$

$$+ k(\beta_x\cos\vartheta_x + \beta_y\cos\vartheta_y)\frac{\cos\vartheta_z}{\sin^2\vartheta_z}\Big]\exp\Big(i\omega t + i\frac{\omega}{c}R\Big)dt, \qquad (17.5)$$

where

$$\Pi_{\omega\parallel} = \Pi'_{\omega2\parallel} + \Pi''_{\omega2\parallel}, \quad \Pi_{\omega\perp} = \Pi'_{\omega2\perp} + \Pi''_{\omega2\perp}.$$

When the particle motion follows a harmonic law, the Hertz vector of the radiation field must be calculated from Eqs. (17.4) and (17.5). To this end, the integration must be performed over a period of the vibrational motion and the result multiplied by $\omega e^{-i\omega t}$. When we assume $\mathbf{v} = \mathbf{v}_0 \cos \omega t$, we obtain from Eqs. (17.4) and (17.5):

$$\Pi_\parallel = \frac{iek}{\omega R} \frac{\cos \vartheta_z}{\sqrt{\varepsilon_1 - \sin^2 \vartheta_z} + \varepsilon_1 \cos \vartheta_z} \left[\frac{\varepsilon_1}{\varepsilon_2} v_{0z} - (v_{0x} \cos \vartheta_x + v_{0y} \cos \vartheta_y) \frac{\sqrt{\varepsilon_1 - \sin^2 \vartheta_z}}{\sin^2 \vartheta_z} \right] e^{i \frac{\omega}{c} R - i\omega t}, \quad (17.6)$$

$$\Pi_\perp = \frac{ie}{\omega R} \frac{\cos \vartheta_z}{\sqrt{\varepsilon_1 - \sin^2 \vartheta_z} + \cos \vartheta_z} \left[\mathbf{i} v_{0x} + \mathbf{j} v_{0y} + \mathbf{k}(v_{0x} \cos \vartheta_x + v_{0y} \cos \vartheta_y) \frac{\cos \vartheta_z}{\sin^2 \vartheta_z} \right] e^{i \frac{\omega}{c} R - i\omega t}, \quad (17.7)$$

where v_{0x}, v_{0y}, and v_{0z} denote the maximum values of the particle's velocity components in the direction of the x, y, and z axes, respectively.

When the probability of oscillator orientation is the same for all directions, we obtain the average radiation energy contributed by waves of various polarizations per unit solid angle and per unit time ($\vartheta \equiv \vartheta_z$) in the form

$$\langle W_{n\parallel} \rangle = \frac{e^2 \omega^2 v_0^2}{4\pi c^3} \frac{|\varepsilon_1|^2 \sin^2 \vartheta + |\varepsilon_2^2 (\varepsilon_1 - \sin^2 \vartheta)|}{|\varepsilon_2 (\sqrt{\varepsilon_1 - \sin^2 \vartheta} + \varepsilon_1 \cos \vartheta)|^2}, \quad (17.8)$$

$$\langle W_{n\perp} \rangle = \frac{e^2 \omega^2 v_0^2}{4\pi c^3} \frac{\cos^2 \vartheta}{|\sqrt{\varepsilon_1 - \sin^2 \vartheta} + \cos \vartheta|^2}. \quad (17.9)$$

This is also the angular distribution of the luminescence radiation in a thin layer of a substance deposited on a thick substrate. These formulas help us to calculate the degree of polarization of the luminescence radiation. We find

$$P = \frac{\langle W_{n\parallel} \rangle - \langle W_{n\perp} \rangle}{\langle W_{n\parallel} \rangle + \langle W_{n\perp} \rangle} = \frac{|\varepsilon_1|^2 \sin^2 \vartheta + |\varepsilon_2^2 (\varepsilon_1 - \sin^2 \vartheta)| - |\varepsilon_2 (\sin^2 \vartheta + \cos \vartheta \sqrt{\varepsilon_1 - \sin^2 \vartheta})|^2}{|\varepsilon_1|^2 \sin^2 \vartheta + |\varepsilon_2^2 (\varepsilon_1 - \sin^2 \vartheta)| + |\varepsilon_2 (\sin^2 \vartheta + \cos \vartheta \sqrt{\varepsilon_1 - \sin^2 \vartheta})|^2}. \quad (17.10)$$

When we set $\varepsilon_1 = 1$ in Eqs. (17.8)–(17.10), we obtain the angular distribution of the radiation intensity generated by waves of various polarizations, and the degree of polarization of the luminescence radiation from a thin isolated film. In this case, Eqs. (17.8)–(17.10) assume the following form:

$$\langle W_{n\parallel} \rangle = \frac{e^2 \omega^2 v_0^2}{16\pi c^3} \frac{\sin^2 \vartheta + |\varepsilon_2|^2 \cos^2 \vartheta}{|\varepsilon_2|^2}, \quad (17.11)$$

$$\langle W_{n\perp} \rangle = e^2 \omega^2 v_0^2 / 16\pi c^3; \quad (17.12)$$

$$P = \frac{1 - |\varepsilon_2|^2}{1 + |\varepsilon_2|^2}. \quad (17.13)$$

These results indicate that the angular distribution and the degree of polarization of the luminescence radiation from thin layers are strongly dependent upon the dielectric constants of both the film and the substrate. The above results can be used in research on the luminescence radiation emitted by thin films, e.g., for an experimental verification of the theory predicting the disappearance of the density effect.

§ 18.   Radiation of a Charged Particle Moving
Uniformly along a Straight Line through
Two Boundaries

Equations (16.1)–(16.12) are the basis for calculating the radiation field of a particle moving according to a certain law, when two boundaries are present. In the case of a uniform motion along a straight line, time integration is easily performed. For the purpose of simplifying the results, we adopt a coordinate system such that the trajectory of the particle is on the yz plane ($\mathbf{r} = \mathbf{j}v_y t + \mathbf{k}v_z t$). Let us perform the integration along the particle trajectory in Eqs. (16.1)–(16.12):

$$\mathbf{\Pi}_{\omega 1\parallel} = \frac{iek}{2\pi\omega R\beta}\, p_{1,2;\,\parallel}\alpha_\parallel p_{2,3;\,\parallel}\left(\beta_z - \beta_y \cos\vartheta_{3v}\frac{\cos\vartheta_{3z}}{\sin^2\vartheta_{3z}}\right) \times$$

$$\times \int_{-\infty}^{-av/v_z} \exp\left[i\frac{\zeta}{v}(\omega - \mathbf{k}_1\mathbf{v}) + ia(k_{2z} - k_{1z}) + ik_3 R\right] d\zeta, \qquad (18.1)$$

$$\mathbf{\Pi}_{\omega 1\perp} = \frac{ie\beta_y}{2\pi\omega R\beta}\, p_{1,2;\,\perp}\alpha_\perp p_{2,3;\,\perp}\left(\mathbf{j} + \mathbf{k}\cos\vartheta_{3y}\frac{\cos\vartheta_{3z}}{\sin^2\vartheta_{3z}}\right)\int_{-\infty}^{-av/v_z} \times$$

$$\times \exp\left[i\frac{\zeta}{v}(\omega - \mathbf{k}_1\mathbf{v}) + ia(k_{2z} - k_{1z}) + ik_3 R\right] d\zeta, \qquad (18.2)$$

$$\mathbf{\Pi}'_{\omega 2\parallel} = \frac{iek}{2\pi\omega R\beta}\,\alpha_\parallel p_{2,3;\,\parallel}\left(\beta_z - \beta_y \cos\vartheta_{3y}\frac{\cos\vartheta_{3z}}{\sin^2\vartheta_{3z}}\right)\int_{-av/v_z}^{0} \exp\left[i\frac{\zeta}{v}(\omega - \mathbf{k}_2\mathbf{v}) + ik_2 R\right] d\zeta, \qquad (18.3)$$

$$\mathbf{\Pi}'_{\omega 2\perp} = \frac{ie\beta_y}{2\pi\omega R\beta}\,\alpha_\perp p_{2,3;\,\perp}\left(\mathbf{j} + \mathbf{k}\cos\vartheta_{3y}\frac{\cos\vartheta_{3z}}{\sin^2\vartheta_{3z}}\right)\int_{-av/v_z}^{0} \exp\left[i\frac{\zeta}{v}(\omega - \mathbf{k}_2\mathbf{v}) + ik_3 R\right] d\zeta, \qquad (18.4)$$

$$\mathbf{\Pi}''_{\omega 2\parallel} = \frac{iek}{2\pi\omega R\beta}\, r_{2,1;\,\parallel}\alpha_\parallel p_{2,3;\,\parallel}\left(\beta_z - \beta_y \cos\vartheta_{3y}\frac{\cos\vartheta_{3z}}{\sin^2\vartheta_{3z}}\right)\int_{-av/v_z}^{v} \times$$

$$\times \exp\left[i\frac{\zeta}{v}(\omega + k_{2z}v_z - k_{2y}v_y) + i2ak_{2z} + ik_3 R\right] d\zeta, \qquad (18.5)$$

$$\mathbf{\Pi}''_{\omega 2\perp} = \frac{ie\beta_y}{2\pi\omega R\beta}\, r_{2,1;\,\perp}\alpha_\perp p_{2,3;\,\perp}\left(\mathbf{j} + \mathbf{k}\cos\vartheta_{3y}\frac{\cos\vartheta_{3z}}{\sin^2\vartheta_{3z}}\right)\int_{-av/v_z}^{0} \times$$

$$\times \exp\left[i\frac{\zeta}{v}(\omega + k_{2z}v_z - k_{2y}v_y) + i2ak_{2z} + ik_3 R\right] d\zeta, \qquad (18.6)$$

$$\mathbf{\Pi}'_{\omega 3\parallel} = \frac{iek}{2\pi\omega R\beta}\left(\beta_z - \beta_y \cos\vartheta_{3y}\frac{\cos\vartheta_{3z}}{\sin^2\vartheta_{3z}}\right)\int_{0}^{\infty} \exp\left[i\frac{\zeta}{v}(\omega - \mathbf{k}_3\mathbf{v}) + ik_3 R\right] d\zeta, \qquad (18.7)$$

$$\mathbf{\Pi}'_{\omega 3\perp} = \frac{ie\beta_y}{2\pi\omega R\beta}\left(\mathbf{j} + \mathbf{k}\cos\vartheta_{3y}\frac{\cos\vartheta_{3z}}{\sin^2\vartheta_{3z}}\right)\int_{0}^{\infty} \exp\left[i\frac{\zeta}{v}(\omega - \mathbf{k}_3\mathbf{v}) + ik_3 R\right] d\zeta, \qquad (18.8)$$

$$\mathbf{\Pi}''_{\omega 3\parallel} = \frac{iek}{2\pi\omega R\beta}\, r_{3,2;\,\parallel}\left(\beta_z - \beta_y \cos\vartheta_{3y}\frac{\cos\vartheta_{3z}}{\sin^2\vartheta_{3z}}\right)\int_{0}^{\infty} \exp\left[i\frac{\zeta}{v}(\omega + k_{3z}v_z - k_{3y}v_y) + ik_3 R\right] d\zeta; \qquad (18.9)$$

$$\Pi'_{\omega 3\perp} = \frac{ie\beta_y}{2\pi\omega R\beta}\, r_{3,\,2;\,\perp}\left(\mathbf{j} + \mathbf{k}\cos\vartheta_{3y}\frac{\cos\vartheta_{3z}}{\sin^2\vartheta_{3z}}\right)\int\limits_0^\infty \exp\left[i\frac{\zeta}{v}(\omega + k_{3z}v_z - k_{3y}v_y) + ik_3R\right]d\zeta, \qquad (18.10)$$

$$\Pi''_{\omega 3\|} = \frac{iek}{2\pi\omega R\beta}\, p_{3,\,2;\,\|}\, r_{2,\,1;\,\|}\alpha_\| p_{2,\,3;\,\|}\left(\beta_z - \beta_y\cos\vartheta_{3y}\frac{\cos\vartheta_{3z}}{\sin^2\vartheta_{3z}}\right)\times$$
$$\times \int\limits_0^\infty \exp\left[i\frac{\zeta}{v}(\omega + k_{3z}v_z - k_{3y}v_y) + i2ak_{2z} + ik_3R\right]d\zeta, \qquad (18.11)$$

$$\Pi''_{\omega 3\perp} = \frac{ie\beta_y}{2\pi\omega R\beta}\, p_{3,\,2;\,\perp}r_{2,\,1;\,\perp}\alpha_\perp p_{2,\,3;\,\perp}\left(\mathbf{j} + \mathbf{k}\cos\vartheta_{3y}\frac{\cos\vartheta_{3z}}{\sin^2\vartheta_{3z}}\right)\int\limits_0^\infty \times$$
$$\times \exp\left[i\frac{\zeta}{v}(\omega + k_{3z}v_z - k_{3y}v_y) + i2ak_{2z} + ik_3R\right]d\zeta. \qquad (18.12)$$

We set $\varepsilon_3 = 1$, $\mu_1 = \mu_2 = \mu_3 = 1$ and omit the subscript 3 at the angles of the direction cosines of the vector $\mathbf{k}_3$; then, the results of the ζ integration can be written as follows:

$$\Pi_{\omega 1\|} = \frac{2eck}{\pi\omega^2 R}\, \frac{A_\|\varepsilon_2\sqrt{\varepsilon_2 - \sin^2\vartheta_z}\cos\vartheta_z}{1 - \beta_y\cos\vartheta_y - \beta_z\sqrt{\varepsilon_1 - \sin^2\vartheta_z}}\left(\beta_z - \beta_y\cos\vartheta_y\frac{\sqrt{\varepsilon_1 - \sin^2\vartheta_z}}{\sin^2\vartheta_z}\right)\times$$
$$\times \exp\left[-i\frac{a\omega}{v_z}(1 - \beta_y\cos\vartheta_y) + i\frac{\omega}{c}R\right], \qquad (18.13)$$

$$\Pi_{\omega 1\perp} = \frac{2ec\beta_y}{\pi\omega^2 R}\, \frac{A_\perp\sqrt{\varepsilon_2 - \sin^2\vartheta_z}\cos\vartheta_z}{1 - \beta_y\cos\vartheta_y - \beta_z\sqrt{\varepsilon_1 - \sin^2\vartheta_z}}\left(\mathbf{j} + \mathbf{k}\cos\vartheta_y\frac{\cos\vartheta_z}{\sin^2\vartheta_z}\right)\times$$
$$\times \exp\left[-i\frac{a\omega}{v_z}(1 - \beta_y\cos\vartheta_y) + i\frac{\omega}{c}R\right], \qquad (18.14)$$

$$\Pi'_{\omega 2\|} = \frac{ec\,\mathbf{k}}{\pi\omega^2 R}\, \frac{(\varepsilon_1\sqrt{\varepsilon_2 - \sin^2\vartheta_z} + \varepsilon_2\sqrt{\varepsilon_1 - \sin^2\vartheta_z})\,A_\|\cos\vartheta_z}{1 - \beta_y\cos\vartheta_y - \beta_z\sqrt{\varepsilon_2 - \sin^2\vartheta_z}}\times$$
$$\times \left(\beta_z - \beta_y\cos\vartheta_y\frac{\sqrt{\varepsilon_2 - \sin^2\vartheta_z}}{\sin^2\vartheta_z}\right)\left\{\exp\left(-i\frac{a\omega}{c}\sqrt{\varepsilon_2 - \sin^2\vartheta_z}\right) -\right.$$
$$\left. - \exp\left[-i\frac{a\omega}{v_z}(1 - \beta_y\cos\vartheta_y)\right]\right\}\exp\left(i\frac{\omega}{c}R\right), \qquad (18.15)$$

$$\Pi'_{\omega 2\perp} = \frac{ec\beta_y}{\pi\omega^2 R}\, \frac{(\sqrt{\varepsilon_2 - \sin^2\vartheta_z} + \sqrt{\varepsilon_1 - \sin^2\vartheta_z})\,A_\perp\cos\vartheta_z}{1 - \beta_y\cos\vartheta_y - \beta_z\sqrt{\varepsilon_2 - \sin^2\vartheta_z}}\left(\mathbf{j} + \mathbf{k}\cos\vartheta_y\frac{\cos\vartheta_z}{\sin^2\vartheta_z}\right)\times$$
$$\times \left\{\exp\left(-i\frac{a\omega}{c}\sqrt{\varepsilon_2 - \sin^2\vartheta_z}\right) - \exp\left[-i\frac{a\omega}{v_z}(1 - \beta_y\cos\vartheta_y)\right]\right\}\exp\left(i\frac{\omega}{c}R\right), \qquad (18.16)$$

$$\Pi''_{\omega 2\|} = \frac{ec\,\mathbf{k}}{\pi\omega^2 R}\, \frac{A_\|(\varepsilon_1\sqrt{\varepsilon_2 - \sin^2\vartheta_z} - \varepsilon_2\sqrt{\varepsilon_1 - \sin^2\vartheta_z})\cos\vartheta_z}{1 - \beta_y\cos\vartheta_y + \beta_z\sqrt{\varepsilon_2 - \sin^2\vartheta_z}}\times$$
$$\times \left(\beta_z + \beta_y\cos\vartheta_y\frac{\sqrt{\varepsilon_2 - \sin^2\vartheta_z}}{\sin^2\vartheta_z}\right)\left\{\exp\left(i\frac{a\omega}{c}\sqrt{\varepsilon_2 - \sin^2\vartheta_z}\right) -\right.$$
$$\left. - \exp\left[-i\frac{a\omega}{v_z}(1 - \beta_y\cos\vartheta_y)\right]\right\}\exp\left(i\frac{\omega}{c}R\right), \qquad (18.17)$$

$$\Pi'_{\omega 2\perp} = \frac{ec\beta_y}{\pi\omega^2 R}\, \frac{A_\perp(\sqrt{\varepsilon_2 - \sin^2\vartheta_z} - \sqrt{\varepsilon_1 - \sin^2\vartheta_z})\cos\vartheta_z}{1 - \beta_y\cos\vartheta_y + \beta_z\sqrt{\varepsilon_2 - \sin^2\vartheta_z}}\times$$
$$\times \left(\mathbf{j} + \mathbf{k}\cos\vartheta_y\frac{\cos\vartheta_z}{\sin^2\vartheta_z}\right)\left\{\exp\left(i\frac{a\omega}{c}\sqrt{\varepsilon_2 - \sin^2\vartheta_z}\right) - \exp\left[-i\frac{a\omega}{v_z}(1 - \beta_y\cos\vartheta_y)\right]\right\}\exp\left(i\frac{\omega}{c}R\right), \qquad (18.18)$$

$$\Pi'_{\omega 3\parallel} = -\frac{ec\,\mathbf{k}}{2\pi\omega^2 R}\Big(\beta_z - \beta_y \cos\vartheta_y\frac{\cos\vartheta_z}{\sin^2\vartheta_z}\Big)\frac{\exp\Big(i\,\dfrac{\omega}{c}\,R\Big)}{1-\beta_y\cos\vartheta_y-\beta_z\cos\vartheta_z};\tag{18.19}$$

$$\Pi'_{\omega 3\perp} = -\frac{ec\beta_y}{2\pi\omega^2 R}\,\frac{\exp\Big(i\,\dfrac{\omega}{c}\,R\Big)}{1-\beta_y\cos\vartheta_y-\beta_z\cos\vartheta_z}\Big(\mathbf{j}+\mathbf{k}\cos\vartheta_y\frac{\cos\vartheta_z}{\sin^2\vartheta_z}\Big),\tag{18.20}$$

$$\Pi''_{\omega 3\parallel} = -\frac{ec\,\mathbf{k}}{2\pi\omega^2 R}\,\frac{\varepsilon_2\cos\vartheta_z-\sqrt{\varepsilon_2-\sin^2\vartheta_z}}{\varepsilon_2\cos\vartheta_z+\sqrt{\varepsilon_2-\sin^2\vartheta_z}}\Big(\beta_z+\beta_y\cos\vartheta_y\frac{\cos\vartheta_z}{\sin^2\vartheta_z}\Big)\frac{\exp\Big(i\,\dfrac{\omega}{c}\,R\Big)}{1-\beta_y\cos\vartheta_y+\beta_z\cos\vartheta_z},\tag{18.21}$$

$$\Pi''_{\omega 3\perp} = -\frac{ec\beta_y}{2\pi\omega^2 R}\,\frac{\cos\vartheta_z-\sqrt{\varepsilon_2-\sin^2\vartheta_z}}{\cos\vartheta_z+\sqrt{\varepsilon_2-\sin^2\vartheta_z}}\,\frac{\exp\Big(i\,\dfrac{\omega}{c}\,R\Big)}{1-\beta_y\cos\vartheta_y+\beta_z\cos\vartheta_z}\Big(\mathbf{j}+\mathbf{k}\cos\vartheta_y\frac{\cos\vartheta_z}{\sin^2\vartheta_z}\Big),\tag{18.22}$$

$$\Pi'''_{\omega 3\parallel} = -\frac{2ec\,\mathbf{k}}{\pi\omega^2 R}\,\frac{A_\parallel\varepsilon_2\sqrt{\varepsilon_2-\sin^2\vartheta_z}\cos\vartheta_z}{\sqrt{\varepsilon_2-\sin^2\vartheta_z}+\varepsilon_2\cos\vartheta_z}\,\frac{\varepsilon_1\sqrt{\varepsilon_2-\sin^2\vartheta_z}-\varepsilon_2\sqrt{\varepsilon_1-\sin^2\vartheta_z}}{1-\beta_y\cos\vartheta_y+\beta_z\cos\vartheta_z}\times$$
$$\times\Big(\beta_z+\beta_y\cos\vartheta_y\frac{\cos\vartheta_z}{\sin^2\vartheta_z}\Big)\exp\Big(i\,\frac{\omega}{c}\,R\Big)\exp\Big(i\,\frac{a\omega}{c}\,\sqrt{\varepsilon_2-\sin^2\vartheta_z}\Big),\tag{18.23}$$

$$\Pi'''_{\omega 3\perp} = -\frac{2ec\beta_y}{\pi\omega^2 R}\,\frac{A_\perp\sqrt{\varepsilon_2-\sin^2\vartheta_z}\cos\vartheta_z}{\sqrt{\varepsilon_2-\sin^2\vartheta_z}+\cos\vartheta_z}\,\frac{\sqrt{\varepsilon_2-\sin^2\vartheta_z}-\sqrt{\varepsilon_1-\sin^2\vartheta_z}}{1-\beta_y\cos\vartheta_y+\beta_z\cos\vartheta_z}\times$$
$$\times\Big(\mathbf{j}+\mathbf{k}\cos\vartheta_y\frac{\cos\vartheta_z}{\sin^2\vartheta_z}\Big)\exp\Big(i\,\frac{a\omega}{c}\,\sqrt{\varepsilon_2-\sin^2\vartheta_z}\Big)\exp\Big(i\,\frac{\omega}{c}\,R\Big),\tag{18.24}$$

where $A_\parallel$ and $A_\perp$ are defined by Eqs. (16.29) and (16.30).

These equations help us to determine the spectral density of the radiation energy generated by waves of the corresponding polarizations per unit solid angle when a charged particle passes through a dielectric film ($\varepsilon_1 = 1$, $\varepsilon_2 \equiv \varepsilon$):*

$$W_{n\omega\parallel} = \frac{e^2\beta_z^2}{\pi^3 c}\,\frac{\cos^2\vartheta_z}{\sin^2\vartheta_z}\,|1-\varepsilon|^2\,[(1-\beta_y\cos\vartheta_y)^2-\beta_z^2\cos^2\vartheta_z]^{-2}\cdot$$
$$\cdot|\,[(1-\beta_z\sqrt{\varepsilon-\sin^2\vartheta_z}-\beta_z^2-\beta_y\cos\vartheta_y)\sin^2\vartheta_z+\beta_y\beta_z\cos\vartheta_y\sqrt{\varepsilon-\sin^2\vartheta_z}]\cdot$$
$$\cdot(1-\beta_y\cos\vartheta_y+\beta_z\sqrt{\varepsilon-\sin^2\vartheta_z})(\sqrt{\varepsilon-\sin^2\vartheta_z}+\varepsilon\cos\vartheta_z)e^{-i\frac{a\omega}{c}\sqrt{\varepsilon-\sin^2\vartheta_z}}+$$
$$+[(1+\beta_z\sqrt{\varepsilon-\sin^2\vartheta_z}-\beta_z^2-\beta_y\cos\vartheta_y)\sin^2\vartheta_z-\beta_y\beta_z\cos\vartheta_y\sqrt{\varepsilon-\sin^2\vartheta_z}]\cdot$$
$$\cdot(1-\beta_y\cos\vartheta_y-\beta_z\sqrt{\varepsilon-\sin^2\vartheta_z})(\sqrt{\varepsilon-\sin^2\vartheta_z}-\varepsilon\cos\vartheta_z)e^{i\frac{a\omega}{c}\sqrt{\varepsilon-\sin^2\vartheta_z}}-$$
$$-2\sqrt{\varepsilon-\sin^2\vartheta_z}\,[(1-\beta_y\cos\vartheta_y+\varepsilon\beta_z\cos\vartheta_z)(1-\beta_z^2-\beta_y\cos\vartheta_y)\sin^2\vartheta_z+$$
$$+(\beta_y\cos\vartheta_y-\sin^2\vartheta_z)(\varepsilon\cos\vartheta_z-\varepsilon\beta_y\cos\vartheta_y\cos\vartheta_z+\varepsilon\beta_z-\beta_z\sin^2\vartheta_z)\beta_z]\times$$
$$\times e^{-i\frac{a\omega}{v_z}(1-\beta_y\cos\vartheta_y)}\Big|^2\,\Big|(\sqrt{\varepsilon-\sin^2\vartheta_z}+\varepsilon\cos\vartheta_z)^2 e^{-i\frac{a\omega}{c}\sqrt{\varepsilon-\sin^2\vartheta_z}}-$$
$$-(\sqrt{\varepsilon-\sin^2\vartheta_z}-\varepsilon\cos\vartheta_z)^2 e^{i\frac{a\omega}{c}\sqrt{\varepsilon-\sin^2\vartheta_z}}\Big|^{-2}\,|(1-\beta_y\cos\vartheta_y)^2-\beta_z^2(\varepsilon-\sin^2\vartheta_z)|^{-2},\tag{18.25}$$

---

* Recently, similar results were reported in [47].

$$W_{n\omega\perp} = \frac{e^2\beta_y^2\beta_z^4}{\pi^2 c}\frac{\cos^2\vartheta_x\cos^2\vartheta_z}{\sin^2\vartheta_z}\,|\,1-\varepsilon\,|^2\,[(1-\beta_y\cos\vartheta_y)^2-\beta_z^2\cos^2\vartheta_z]^{-2}\cdot$$

$$\cdot\Big|(1-\beta_y\cos\vartheta_y+\beta_z\sqrt{\varepsilon-\sin^2\vartheta_z})(\sqrt{\varepsilon-\sin^2\vartheta_z}+\cos\vartheta_z)e^{-i\frac{a\omega}{c}\sqrt{\varepsilon-\sin^2\vartheta_z}}+$$

$$+(1-\beta_y\cos\vartheta_y-\beta_z\sqrt{\varepsilon-\sin^2\vartheta_z})(\sqrt{\varepsilon-\sin^2\vartheta_z}-\cos\vartheta_z)e^{i\frac{a\omega}{c}\sqrt{\varepsilon-\sin^2\vartheta_z}}-$$

$$-2\sqrt{\varepsilon-\sin^2\vartheta_z}\,(1-\beta_y\cos\vartheta_z+\beta_z\cos\vartheta_z)\,e^{-i\frac{a\omega}{v_z}(1-\beta_y\cos\vartheta_y)}\Big|^2\cdot$$

$$\cdot\Big|(\sqrt{\varepsilon-\sin^2\vartheta_z}+\cos\vartheta_z)^2 e^{-i\frac{a\omega}{c}\sqrt{\varepsilon-\sin^2\vartheta_z}}-$$

$$-(\sqrt{\varepsilon-\sin^2\vartheta_z}-\cos\vartheta_z)^2 e^{i\frac{a\omega}{c}\sqrt{\varepsilon-\sin^2\vartheta_z}}\Big|^{-2}\cdot|(1-\beta_y\cos\vartheta_y)^2-\beta_z^2(\varepsilon-\sin^2\vartheta_z)|^{-2}\tag{18.26}$$

$$(\mathrm{Im}\,\varepsilon>0,\quad \mathrm{Im}\,\sqrt{\varepsilon-\sin^2\vartheta_z}>0).$$

Taking into account that $\mathrm{Im}\,(\varepsilon-\sin^2\vartheta_z)^{1/2}>0$ for $a\to\infty$, we obtain equations for the transition radiation in the case of inclined particle incidence:

$$W_{n\omega\parallel}=\frac{e^2\beta_z^2\cos^2\vartheta_z\,|\,1-\varepsilon\,|^2}{\pi^2 c\,[(1-\beta_y\cos\vartheta_y)^2-\beta_z^2\cos^2\vartheta_z]^2\sin^2\vartheta_z}\times$$

$$\times\left|\frac{(1-\beta_z\sqrt{\varepsilon-\sin^2\vartheta_z}-\beta_z^2-\beta_y\cos\vartheta_y)\sin^2\vartheta_z+\beta_y\beta_z\cos\vartheta_y\sqrt{\varepsilon-\sin^2\vartheta_z}}{(1-\beta_y\cos\vartheta_y-\beta_z\sqrt{\varepsilon-\sin^2\vartheta_z})(\varepsilon\cos\vartheta_z+\sqrt{\varepsilon-\sin^2\vartheta_z})}\right|^2,\tag{18.27}$$

$$W_{n\omega\perp}=\frac{e^2\beta_y^2\beta_z^4\cos^2\vartheta_x\cos^2\vartheta_z\,|\,1-\varepsilon\,|^2}{\pi^2 c\,[(1-\beta_y\cos\vartheta_y)^2-\beta_z^2\cos^2\vartheta_z]^2\sin^2\vartheta_z}\times$$

$$\times\,|\,(1-\beta_y\cos\vartheta_y-\beta_z\sqrt{\varepsilon-\sin^2\vartheta_z})(\cos\vartheta_z+\sqrt{\varepsilon-\sin^2\vartheta_z})\,|^{-2}.\tag{18.28}$$

The results stated in Eqs. (18.25) and (18.26) are simplified considerably for a nonrelativistic particle and assume the form:

$$W_{n\omega\parallel}=\frac{e^2\beta_z^2}{\pi^2 c}\sin^2\vartheta_z\cos^2\vartheta_z|\,1-\varepsilon\,|^2|(\sqrt{\varepsilon-\sin^2\vartheta_z}+\varepsilon\cos\vartheta_z)e^{-i\frac{a\omega}{c}\sqrt{\varepsilon-\sin^2\vartheta_z}}+$$

$$+(\sqrt{\varepsilon-\sin^2\vartheta_z}-\varepsilon\cos\vartheta_z)e^{i\frac{a\omega}{c}\sqrt{\varepsilon-\sin^2\vartheta_z}}-2\sqrt{\varepsilon-\sin^2\vartheta_z}\,e^{-i\frac{a\omega}{v_z}(1-\beta_y\cos\vartheta_y)}|^2\times$$

$$\times\Big|(\sqrt{\varepsilon-\sin^2\vartheta_z}+\varepsilon\cos\vartheta_z)^2\,e^{-i\frac{a\omega}{c}\sqrt{\varepsilon-\sin^2\vartheta_z}}-(\sqrt{\varepsilon-\sin^2\vartheta_z}-\varepsilon\cos\vartheta_z)^2\,e^{i\frac{a\omega}{c}\sqrt{\varepsilon-\sin^2\vartheta_z}}\Big|^{-2},\tag{18.29}$$

$$W_{n\omega\perp}=\frac{e^2\beta_y^2\beta_z^4}{\pi^2 c}\frac{\cos^2\vartheta_x\cos^2\vartheta_z}{\sin^2\vartheta_z}|\,1-\varepsilon\,|^2|(\sqrt{\varepsilon-\sin^2\vartheta_z}+\cos\vartheta_z)e^{-i\frac{a\omega}{c}\sqrt{\varepsilon-\sin^2\vartheta_z}}+$$

$$+(\sqrt{\varepsilon-\sin^2\vartheta_z}-\cos\vartheta_z)e^{i\frac{a\omega}{c}\sqrt{\varepsilon-\sin^2\vartheta_z}}-2\sqrt{\varepsilon-\sin^2\vartheta_z}\,e^{-i\frac{a\omega}{v_z}(1-\beta_y\cos\vartheta_y)}|^2\times$$

$$\times\Big|(\sqrt{\varepsilon-\sin^2\vartheta_z}+\cos\vartheta_z)^2 e^{-i\frac{a\omega}{c}\sqrt{\varepsilon-\sin^2\vartheta_z}}-(\sqrt{\varepsilon-\sin^2\vartheta_z}-\cos\vartheta_z)^2 e^{i\frac{a\omega}{c}\sqrt{\varepsilon-\sin^2\vartheta_z}}\Big|^{-2}.\tag{18.30}$$

From these results we obtain in the case of a thin film $\left(\frac{a\omega}{c}\left|\sqrt{\varepsilon-\sin^2\vartheta_z}\right|\ll 1\right)$:

$$W_{n\omega\parallel}=\frac{e^2\beta_z^2}{\pi^2 c}\left|\frac{1-\varepsilon}{\varepsilon}\right|^2\sin^2\vartheta_z\sin^2\left[\frac{a\omega}{2v_z}(1-\beta_y\cos\vartheta_y)\right],\tag{18.31}$$

$$W_{n\omega\perp} = \frac{e^2\beta_y^2\beta_z^4}{\pi^2 c} \mid 1 - \varepsilon \mid^2 \frac{\cos^2\vartheta_x}{\sin^2\vartheta_z} \sin^2\left[\frac{a\omega}{2v_z}(1 - \beta_y\cos\vartheta_y)\right]. \tag{18.32}$$

In these equations, the squares of the sine functions account for the interference of two flashes of transition radiation generated at the first and the second boundary. The portion of the argument which depends upon the particle's velocity component in the y direction determines the relative phase shift of the waves due to the displacement of the particle relative to the normal to the boundary, along the particle's path in the film. In the event that the phase shift related to this displacement is small, i.e., if we have $\frac{a\omega}{v_z}\beta_y\cos\vartheta_y \ll 1$, the results of Eqs. (18.31) and (18.32) transform into

$$W_{n\omega\parallel} = \frac{e^2\beta_z^2}{\pi^2 c}\left|\frac{1-\varepsilon}{\varepsilon}\right|^2 \sin^2\vartheta_z \sin^2\frac{a\omega}{2v_z}, \tag{18.33}$$

$$W_{n\omega\perp} = \frac{e^2\beta_y^2\beta_z^4}{\pi^2 c} \mid 1 - \varepsilon\mid^2 \frac{\cos^2\vartheta_x}{\sin^2\vartheta_z} \sin^2\frac{a\omega}{2v_z}. \tag{18.34}$$

If the phase shift is small, which results from the phase lag of the particle relative to the wave emitted along the path in the film, i.e., if $a\omega/v_z \ll 1$, we obtain for a nonrelativistic particle from Eqs. (18.31) and (18.32)

$$W_{n\omega\parallel} = \frac{e^2 a^2 \omega^2}{4\pi^2 c^3}\left|\frac{1-\varepsilon}{\varepsilon}\right|^2 \sin^2\vartheta_z, \tag{18.35}$$

$$W_{n\omega\perp} = \frac{e^2 a^2 \omega^2 \beta_y^2 \beta_z^4}{4\pi^2 c^3} \mid 1 - \varepsilon\mid^2 \frac{\cos^2\vartheta_x}{\sin^2\vartheta_z}. \tag{18.36}$$

These are the very simple consequences of the results obtained above.

Let us now consider the radiation generated when a charged particle moves uniformly along a straight line perpendicular to the boundary. In this particular case, the radiation is polarized in the plane of incidence, and it follows from Eqs. (18.1)–(18.12) that the radiation is given by the sum of the following Hertz vectors:

$$\Pi_{\omega 1} = \frac{evk}{2\pi\omega^2 R}\frac{p_{1,2}\alpha p_{2,3}}{1-\beta n_1\cos\vartheta_1}\exp\left(-i\frac{\omega}{v}a + i\frac{\omega}{c}an_2\cos\vartheta_2 + i\frac{\omega}{c}n_3 R\right), \tag{18.37}$$

$$\Pi_{\omega 2} = \frac{evk}{2\pi\omega^2 R}\alpha p_{2,3}\left\{\frac{1}{1-\beta n_2\cos\vartheta_2}\left[\exp\left(-i\frac{\omega}{c}an_2\cos\vartheta_2\right) - \exp\left(-i\frac{\omega}{v}a\right)\right] + \right.$$

$$\left. + \frac{r_{2,1}}{1+\beta n_2\cos\vartheta_2}\left[\exp\left(i\frac{\omega}{c}an_2\cos\vartheta_2\right) - \exp\left(-i\frac{\omega}{v}a\right)\right]\right\}\exp\left(i\frac{\omega}{c}an_2\cos\vartheta_2 + i\frac{\omega}{c}n_3 R\right), \tag{18.38}$$

$$\Pi_{\omega 3} = -\frac{evk}{2\pi\omega^2 R}\left\{\frac{1}{1-\beta n_3\cos\vartheta_3} + \frac{\alpha}{1+\beta n_3\cos\vartheta_3}\left[r_{3,2} + r_{2,1}\exp\left(2i\frac{\omega}{c}an_2\cos\vartheta_2\right)\right]\right\}\exp\left(i\frac{\omega}{c}n_3 R\right). \tag{18.39}$$

Here, as well as in the following discussion, we omit the symbol $\parallel$ which indicates the polarization of the waves. We also omit the subscript z in the direction cosines of the angles which the wave includes with the z axis in the various regions.

With the motion which we consider (particles moving along the normal to the boundary) we must set $\beta_x = \beta_y = 0$ in expressions (2.7)–(2.10) which were found in the case of arbitrary particle motion for the coefficients r_{jk} and p_{jk}. In deriving Eq. (18.39), we used the relation

$$p_{jk}p_{hj} - r_{jk}r_{hj} = 1,$$ (18.40)

which holds in this case.

The spectral density of the radiation energy per unit solid angle is given by the equation

$$W_{n\omega} = \frac{\omega^4}{c^3} R^2 |\Pi_{\omega 1} + \Pi_{\omega 2} + \Pi_{\omega 3}|^2 n_3' \sin^2 \vartheta_3.$$ (18.41)

n_3' denotes the real part of the refractive index of medium 3.

Let us consider some examples for the application of Eq. (18.41) [20].

1. Assume that layer 2 is so thin that the inequality

$$\frac{a\omega}{c} |n_2 \cos \vartheta_2| \ll 1$$ (18.42)

holds. In the limit, we denote the quantity α by α_0:

$$\alpha_0 = (1 - r_{2,1}r_{2,3})^{-1}$$ (18.43)

and recall that

$$p_{1,2}\alpha_0 p_{2,3} = p_{1,3}, \quad \alpha_0 (r_{3,2} + r_{2,1}) = r_{3,1}.$$ (18.44)

Then, Eqs. (18.37)-(18.39) assume the following form:

$$\Pi_{\omega 1} = \frac{ev\mathbf{k}}{2\pi\omega^2 R} \frac{p_{1,3}}{1 - \beta n_1 \cos \vartheta_1} \exp\left(-i\frac{\omega}{v} a + i\frac{\omega}{c} n_3 R\right),$$ (18.45)

$$\Pi_{\omega 2} = \frac{ev\mathbf{k}}{2\pi\omega^2 R} \alpha_0 p_{2,3} \left(\frac{1}{1 - \beta n_2 \cos \vartheta_2} + \frac{r_{2,1}}{1 + \beta n_2 \cos \vartheta_2}\right) \left[1 - \exp\left(-i\frac{\omega}{v} a\right)\right] \exp\left(i\frac{\omega}{c} n_3 R\right),$$ (18.46)

$$\Pi_{\omega 3} = -\frac{ev\mathbf{k}}{2\pi\omega^2 R} \left(\frac{1}{1 - \beta n_3 \cos \vartheta_3} + \frac{r_{3,1}}{1 + \beta n_3 \cos \vartheta_3}\right) \exp\left(i\frac{\omega}{c} n_3 R\right).$$ (18.47)

When the condition (18.42) holds, the inequality $\omega a \ll v$ is fulfilled for a relativistic particle. Therefore, Eq. (18.46), which includes the factor $[1 - \exp(-i\omega a/v)]$ (which in this case is much smaller than unity) has an absolute value which is much smaller than that of Eqs. (18.45) and (18.47). The conclusion is that a layer whose thickness is much smaller than the wavelength does not influence the transition radiation generated by charged relativistic particles.

Let us consider the radiation of a nonrelativistic particle. In this case, the quantities which contain the factor β in the denominators of Eq. (18.45)-(18.47) can be ignored and the summation of these equations gives us (for $\mu_1 = \mu_2 = \mu_3 = 1$):

$$\Pi_\omega = \frac{ev\mathbf{k}p_{1,3}}{2\pi\omega^2 R} \left\{\left(1 - \frac{n_1^2}{n_3^2}\right) + \left(\frac{n_1^2}{n_2^2} - 1\right)\left[1 - \exp\left(-i\frac{\omega}{v} a\right)\right]\right\} \exp\left(i\frac{\omega}{c} n_3 R\right).$$ (18.48)

The first term in the braces of this expression describes the transition radiation at one boundary (between medium 1 and medium 3). The second term results from the intermediate layer 2. When the particle velocity is small so that the time for passing through layer 2 (i.e., a/v) is not very small (or simply not small) relative to the period $2\pi/\omega$ of the light wave, then layer 2 gives a noticeable contribution to the transition radiation. This means that the properties of a thin film can be established from its transition radiation. It is obvious that for $n_1 = n_2$, as well as for $n_2 = n_3$, Eq. (18.48) transforms into the usual formula for the transition radiation resulting from a single boundary, as we could expect.

2. "Translucent optics" is of particular interest when we consider the superposition of the transition radiation from several boundaries. Translucence is obtained by applying a thin layer which suppresses reflected light rays. The second term in the braces of Eq. (18.39) corresponds to the reflected light passing from medium 3 into medium 1. When this term vanishes, we have the translucency case. When we set the factor in the brackets in the second term of Eq. (18.39) equal to zero, we obtain the translucency condition:

$$r_{3,2} + r_{2,1} \exp\left(2i\,\frac{\omega}{c}\,an_2\cos\vartheta_2\right) = 0. \tag{18.49}$$

In the most interesting case of transparent media and for real $r_{3,2}$ and $r_{2,1}$ this equation can be fulfilled if the exponent is equal to ± 1. When the exponent is equal to unity, we obtain from Eq. (18.49) $r_{3,2} = -r_{2,1}$, and, since $r_{2,1} = -r_{1,2}$, we have consequently $r_{3,2} = r_{1,2}$, i.e., the refractive indices of media 1 and 3 must be the same. This corresponds to an isolated film and we will not dwell on this problem.

Let us consider the consequences of the translucency condition if we have

$$\exp\left(2i\,\frac{\omega}{c}\,an_2\cos\vartheta_2\right) = -1. \tag{18.50}$$

Then $r_{3,2} = r_{2,1}$ must hold, which is equivalent to the condition

$$n_2^2 \cos\vartheta_1 \cos\vartheta_3 = n_1 n_3 \cos^2\vartheta_2. \tag{18.51}$$

The above condition is symmetric relative to transposition of the subscripts 1 and 3, which means that we simultaneously have transparency in the forward direction (for a wave passing from medium 1 into medium 3) as well as in the backward direction (for a wave passing from medium 3 into medium 1). This treatment can be easily generalized to a stack of translucent plates (see below).

When Eqs. (18.50) and (18.51) hold, we have

$$r_{3,2} = r_{2,1} = \frac{\sqrt{n_1\cos\vartheta_3} - \sqrt{n_3\cos\vartheta_1}}{\sqrt{n_1\cos\vartheta_3} + \sqrt{n_3\cos\vartheta_1}}, \qquad \alpha = \frac{(\sqrt{n_1\cos\vartheta_3} + \sqrt{n_3\cos\vartheta_1})^2}{4\sqrt{n_1 n_3}\cos\vartheta_1\cos\vartheta_3},$$

$$p_{1,2} = \frac{2\sqrt{n_1\cos\vartheta_3}}{\sqrt{n_1\cos\vartheta_3} + \sqrt{n_3\cos\vartheta_1}}\frac{n_2^2}{n_1^2}, \qquad P_{2,3} = \frac{2\sqrt{n_1\cos\vartheta_3}}{\sqrt{n_1\cos\vartheta_3} + \sqrt{n_3\cos\vartheta_1}}\frac{n_3^2}{n_2^2}. \tag{18.52}$$

Thus, with translucency the Hertz vectors defined by Eqs. (18.37)–(18.39) assume the following form:

$$\mathbf{\Pi}_{\omega 1} = \frac{evk}{2\pi\omega^2 R}\frac{n_3}{n_1}\left(\frac{n_3\cos\vartheta_3}{n_1\cos\vartheta_1}\right)^{1/2}\frac{i}{1-\beta n_1\cos\vartheta_1}\exp\left(-i\,\frac{\omega}{v}\,a + i\,\frac{\omega}{c}\,n_3 R\right), \tag{18.53}$$

$$\mathbf{\Pi}_{\omega 2} = \frac{evk}{2\pi\omega^2 R}\frac{n_3^2}{n_2^2}\frac{\sqrt{n_1\cos\vartheta_3} + \sqrt{n_3\cos\vartheta_1}}{2\sqrt{n_3\cos\vartheta_1}}\left\{\frac{1}{1-\beta n_2\cos\vartheta_2}\left[1 - i\exp\left(-i\,\frac{\omega}{v}\,a\right)\right] - \right.$$

$$\left. - \frac{\sqrt{n_1\cos\vartheta_3} - \sqrt{n_3\cos\vartheta_1}}{\sqrt{n_1\cos\vartheta_3} + \sqrt{n_3\cos\vartheta_1}}\frac{1}{1+\beta n_2\cos\vartheta_2}\left[1 + i\exp\left(-i\,\frac{\omega}{v}\,a\right)\right]\right\}\exp\left(i\,\frac{\omega}{c}\,n_3 R\right), \tag{18.54}$$

$$\mathbf{\Pi}_{\omega 3} = -\frac{ev\mathbf{k}}{2\pi\omega^2 R}\frac{1}{1-\beta n_3\cos\vartheta_3}\exp\left(i\frac{\omega}{c}n_3 R\right). \tag{18.55}$$

§19. Radiation of a Particle Passing in the Normal Direction through a Plate

The spherical wave of a radiation field generated by a charged particle passing through a dielectric plate in vacuum normal to the plate surface can be obtained from the more general equations (16.17)-(16.28) when we set $\varepsilon_1 = 1$, $\beta_y = 0$. In this particular case, the Hertz vectors defined by Eqs. (16.18), (16.20), (16.22), (16.24), (16.26), and (16.28) (these Hertz vectors describe waves polarized in a plane perpendicular to the plane of incidence) vanish. The result simplifies considerably. The result is given by the sum of the vectors stated in Eqs. (16.17), (16.19), (16.21), (16.23), (16.25), and (16.27), wherein we have to set $\beta_y = 0$. For $\varepsilon_1 = 1$ we can write these vectors in the form ($\varepsilon_2 \equiv \varepsilon$, $\beta_z \equiv \beta$, $\vartheta_z \equiv \vartheta$)

$$\mathbf{\Pi}_{\omega 1} = \frac{2ev\mathbf{k}}{\pi\omega^2 R}\,A\,\frac{\varepsilon\sqrt{\varepsilon-\sin^2\vartheta}}{1-\beta\cos\vartheta}\cos\vartheta\exp\left(-i\frac{a\omega}{v}+i\frac{\omega}{c}R\right), \tag{19.1}$$

$$\mathbf{\Pi}'_{\omega 2} = \frac{ev\mathbf{k}}{\pi\omega^2 R}\,A\,\frac{\sqrt{\varepsilon-\sin^2\vartheta}+\varepsilon\cos\vartheta}{1-\beta\sqrt{\varepsilon-\sin^2\vartheta}}\cos\vartheta\left[\exp\left(-i\frac{a\omega}{c}\sqrt{\varepsilon-\sin^2\vartheta}\right)-\exp\left(-i\frac{a\omega}{v}\right)\right]\exp\left(i\frac{\omega}{c}R\right), \tag{19.2}$$

$$\mathbf{\Pi}''_{\omega 2} = \frac{ev\mathbf{k}}{\pi\omega^2 R}\,A\,\frac{\sqrt{\varepsilon-\sin^2\vartheta}-\varepsilon\cos\vartheta}{1+\beta\sqrt{\varepsilon-\sin^2\vartheta}}\cos\vartheta\left[\exp\left(i\frac{a\omega}{c}\sqrt{\varepsilon-\sin^2\vartheta}\right)-\exp\left(-i\frac{a\omega}{v}\right)\right]\exp\left(i\frac{\omega}{c}R\right), \tag{19.3}$$

$$\mathbf{\Pi}'_{\omega 3} = -\frac{ev\mathbf{k}}{2\pi\omega^2 R}\frac{1}{1-\beta\cos\vartheta}\exp\left(i\frac{\omega}{c}R\right), \tag{19.4}$$

$$\mathbf{\Pi}''_{\omega 3} = -\frac{ev\mathbf{k}}{2\pi\omega^2 R}\frac{\varepsilon\cos\vartheta-\sqrt{\varepsilon-\sin^2\vartheta}}{\varepsilon\cos\vartheta+\sqrt{\varepsilon-\sin^2\vartheta}}\frac{\exp\left(i\frac{\omega}{c}R\right)}{1+\beta\cos\vartheta}, \tag{19.5}$$

$$\mathbf{\Pi}'''_{\omega 3} = -\frac{2ev\mathbf{k}}{\pi\omega^2 R}\,A\,\frac{\varepsilon\sqrt{\varepsilon-\sin^2\vartheta}\cos\vartheta}{\varepsilon\cos\vartheta+\sqrt{\varepsilon-\sin^2\vartheta}}\frac{\sqrt{\varepsilon-\sin^2\vartheta}-\varepsilon\cos\vartheta}{1+\beta\cos\vartheta}\exp\left(i\frac{a\omega}{c}\sqrt{\varepsilon-\sin^2\vartheta}+i\frac{\omega}{c}R\right), \tag{19.6}$$

where

$$A = \left[(\sqrt{\varepsilon-\sin^2\vartheta}+\varepsilon\cos\vartheta)^2\exp\left(-i\frac{a\omega}{c}\sqrt{\varepsilon-\sin^2\vartheta}\right)-\right.$$
$$\left.-(\sqrt{\varepsilon-\sin^2\vartheta}-\varepsilon\cos\vartheta)^2\exp\left(i\frac{a\omega}{c}\sqrt{\varepsilon-\sin^2\vartheta}\right)\right]^{-1}. \tag{19.7}$$

With these formulas we can calculate the angular distribution of the spectral density of the radiation energy in the forward direction (radiation emitted at an acute angle relative to the velocity vector of the particle):

$$W_{n\omega} = \frac{e^2\beta^2}{\pi^2 c}\sin^2\vartheta\cos^2\vartheta\,|1-\varepsilon|^2(1-\beta^2\cos^2\vartheta)^{-2}|(1-\beta\sqrt{\varepsilon-\sin^2\vartheta}-\beta^2)\times$$
$$+(1+\beta\sqrt{\varepsilon-\sin^2\vartheta}-\beta^2)(1-\beta\sqrt{\varepsilon-\sin^2\vartheta})\times$$
$$\times(1+\beta\sqrt{\varepsilon-\sin^2\vartheta})(\sqrt{\varepsilon-\sin^2\vartheta}+\varepsilon\cos\vartheta)\exp\left(-i\frac{a\omega}{c}\sqrt{\varepsilon-\sin^2\vartheta}\right)+$$
$$\times(\sqrt{\varepsilon-\sin^2\vartheta}-\varepsilon\cos\vartheta)\exp\left(i\frac{a\omega}{c}\sqrt{\varepsilon-\sin^2\vartheta}\right)-2\sqrt{\varepsilon-\sin^2\vartheta}\times \tag{19.8}$$

$$\times\,(1-\beta^2\cos^2\vartheta - \varepsilon\beta^2 - \varepsilon\beta^3\cos\vartheta)\exp\left(-\,i\,\frac{a\omega}{c}\right)\Big|^2\Big|\,(\sqrt{\varepsilon - \sin^2\vartheta} + \varepsilon\cos\vartheta)^2\times$$

$$\times\,\exp\left(-\,i\,\frac{a\omega}{c}\,\sqrt{\varepsilon - \sin^2\vartheta}\right) - (\sqrt{\varepsilon - \sin^2\vartheta} - \varepsilon\cos\vartheta)^2\,\times$$

$$\times\,\exp\left(i\,\frac{a\omega}{c}\,\sqrt{\varepsilon - \sin^2\vartheta}\right)\Big|^2\,|\,1 - \beta^2(\varepsilon - \sin^2\vartheta)\,|^{-2}. \qquad (19.8)$$

When the direction of motion of the particle is reversed, the angular distribution of the spectral density of the radiation energy in this portion of the half-space is given by Eq. (19.8), in which the sign of the velocity must be reversed.

The author obtained this result ten years ago [7]. The radiation of a charged particle passing through a plate in the normal direction has been treated in [48]-[55]. The radiation resulting from the passage of a particle through an isotropic ferrodielectric, crystalline plate was discussed in [9] and [11]; in [56], the radiation resulting from the passage of a particle through a layer with spatial dispersion was described.

By now, many data from experimental studies of the transition radiation at plates have been accumulated. Obviously, the authors of the first experimental papers [57, 59] did not know of the existence of transition-radiation theory. Their work was initiated by Ferrell [48], who attempted to treat the transformation of longitudinal waves generated in a thin metal layer by a passing charge, into transverse waves. In principle, this could be detected experimentally. But, in [48] the particular circumstances resulting from the electrodynamics of media with spatial dispersion were not taken into account, which is necessary for the treatment of this type of wave transformation. Silin and Fetisov [56] discussed this problem for the first time. They could show that the excitation and propagation of longitudinal waves under the particular experimental conditions is immaterial and that the experimentally observed radiation can be quite adequately described by the theory of transition radiation [7]. This conclusion was confirmed by the results of subsequent research work [60]-[65].

Let us outline some consequences of the theory.

When the particle velocity exceeds the phase velocity of light in the plate, Cerenkov radiation develops which leaves the plate, unless total reflection can occur. The refraction angle of the Cerenkov radiation satisfies the equation

$$\beta\,\sqrt{\varepsilon - \sin^2\vartheta} = 1. \qquad (19.9)$$

The denominator of Eq. (19.8) vanishes for this direction. A careful analysis shows that the numerator also vanishes. The resulting undetermined expression is of the form

$$\frac{\sin^2\left[\dfrac{a\omega}{2v}\,(1 - \beta\,\sqrt{\varepsilon - \sin^2\vartheta})\right]}{(1 - \beta\,\sqrt{\varepsilon - \sin^2\vartheta})^2}. \qquad (19.10)$$

This means that for

$$a\omega/v \gg 1, \qquad (19.11)$$

Eq. (19.8) describes a sharp peak of the radiation intensity. A peak of diffracted Cerenkov radiation occurs before the plate, too. This peak is obviously the result of the reflection of the Cerenkov radiation at the front face of the plate. This peak must vanish when a radiation with a velocity greater than that of light is generated under the Brewster angle. As a matter of fact, it is known from optics that the intensity of a reflected wave whose electric vector coincides with the plane of incidence vanishes, and the wave emitted in this case is polarized in the plane of incidence.

The intensity ratio of the Cerenkov radiation emitted in the forward direction (W_B) to the radiation emitted in the backward direction (W_H) is, according to Eq. (19.8), equal to

$$\frac{W_B}{W_H} = \frac{(1 + \beta\varepsilon\cos\vartheta_r)^2}{(1 - \beta\varepsilon\cos\vartheta_r)^2} = \frac{(1 + \varepsilon\sqrt{1 + \beta^2 - \varepsilon\beta^2})^2}{(1 - \varepsilon\sqrt{1 + \beta^2 - \varepsilon\beta^2})^2} . \tag{19.12}$$

We recall that when the wave is incident upon the boundary under the Brewster angle, the reflected ray includes a right angle with the diffracted ray. This gives us a condition for the generation of Cerenkov radiation under the Brewster angle:

$$\beta^2\varepsilon^2 = 1 + \varepsilon . \tag{19.13}$$

When this condition holds, the denominator in Eq. (19.12) vanishes, and this means that the intensity peak of the radiation before the plate vanishes.

Equation (19.8) holds at sufficiently large distances from the film in the regions in which Eqs. (19.1)-(19.7) are applicable, i.e., where the entire radiation field transforms into a spherical wave. It is known that at small distances from the particle trajectory the Cerenkov radiation is given by a cylindrical wave. This is the reason why near the plate the diffracted Cerenkov radiation which passes into the vacuum has the form of a cylindrical wave. If the radiation is not generated under the Brewster angle, a second, third, etc., cylindrical wave develops from multiple reflections of the Cerenkov radiation from the walls of the plate, in addition to the first diffracted cylindrical wave. Obviously, these waves must be separated in space by conical surfaces.

Let us consider diffracted cylindrical waves and the regions of space in which they exist. We write the incident cylindrical wave of the Cerenkov radiation of Eq. (6.12):

$$\Pi'_{\omega,0} = \frac{-ev}{\sqrt{2\pi vR\sin\vartheta}} \frac{\sqrt{-i\omega}}{\omega^2(\varepsilon\beta^2 - 1)^{1/4}} \exp\left[i\omega\frac{R}{v}(\sqrt{\varepsilon\beta^2 - 1}\sin\vartheta + \cos\vartheta)\right] \tag{19.14}$$

and the first diffracted wave defined by Eq. (8.25):

$$\Pi_{\omega,1} = -\frac{2ev\sqrt{-i\omega}\exp\left[i\omega\frac{R}{v}(\sqrt{\varepsilon\beta^2 - 1}\sin\vartheta + \sqrt{1 + \beta^2 - \varepsilon\beta^2}\cos\vartheta)\right]}{\sqrt{2\pi vR\sin\vartheta}\omega^2(\varepsilon\beta^2 - 1)^{1/4}(\varepsilon\sqrt{1 + \beta^2 - \varepsilon\beta^2} + 1)} . \tag{19.15}$$

With these results, we obtain the first reflected wave in the form

$$\Pi'_{\omega,1} = \frac{ev\sqrt{-i\omega}(\varepsilon\sqrt{1 + \beta^2 - \varepsilon\beta^2} - 1)\exp\left[i\omega\frac{R}{v}(\sqrt{\varepsilon\beta^2 - 1}\sin\vartheta - \cos\vartheta)\right]}{\sqrt{2\pi vR\sin\vartheta}\omega^2(\varepsilon\beta^2 - 1)^{1/4}(\varepsilon\sqrt{1 + \beta^2 - \varepsilon\beta^2} + 1)} . \tag{19.16}$$

In the following discussion we will denote the m-th diffracted wave by the subscript m, and the m-th reflected wave by an additional prime. With the self-evident relations for the wave amplitudes

$$\Pi'_{\omega,m-1}/\Pi_{\omega,m} = \Pi'_{\omega,m}/\Pi_{\omega,m+1}, \qquad \Pi_{\omega,m-1}/\Pi_{\omega,m} = \Pi'_{\omega,m}/\Pi'_{\omega,m+1}, \tag{19.17}$$

we can express the amplitude of the m-th diffracted wave by the amplitudes of the primary, first reflected, and first diffracted waves:

$$\Pi_{\omega,m} = \Pi_{\omega,1}\frac{(\Pi'_{\omega,1})^{m-1}}{(\Pi'_{\omega,0})^{m-1}} . \tag{19.18}$$

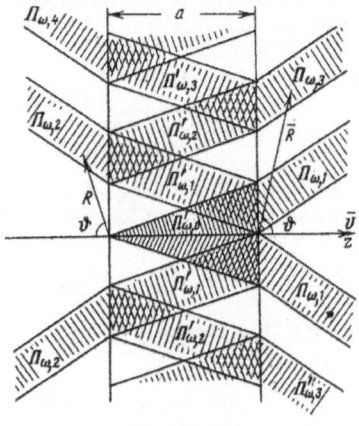

Fig. 18

When we take into account the phase factor which results from the increase in the optical path of the wave with each reflection, and use Eqs. (19.14), (19.15), (19.16), and (19.18), we obtain the Hertz vector of the m-th diffracted wave:

$$\Pi_{\omega,m} = \frac{2ev}{\sqrt{2\pi v R \sin\vartheta}} \frac{\sqrt{i\omega}\,(\varepsilon\sqrt{1+\beta^2-\varepsilon\beta^2}-1)^{m-1}}{\omega^2\,(\varepsilon\sqrt{1+\beta^2-\varepsilon\beta^2}+1)^m\,(\varepsilon\beta^2-1)^{1/4}} \times$$

$$\times \exp\left\{i\omega\left[\frac{R}{v}\sqrt{\varepsilon\beta^2-1}\sin\vartheta + (-1)^{m-1}\frac{R}{v}\sqrt{1+\beta^2-\varepsilon\beta^2}\cos\vartheta + \right.\right.$$

$$\left.\left. + (m-1)\frac{aev}{c^2}\right]\right\}. \qquad (19.19)$$

This wave passes into the vacuum through the annular area

$$(m-1)a\sqrt{\varepsilon\beta^2-1} < \rho < ma\sqrt{\varepsilon\beta^2-1} \qquad (19.20)$$

and propagates between conical surfaces in the region

$$\frac{R\sqrt{\varepsilon\beta^2-1}\cos\vartheta}{\sqrt{1+\beta^2-\varepsilon\beta^2}} + (m-1)a\sqrt{\varepsilon\beta^2-1} < R\sin\vartheta < \frac{R\sqrt{\varepsilon\beta^2-1}\cos\vartheta}{\sqrt{1+\beta^2-\varepsilon\beta^2}} + ma\sqrt{\varepsilon\beta^2-1}. \qquad (19.21)$$

In this equation, as well as in Eq. (19.19), in the case of even m, ϑ denotes the angle between the negative z axis and the radius vector from the point at which the particle enters into the plate to the point of observation; in the case of odd m, ϑ denotes the angle between the positive z axis and the radius vector from the point at which the particle passes into the vacuum to the point of observation (Fig. 18); R denotes the absolute value of the corresponding radius vectors.

The nonvanishing components of the electromagnetic field of the m-th diffracted wave in a cylindrical z, ρ, φ coordinate system (φ reckoned counterclockwise when observations from a point with z > 0 are made) can be expressed by the Hertz vector of Eq. (19.19) in the following form:

$$H_{\omega,m\varphi} = \frac{\omega^2}{vc}\sqrt{\varepsilon\beta^2-1}\,\Pi_{\omega,m},$$

$$E_{\omega,mz} = \frac{\omega^2}{v^2}(\varepsilon\beta^2-1)\,\Pi_{\omega,m},$$

$$E_{\omega,m\rho} = (-1)^m\frac{\omega^2}{v^2}\sqrt{\varepsilon\beta^2-1}\sqrt{1+\beta^2-\varepsilon\beta^2}\,\Pi_{\omega,m}. \qquad (19.22)$$

Let us calculate the spectral density of the energy flow of the m-th wave beyond the plate:

$$W_{\omega,m} = c\int\limits_{(m-1)a\sqrt{\varepsilon\beta^2-1}}^{ma\sqrt{\varepsilon\beta^2-1}}\rho\,d\rho\int\limits_{-\infty}^{+\infty}E_{\omega,m\varphi}H_{\omega',m\varphi}e^{i(\omega+\omega')t}\,dt, \qquad (19.23)$$

where we have to set $\vartheta = \pi/2$ in the expressions for $E_{\omega,m\rho}$ and $H_{\omega,m\rho}$. When we use Eqs. (19.19) and (19.22) and integrate in Eq. (19.23) over both the time and ρ, we obtain

$$W_{\omega,m} = \frac{4(-1)^{m-1}e^2a\,(\varepsilon\beta^2-1)\sqrt{1+\beta^2-\varepsilon\beta^2}\,(\varepsilon\sqrt{1+\beta^2-\varepsilon\beta^2}-1)^{2m-2}}{v^2\,(\varepsilon\sqrt{1+\beta^2-\varepsilon\beta^2}+1)^{2m}}\,\omega. \qquad (19.24)$$

The sign of this expression determines the direction of the energy flow. When we sum the energy flows separately for even and odd m values and determine the ratio of the radiation energy in the half-space before the plate to the radiation energy passing through the front wall of the plate, we find Eq. (19.12), as we could expect. The energy of the Cerenkov radiation which passes through the two boundaries is

$$W_\omega = \sum_{m=1}^\infty W_{\omega, m} (-1)^{m-1} = \frac{e^2 a}{c^2}\left(1 - \frac{1}{\varepsilon\beta^2}\right)\omega \qquad (19.25)$$

in agreement with the fact that the entire energy of the Cerenkov radiation which does not undergo total reflection leaves the plate.

Similarly to the cylindrical waves discussed in §6 for a finite particle path in a medium, we can conclude that these cylindrical waves do not come close to the conical surfaces. The angular dimensions of the intermediate region are of the order $\Delta\vartheta_r \sim \sqrt{c/\omega R_1}$ ($\Delta\vartheta_r \ll 1$), where R_1 denotes the distance along the corresponding cone (not taking into account the diffraction at the boundary) to the particle trajectory. We note that for $R_1 \gg a$, the angular dimensions of the region between the cones are:

$$\Delta\vartheta_1 = \frac{a\beta}{R_1}\sin\vartheta_r\cos\vartheta_r \quad (\vartheta_r = \arcsin\frac{1}{\beta}\sqrt{\varepsilon\beta^2 - 1}). \qquad (19.26)$$

When we have $\Delta\vartheta_r > \Delta\vartheta_1$, the cylindrical wave in the intermediate space between the cones is missing. After inserting $\Delta\vartheta_r$ and $\Delta\vartheta_1$, we can write this inequality in the form

$$\sqrt{R_1 c/\omega} > a\beta\sin\vartheta_r\cos\vartheta_r. \qquad (19.27)$$

We recall that $R\sin\vartheta = R_1\sin\vartheta_r$, and therefore the inequality (19.27) can be rewritten in the form

$$R\sin\vartheta > \frac{\omega}{c}a^2\beta^2\sin^3\vartheta_r\cos^2\vartheta_r. \qquad (19.28)$$

It follows that beyond the cylinder we have

$$\rho = \frac{\omega}{c}a^2\beta^2\sin^3\vartheta_r\cos^2\vartheta_r, \qquad (19.29)$$

i.e., the diffracted waves of the Cerenkov radiation are not cylindrical waves. When we compare this result with inequality (19.21) defining the conical regions in space in which only cylindrical waves can exist, we come to the conclusion that the m-th diffracted wave passes into the vacuum, but is no longer a cylindrical wave if

$$m > \frac{\omega}{c}a\beta\sin^2\vartheta_r\cos^2\vartheta_r. \qquad (19.30)$$

Since we may speak of a radiation field in space only when the linear dimensions of the field are much greater than the wavelength, then, if the condition (19.27) is formally fulfilled only for $R_1 < c/\omega$, i.e., if

$$\frac{\omega}{c}a\beta\sin\vartheta_r\cos\vartheta_r < 1, \qquad (19.31)$$

no cylindrical waves develop. It is interesting to compare this result with the diffraction of a plane wave at a slit. When the slit width is smaller than the wavelength, no region of space in which the wave is a plane wave exists behind the slit. In our case, the distance between neighboring conical surfaces plays the role of the slit width. This distance is equal to $a\beta\sin\vartheta_r\cos\vartheta_r$. We see from this comparison with the slit, that the transformation of the cylindrical wave obeys the laws of diffraction.

When the inverse inequality (19.31) is fulfilled, the diffracted waves of the Cerenkov radiation are cylindrical waves near the plate. At very large distances from the point at which the particle leaves the plate and enters into the vacuum, or from the point at which the particle passes into the plate, first diffracted waves of the Cerenkov radiation propagate (remaining cylindrical waves) for which $\sin \vartheta \approx \sin \vartheta_r$. Thus, if we set in Eq. (19.28) $\sin \vartheta = \sin \vartheta_r$, we obtain that at the distances

$$R > \frac{\omega}{c} a^2 \beta^2 \sin^2 \vartheta_r \cos^2 \vartheta_r \qquad (19.32)$$

no cylindrical waves exist. When we recall that the projection of the apparent size of the source (created by diffraction) upon the direction of propagation of the diffracted waves in the vacuum is equal to

$$s = a \beta \sin \vartheta_r \cos \vartheta_r, \qquad (19.33)$$

we see from a comparison with the condition (6.5) that the field of a finite source be converted into a spherical wave, that Eq. (19.32) agrees with condition (6.5), provided that in (6.5) the refractive index is assumed equal to unity for the emission into the vacuum. Consequently, at distances

$$R \gg \frac{\omega}{c} a^2 \beta^2 \sin^2 \vartheta_r \cos^2 \vartheta_r \qquad (19.34)$$

the Cerenkov radiation transforms into a spherical wave.

We assume that the plate is completely transparent. Let us consider the diffracted cylindrical waves in the case of weak absorption: $\varepsilon = \varepsilon' + i\varepsilon''$ ($\varepsilon'' \ll \varepsilon'$). In the exponent of the exponential function we must take into account that

$$\sqrt{\varepsilon\beta^2 - 1} \approx \sqrt{\varepsilon'\beta^2 - 1} + \frac{i\varepsilon''\beta^2}{2\sqrt{\varepsilon'\beta^2 - 1}},$$

$$\sqrt{1 + \beta^2 - \varepsilon\beta^2} \approx \sqrt{1 + \beta^2 - \varepsilon'\beta^2} - \frac{i\varepsilon''\beta^2}{2\sqrt{1 + \beta^2 - \varepsilon'\beta^2}}. \qquad (19.35)$$

In this case the exponent of the exponential function assumes the form

$$i \frac{\omega}{c} R \left[\cos(\vartheta - \vartheta_r) + i\varepsilon'' \frac{\sin(\vartheta - \vartheta_r)}{\sin 2\vartheta_r} + (m-1) \frac{\varepsilon' a\beta}{R} + i(m-1) \frac{\varepsilon'' a\beta}{R} \right]$$

$$\left(\vartheta_r = \arcsin \frac{\sqrt{\varepsilon'\beta^2 - 1}}{\beta} \right). \qquad (19.36)$$

If $\varepsilon'' a\beta\omega \gg 1$ (nontransparent plate), only the first diffracted wave (m = 1) passes into the vacuum. But, if $\varepsilon'' a\beta\omega \ll 1$ (transparent plate), diffracted waves, for which the condition

$$m\varepsilon'' a\beta\omega < 1 \qquad (19.37)$$

is satisfied, run into the vacuum.

The wave attenuation is determined by the factor

$$\exp\left[-\frac{\omega}{c} \frac{\rho\varepsilon'' \sin(\vartheta - \vartheta_r)}{\sin\vartheta \sin 2\vartheta_r} - (m-1)\frac{\omega}{c} \varepsilon'' a\beta \right]. \qquad (19.38)$$

Equation (19.38) helps us to determine the spectral energy density of the radiation provided by the m-th wave in the diffracted Cerenkov radiation in the attenuated case:

$$W_{\omega, m} = \frac{4e^2c^3}{v^3\varepsilon'} \frac{(\varepsilon'\beta^2 - 1)\sqrt{1 + \beta^2 - \varepsilon'\beta^2}(\varepsilon'\sqrt{1 + \beta^2 - \varepsilon'\beta^2} - 1)^{2m-2}}{(\varepsilon'\sqrt{1 + \beta^2 - \varepsilon'\beta^2} - 1)^{2m}} \times$$

$$\times \left[1 - \exp\left(-\frac{\omega}{c} \varepsilon'' a\beta \right) \right] \exp\left[-3(m-1)\frac{\omega}{c} \varepsilon'' a\beta \right]. \qquad (19.39)$$

A consequence of this equation is that in the case of a nontransparent plate $(\varepsilon^n a\beta\omega \gg 1)$ the energy of the Cerenkov radiation passing into the vacuum is determined by the first diffracted cylindrical wave:

$$W_\omega = \frac{4e^2c^2}{v^3\varepsilon''}\frac{(\varepsilon'\beta^2-1)\sqrt{1+\beta^2-\varepsilon'\beta^2}}{(\varepsilon'\sqrt{1+\beta^2-\varepsilon'\beta^2}+1)^2}.$$

(19.40)

§20. Energy Losses of a Charged Relativistic Particle in a Thin Dielectric Plate

It is well known that the field of a moving charged particle is shielded at large distances from the particle, due to the polarization of the medium. This helps to explain the so-called density effect, i.e., the ionization losses tend to some constant value with increasing particle energy. The field is not shielded when the material layer is thin enough. It is, therefore, logical to assume that the density effect disappears in the energy losses of a charged relativistic particle which passes through a thin dielectric plate, i.e., to conclude that the energy losses increase logarithmically with the particle energy. This prediction was made in [77] and proved to be correct. However, several related questions required additional considerations, which were made in a publication by the author [13]. This paragraph outlines the work of the author [13].

We start from the field of a particle passing through a plate (thickness a) situated in vacuum [11]:

$$\Pi_k = \frac{ie v}{\pi v}\int_{-\infty}^{+\infty}\frac{e^{-i\omega t}}{\omega}\,d\omega\int_0^{\varkappa_0}\varkappa J_0(\varkappa r)\left\{\frac{e^{i\frac{\omega}{v}z}}{\varkappa^2+\frac{\omega^2}{v^2}(1-\beta^2\varepsilon_k)}+A_k e^{z\beta_k}+B_k e^{-z\beta_k}\right\}d\varkappa,$$

(20.1)

where

$$A_1 = \eta\{(\beta_2-\varepsilon\beta_1)(\beta_2\gamma_2+\varepsilon\gamma_1)e^{-a\beta_2}-(\beta_2+\varepsilon\beta_1)(\beta_2\gamma_2-\varepsilon\gamma_1)e^{a\beta_2}+2\varepsilon\beta_2(\beta_1\gamma_2-\gamma_1)e^{i\frac{\omega}{v}a}\}e^{-i\frac{\omega}{v}a},$$

$$B_3 = \eta\{(\beta_2-\varepsilon\beta_1)(\beta_2\gamma_2-\varepsilon\gamma_1)e^{-a\beta_2}-(\beta_2+\varepsilon\beta_1)(\beta_2\gamma_2+\varepsilon\gamma_1)e^{a\beta_2}+2\varepsilon\beta_2(\beta_1\gamma_2+\gamma_1)e^{-i\frac{\omega}{v}a}\}e^{a\beta_1},$$

$$A_2 = \eta\varepsilon\{(\beta_2-\varepsilon\beta_1)(\beta_1\gamma_2+\gamma_1)e^{-i\frac{\omega}{v}a}+(\beta_2+\varepsilon\beta_1)(\beta_1\gamma_2-\gamma_1)e^{a\beta_1}\}e^{-a\beta_2},$$

$$B_2 = \eta\varepsilon\{(\beta_2-\varepsilon\beta_1)(\beta_1\gamma_2-\gamma_1)e^{-a\beta_2}+(\beta_2+\varepsilon\beta_1)(\beta_1\gamma_2+\gamma_1)e^{-i\frac{\omega}{v}a}\}e^{a\beta_1},$$

$$\eta = (\varepsilon-1)[(\beta_2-\varepsilon\beta_1)^2 e^{-a\beta_2}-(\beta_2+\varepsilon\beta_1)^2 e^{a\beta_2}]^{-1}\left[\varkappa^2+\frac{\omega^2}{v^2}(1-\beta^2\varepsilon)\right]^{-1}\left[\varkappa^2+\frac{\omega^2}{v^2}(1-\beta^2)\right]^{-1}e^{a\beta_1},$$

$$B_1 = 0, \quad A_3 = 0, \quad \beta = v/c,$$

$$\beta_1 = \sqrt{\varkappa^2-\frac{\omega^2}{c^2}}, \quad \beta_2 = \sqrt{\varkappa^2-\varepsilon\frac{\omega^2}{c^2}},$$

$$\gamma_1 = \frac{i\omega}{\varepsilon v}\left[\varkappa^2+\frac{\omega^2}{v^2}(1-\beta^2-\beta^2\varepsilon)\right],$$

$$\gamma_2 = \frac{\omega^2}{c^2}\quad(\operatorname{Im}\varepsilon>0,\quad \operatorname{Im}\beta_{1,2}<0\quad\text{for}\quad\omega>0),$$

$$E_{k\omega} = \frac{1}{\varepsilon_k}\operatorname{grad}\operatorname{div}\Pi_{k\omega}+\frac{\omega^2}{c^2}\Pi_{k\omega},\quad H_{k\omega} = -i\frac{\omega}{c}\operatorname{rot}\Pi_{k\omega}\quad(k=1,2,3),$$

$$k=1, \quad \varepsilon_1=1, \quad z<-a,$$

$$k=2, \quad \varepsilon_2=\varepsilon, \quad -a<z<0, \quad k=3, \quad \varepsilon_3=1, \quad z>0.$$

Naturally, this solution can also be obtained with the image method, but we will not concern ourselves with this approach.

We see from Eq. (20.1) that the electrical field at the point of the particle ($r = 0$, $t = z/v$) is given by

$$E_1 = \frac{ie}{\pi} \int_{-\infty}^{+\infty} \frac{e^{-i\frac{\omega}{v}z}}{\omega} d\omega \int_0^{\varkappa_0} \varkappa^3 A_1 e^{z\beta_1} d\varkappa \quad (z < -a),$$ (20.2)

$$E_2 = \frac{ie}{\pi} \int_{-\infty}^{+\infty} \frac{e^{-i\frac{\omega}{v}z}}{\omega} d\omega \int_0^{\varkappa_0} \frac{\varkappa}{\varepsilon} \left\{ \frac{\left(\frac{\omega^2}{c^2}\varepsilon - \frac{\omega^2}{v^2}\right)e^{i\frac{\omega}{v}z}}{\varkappa^2 + \frac{\omega^2}{v^2}(1-\beta^2\varepsilon)} + \varkappa^2 A_2 e^{z\beta_2} + \varkappa^2 B_2 e^{-z\beta_2} \right\} d\varkappa \quad (-a < z < 0),$$ (20.3)

$$E_3 = \frac{ie}{\pi} \int_{-\infty}^{+\infty} \frac{e^{-i\frac{\omega}{v}z}}{\omega} d\omega \int_0^{\varkappa_0} \varkappa^3 B_3 e^{-z\beta_1} d\varkappa \quad (z > 0),$$ (20.4)

where the integration limit $\varkappa_0$ is obtained from macroscopical considerations: $\omega_0/v \ll \varkappa_0 \ll 1/d$, with ω_0 denoting some average frequency corresponding to the motion of the majority of electrons in an atom, and d denoting the distance between atoms, or (in a condensed medium) the size of the atoms.

We obtain from Eqs. (20.2)-(20.4) the portion of the work done by the field, a portion which increases linearly with the thickness of the plate (starting from thickness zero). This part of the work is equal to the sum of the following expressions, which are obtained by integrating along the path to the plate (W_1), inside the plate (W_2), and beyond the plate (W_3):

$$W_1 = I_1 + I_2,$$ (20.5)

$$W_2 = -\frac{ie^2 a}{\pi v^2}(1-\beta^2) \int_{-\infty}^{\infty} \frac{\omega}{\varepsilon} d\omega \int_0^{\varkappa_0} \left(\varkappa^2 + \frac{\omega^2}{v^2} - \frac{\omega^2}{c^2}\right)^{-1} \varkappa \, d\varkappa,$$ (20.6)

$$W_3 = I_1 - I_2,$$ (20.7)

where

$$I_1 = \frac{ie^2 a}{2\pi v^2} \int_{-\infty}^{+\infty} \frac{(1-\varepsilon)}{\varepsilon}(1-\beta^2-\varepsilon)\omega \, d\omega \int_0^{\varkappa_0} \left(\varkappa^2 + \frac{\omega^2}{v^2} - \frac{\omega^2}{c^2}\right)^{-2} \varkappa^3 \, d\varkappa,$$ (20.8)

$$I_2 = -\frac{e^2 a}{2\pi v} \int_{-\infty}^{+\infty} \frac{(1-\varepsilon)}{\varepsilon} d\omega \int_0^{\varkappa_0} \left(\frac{\omega^2}{v^2} - \frac{\omega^2}{c^2} + \varkappa^2\varepsilon - \frac{\omega^2}{c^2}\varepsilon\right)\left(\varkappa^2 - \frac{\omega^2}{c^2}\right)^{-1/2}\left(\varkappa^2 + \frac{\omega^2}{v^2} - \frac{\omega^2}{c^2}\right)^{-2} \varkappa^3 \, d\varkappa.$$ (20.9)

Integration in Eq. (20.6) leads to the result

$$W_2 = -\frac{4\pi Ne^4 a}{mv^2}(1-\beta^2)\ln\frac{\varkappa_0 v}{\bar{\omega}\sqrt{1-\beta^2}},$$ (20.10)

where $\bar{\omega}$ denotes some average frequency of the motion of the electrons in the atom; this frequency is given by the expression

$$\ln\bar{\omega} = \frac{\displaystyle\int_0^\infty \frac{\omega\varepsilon''(\omega)}{|\varepsilon(\omega)|^2}\ln\omega \, d\omega}{\displaystyle\int_0^\infty \frac{\omega\varepsilon''(\omega)}{|\varepsilon(\omega)|^2} d\omega} = \frac{m}{2\pi^2 Ne^2}\int_0^\infty \frac{\omega\varepsilon''(\omega)}{|\varepsilon(\omega)|^2}\ln\omega \, d\omega,$$ (20.11)

where

$$\varepsilon''(\omega) = \operatorname{Im} \varepsilon(\omega).$$

We note that the integral I_2 diverges logarithmically in the frequency. The particle-charge renormalization results in the appearance of this integral in W_1 and W_3. As one could expect from the symmetry of the problem, the divergent integral I_2 cancels out in the sum $W_1 + W_3$, which describes the work done by the braking force beyond the plate. With the well-known equality

$$\int_0^\infty \frac{\omega \varepsilon''(\omega)}{|\varepsilon(\omega)|^2}\, d\omega = \int_0^\infty \omega \varepsilon''(\omega)\, d\omega = \frac{2\pi^2 Ne^2}{m}, \tag{20.12}$$

we obtain

$$W_1 + W_3 = \frac{4\pi Ne^4 a}{mv^2}\left[(1-\beta^2)\ln\frac{\varkappa_0 v}{\bar\omega \sqrt{1-\beta^2}} - \ln\frac{\varkappa_0 v}{\bar\omega' \sqrt{1-\beta^2}} + \frac{\beta^2}{2}\right], \tag{20.13}$$

where

$$\ln\bar\omega' = \frac{\displaystyle\int_0^\infty \omega\varepsilon''(\omega)\ln\omega\, d\omega}{\displaystyle\int_0^\infty \omega\varepsilon''(\omega)\, d\omega} = \frac{m}{2\pi^2 Ne^2}\int_0^\infty \omega\varepsilon''(\omega)\ln\omega\, d\omega. \tag{20.14}$$

The work which is done by the braking force along the entire path of particle motion is given by the expression

$$W = \frac{ie^2 a}{\pi v^2}\int_{-\infty}^{+\infty}\frac{\omega}{\varepsilon}\, d\omega \int_0^{\varkappa_0}\frac{\varepsilon\varkappa^2(\varepsilon+\beta^2-2) - \frac{\omega^2}{v^2}(1-\beta^2)^2}{\left(\varkappa^2 + \frac{\omega^2}{v^2} - \frac{\omega^2}{c^2}\right)^2}\, \varkappa\, d\varkappa. \tag{20.15}$$

When we integrate over $\varkappa$ and use the notation defined by Eq. (20.14) we obtain

$$W = -\frac{4\pi Ne^4 a}{mv^2}\left(\ln\frac{\varkappa_0 v}{\bar\omega' \sqrt{1-\beta^2}} - \frac{\beta^2}{2}\right); \tag{20.16}$$

in this expression, restrictions for the particle velocity result from the macroscopic approach.*

In the case of a relativistic particle, the work done by the braking force is equal to two integrals I_1. One of these integrals appears in the expression for the work done by the field along the path to the plate; the other results from the work done by the field along the path beyond the plate [13]:

$$W \simeq -\frac{4\pi Ne^4 a}{mc^2}\ln\frac{\varkappa_0 c}{\bar\omega' \sqrt{1-\beta^2}}. \tag{20.17}$$

---

*Garibyan and Lorikyan [67] came to the conclusion that it is necessary to review the energy-loss problem of a fast particle in a thin dielectric plate. Contrary to [77], in [67] it was believed necessary to include the work done by the braking force along the path to the plate, too. In the relativistic case, the authors of [67] obtained a result which agrees with the previously obtained result of the author of this article [13], i.e., they obtained a result coinciding with Eq. (20.16). However, in the case of particles moving with velocities not close to the velocity of light, the result of [67] differs from Eq. (20.16) in the definition of the average of frequency because in [67] an error was made in the integral over $\varkappa$.

It follows from the definition of the frequency $\bar{\omega}'$ in Eq. (20.14) that the particle loses energy in the absorption bands. It is now easy to understand quantitatively that the particle is slowed down along its path beyond the plate. As a matter of fact, due to the discontinuity of the normal component of the electric induction vector, the braking force (which is proportional to the normal component of the electrical field) is (in the absorption bands) stronger beyond the plate than inside the plate. The author is indebted to V. L. Ginzburg for this simple interpretation.

When narrow absorption bands exist, Eq. (20.17) can be written in the form of a sum over all resonance frequencies

$$W = \sum_s W_s \approx - \frac{4\pi e^4 a}{mc^2} \sum_s N_s \ln \frac{\varkappa_0 c}{\omega_s \sqrt{1-\beta^2}}, \qquad (20.18)$$

where N_s denotes a quantity proportional to the strength of the oscillators with resonance frequency ω_s $\left(\sum_s N_s = N\right)$.

The result expresses the energy absorbed in the plate (see below). Therefore, in order to clarify the applicability range of the result, one must compare the calculated energy losses with the energy-absorption–dependent part of the quadratic term of the expansion (over the thickness of the plate) for the work done by the braking force. The emission energy must be separated from the quadratic term.

The spectral energy density of the emission of a relativistic charged particle passing through a thin plate was calculated in [9]. The spectral energy density is proportional to the square of the thickness of the plate:

$$W'_\omega \approx \frac{e^2 \omega^3 a^2}{\pi c^3} |\varepsilon - 1|^2 \ln \frac{1}{\sqrt{1-\beta^2}}. \qquad (20.19)$$

The quadratic term of the expansion for the work done by the braking force does not depend upon the widths of the absorption bands [77], whereas the integral over the frequencies [Eq. (20.19)] increases with decreasing width of the absorption band. When narrow absorption bands exist, this integral is greater than the absolute value of the quadratic term, and the portion of the quadratic term which results from absorption losses becomes comparable to the energy losses by radiation. In this case, Eq. (20.18) describes the losses by absorption, provided that the energy of the emission can be neglected.

When we integrate Eq. (20.19) over the frequency, we obtain in the case of narrow absorption bands:

$$W' \approx \frac{2\pi e^4 a^2}{mc^3} \sum_s N_s \omega_s \varepsilon_s'' \ln \frac{1}{\sqrt{1-\beta^2}}, \qquad \varepsilon_s'' \simeq \frac{4\pi N_s e^2}{m \omega_s \Delta \omega_s}, \qquad (20.20)$$

where ε_s'' denotes the ε'' value at the resonance frequency ω_s. The energy of the emission is small relative to the energy of the losses defined in Eq. (20.18), provided that

$$a \ll c \sum_s N_s \ln \frac{\varkappa_0 c}{\omega_s \sqrt{1-\beta^2}} \Big/ \sum_s N_s \omega_s \varepsilon_s'' \ln \frac{1}{\sqrt{1-\beta^2}}. \qquad (20.21)$$

Since the ratio of the logarithms is of the order of unity, we obtain for an individual absorption band

$$a \ll c/\omega_s \varepsilon_s''. \tag{20.22}$$

These inequalities define the applicability range of the result stated in Eq. (20.18).

B. M. Bolotovskii brought to the author's attention that usually photons behave much like the pseudo-photons of the field which is generated by relativistic particles, and that the condition for weak absorption of a plane electromagnetic wave passing through a plate has the form of Eq. (20.22). In other words, condition (20.22) can be interpreted as the condition for weak absorption of the pseudo-photons carried along by the particle field.

The energy emitted by a particle passing through a thin plate is proportional to the square of the thickness of the plate. More particularly, this means that the energy defined by Eq. (20.17) (which is proportional to the first power of the plate thickness) is absorbed in the plate. When we consider the distribution of the absorbed energy over the various sections of the plate, we must take into account that the particle is slowed down along its path in the vacuum and therefore, draws energy from the outside into the plate, through the boundary. Thus, we have to calculate the energy flow through the boundary. The rather laborious calculations do not pose any major difficulties. As a result, we obtain the energy which flows in through the plate surfaces (through both boundaries); the surfaces have the form of concentric circles with radii r and $r + dr$ and centers at the points at which the particle passes through the boundaries. This energy is given by

$$dW \approx \frac{2e^2 a}{\pi c^2} \frac{dr}{r} \int_0^\infty \omega \varepsilon''(\omega) \, \alpha^2 K_1^2(\alpha) \, d\omega, \quad \alpha = \frac{r\omega}{c} \sqrt{1 - \beta^2}, \tag{20.23}$$

where $K_1(\alpha)$ denotes the Macdonald function. When we calculated the entire absorbed energy from Eq. (20.23), we obtain with an adequate degree of accuracy

$$W \approx \frac{2e^2 a}{\pi c^2} \int_0^\infty \omega \varepsilon''(\omega) \, d\omega \int_{r_1}^{r_2} \frac{dr}{r}, \quad r_1 \sim \frac{1}{\varkappa_0}, \quad r_2 = \frac{c}{\omega \sqrt{1 - \beta^2}}. \tag{20.24}$$

It is easy to see that the result obtained after integrating over r is identical with the result stated in Eq. (20.17), as we could expect.

According to Eq. (20.24), the energy absorbed at various distances from the particle trajectory is proportional to the number of pseudo-photons of the electromagnetic field carried along by the particle. In the s-th resonance band, energy is absorbed in the region $r < c/\omega_s \cdot \sqrt{1 - \beta^2}$, and the energy per unit surface is inversely proportional to the square of the distance from the particle trajectory, but independent of the particle energy. The logarithmic increase in the losses [according to Eq. (20.17)] with the energy results from an increase in the size of the absorbing regions in space. This means that the logarithmic increase is limited by the dimensions of the plate. Thus, in the case of a disk of radius r_0, the losses at the s-th resonance frequency reach their limit for $E \sim r_0 \omega_s mc$ and do not increase further on when the particle energy increases.

CHAPTER 4

Radiation of a Particle Moving in a Layered Medium

The radiation from a charged particle moving in media with periodic discontinuities in space has been discussed in several articles. The exact solution to the problem was given by Fainberg and Khizhnyak [68] for an infinite, layered medium. They discussed in detail the case of inhomogeneities which are small compared to the wavelength of the quanta emitted. The same case, yet with a continuously changing periodical inhomogeneity structure, was considered by Bliokh [69]. The Nobel Lecture of Frank [70] emphasized that in the case of relativistic particles, the intensity of the transition radiation reaches its greatest possible value when the optical properties of the medium along the particle's path change only slightly. In the case of greatest importance for practice, the particle path along which the radiation is generated is much greater than the wavelength. Of particular interest were studies concerning inhomogeneities of large dimensions. Ter-Mikaelyan [71] and Ter-Mikaelyan and Gazazyan [72] discussed the radiation in media with periodically changing inhomogeneities for the case in which the wavelength of radiation is smaller than the period of the structure. In some particular cases, similar results were confirmed in the work of Garibyan and Gol'dman [73] and Amatuni and Korkhmazyan [74]. In [71]-[74], the interest focused on the radiation of waves with frequencies greater than the optical frequencies; for such waves, the dielectric constant of the various parts of a periodically inhomogeneous medium varies slightly and is close to unity. Barsukov and Bolotovskii [75, 76] considered the energy losses of a particle in the two cases of a periodically inhomogeneous stationary and nonstationary medium.

We will determine the radiation field of a particle moving in a limited layered medium whose structure is given by some particular law. We will determine the radiation of a charged particle in the optical frequency range; the radiation is generated in a stack of plates or during the passage of a charged particle through a stack of transparent plates.

§ 21. Radiation from a Charged Particle Moving in Some Fashion in a Layered Medium

In this section, we will determine the radiation field of an arbitrarily moving charged particle in a layered medium and express the radiation field by the law of particle motion.

Let us assume that there exist m plane layers of a substance (with a finite thickness of the first and the last layer) with varying optical properties, i.e., a layered medium with $(m-1)$ separating boundaries. We assume that the z axis is parallel to the normal to the boundaries and that the coordinate origin is situated on the last, $(m-1)$-th boundary. The media will be denoted by the natural numbers with increasing order from left to right, i.e., in the direction of the positive z axis. The thickness of the n-th layer is denoted by a_n (Fig. 19).

Let us determine in the region m the radiation field generated by the particle along its path in the n-th layer.

The Hertz vectors of the spherical waves having various polarizations of the primary radiation in the forward direction (in the direction of the positive z axis) have the form

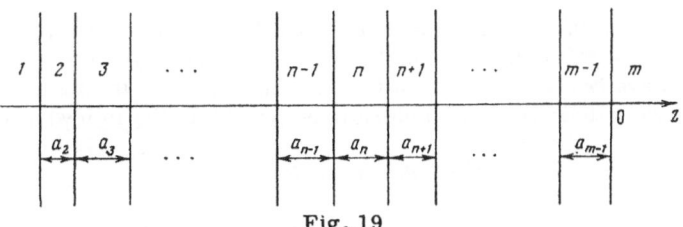

Fig. 19

$$\Pi'_{n,\,\|} = \frac{ieck}{2\pi\omega R}\int p'_{n,\,m,\,\|}\left[\beta_z - (\beta_x\cos\vartheta_{mx} + \beta_y\cos\vartheta_{my})\frac{\cos\vartheta_{mz}}{\sin^2\vartheta_{mz}}\right]$$

$$\times \exp\left(i\omega t - i\mathbf{k}_n\mathbf{r} - ia_nk_{nz} + i\sum_{j=n+1}^{m-1}a_jk_{jz}\right)dt, \tag{21.1}$$

$$\Pi'_{n,\,\perp} = \frac{iec}{2\pi\omega R}\,p'_{n,\,m;\,\perp}\int\left[i\beta_x + j\beta_y + \mathbf{k}(\beta_x\cos\vartheta_{mx} + \right.$$

$$\left. + \beta_y\cos\vartheta_{my})\frac{\cos\vartheta_{mz}}{\sin^2\vartheta_{mz}}\right]\exp\left(i\omega t - i\mathbf{k}_n\mathbf{r} - ia_nk_{nz} + i\sum_{j=n+1}^{m-1}a_jk_{jz}\right)dt. \tag{21.2}$$

The backward radiation, which is reflected from the boundaries separating media 1, 2, 3,..., n − 1 and passes through the media n + 1, n + 2, n + 3,..., m − 1, gives the following contribution in the region m:

$$\Pi''_{n,\,\|} = \frac{ieck}{2\pi\omega R}\int r_{n,\,1;\,\|}\,p'_{m,\,m;\,\|}\left[\beta_z - (\beta_x\cos\vartheta_{mx} + \beta_y\cos\vartheta_{my})\frac{\cos\vartheta_{mz}}{\sin^2\vartheta_{mz}}\right]\times$$

$$\times\exp\left(i\omega t + ik_{nz}z - ik_{nx}x - ik_{ny}y + 2ia_nk_{nz} + i\sum_{j=n+1}^{m-1}a_jk_{jz}\right)dt, \tag{21.3}$$

$$\Pi''_{n,\,\perp} = \frac{iec}{2\pi\omega R}\,r_{n,\,1;\,\perp}p_{n,\,m;\,\perp}\int\left[i\beta_x + j\beta_y + \mathbf{k}(\beta_x\cos\vartheta_{mx} + \beta_y\cos\vartheta_{my})\frac{\cos\vartheta_{mz}}{\sin^2\vartheta_{mz}}\right]\times$$

$$\times\exp\left(i\omega t - ik_{nx}x - ik_{ny}y + ik_{nz}z + 2ia_nk_{nz} + i\sum_{j=n+1}^{m-1}a_jk_{jz}\right)dt, \tag{21.4}$$

where $r_{n,1}$ denotes the reflection coefficient for the media 1, 2, 3,..., n − 1; $p'_{n,m}$ is the transmission coefficient describing the wave transmission from medium n into medium m. The problem is to determine these coefficients.

We determined the coefficients r_{jk} and p_{jk} for contacting media in §2 (one separating boundary). We see from a comparison of Eqs. (21.1)-(21.4) for m = 3 with the results stated in Eqs. (16.1)-(16.12) (which correspond to m = 3) that for example, the coefficients $p'_{1,3}$ and $r_{3,1}$ (describing the wave transmission through the layer and the wave reflection from the layer, respectively) can be expressed by the coefficients for contacting media:

$$p'_{1,3} = p_{1,2}\alpha_{1,\,2,\,3}p_{2,3}, \quad r_{3,1} = r_{3,2} + p_{3,2}r_{2,1}\alpha_{1,\,2,\,3}p_{2,3}\exp(2ia_2k_{2,\,z}). \tag{21.5}$$

In the case of four media, the wave generated in region 1 passes into the third region (which is taken into account by the coefficient $p'_{1,3}$), undergoes multiple reflection (the corresponding coefficient $\alpha_{1,3,4}$ depends on the optical properties of all transgressed layers 1, 2, 3, and 4), and leaves into medium 4 (taken into account by the coefficient $p_{3,4}$). Accordingly, the coefficient describing the wave transmission from the first medium into the fourth medium is $p'_{1,4} = p'_{1,3}\alpha_{1,3,4}p_{3,5}$. When we extend these considerations, we obtain the following recurrence formula

$$p'_{1,j} = p'_{1,j-1}\alpha_{1,j-1,j}p_{j-1,j},$$

(21.6)

where $\alpha_{1,j-1,j}$ denotes the coefficient of multiple reflection in the layer j. This coefficient accounts for the reflection at all separating boundaries preceding this layer and, by definition, this coefficient is given by

$$\alpha_{1,j-1,j} = [1 - r_{j-1,1}r_{j-1,j}\exp{(2ia_{j-1}k_{j-1,z})}]^{-1},$$

(21.7)

where $r_{j-1,j}$ denotes the reflection coefficient of the boundary separating media $j - 1$ and j, and $r_{j-1,1}$ is the reflection coefficient referring to the portion of the layered media situated at the left from the $(j - 1)$-th layer under consideration (to the left from the boundary separating media 1, 2, 3,..., $j - 2$). Equation (21.6) can be rewritten in the form

$$p'_{1,n} = p'_{1,k}\alpha_{1,k,k+1}p_{k,k+1}\alpha_{1,k+1,k+2}p_{k+1,k+2}\cdots\alpha_{1,n-1,n}p_{n-1,n} \quad (k < n).$$

(21.8)

Since the wave which passes from the first layer into the n-th layer runs successively through all intermediate layers, the transmission coefficient of the layered medium is equal to the product of the transmission coefficients of the individual sections of the layered medium. This means that the factor at the transmission coefficient $p'_{1,k}$ referring to the transmission of the waves from the first into the k-th layer at the right side of Eq. (21.8) is identical to the coefficient for wave transmission from the k-th to the n-th layer. We obtain

$$p'_{k,n} = \alpha_{1,k,k+1}p_{k,k+1}\alpha_{1,k+1,k+2}p_{k+1,k+2}\cdots\alpha_{1,n-1,n}p_{n-1,n} \quad (k < n).$$

(21.9)

In order to obtain the transmission coefficient for a wave moving in the opposite direction (for k > n) we must relabel all the media in reversed order, i.e.,

$$p'_{k,n} = \alpha_{m,k,k-1}p_{k,k-1}\alpha_{m,k-1,k-2}p_{k-1,k-2}\cdots\alpha_{m,n+1,n}p_{n+1,n} \quad (k > n),$$

(21.10)

where

$$\alpha_{m,j,j-1} = [1 - r_{j,m}r_{j,j-1}\exp{(2ia_jk_{j,z})}]^{-1}.$$

(21.11)

The coefficients $r_{n,k}$ can be found similarly. According to the second equation of Eqs. (21.5), when three media 2, 3, and 4 are present (two separating boundaries) the reflection coefficient for the wave propagating in the fourth medium is

$$r_{4,2} = r_{4,3} + p_{4,3}r_{3,2}\alpha_{2,3,4}p_{3,4}\exp{(2ia_2k_{2,z})}.$$

(21.12)

We note that the coefficient of reflection from two separating boundaries is equal to the sum of the reflection coefficients for each of the two boundaries. In the case of four media, 1, 2, 3, and 4, another term must be added which takes into account the reflection from the third (along the path of the ray) separating boundary (separating medium 1 from medium 2). The reflection from the third boundary is obtained as follows: the wave passes into medium 2 $(p'_{4,2})$, is reflected from the boundary separating medium 1 from medium 2 $(r_{2,1})$, is multiply reflected $(\alpha_{1,2,4})$, and passes into layer 4 in the opposite direction. Accordingly, the reflection coefficient resulting from the third separating boundary is $p'_{4,2}r_{2,1}\alpha_{1,2,4}p'_{2,4}$. When we introduce the time lag of the wave relative to the wave reflected from the first boundary separating medium 3 from medium 4, we obtain

$$r_{4,1} = r_{4,2} + p'_{4,2}r_{2,1}\alpha_{1,2,4}p'_{2,4}\exp{(2ia_2k_{2,z} + 2ia_3k_{3,z})}.$$

(21.13)

Finally, we obtain the following recurrence formulas:

$$r_{n,\,k} = r_{\tilde{n},\,k+1} + p'_{n,\,k+1} r_{k+1,\,k} \alpha_{n,\,k+1,\,l} p'_{k+1,\,n} \exp\left(2i \sum_{j=k+1}^{n-1} a_j k_{j,\,z}\right) \quad (k < n), \tag{21.14}$$

$$r_{n,\,k} = r_{n,\,k-1} + p'_{n,\,l-1} r_{k-1,\,k} \alpha_{n,\,k-1,\,k} p'_{l-1,\,n} \exp\left(2i \sum_{j=n+1}^{k-1} a_j k_{j,\,z}\right) \quad (k > n). \tag{21.15}$$

We obtain from the above equations

$$r_{n,\,k} = r_{n,\,n-1} + p_{n,\,n-1} r_{n-1,\,n-2} \alpha_{n,\,n-1,\,n-2} p_{n-1,\,n} \exp\left(2i a_{n-1} k_{n-1,\,z}\right) +$$
$$+ p'_{n,\,n-2} r_{n-2,\,n-3} \alpha_{n,\,n-2,\,n-3} p'_{n-2,\,n} \exp\left(2i a_{n-1} k_{n-1,\,z} + 2i a_{n-2} k_{n-2,\,z}\right) + \dots$$
$$\dots + p'_{n,\,k+1} r_{k+1,\,k} \alpha_{n,\,k+1,\,k} p'_{k+1,\,n} \exp\left(2i \sum_{j=k+1}^{n-1} a_j k_{j,\,z}\right) \quad (k < n), \tag{21.16}$$

$$r_{n,\,k} = r_{n,\,n+1} + p_{n,\,n+1} r_{n+1,\,n+2} \alpha_{n,\,n+1,\,n+2} p_{n+1,\,n} \exp\left(2i a_{n+1} k_{n+1,\,z}\right) +$$
$$+ p'_{n,\,n+2} r_{n+2,\,n+3} \alpha_{n,\,n+2,\,n+3} p'_{n+2,\,n} \exp\left(2i a_{n+1} k_{n+1,\,z} + 2i a_{n+2} k_{n+2,\,z}\right) + \dots$$
$$\dots + p'_{n,\,k-1} r_{k-1,\,k} \alpha_{n,\,k-1,\,k} p'_{k-1,\,n} \exp\left(2i \sum_{j=n+1}^{k-1} a_j k_{j,\,z}\right) \quad (k > n). \tag{21.17}$$

Equations (21.7), (21.9)–(21.11), (21.16), and (21.17) fully determine the solution to Eqs. (21.1)–(21.4).

§ 22. Radiation Generated by a Charged Particle in a Stack of Plates

Let us use the results of the preceding section to calculate the radiation field generated by a charged particle moving through a stack of plates. We assume that the plates are separated by vacuum gaps and that the plates have identical optical properties. Thus, we have $\varepsilon_k = \varepsilon$ and $\mu_k = \mu$ for even k, and $\varepsilon_k = 1$, $\mu_k = 1$ for odd k.

It follows from Eqs. (2.7)–(2.10) that for a boundary separating vacuum (subscript 1) from a medium (subscript 2) the reflection coefficient equals the transmission coefficient (in the following discussion we omit the subscripts at the angles of the direction cosines of the wave vector referring to the vacuum region). We obtain

$$r_{1,2;\,\perp} = \frac{\mu \cos \vartheta_z - \sqrt{\varepsilon\mu - \sin^2 \vartheta_z}}{\mu \cos \vartheta_z + \sqrt{\varepsilon\mu - \sin^2 \vartheta_z}}, \qquad r_{2,1;\,\perp} = -\frac{\mu \cos \vartheta_z - \sqrt{\varepsilon\mu - \sin^2 \vartheta_z}}{\mu \cos \vartheta_z + \sqrt{\varepsilon\mu - \sin^2 \vartheta_z}}, \tag{22.1}$$

$$p_{2,1;\,\perp} = \frac{2\mu \cos \vartheta_z}{\mu \cos \vartheta_z + \sqrt{\varepsilon\mu - \sin^2 \vartheta_z}}, \qquad p_{1,2;\,\perp} = \frac{2\sqrt{\varepsilon\mu - \sin^2 \vartheta_z}}{\mu \cos \vartheta_z + \sqrt{\varepsilon\mu - \sin \vartheta_z}}, \tag{22.2}$$

$$r_{1,2;\,\parallel} = \frac{\varepsilon \cos \vartheta_z - \sqrt{\varepsilon\mu - \sin^2 \vartheta_z}}{\varepsilon \cos \vartheta_z + \sqrt{\varepsilon\mu - \sin^2 \vartheta_z}} \frac{\beta_z \sin^2 \vartheta_z - (\beta_x \cos \vartheta + \beta_y \cos \vartheta_y) \cos \vartheta_z}{\beta_z \sin^2 \vartheta_z + (\beta_- \cos \vartheta_y + \beta_y \cos \vartheta_y) \cos \vartheta_z},$$

$$\tag{22.3}$$

$$r_{2,1;\,\parallel} = -\frac{\varepsilon \cos \vartheta_z - \sqrt{\varepsilon\mu - \sin^2 \vartheta_z}}{\varepsilon \cos \vartheta_z + \sqrt{\varepsilon\mu - \sin^2 \vartheta_z}} \frac{\beta_z \sin^2 \vartheta_z + (\beta_x \cos \vartheta_x + \beta_y \cos \vartheta_y) \sqrt{\varepsilon\mu - \sin^2 \vartheta_z}}{\beta_z \sin^2 \vartheta_z - (\beta_x \cos \vartheta_x + \beta_y \cos \vartheta_y) \sqrt{\varepsilon\mu - \sin^2 \vartheta_z}},$$

$$p_{2,1;\,\parallel} = \frac{2\cos\vartheta_z}{\varepsilon\cos\vartheta_z + \sqrt{\varepsilon\mu - \sin^2\vartheta_z}}\;\frac{\beta_z\sin^2\vartheta_z + (\beta_x\cos\vartheta_x + \beta_y\cos\vartheta_y)\sqrt{\varepsilon\mu - \sin^2\vartheta_z}}{\beta_z\sin^2\vartheta_z + (\beta_x\cos\vartheta_x + \beta_y\cos\vartheta_y)\cos\vartheta_z},$$

$$p_{1,2;\,\parallel} = \frac{2\varepsilon\sqrt{\varepsilon\mu - \sin^2\vartheta_z}}{\varepsilon\cos\vartheta_z + \sqrt{\varepsilon\mu - \sin^2\vartheta_z}}\;\frac{\beta_z\sin^2\vartheta_z - (\beta_x\cos\vartheta_x + \beta_y\cos\vartheta_y)\cos\vartheta_z}{\beta_z\sin^2\vartheta_z - (\beta_x\cos\vartheta_x + \beta_y\cos\vartheta_y)\sqrt{\varepsilon\mu - \sin^2\vartheta_z}}.$$

$$(22.4)$$

We obtain from Eqs. (2.7)–(2.10) for the boundary separating the first layer (subscript 2) from the first vacuum gap (subscript 3)

$$r_{3,2;\,\perp} = \frac{\mu\cos\vartheta_z - \sqrt{\varepsilon\mu - \sin^2\vartheta_z}}{\mu\cos\vartheta_z + \sqrt{\varepsilon\mu - \sin^2\vartheta_z}}, \qquad r_{2,3;\,\perp} = -\frac{\mu\cos\vartheta_z - \sqrt{\varepsilon\mu - \sin^2\vartheta_z}}{\mu\cos\vartheta_z + \sqrt{\varepsilon\mu - \sin^2\vartheta_z}},$$

$$(22.5)$$

$$p_{2,3;\,\perp} = \frac{2\mu\cos\vartheta_z}{\mu\cos\vartheta_z + \sqrt{\varepsilon\mu - \sin^2\vartheta_z}}, \qquad p_{3,2;\,\perp} = \frac{2\sqrt{\varepsilon\mu - \sin^2\vartheta_z}}{\mu\cos\vartheta_z - \sqrt{\varepsilon\mu - \sin^2\vartheta_z}},$$

$$(22.6)$$

$$r_{3,2;\,\parallel} = \frac{\varepsilon\cos\vartheta_z - \sqrt{\varepsilon\mu - \sin^2\vartheta_z}}{\varepsilon\cos\vartheta^3 + \sqrt{\varepsilon\mu - \sin^2\vartheta_z}}\;\frac{\beta_z\sin^2\vartheta_z + (\beta_x\cos\vartheta_x + \beta_y\cos\vartheta_y)\cos\vartheta_z}{\beta_z\sin^2\vartheta_z - (\beta_x\cos\vartheta_x + \beta_y\cos\vartheta_y)\cos\vartheta_z},$$

$$r_{2,3;\,\parallel} = -\frac{\varepsilon\cos\vartheta_z - \sqrt{\varepsilon\mu - \sin^2\vartheta_z}}{\varepsilon\cos\vartheta_z + \sqrt{\varepsilon\mu - \sin^2\vartheta_z}}\;\frac{\beta_z\sin^2\vartheta_z - (\beta_x\cos\vartheta_x + \beta_y\cos\vartheta_y)\sqrt{\varepsilon\mu - \sin^2\vartheta_z}}{\beta_z\sin^2\vartheta_z + (\beta_x\cos\vartheta_x + \beta_y\cos\vartheta_y)\sqrt{\varepsilon\mu - \sin^2\vartheta_z}},$$

$$(22.7)$$

$$p_{2,3;\,\parallel} = \frac{2\cos\vartheta_z}{\varepsilon\cos\vartheta_z + \sqrt{\varepsilon\mu - \sin^2\vartheta_z}}\;\frac{\beta_z\sin^2\vartheta_z - (\beta_x\cos\vartheta_x + \beta_y\cos\vartheta_y)\sqrt{\varepsilon\mu - \sin^2\vartheta_z}}{\beta_z\sin^2\vartheta_z - (\beta_x\cos\vartheta_x + \beta_y\cos\vartheta_y)\cos\vartheta_z},$$

$$p_{3,2;\,\parallel} = \frac{2\varepsilon\sqrt{\varepsilon\mu - \sin^2\vartheta_z}}{\varepsilon\cos\vartheta_z + \sqrt{\varepsilon\mu - \sin^2\vartheta_z}}\;\frac{\beta_z\sin^2\vartheta_z + (\beta_x\cos\vartheta_x + \beta_y\cos\vartheta_y)\cos\vartheta_z}{\beta_z\sin^2\vartheta_z + (\beta_x\cos\vartheta_x + \beta_y\cos\vartheta_y)\sqrt{\varepsilon\mu - \sin^2\vartheta_z}}.$$

$$(22.8)$$

In the case of alternating, similar media, we obtain

$$r_{2k+i,\,2k+j} = r_{i,j}, \qquad p_{2k+i,\,2k+j} = p_{i,j} \qquad (k = 1, 2, 3, \ldots).$$

$$(22.9)$$

These equations refer to coefficients corresponding to waves which are polarized either in the plane of propagation or perpendicular to that plane. Since it is always possible to select an even number such that by subtracting the number from the subscripts of some coefficient, one of the subscripts appearing in Eqs. (22.1)–(22.8) is obtained, any coefficient for a boundary in the stack is available from the coefficients defined by Eqs. (22.1)–(22.8), when Eq. (22.9) is employed.

In the simplest case, in which the particle moves linearly and uniformly in a direction perpendicular to the plates, rather laborious transformations and calculations lead to the following formula for the spectral density of the radiation energy in the forward direction, i.e., under acute angles relative to the velocity vector of the particle (energy density per unit solid angle) [14]:

$$W_{n\omega} = \frac{e^2 v^2}{\pi^2 c^3}\;\frac{\sin^2\vartheta\cos^2\vartheta}{(1 - \beta^2\cos^2\vartheta)^2}\,|f(\omega,\vartheta)|^2,$$

$$f(\omega,\vartheta) = \Big\{ (e^{im\psi} - e^{-im\psi})\,h_1 e^{-i\frac{\omega}{c}b\cos\vartheta} + (h_1 e^{-i\frac{\omega}{c}b\cos\vartheta} + h_2 e^{i\frac{\omega}{v}b}) \times$$

$$\times \Big[\frac{1 - e^{-i(m-1)(\varphi+\psi)}}{1 - e^{-i(\varphi+\psi)}}\,e^{i(m-1)\psi} - \frac{1 - e^{-i(m-1)(\varphi-\psi)}}{1 - e^{-i(\varphi-\psi)}}\,e^{-i(m-1)\varphi} \Big] e^{-i\varphi} \Big\}\,q^{-1},$$

$$q = \{[(\chi + \eta)^2 e^{-i\frac{\omega}{c}a\chi} - (\chi - \eta)^2 e^{i\frac{\omega}{c}a\chi}](e^{im\psi} - e^{-im\psi})e^{-i\frac{\omega}{c}b\cos\vartheta} -$$

$$- 4\chi\eta(e^{i(m-1)\psi} - e^{-i(m-1)\psi})\}(1 - \beta^2\chi^2),$$

$$h_{1,2} = (\chi + \eta)(1 + \beta\chi)[(\varepsilon - 1)(1 - \beta\chi) - \beta^2(\varepsilon\mu - 1)]e^{\mp i\frac{\omega}{c}a\chi} +$$

$$+ (\chi - \eta)(1 - \beta\chi)[(\varepsilon - 1)(1 + \beta\chi) - \beta^2(\varepsilon\mu - 1)]e^{\pm i\frac{\omega}{c}a\chi} -$$

$$- 2\chi[(\varepsilon - 1)(1 - \beta^2\chi^2) - \beta^2(\varepsilon\mu - 1)(1 + \beta\eta)]e^{\mp i\frac{\omega}{v}a},$$

$$\psi = \arccos\left[\cos\left(\frac{\omega}{c}a\chi\right)\cos\left(\frac{\omega}{c}b\cos\vartheta\right) - \frac{\chi^2 + \eta^2}{2\chi\eta}\sin\left(\frac{\omega}{c}a\chi\right)\sin\left(\frac{\omega}{c}b\cos\vartheta\right)\right],$$

$$\varphi = \frac{\omega}{v}(a + b) \quad \chi = \sqrt{\varepsilon\mu - \sin^2\vartheta}, \quad \eta = \varepsilon\cos\vartheta,$$

(22.10)

where m denotes the number of plates; a and b are the thickness of the plates and the vacuum gaps, respectively; ϑ is the angle between the normal to the plates and the direction of observation; and β is the ratio of particle velocity v to velocity c of light in vacuum. We note that in the case of wave attenuation, we have Im $(\varepsilon, \mu) > 0$.

The angular distribution of the spectral density of the radiation in the backward direction before the plates is obtained from Eq. (22.10) by changing the sign before the particle velocity everywhere.

It follows from Eq. (22.10) that for a large number of plates the electromagnetic waves are emitted in certain directions for which the denominators of the terms in the square brackets vanish, i.e., for

$$\varphi \pm \psi = 2\pi r,$$

$$\frac{\omega}{v}(a + b) \pm \arccos\left[\cos\left(\frac{\omega}{c}a\chi\right)\cos\left(\frac{\omega}{c}b\cos\vartheta\right) - \frac{\chi^2 + \eta^2}{2\chi\eta}\sin\left(\frac{\omega}{c}a\chi\right)\sin\left(\frac{\omega}{c}b\cos\vartheta\right)\right] = 2\pi r,$$

(22.11)

where r denotes an integer. The upper sign corresponds to the radiation in the backward direction or to reflected waves; the lower sign refers to the forward direction. These equations for dielectric plates $(\mu = 1)$ are identical to the well-known dispersion equations obtained and discussed in [68] and in [71, 72]. These equations are a generalization to the case of a periodical structure composed of ferrodielectric plates. We find near the Brewster angle

$$(\chi - \eta)^2 \ll \chi\eta, \quad \psi \approx \frac{\omega}{c}(a\chi + b\cos\vartheta),$$

(22.12)

and Eq. (22.11) assumes the following simple form:

$$a(1 \pm \beta\chi) + b(1 \pm \beta\cos\vartheta) = 2\pi r v/\omega.$$

(22.13)

When this equality holds, the wave lags behind the particle, or runs before the particle, and along one period of the layered medium the phase shift is $2\pi r$. The result is that waves emitted in corresponding directions interfere and amplify. Multiple reflections must be taken into account in order to interpret the more general relations stated in Eq. (22.11).

It is interesting to consider the dispersion equation in the case in which the group velocity in the various layers of the periodic structure is characterized by different signs. In principle, this case can occur in ferrodielectric plates. We have Re $\chi < 0$ and Re $\eta < 0$ in the corresponding frequency regions. For example, near the Brewster angle, the dispersion equation can be rewritten in the form

$$a(1 \mp |\sqrt{\varepsilon\mu - \sin^2\vartheta}|) + b(1 \pm \beta\cos\vartheta) = 2\pi r v/\omega,$$

(22.14)

where the lower and upper signs refer to emission in the forward and backward direction, respectively. The dispersion equation expresses the fact that the projections of the wave vectors of a wave transferring energy in a certain direction upon the velocity vector of the particle in vacuum and upon the velocity vector of the particle in the plate are characterized by different signs.

Let us use our results and consider the radiation near the Brewster angle. When we assume that the conditions of (22.12) are satisfied and when we neglect in Eq. (22.10) terms with the factor $\chi - \eta$, we obtain

$$W_{n\omega} = \frac{e^2 v^2}{\pi^2 c^3} \frac{\sin^2 \vartheta \cos^2 \vartheta}{(1 - \beta^2 \cos^2 \vartheta)^2} \frac{[(\varepsilon - 1)(1 - \beta \sqrt{\varepsilon\mu - \sin^2 \vartheta}) - \beta^2(\varepsilon\mu - 1)]^2}{(\varepsilon\mu - \sin^2 \vartheta)} \times$$

$$\times \frac{\sin^2 \left[\frac{\omega a}{2v}(1 - \beta\sqrt{\varepsilon\mu - \sin^2 \vartheta})\right]}{(1 - \beta\sqrt{\varepsilon\mu - \sin^2 \vartheta})^2} \frac{\sin^2 \left[\frac{m\omega a}{2v}(1 - \beta\sqrt{\varepsilon\mu - \sin^2 \vartheta}) + \frac{m\omega b}{2v}(1 - \beta \cos \vartheta)\right]}{\sin^2 \left[\frac{\omega a}{2v}(1 - \beta\sqrt{\varepsilon\mu - \sin^2 \vartheta}) + \frac{\omega b}{2v}(1 - \beta \cos \vartheta)\right]} ; \qquad (22.15)$$

it was assumed that no wave attenuation occurs and that $\varepsilon\mu > \sin^2 \vartheta$.

It is generally accepted that electromagnetic waves polarized in the plane of incidence pass completely through a boundary when they are incident, at the Brewster angle, onto a plane defined by the normal to the separating boundary and by the wave vector. In this particular case, the polarization is such that no reflected waves are generated, and therefore simplifications can be made. This particular condition makes it possible to obtain Eq. (22.15) directly from the result for a single plate. We must keep in mind that the waves generated at neighboring plates differ only by the phase factor

$$\exp\left[i\frac{\omega a}{v}(1 - \beta\sqrt{\varepsilon\mu - \sin^2 \vartheta}) + i\frac{\omega b}{v}(1 - \beta \cos \vartheta)\right], \qquad (22.16)$$

which results from the relative lag of the waves emitted in a certain direction from the ends of a period of the layered medium.

The angular distribution of the radiation intensity in the backward direction is given by Eq. (22.15) near the Brewster angle, but the sign of the particle velocity must be reversed. When we consider in Eq. (22.15) $(\varepsilon\mu - \sin^2 \vartheta)^{\frac{1}{2}}$ as the positive root, then in the frequency region involving negative group velocities, Eq. (22.15) assumes the following form:

$$W_{n\omega} = \frac{e^2 v^2}{\pi^2 c^3} \frac{\sin^2 \vartheta \cos^2 \vartheta}{(1 - \beta^2 \cos^2 \vartheta)^2} \frac{[(\varepsilon - 1)(1 + \beta\sqrt{\varepsilon\mu - \sin^2 \vartheta}) - \beta^2(\varepsilon\mu - 1)]^2}{(\varepsilon\mu - \sin^2 \vartheta)} \times$$

$$\times \frac{\sin^2 \left[\frac{\omega a}{2v}(1 + \beta\sqrt{\varepsilon\mu - \sin^2 \vartheta})\right]}{(1 + \beta\sqrt{\varepsilon\mu - \sin^2 \vartheta})^2} \frac{\sin^2 \left[\frac{m\omega a}{2v}(1 + \beta\sqrt{\varepsilon\mu - \sin^2 \vartheta}) + \frac{m\omega b}{2v}(1 - \beta \cos \vartheta)\right]}{\sin^2 \left[\frac{\omega a}{2v}(1 + \beta\sqrt{\varepsilon\mu - \sin^2 \vartheta}) + \frac{\omega b}{2v}(1 - \beta \cos \vartheta)\right]} , \qquad (22.17)$$

where ϑ denotes the angle between the wave vector in vacuum and the direction of motion of the particle. After reversing the sign of the velocity, this formula describes the emission in the backward direction, under an obtuse angle relative to the velocity vector of the particle. In this case, ϑ denotes the angle between the wave vector and the negative direction of the particle motion ($\vartheta < \pi/2$).

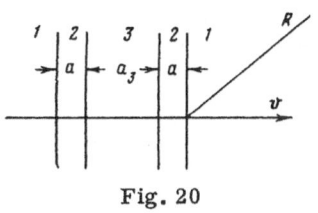

Fig. 20

§23. Radiation Generated by a Particle Passing through a Stack of Transparent Plates

The solution obtained in §18 for the passage of a particle through a transparent boundary can be easily generalized to the case of a stack of transparent plates (because no reflected waves occur) [20].

First, let us determine the radiation field for one transparent plate, i.e., for a plane layer of a medium 3 (whose thickness is denoted by a_3), which is surrounded on both sides by layers of a medium 2 (thickness a; see Fig. 20). We denote by R the distance between the point of observation and the point at which the particle leaves the second layer of medium 2.

The Hertz vector for the radiation field generated along the path in medium 1 to the first boundary separating medium 1 from medium 2 differs from the Hertz vector of Eq. (18.53) by the factor

$$i \exp \left(i \frac{\omega}{c} a_3 n_3 \cos \vartheta_3 - i \frac{\omega}{v} a - i \frac{\omega}{v} a_3 \right), \tag{23.1}$$

which accounts for the delay of the wave on its path through medium 3 and the second layer 2. Furthermore, we must omit the factor $\frac{n_3}{n_1} \frac{\sqrt{n_3 \cos \vartheta_3}}{\sqrt{n_1 \cos \vartheta_1}}$, which resulted in Eq. (18.53) from the transition of the wave from medium 1 into medium 3. Consequently, the field generated along the path to the first separating boundary is given by the equation

$$\Pi_{\omega 1} = -\frac{evk}{2\pi\omega^2 R} \frac{1}{1 - \beta n_1 \cos \vartheta_1} \exp \left(i \frac{\omega}{c} a_3 n_3 \cos \vartheta_3 - 2i \frac{\omega}{v} a - i \frac{\omega}{v} a_3 + i \frac{\omega}{c} n_1 R \right), \tag{23.2}$$

in which we assume that Eqs. (18.50) and (18.51) are satisfied.

The delay of the radiation field generated in the first layer of medium 2, and the exit of the wave field into medium 1 are calculated in a similar fashion. The result is expressed by the Hertz vector of Eq. (18.54), which must be multiplied by

$$i \frac{n_1}{n_3} \frac{\sqrt{n_1 \cos \vartheta_1}}{\sqrt{n_3 \cos \vartheta_3}} \exp \left(i \frac{\omega}{c} a_3 n_3 \cos \vartheta_3 - i \frac{\omega}{v} a - i \frac{\omega}{v} a_3 \right). \tag{23.3}$$

We obtain

$$\Pi_{\omega 2} = i \frac{evk}{2\pi\omega^2 R} \frac{n_1^2}{n_2^2} \frac{\sqrt{n_1 \cos \vartheta_3} + \sqrt{n_3 \cos \vartheta_1}}{2 \sqrt{n_1 \cos \vartheta_3}} \left\{ \frac{1}{1 - \beta n_2 \cos \vartheta_2} \times \right.$$
$$\times \left[1 - i \exp \left(-i \frac{\omega}{v} a \right) \right] - \frac{\sqrt{n_1 \cos \vartheta_3} - \sqrt{n_3 \cos \vartheta_1}}{\sqrt{n_1 \cos \vartheta_3} + \sqrt{n_3 \cos \vartheta_1}} \frac{1}{1 + \beta n_2 \cos \vartheta_2} \times$$
$$\left. \times \left[1 + i \exp \left(-i \frac{\omega}{v} a \right) \right] \right\} \exp \left(i \frac{\omega}{c} a_3 n_3 \cos \vartheta_3 - i \frac{\omega}{v} a - i \frac{\omega}{v} a_3 + i \frac{\omega}{c} n_1 R \right). \tag{23.4}$$

The radiation along the path in layer 3 is defined by the Hertz vector

$$\Pi_{\omega 3} = -\frac{evk}{2\pi\omega^2 R} \frac{i}{1 - \beta n_3 \cos \vartheta_3} \frac{n_1}{n_3} \frac{\sqrt{n_1 \cos \vartheta_1}}{\sqrt{n_3 \cos \vartheta_3}} \times$$
$$\times \left[\exp \left(i \frac{\omega}{c} a_3 n_3 \cos \vartheta_3 - i \frac{\omega}{v} a_3 \right) - 1 \right] \exp \left(-i \frac{\omega}{v} a + i \frac{\omega}{c} n_1 R \right). \tag{23.5}$$

The wave generated along the path in the second layer of medium 2 is given by an equation which results from Eq. (18.54) by transposition of the subscripts 1 and 3:

$$\Pi'_{\omega 2} = \frac{evk}{2\pi\omega^2 R} \frac{n_1^2}{n_2^2} \frac{\sqrt{n_1 \cos \vartheta_3} + \sqrt{n_3 \cos \vartheta_1}}{2\sqrt{n_1 \cos \vartheta_3}} \left\{ \frac{1}{1 - \beta n_2 \cos \vartheta_2} \times \right.$$

$$\times \left[1 - i \exp\left(-i\frac{\omega}{v} a\right) \right] + \frac{\sqrt{n_1 \cos \vartheta_3} - \sqrt{n_3 \cos \vartheta_1}}{\sqrt{n_1 \cos \vartheta_3} + \sqrt{n_3 \cos \vartheta_1}} \times$$

$$\times \left. \frac{1}{1 + \beta n_2 \cos \vartheta_2} \left[1 + i \exp\left(-i\frac{\omega}{v} a\right) \right] \right\} \exp\left(i\frac{\omega}{c} n_1 R\right). \tag{23.6}$$

Finally, the Hertz vector of the radiation field generated along the path in medium 1 (after passage of the particle through the last boundary) is

$$\Pi'_{\omega 1} = -\frac{evk}{2\pi\omega^2 R} \frac{1}{1 - \beta n_1 \cos \vartheta_1} \exp\left(i\frac{\omega}{c} n_1 R\right). \tag{23.7}$$

The radiation field of a particle which passes through a transparent plate is given by the sum of the Hertz vectors of Eqs. (23.2) and (23.4)-(23.7). We obtain

$$\Pi_\omega = -\frac{evk}{2\pi\omega^2 R} \left\{ \frac{1}{1 - \beta n_1 \cos \vartheta_1} \left[\exp\left(i\frac{\omega}{c} a_3 n_3 \cos \vartheta_3 - 2i\frac{\omega}{v} a - i\frac{\omega}{v} a_3\right) + 1 \right] - \right.$$

$$- \frac{n_1^2}{n_2^2} \frac{1}{(1 - \beta^2 a_2^2 \cos^2 \vartheta_2)} \left[i \left(\frac{\sqrt{n_3 \cos \vartheta_1}}{\sqrt{n_1 \cos \vartheta_3}} + \beta n_2 \cos \vartheta_2 \right) + \left(1 + \beta n_2 \cos \vartheta_2 \frac{\sqrt{n_3 \cos \vartheta_1}}{\sqrt{n_1 \cos \vartheta_3}} \right) e^{-i\frac{\omega}{v} a} \right] \times$$

$$\times \exp\left(i\frac{\omega}{c} a_3 n_3 \cos \vartheta_3 - i\frac{\omega}{v} a - i\frac{\omega}{v} a_3\right) - \frac{n_1^2}{n_2^2} \frac{1}{(1 - \beta^2 n_2^2 \cos^2 \vartheta_2)} \left[\left(1 + \beta n_2 \cos \vartheta_2 \frac{\sqrt{n_3 \cos \vartheta_1}}{\sqrt{n_1 \cos \vartheta_3}} \right) - \right.$$

$$- i \left(\frac{\sqrt{n_3 \cos \vartheta_1}}{\sqrt{n_1 \cos \vartheta_3}} + \beta n_2 \cos \vartheta_2 \right) e^{-i\frac{\omega}{v} a} \right] + i \frac{n_1}{n_3} \frac{\sqrt{n_1 \cos \vartheta_1}}{\sqrt{n_3 \cos \vartheta_3}} \frac{1}{1 - \beta n_3 \cos \vartheta_3} \times$$

$$\times \left. \left[\exp\left(i\frac{\omega}{c} a_3 n_3 \cos \vartheta_3 - i\frac{\omega}{v} a_3\right) - 1 \right] e^{-i\frac{\omega}{v} a} \right\} \exp\left(i\frac{\omega}{c} n_1 R\right). \tag{23.8}$$

Let us assume that $a_3 = 0$. Then we obtain from Eq. (23.8)

$$\Pi_\omega = -\frac{evk}{2\pi\omega^2 R} \frac{(1 + e^{-2i\frac{\omega}{v} a})(1 - \beta n_1 \cos \vartheta_1 - \beta^2 n_2^2)}{n_2^2 (1 - \beta n_1 \cos \vartheta_1)(1 - \beta^2 n_2^2 \cos^2 \vartheta_2)} (n_2^2 - n_1^2) e^{i\frac{\omega}{c} n_1 R} \tag{23.9}$$

As could be anticipated, the dependence of the radiation upon the optical properties of medium 3 cancels. This case corresponds to an isolated plate of thickness $2a$ which is placed into a medium with refractive index n_1. Equation (23.9), which was obtained from Eq. (23.8) for a transparent plate, corresponds to a transparent optical system. The same result is obtained from the translucency condition of a single separating boundary [see Eq. (18.50)] — a condition not discussed in detail in §18.

Since no reflected waves occur and since the radiation energy passes completely through the translucent boundaries, the solution stated in Eq. (23.8) can be easily generalized to the case of a stack of translucent plates. For example, in the case of two transparent plates, one must add a Hertz vector of Eq. (23.8) multiplied by the exponential function $e^{-i\varphi_1}$ to the solution stated in Eq. (23.8); we have

$$\varphi_1 = (\omega / v) \left[a_1(1 - \beta n_1 \cos \vartheta_1) + 2a(1 - \beta n_2 \cos \vartheta_2) + a_3(1 - \beta n_3 \cos \cdot \vartheta_3) \right], \tag{23.10}$$

which takes into account the relative phase shift of the waves emitted at the plates. The thickness of the interlayer between the plates is denoted by a_1; the interlayer is assumed to consist of a medium with refractive index n_1. In the case of three equidistant plates, one must add a Hertz vector of Eq. (23.8) multiplied by $e^{-2i\varphi_1}$, etc.

Thus, in the case of m transparent plates, the result is given by a Hertz vector of Eq. (23.8), which is multiplied by the following sum:

$$1 + e^{-i\varphi_1} + e^{-2i\varphi_1} + e^{-3i\varphi_1} + \dots + e^{-i(m-1)\varphi_1} = \frac{1 - e^{-im\varphi_1}}{1 - e^{-i\varphi_1}}. \qquad (23.11)$$

With Eq. (18.51) we obtain the final result

$$\Pi_{\omega m} = \Pi_\omega \frac{1 + (-1)^{m-1} e^{-im\varphi}}{1 + e^{-i\varphi}}, \qquad (23.12)$$

where

$$\varphi = (\omega / v)[a_1(1 - \beta n_1 \cos \vartheta_1) + 2a + a_3(1 - \beta n_3 \cos \vartheta_3)]. \qquad (23.13)$$

Π_ω is defined by Eq. (23.8).

CHAPTER 5

Radiation with Scattering

§24. Radiation with Single Scattering of a Nonrelativistic Particle near a Surface

In order to clarify the effect of particle scattering in the medium upon the radiation, it is convenient to start from the simplest case, i.e., to consider the influence of a single scattering process near the boundary upon the radiation emission.

When in the yz plane a nonrelativistic particle is incident at a certain angle to the boundary, we obtain from the general result stated in Eqs. (4.14)-(4.16):

$$\Pi_\omega = \frac{ec \cos \vartheta_z}{\pi \omega^2 R} \{[i\beta'_x + j (\beta'_y - \beta_y)](\sin^2 \vartheta_z + \cos \vartheta_z \sqrt{\varepsilon - \sin^2 \vartheta_z}) + k [\beta'_z - \varepsilon\beta_z + (\beta'_x \cos \vartheta_x + \beta'_y \cos \vartheta_y -$$
$$- \beta_y \cos \vartheta_y)(\cos \vartheta_z - \sqrt{\varepsilon - \sin^2 \vartheta_z})]\} (\varepsilon \cos \vartheta_z + \sqrt{\varepsilon - \sin^2 \vartheta_z})^{-1} e^{i\omega R'c}, \qquad (24.1)$$

where β_y and β_z ($\beta_x = 0$, $\beta_z < 0$) denote the components of the velocity vector of the particle in units of the velocity of light along the path in vacuum; β'_x, β'_y, and β'_z ($\beta'_z < 0$) denote the components of the velocity vector along the path in the medium (see Fig. 21):

$$\beta = \beta(j \sin \theta_0 + k \cos \theta_0),$$
$$\beta' = \beta(i \cos \theta_x + j \cos \theta_y + k \cos \theta_z) \ (\theta_0 > \pi /2, \ \theta_z > \pi / 2). \qquad (24.2)$$

When we omit in Eq. (24.1) all terms containing components of the velocity vector of the particle in the medium (β'), the result corresponds to the radiation from a particle stopped at the separating boundary:

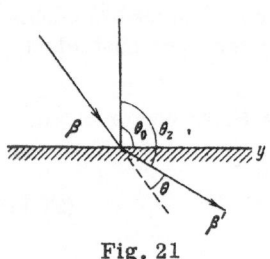

Fig. 21

$$\mathbf{\Pi}_\omega = -\frac{ec\cos\vartheta_z}{\pi\omega^2 R}\{\mathbf{j}\beta_y(\sin^2\vartheta_z + \cos\vartheta_z\sqrt{\varepsilon-\sin^2\vartheta_z}) +$$

$$+ \mathbf{k}[\varepsilon\beta_z + \beta_y\cos\vartheta_y(\cos\vartheta_z - \sqrt{\varepsilon-\sin^2\vartheta_z})]\} \times$$

$$\times (\varepsilon\cos\vartheta_z + \sqrt{\varepsilon-\sin^2\vartheta_z})^{-1}e^{i\omega R/c}. \tag{24.3}$$

After omitting all terms which contain vector components of β, we obtain the radiation resulting from the ejection of the particle into the medium:

$$\mathbf{\Pi}_\omega = \frac{ec\cos\vartheta_z}{\pi\omega^2 R}\{(\mathbf{i}\beta_x' + \mathbf{j}\beta_y')(\sin^2\vartheta_z + \cos\vartheta_z\sqrt{\varepsilon-\sin^2\vartheta_z}) +$$

$$+ \mathbf{k}[\beta_z' + (\beta_x'\cos\vartheta_x + \beta_y'\cos\vartheta_y)(\cos\vartheta_z - \sqrt{\varepsilon-\sin^2\vartheta_z})]\}(\varepsilon\cos\vartheta_z + \sqrt{\varepsilon-\sin^2\vartheta_z})^{-1}e^{i\omega R/c}. \tag{24.4}$$

With a completely transparent medium and $\varepsilon > 1$, the radiation is linearly polarized in the entire angular range. For $\varepsilon < 1$, the radiation is elliptically polarized in the angular range in which $\sin^2\vartheta_z > \varepsilon$. When the medium absorbs or when $\varepsilon < 0$, the radiation is elliptically polarized in all directions. The parameters of the polarization ellipse are determined by the normal component (normal to the separating boundary) and the tangential component of the Hertz vector defined by the formulas of §11.

We note that the results of Eqs. (24.1)-(24.4) can be used when the particle moves along the boundary ($\beta_z = \beta_z' = 0$). For example, when we assume in Eq. (24.3) that $\beta_x' = \beta_z' = 0$, we obtain the radiation resulting from the ejection of the particle from the plane of the boundary (in the direction of the y axis):

$$\mathbf{\Pi}_\omega = \frac{ev\cos\vartheta_z}{\pi\omega^2 R}\frac{\mathbf{j}(\sin^2\vartheta_z + \cos\vartheta_z\sqrt{\varepsilon-\sin^2\vartheta_z}) + \mathbf{k}\cos\vartheta_y(\cos\vartheta_z - \sqrt{\varepsilon-\sin^2\vartheta_z})}{\varepsilon\cos\vartheta_z + \sqrt{\varepsilon-\sin^2\vartheta_z}}e^{i\omega R/c}. \tag{24.5}$$

According to this result, the Hertz vectors of waves polarized in the plane of propagation or perpendicular to that plane are:

$$\mathbf{\Pi}_\parallel = -\frac{\mathbf{k}ev\cos\vartheta_y\cos\vartheta_z\sqrt{\varepsilon-\sin^2\vartheta_z}}{\pi\omega^2 R\sin^2\vartheta_z(\varepsilon\cos\vartheta_z + \sqrt{\varepsilon-\sin^2\vartheta_z})}e^{i\omega R/c}, \tag{24.6}$$

$$\mathbf{\Pi}_\perp = \frac{ev\cos\vartheta_z(\mathbf{j}\sin^2\vartheta_z + \mathbf{k}\cos\vartheta_y\cos\vartheta_z)}{\pi\omega^2 R\sin^2\vartheta_z(\cos\vartheta_z + \sqrt{\varepsilon-\sin^2\vartheta_z})}e^{i\omega R/c}. \tag{24.7}$$

The angular distribution of the radiation energy of these waves is given by the formulas

$$W_{n\omega\parallel} = \frac{e^2v^2}{\pi^2 c^3}\frac{|\varepsilon-\sin^2\vartheta_z|\cos^2\vartheta_y\cos^2\vartheta_z}{|\varepsilon\cos\vartheta_z + \sqrt{\varepsilon-\sin^2\vartheta_z}|^2\sin^2\vartheta_z}, \tag{24.8}$$

$$W_{n\omega\perp} = \frac{e^2v^2}{\pi^2 c^3}\frac{\cos^2\vartheta_x\cos^2\vartheta_z}{|\cos\vartheta_z + \sqrt{\varepsilon-\sin^2\vartheta_z}|^2\sin^2\vartheta_z}. \tag{24.9}$$

It is convenient to use the expressions for the electrical field in the subsequent discussion. We obtain from Eqs. (24.6), (24.7), and (4.7) the Fourier components of the electrical field of the waves polarized in the plane of propagation and perpendicular to that plane:

$$\mathbf{E}_{\omega\parallel} = -\frac{ev\cos\vartheta_z}{\pi Rc^2\sin^2\vartheta_z}(\mathbf{i}\cos\vartheta_x\cos\vartheta_z + \mathbf{j}\cos\vartheta_y\cos\vartheta_z - \mathbf{k}\sin^2\vartheta_z) \times$$

$$\times [(\cos\vartheta_z - \varepsilon\cos\theta_0)\sin^2\vartheta_z - (\cos\vartheta_x\cos\vartheta_x + \cos\vartheta_y\cos\vartheta_y -$$

$$- \sin\theta_0\cos\vartheta_y)\sqrt{\varepsilon-\sin^2\vartheta_z}](\varepsilon\cos\vartheta_z + \sqrt{\varepsilon-\sin^2\vartheta_z})^{-1}e^{i\omega R/c}. \tag{24.10}$$

$$\mathbf{E}_{\omega \perp} = -\frac{ev \cos \vartheta_z}{\pi R c^2 \sin^2 \vartheta_z} (\mathbf{i} \cos \vartheta_y - \mathbf{j} \cos \vartheta_x)(\cos \theta_x \cos \vartheta_y +$$

$$+ \cos \theta_y \cos \vartheta_x - \sin \theta_0 \cos \vartheta_x)(\cos \vartheta_z + \sqrt{\varepsilon - \sin^2 \vartheta_z}) e^{i\omega R/c}. \qquad (24.11)$$

The spectral density of the radiation energy per unit solid angle is expressed by these vectors according to the formulas

$$W_{n\omega \parallel} = cR^2 |\mathbf{E}_{\parallel}|^2, \quad W_{n\omega \perp} = cR^2 |\mathbf{E}_{\perp}|^2. \qquad (24.12)$$

When we omit in Eqs. (24.10) and (24.11) all terms containing direction cosines of the particle's velocity vector in the medium ($\cos \theta_x$, $\cos \theta_y$, and $\cos \theta_z$), we obtain the radiation field resulting from the instantaneous arrest of the particle at the boundary. When we omit all terms containing $\sin \theta_0$ and $\cos \theta_0$, we obtain the radiation generated by the ejection of the particle into the medium.

Let us consider several consequences of these results. When the particle is perpendicularly incident on the boundary ($\theta_0 = \pi$), we obtain from Eq. (24.10)-(24.12):

$$W_{n\omega} = \frac{e^2 v^2 \cos^2 \vartheta_z}{\pi^2 c^3 \sin^2 \vartheta_z} \{ |(\cos \theta_z + \varepsilon) \sin^2 \vartheta_z - (\cos \theta_x \cos \vartheta_x + \cos \theta_y \cos \vartheta_y) \times$$

$$\times \sqrt{\varepsilon - \sin^2 \vartheta_z}|^2 + |\sin^2 \vartheta_z + \cos \vartheta_z \sqrt{\varepsilon - \sin^2 \vartheta_z}|^2 (\cos \theta_x \cos \vartheta_y + \cos \theta_y \cos \vartheta_x)^2 \} |\varepsilon \cos \vartheta_z + \sqrt{\varepsilon - \sin^2 \vartheta_z}|^{-2}, \qquad (24.13)$$

$$\frac{W_{n\omega \parallel}}{W_{n\omega \perp}} = \frac{|(\cos \theta_z + \varepsilon) \sin^2 \vartheta_z - (\cos \theta_x \cos \vartheta_x + \cos \theta_y \cos \vartheta_y) \sqrt{\varepsilon - \sin^2 \vartheta_z}|^2}{|\sin^2 \vartheta_z + \cos \vartheta_z \sqrt{\varepsilon - \sin^2 \vartheta_z}|^2 (\cos \theta_x \cos \vartheta_y + \cos \theta_y \cos \vartheta_x)^2}. \qquad (24.14)$$

When, after the scattering the particle moves in the xz plane ($\cos \theta_y = 0$), Eqs. (24,13) and (24.14) assume the following form in the yz plane which is perpendicular to the plane over which the particle moves ($\cos \vartheta_x = 0$):

$$W_{n\omega} = \frac{e^2 v^2 \cos^2 \vartheta_z}{\pi^2 c^3} \{ |\varepsilon - \cos \theta|^2 \sin^2 \vartheta_z + |\sin^2 \vartheta_z +$$

$$+ \cos \vartheta_z \sqrt{\varepsilon - \sin^2 \vartheta_z}|^2 \sin^2 \theta \} |\varepsilon \cos \vartheta_z + \sqrt{\varepsilon - \sin^2 \vartheta_z}|^{-2}, \qquad (24.15)$$

$$\frac{W_{n\omega \parallel}}{W_{n\omega \perp}} = \frac{|\varepsilon - \cos \theta|^2 \sin^2 \vartheta_z}{|\sin^2 \vartheta_z + \cos \vartheta_z \sqrt{\varepsilon - \sin^2 \vartheta_z}|^2 \sin^2 \theta} \quad (\theta = \hat{\beta}\hat{\beta}'). \qquad (24.16)$$

With these equations we obtain for $\varepsilon = 1$

$$W_{n\omega} = \frac{e^2 v^2}{4\pi^2 c^3} [(1 - \cos \theta)^2 \sin^2 \vartheta_z + \sin^2 \theta], \qquad (24.17)$$

$$\frac{W_{n\omega \parallel}}{W_{n\omega \perp}} = \frac{(1 - \cos \theta)^2 \sin^2 \vartheta_z}{\sin^2 \theta}. \qquad (24.18)$$

When the scattering angle is small, these equations assume the form

$$W_{n\omega} = \frac{e^2 v^2 \theta^2}{4\pi^2 c^3} \left(\frac{\theta^2}{4} \sin^2 \vartheta_z + 1 \right), \qquad (24.19)$$

$$\frac{W_{n\omega \parallel}}{W_{n\omega \perp}} = \frac{\theta^2}{4} \sin^2 \vartheta_z. \qquad (24.20)$$

When the particle is ejected from the boundary into the medium, the angular and spectral distributions of the radiation intensity are given by the following equations:

$$W_{n\omega} = \frac{e^2 v^2 \cos^2 \vartheta_z}{\pi^2 c^3 \sin^2 \vartheta_z} \{ |\cos \theta_z \sin^2 \vartheta_z - (\cos \theta_x \cos \vartheta_x +$$

$$+ \cos \theta_y \cos \vartheta_y) \sqrt{\varepsilon - \sin^2 \vartheta_z} |^2 + | \sin^2 \vartheta_z + \cos \vartheta_z \sqrt{\varepsilon - \sin^2 \vartheta_z} |^2 \times$$

$$\times (\cos \theta_x \cos \vartheta_y + \cos \theta_y \cos \vartheta_x)^2 \} |\varepsilon \cos \vartheta_z + \sqrt{\varepsilon - \sin^2 \vartheta_z} |^{-2}, \tag{24.21}$$

$$\frac{W_{n\omega \parallel}}{W_{n\omega \perp}} = \frac{|\cos \theta_z \sin^2 \vartheta_z - (\cos \theta_x \cos \vartheta_x + \cos \theta_y \cos \vartheta_y) \sqrt{\varepsilon - \sin^2 \vartheta_z} |^2}{|\sin^2 \vartheta_z + \cos \vartheta_z \sqrt{\varepsilon - \sin^2 \vartheta_z} |^2 (\cos \theta_x \cos \vartheta_y + \cos \theta_y \cos \vartheta_x)^2}. \tag{24.22}$$

In the plane which is perpendicular to the plane over which the particle moves, we obtain

$$W_{n\omega} = \frac{e^2 v^2}{\pi^2 c^3} \cos^2 \vartheta_z \{\cos^2 \theta \sin^2 \vartheta_z + |\sin^2 \vartheta_z + \cos \vartheta_z \sqrt{\varepsilon - \sin^2 \vartheta_z} |^2 \sin^2 \theta \} |\varepsilon \cos \vartheta_z + \sqrt{\varepsilon - \sin^2 \vartheta_z} |^{-2}, \tag{24.23}$$

$$\frac{W_{n\omega \parallel}}{W_{n\omega \perp}} = \frac{\cos^2 \theta \sin^2 \vartheta_z}{|\sin^2 \vartheta_z + \cos \vartheta_z \sqrt{\varepsilon - \sin^2 \vartheta_z} |^2 \sin^2 \theta}. \tag{24.24}$$

We obtain with these equations for $\varepsilon = 1$

$$W_{n\omega} = \frac{e^2 v^2}{4\pi^2 c^3} (\cos^2 \theta \sin^2 \vartheta_z + \sin^2 \theta), \tag{24.25}$$

$$\frac{W_{n\omega \parallel}}{W_{n\omega \perp}} = \frac{\cos^2 \theta \sin^2 \vartheta_z}{\sin^2 \theta}. \tag{24.26}$$

When the angle θ is small, we find

$$W_{n\omega} = \frac{e^2 v^2}{4\pi^2 c^3} (\sin^2 \vartheta_z + \theta^2), \tag{24.27}$$

$$\frac{W_{n\omega \parallel}}{W_{n\omega \perp}} = \frac{\sin^2 \vartheta_z}{\theta^2}. \tag{24.28}$$

Let us calculate the average values of the angular distribution of the radiation intensity when the scattering has azimuthal symmetry. We express the direction cosines of the vector $\boldsymbol{\beta}'$ by the azimuthal scattering angle φ referred to the plane which is defined by the x axis and the direction of motion of the particle in vacuum:

$$\cos \theta_x = \sin \theta \cos \varphi,$$
$$\cos \theta_y = \cos \theta \sin \theta_0 + \sin \theta \cos \theta_0 \sin \varphi, \tag{24.29}$$
$$\cos \theta_z = \cos \theta \cos \theta_0 - \sin \theta \sin \theta_0 \sin \varphi,$$

where

$$\theta = \widehat{\boldsymbol{\beta}\boldsymbol{\beta}'}. \tag{24.30}$$

With these expressions we can find the average values of both the direction cosines of the vector $\boldsymbol{\beta}'$ and the products of these direction cosines for scattering with azimuthal symmetry

$$\overline{\cos \theta_x} = 0, \qquad \overline{\cos \theta_y} = \cos \theta \sin \theta_0, \qquad \overline{\cos \theta_z} = \cos \theta \cos \theta_0,$$

$$\overline{\cos^2 \theta_x} = \frac{1}{2} \sin^2 \theta, \qquad \overline{\cos^2 \theta_y} = \cos^2 \theta \sin^2 \theta_0 + \frac{1}{2} \sin^2 \theta \cos^2 \theta_0,$$

$$\overline{\cos^2 \theta_z} = \cos^2 \theta \cos^2 \theta_0 + \tfrac{1}{2} \sin^2 \theta \sin^2 \theta_0, \quad \overline{\cos \theta_x \cos \theta_y} = 0,$$

$$\overline{\cos \theta_x \cos \theta_z} = 0, \qquad \overline{\cos \theta_y \cos \theta_z} = \tfrac{1}{2}(3 \cos^2 \theta - 1) \sin \theta_0 \cos \theta_0. \tag{24.31}$$

In addition, we can determine the average values of the square of the absolute value of the electric vectors defined by Eqs. (24.10) and (24.11)

$$\overline{|\mathbf{E}_{\omega \parallel}|^2} = \frac{e^2 v^2 \cos^2 \vartheta_z}{\pi^2 R^2 c^4 \sin^2 \vartheta_z} \Big\{ | \sqrt{\varepsilon - \sin^2 \vartheta_z} \, (1 - \cos \theta) \sin \theta_0 \cos \vartheta_y -$$

$$- (\varepsilon - \cos \theta) \cos \theta_0 \sin^2 \vartheta_z |^2 + \tfrac{1}{2} \sin^2 \theta \, [| \sqrt{\varepsilon - \sin^2 \vartheta_z} \cos \theta_0 \cos \vartheta_y +$$

$$+ \sin \theta_0 \sin^2 \vartheta_z |^2 + | \varepsilon - \sin^2 \vartheta_z \, | \cos^2 \vartheta_x |] \} | \varepsilon \cos \vartheta_z + \sqrt{\varepsilon - \sin^2 \vartheta_z} |^{-2}, \tag{24.32}$$

$$\overline{|\mathbf{E}_{\omega \perp}|^2} = \frac{e^2 v^2 \cos^2 \vartheta_z}{\pi^2 R^2 c^4 \sin^2 \vartheta_z} [(1 - \cos \theta)^2 \sin^2 \theta_0 \cos^2 \vartheta_x +$$

$$+ \tfrac{1}{2} \sin^2 \theta \, (\cos^2 \vartheta_y + \cos^2 \theta_0 \cos^2 \vartheta_x)] | \cos \vartheta_z + \sqrt{\varepsilon - \sin^2 \vartheta_z} |^{-2}. \tag{24.33}$$

The angular distribution of the radiation intensity and the degree of polarization can be expressed with these quantities according to the following formulas:

$$W_{n\omega} = cR^2 [\overline{|\mathbf{E}_{\omega \parallel}|^2} + \overline{|\mathbf{E}_{\omega \perp}|^2}], \tag{24.34}$$

$$P = \frac{\overline{|\mathbf{E}_{\omega \parallel}|^2} - \overline{|\mathbf{E}_{\omega \perp}|^2}}{\overline{|\mathbf{E}_{\omega \parallel}|^2} + \overline{|\mathbf{E}_{\omega \perp}|^2}}. \tag{24.35}$$

When the particle is incident normally to the boundary ($\theta_0 = \pi$), the results simplify considerably and assume the form ($\vartheta \equiv \vartheta_z$):

$$W_{n\omega} = \frac{e^2 v^2}{2\pi^2 c^3} \cos^2 \vartheta \, (2 |\varepsilon - \cos \theta|^2 \sin^2 \vartheta + |\varepsilon - \sin^2 \vartheta| \sin^2 \theta +$$

$$+ |\sin^2 \vartheta + \cos \vartheta \sqrt{\varepsilon - \sin^2 \vartheta}|^2 \sin^2 \theta) | \varepsilon \cos \vartheta + \sqrt{\varepsilon - \sin^2 \vartheta} |^{-2}, \tag{24.36}$$

$$\frac{W_{n\omega \parallel}}{W_{n\omega \perp}} = \frac{2 |\varepsilon - \cos \theta|^2 \sin^2 \vartheta + |\varepsilon - \sin^2 \vartheta| \sin^2 \theta}{|\sin^2 \vartheta + \cos \vartheta \sqrt{\varepsilon - \sin^2 \vartheta}|^2 \sin^2 \theta}. \tag{24.37}$$

We obtain for small scattering angles

$$W_{n\omega \parallel} = \frac{e^2 \beta^2}{\pi^2 c} \frac{|\varepsilon - 1|^2 \sin^2 \vartheta \cos^2 \vartheta}{|\varepsilon \cos \vartheta + \sqrt{\varepsilon - \sin^2 \vartheta}|^2} + \frac{e^2 \beta^2}{2\pi^2 c} \frac{|\varepsilon - \sin^2 \vartheta| \theta^2 \cos^2 \vartheta}{|\varepsilon \cos \vartheta + \sqrt{\varepsilon - \sin^2 \vartheta}|^2}, \tag{24.38}$$

$$W_{n\omega \perp} = \frac{e^2 \beta^2}{2\pi^2 c} \frac{\theta^2 \cos^2 \vartheta}{|\cos \vartheta + \sqrt{\varepsilon - \sin^2 \vartheta}|^2}. \tag{24.39}$$

When we insert into Eqs. (24.38) and (24.39) instead of an arbitrary angle θ the most likely scattering angles along the path of wave absorption

$$\frac{c}{\omega \, \mathrm{Im} \, \sqrt{\varepsilon - \sin^2 \vartheta}}, \tag{24.40}$$

the results provide us with a qualitative description of the effect of the scattering upon the radiation. We note that the nonpolarized radiation component resulting from the inclined incidence is asymmetric relative to the normal of the boundary. This typical asymmetry can be important when we have to determine the relative contribution of the bremsstrahlung to an experimentally observed, partially polarized radiation.

Let us consider in detail the radiation generated in the very simple case in which a particle is incident in normal direction on a target and scattered once at the boundary. We note that Eqs. (24.36) and (24.37) are applicable, provided that the particle does not leave into the vacuum after the scattering, i.e., provided that the scattering angle θ is smaller than a right angle. In the limit of scattering under a right angle ($\theta = \pi/2$), Eqs. (24.36) and (24.37) assume the form

$$W_{n\omega} = \frac{e^2 v^2}{2\pi^2 c^3} \cos^2 \vartheta \left(2 \, | \varepsilon |^2 \sin^2 \vartheta + | \varepsilon - \sin^2 \vartheta | + \right.$$

$$+ | \sin^2 \vartheta + \cos \vartheta \, \sqrt{\varepsilon - \sin^2 \vartheta} \, |^2) | \varepsilon \cos \vartheta + \sqrt{\varepsilon - \sin^2 \vartheta} \, |^{-2}, \tag{24.41}$$

$$\frac{W_{n\omega \parallel}}{W_{n\omega \perp}} = \frac{2 \, | \varepsilon |^2 \sin^2 \vartheta + | \varepsilon - \sin^2 \vartheta |}{| \sin^2 \vartheta + \cos \vartheta \, \sqrt{\varepsilon - \sin^2 \vartheta} \, |^2} \ . \tag{24.42}$$

We see from Eqs. (24.36) and (24.37) that the region in which the scattering substantially affects the radiation decreases with increasing absolute value of the dielectric constant. The region in which the scattering strongly affects the radiation is close to the angle $\vartheta = 0$. The size of the region is greatest for scattering at the angle $\theta = \pi/2$ and is given by the condition

$$W_{n\omega \parallel} < W_{n\omega \perp}. \tag{24.43}$$

It follows from Eqs. (24.41) and (24.42) that the maximum dimensions of the region must be proportional to the absolute value of the complex index of refraction of the medium

$$\Delta \vartheta \sim | \sqrt{\varepsilon} \, |^{-1}. \tag{24.44}$$

In the center of the region, i.e., at $\vartheta = 0$, the radiation is not polarized and is completely determined by the scattering process. According to Eq. (24.36) the radiation intensity in the direction $\vartheta = 0$ is given by

$$W_{n\omega} = \frac{e^2 v^2}{\pi^2 c^3} \frac{\sin^2 \theta}{| 1 + \sqrt{\varepsilon} \, |^2} \ . \tag{24.45}$$

The radiation intensity increases with increasing scattering angle and assumes its maximum value for the angle $\theta = \pi/2$:

$$W_{n\omega} = \frac{c^2 v^2}{\pi^2 c^3} \frac{1}{| 1 + \sqrt{\varepsilon} \, |^2} \ . \tag{24.46}$$

Of particular interest is another case, namely, the radiation at frequencies for which the absolute value of the dielectric constant is much smaller than unity. Except for a small region of angles near $\vartheta = 0$, in this case the results are satisfactorily described by Eqs. (24.36) and (24.37) when we set in these equations $\varepsilon = 0$. Then, these equations assume the form

$$W_{n\omega} = \frac{e^2 v^2}{\pi^2 c^3} \cos^2 \vartheta, \tag{24.47}$$

$$\frac{W_{n\omega \parallel}}{W_{n\omega \perp}} = \frac{1 + \cos^2 \theta}{\sin^2 \theta} \ . \tag{24.48}$$

Interestingly enough, the angular distribution of the radiation intensity is independent of the scattering angle in this particular case, i.e., for $\varepsilon = 0$. However, according to Eq. (24.48), the scattering has a strong influence upon the degree of polarization of the radiation. The degree of polarization is given by

$$P = (1 + \cos 2\theta) / 2. \tag{24.49}$$

The degree of polarization is independent of the direction of the waves emitted and decreases monotonically from unit to zero with increasing scattering angles. Thus, the radiation is non-polarized in right-angle scattering. In order to understand this unexpected result, we must consider the part of the radiation which is generated along the path in the medium.

Let us consider the radiation generated in the ejection of the particle from the boundary into the medium. When we use Eq. (24.4) and average the radiation intensity in the case in which the particle ejection has axial symmetry relative to the normal to the boundary, we find

$$W_{n\omega} = \frac{e^2 v^2}{2\pi^2 c^3} \cos^2 \vartheta \, (2\cos^2 \theta \sin^2 \vartheta + |\varepsilon - \sin^2 \vartheta| \sin^2 \theta +$$

$$+ |\sin^2 \vartheta + \cos \vartheta \sqrt{\varepsilon - \sin^2 \vartheta}|^2 \sin^2 \theta) |\varepsilon \cos \vartheta + \sqrt{\varepsilon - \sin^2 \vartheta}|^{-2}, \tag{24.50}$$

$$\frac{W_{n\omega\parallel}}{W_{n\omega\perp}} = \frac{2\cos^2 \theta \sin^2 \vartheta + |\varepsilon - \sin^2 \vartheta| \sin^2 \theta}{|\sin^2 \vartheta + \cos \vartheta \sqrt{\varepsilon - \sin^2 \vartheta}|^2 \sin^2 \theta}. \tag{24.51}$$

It follows from these equations that for large values of the absolute value of the dielectric constant and for not very small θ values, the following relations are valid (except for a small angular region near $\vartheta = \pi/2$):

$$W_{n\omega} = \frac{e^2 v^2}{2\pi^2 c^3} \frac{\sin^2 \theta}{|\varepsilon|} (1 + \cos^2 \vartheta), \tag{24.52}$$

$$\frac{W_{n\omega\parallel}}{W_{n\omega\perp}} = \frac{1}{\cos^2 \vartheta}. \tag{24.53}$$

For particles scattered parallel to the separating boundary ($\theta = \pi/2$), we obtain from Eqs. (24.50) and (24.51)

$$W_{n\omega} = \frac{e^2 v^2}{2\pi^2 c^3} \cos^2 \vartheta \, (|\varepsilon - \sin^2 \vartheta| + |\sin^2 \vartheta + \cos \vartheta \sqrt{\varepsilon - \sin^2 \vartheta}|^2) |\varepsilon \cos \vartheta + \sqrt{\varepsilon - \sin^2 \vartheta}|^{-2}, \tag{24.54}$$

$$\frac{W_{n\omega\parallel}}{W_{n\omega\perp}} = \frac{|\varepsilon - \sin^2 \vartheta|}{|\sin^2 \vartheta + \cos \vartheta \sqrt{\varepsilon - \sin^2 \vartheta}|^2}. \tag{24.55}$$

Let us consider the radiation in the frequency region in which the absolute value of the dielectric constant is much smaller than unity. We assume in Eqs. (24.50) and (24.51) that $\varepsilon = 0$ and obtain

$$W_{n\omega} = \frac{e^2 v^2}{\pi^2 c^3} \cos^2 \vartheta, \tag{24.56}$$

$$\frac{W_{n\omega\parallel}}{W_{n\omega\perp}} = \frac{1 + \cos^2 \theta}{\sin^2 \theta}. \tag{24.57}$$

These results agree with the results stated in Eqs. (24.47) and (24.48), in which we included the radiation of the particle along its path in vacuum. Thus, the section of the particle trajectory in vacuum has no influence upon the resulting radiation: a particle which is suddenly stopped at the boundary does not emit radiation in the case $\varepsilon = 0$. In this case, the image of the charged particle moving in vacuum normal to the boundary generates the same radiation field as a particle emitted into the medium.

§ 25. Effect of Multiple Scattering upon the Transition Radiation

Transition radiation is usually treated under the assumption that the particle motion is uniform and linear. When this assumption is made, the energy of the transition radiation increases proportionally to the energy of the particle in the relativistic case [77, 78]. Since quanta of the transition radiation are generated along a path section which is large compared to the wavelength, and since the entire radiation is emitted in forward direction at small angles relative to the velocity vector [of the order $(1 - \beta^2)^{\frac{1}{2}}$], multiple scattering can substantially modify both the spectrum and the energy of the transition radiation. This point was mentioned in [79]. First, we will evaluate both the spectrum and the energy of the transition radiation (see [10]); after that, a quantitative treatment of the problem will follow (see [16]).

Let us consider the radiation emitted by a particle which moves through a boundary separating a medium from vacuum. According to the preceding discussion, when the angle at which the particle trajectory is inclined relative to the boundary, is much greater than $(1 - \beta^2)^{\frac{1}{2}}$, the spherical wave amplitude of the transition-radiation field generated by a relativistic particle at small angles ϑ relative to the trajectory is the same as in the case of perpendicular particle incidence, and the spherical wave amplitude is proportional to

$$H \sim \vartheta \left[\left(1 - \beta + \frac{\vartheta^2}{2}\right)^{-1} - \left(1 - n\beta + \frac{\vartheta^2}{2}\right)^{-1} \right] \tag{25.1}$$

provided that the refractive index of the medium is close to unity: $n = 1 - \omega_0^2/2\omega^2$, where $\omega \gg \omega_0$, $\omega_0^2 = 4\pi e^2 N/m$. The first term of Eq. (25.1) corresponds to the field generated along the particle path in vacuum, the second term to the field generated along the particle path in the medium. It follows from Eq. (25.1) that the amplitude of the field is proportional to the path difference of coherent interaction between the particle and the wave in vacuum (s_v) and in the medium (s_m). The path lengths depend upon the angle between the direction of wave propagation and the velocity vector of the particle and are given by $s_v \sim \frac{c}{\omega}\left(1 - \beta + \frac{\vartheta^2}{2}\right)^{-1}$, $s_m \sim \frac{c}{\omega}\left(1 - n\beta + \frac{\vartheta^2}{2}\right)^{-1}$. In the direction in which the energy density of the radiation emitted per unit solid angle is maximal, we have $\left(\vartheta \approx \frac{\mu c^2}{E} \ll 1\right) s_v \sim \frac{c}{\omega}\left(\frac{E}{\mu c^2}\right)^2$, $s_m \sim \frac{c}{\omega}\left[\left(\frac{\mu c^2}{E}\right)^2 + \frac{\omega_0^2}{2\omega^2}\right]^{-1}$, where E and μ denote the total energy and the rest mass of the particle, respectively. At frequencies below a critical frequency

$$(\omega < \omega_{cr} = \omega_0 E/\mu c^2),$$

we have $s_v \gg s_m$, and therefore, the field of the transition radiation is generated mainly along the particle path in vacuum. Multiple scattering in the medium affects the small term in Eq. (25.1), and this is the reason why multiple scattering can decrease the probability for the emission of transition-radiation quanta. Furthermore, when scattering is ignored, transition-radiation quanta are generated in other directions ($\vartheta \neq \mu c^2/E$) only in the frequency range in which the second term of Eq. (25.1) is small compared with the first term.

At frequencies $\omega > \omega_{cr}$, we have $s_v \approx s_m$ for unscattered particles, and therefore, energy is not emitted. Multiple scattering plays an important role if along the path $s \approx \frac{c}{\omega}\left(\frac{E}{\mu c^2}\right)^2$ the particle passes beyond the angular region defined by the angle $\vartheta \approx \mu c^2/E$. When this is the case, the path of coherent interaction between the particle and the wave in the medium is smaller than without scattering, and the field which corresponds to the second term in Eq. (25.1) and causes the cutoff in the spectrum is not generated. Then, new frequencies appear in the radiation spectrum. Thus, the relation $\sqrt{\langle\vartheta^2\rangle_s} > \mu c^2/E$ must hold where $\langle\vartheta^2\rangle_s$ denotes the mean square of the scattering angle along the path s: $\langle\vartheta^2\rangle_s = (E_s/E)^2 \frac{s}{L}$, $E_s = 21 \cdot 10^6$ eV, and L denotes the unit radiation length. We rewrite the last inequality in the form

$$\frac{E_s E}{(\mu c^2)^2} \sqrt{\frac{c}{\omega L}} > 1. \tag{25.2}$$

When this inequality holds for the frequency ω_{cr}, i.e., if

$$\frac{E_s}{\mu c^2} \sqrt{\frac{E}{\mu c^2} \frac{c}{\omega_0 L}} > 1, \tag{25.3}$$

new frequencies $\omega_{cr} < \omega \lesssim \omega_{cr}^*$ appear in the spectrum of the transition radiation and the relation (25.2) holds for these frequencies so that

$$\overset{\cdot}{\omega}_{cr} = \frac{E_s^2 E^2}{(\mu c^2)^4} \frac{c}{L}. \tag{25.4}$$

Let us denote by E' the particle energy for which $\omega_{cr}^* = \omega_{cr}$: $E' = \frac{\omega_0 L}{c} \frac{(\mu c^2)^3}{E_s^2}$. For E < E', the usual formula for the energy of the transition radiation is valid: $W = \frac{e^2\omega_0}{3c} \frac{E}{\mu c^2}$ [77, 78]. If the scattering would not change the cutoff frequency in the spectrum, the energy of the transition radiation could be obtained for E > E' from the last equation by replacing $\omega_0 E/\mu c^2$ with ω_{cr}^*. However, the maximum frequency of the radiation drops off faster with increasing angle when the scattering is taken into account. As a matter of fact, disregarding the scattering, in any direction frequencies are emitted for which the path of coherent interaction between the particle and the field differs strongly for the two sides of the separating boundary. We can write

an approximate radiation condition $(s_v > 2 s_m)$: $\sqrt{1 + \frac{\vartheta^2}{1-\beta^2}} < \frac{\omega_{cr}}{\omega}$. Due to the scattering, the

field in the angular interval below a certain ϑ value [the field corresponding to the second term of Eq. (25.1)] is not generated if, along the path s_m of coherent interaction (disregarding coherent interaction), the particle passes beyond that angle. In the range of frequncies and angles in which $s_v \sim s_m \sim 2c/\omega(1 + \beta^2 + \vartheta^2)$, multiple scattering decreases the path of coherent interaction more than twice in the direction $\vartheta > \sqrt{1-\beta^2}$ whenever $\left(\frac{E_s}{E}\right)\frac{s_m}{L} > 2\vartheta^2$. When we consider this condition to be an additional radiation condition (which is correct for any direction as far as the order of magnitude is concerned), we can rewrite it in the form $\frac{\vartheta^4}{1-\beta^2}\left(1 + \frac{\vartheta^2}{1-\beta^2}\right) < \frac{\omega_{cr}}{\omega}$.

For example, in the region $\vartheta > \sqrt{1-\beta^2}$ the maximum frequency emitted is inversely proportional to the fourth power of ϑ, whereas without scattering the maximum frequency is proportional to the first power of ϑ.

In our calculations we will make use of the fact that the field is generated mainly in the vacuum and determined by the first term of Eq. (25.1). The corresponding equation for the radiation energy per unit polar angle and per unit frequency interval has the form

$$\frac{dW_\omega}{d\vartheta} = \frac{2e^2}{\pi c} \frac{\vartheta^3}{(1-\beta^2+\vartheta^2)^2}. \tag{25.5}$$

The integration must be extended over regions in which one of the following inequalities is satisfied:

$$\frac{\vartheta^2}{1-\beta^2}\left(1+\frac{\vartheta^2}{1-\beta^2}\right) < \frac{\overset{\bullet}{\omega}_{cr}}{\omega}, \qquad (25.6a)$$

$$\sqrt{1+\frac{\vartheta^2}{1-\beta^2}} < \frac{\omega_{cr}}{\omega}. \qquad (25.6b)$$

The first inequality provides a small correction at $E < E'$. When we integrate Eq. (25.5) over the region defined by (25.6), we obtain $W \approx 4e^2\omega_{cr}/3\pi c$, i.e., we obtain a well-known result [77, 78]. A small contribution results from inequality (25.6b) for $E \gg E'$. Integration of Eq. (25.5) over the region defined by Eq. (25.6a) renders the spectral density of the radiation energy

$$W_\omega \approx \frac{e^2}{\pi c}\left(\ln\frac{\sqrt{\omega}+\sqrt{\omega+4\overset{\bullet}{\omega}_{cr}}}{2\sqrt{\omega}} + \frac{2\sqrt{\omega}}{\sqrt{\omega}+\sqrt{\omega+4\overset{\bullet}{\omega}_{cr}}} - 1\right), \qquad (25.7)$$

the density of the radiation energy per unit polar angle

$$\frac{dW}{d\vartheta} \approx 2\,(1-\beta^2)^2\,e^2\,\overset{\bullet}{\omega}_{cr}\,\vartheta/\pi c\,(1-\beta^2+\vartheta^2)^3$$

and the total radiation energy

$$W \approx e^2\overset{\bullet}{\omega}_{cr}/2\pi c,$$

where ω_{cr}^* is defined by Eq. (25.4). Thus, the energy of the transition radiation is of the order of magnitude

$$W \approx \frac{e^2 E_s^2}{2\pi L(\mu c^2)^3}\left(\frac{E}{\mu c^2}\right)^2 \qquad (25.8)$$

for $E > E'$. The energy of the transition radiation increases proportional to the square of the particle energy.

For $E \sim E'$ the cutoff of the spectrum in the region which is essential for the integration is partially determined by multiple scattering and partially by the polarization of the medium. Due to the differences in the cutoff, the linear increase in the radiation energy is replaced by a quadratic law in a certain finite interval near E' where the result of Eq. (25.8) is too small. We note that an exact integration over the region defined by (25.6) does not exceed the value

$$W = \frac{e^2}{\pi c}\left(\frac{4}{3}\omega_{cr} + \frac{1}{2}\overset{\bullet}{\omega}_{cr}\right). \qquad (25.9)$$

Since no quanta with energies exceeding the particle energy are emitted, Eq. (25.8) remains valid for $\hbar\omega_{cr}^* \gtrsim E$ $(E \gtrsim L(\mu c^2)^4/E_s^2\hbar c)$. If $E \gg L(\mu c^2)^4/E_s^2\hbar c$, the radiation energy is of the order of magnitude

$$W \approx \int_0^{E/\hbar} W_\omega d\omega \approx \frac{e^2 E}{2\pi\hbar c}\left(\ln\frac{E_s^2 E\hbar c}{L(\mu c^2)^4} - 1\right). \qquad (25.10)$$

So far, this is our estimate of the effect.

Multiple scattering and its effect on transition radiation have been discussed in several publications [80-82]. A careful review reveals that the results obtained in those papers for the spectral density comprise negative values, and due to the neglect of quantum effects and of the polarization of the medium, the integral result over all frequencies vanishes. This discrepancy in our estimate of the effect resides in the fact that the splitting of the total radiation into transition radiation and bremsstrahlung is, to some extent arbitrary since both types of radiation are interrelated. In order to calculate one type of radiation in a sensible form, one must accurately define what one understands under the second type of radiation. We will outline a method for calculating the transition radiation, in which this type of separating the radiations is employed.

When the energy losses are small in a homogeneous medium, the spectral density of the radiation energy per unit length is constant along the particle path and, consequently, along the path L' the spectral density of the radiation energy is equal to some value $W''_\omega L'$. We assume that the entire space is filled with a medium, except for a layer L; we assume that L is large relative to the length of coherent interaction between the particles and the electromagnetic wave (for the frequency range under consideration). Then the radiation generated by the particle along a path L' + L (intersecting the vacuum gap of width L in the middle of the trajectory) consists of two noninterfering contributions, and the dependence upon L cancels. We split the intensity of each of these contributions into $W''_\omega L'/2$ and a remainder which represents the transition radiation. Since W''_ω is known, the calculation of the rest helps to find the total radiation.

We consider the radiation of waves with frequencies much greater than optical frequencies. Both the reflection and the diffraction of these waves at boundaries can be ignored. At large distances, the radiation is entirely determined by the particle trajectory. When the energy losses are small, the particle trajectory is determined by the same type of scattering before and after passage through the vacuum gap. The implication is that the intensity of the transition radiation is independent of whether the particle moves from the medium into the vacuum or in the opposite direction. We note that the radiation generated along the various path sections can be added independently because no radiation interference occurs. Below, we will use this fact to calculate the angles of quanta emission from the possible trajectories as functions of the actual direction of motion of the particle in the vacuum gap.

Averaging the radiation intensity over all trajectories is equivalent to this type of summation and corresponds to an averaging over a set of trajectories which are shifted and rotated in space so that both the coordinates and the velocity of the particle are uniquely determined in the vacuum gap. We will call this "averaging over the actual direction of motion of the particle in vacuum." The "distribution function" of the set of trajectories shifted in this fashion along the path after the particle's passage through the gap satisfies the usual kinetic equation. Before the passage, the distribution function satisfies the equation which is obtained from the kinetic equation by reversing the direction of the time axis.

The distribution function is symmetric relative to the vacuum gap and this means that the intensities of the noninterfering contributions to the radiation are equal (at large L). We denote the spectral density of the radiation energy per unit solid angle of one of these contributions by

$$W'_{n\omega} = c \left| \int_{-\infty}^{0} \mathbf{A} \, dt + \int_{0}^{\infty} \tilde{\mathbf{A}} \, dt \right|^2, \tag{25.11}$$

for L' → ∞; the notations are interpreted as follows:

$$\mathbf{A} = \frac{e\omega}{2\pi c^2} [\mathbf{n}\mathbf{v}] \, e^{i(\mathbf{k}\mathbf{r} - \omega t)}; \tag{25.12}$$

r and **v** are the coordinates and the velocity of the particle at the time t, respectively; **k** denotes the wave vector of the photon in the medium or in the vacuum, depending upon the location of the particle at the particular time; and **n** = **k**/k. We denote by the tilde the vector **A** defining the radiation field generated along the path in the medium for a particular scattering trajectory. It is assumed in Eq. (25.11) that the particle moves from the vacuum into the medium and intersects the boundary at the time t = 0.

The influence of the boundary upon the radiation is found by comparing the radiation with that in a uniform medium, with the gap tending to zero. Since Eq. (25.11) corresponds to half the radiation density at large L values, the comparison with Eq. (25.11) must be made with half the radiation intensity in the limit L → 0:

$$W_{n\omega}^{\sim} = \frac{c}{2} \left| \overline{\int_{-\infty}^{0} \widetilde{\mathbf{A}} \, dt + \int_{0}^{\infty} \widetilde{\mathbf{A}} \, dt} \right|^{2}. \qquad (25.13)$$

We see that the transition radiation is given by the difference $W_{n\omega} = W_{n\omega}^{!} - W_{n\omega}^{n}$. Let us write this difference in the form

$$W_{n\omega} = c \left| \int_{-\infty}^{0} \mathbf{A} \, dt \right|^{2} + c \overline{\int_{-\infty}^{0} \mathbf{A} \, dt \int_{0}^{\infty} \widetilde{\mathbf{A}}^{\cdot} \, dt} + c \overline{\int_{-\infty}^{0} \mathbf{A}^{\cdot} \, dt \int_{0}^{\infty} \widetilde{\mathbf{A}} \, dt} +$$

$$+ c \left| \int_{0}^{\infty} \widetilde{\mathbf{A}} \, dt \right|^{2} - \frac{c}{2} \left| \int_{-\infty}^{0} \widetilde{\mathbf{A}} \, dt \right|^{2} - \frac{c}{2} \overline{\int_{-\infty}^{0} \widetilde{\mathbf{A}} \, dt \int_{0}^{\infty} \widetilde{\mathbf{A}}^{\cdot} dt} - \frac{c}{2} \overline{\int_{-\infty}^{0} \widetilde{\mathbf{A}}^{\cdot} dt \int_{0}^{\infty} \widetilde{\mathbf{A}} \, dt} - \frac{c}{2} \left| \int_{0}^{\infty} \widetilde{\mathbf{A}} \, dt \right|^{2}. \qquad (25.14)$$

We note that when the particle moves from the vacuum into the medium (this direction was assumed in Eq. (25.11) only to obtain definite conditions, but the effect is independent of the direction of motion), no braking forces resulting in the emission of waves are experienced by the particle along its path in the vacuum far from the other boundary, because the waves are not reflected from the boundary. This means that the generation of transition radiation results in this case from another prehistory of the motion (relative to the time t = 0) as in the case of a motion in a homogeneous medium. The result of the integration of Eq. (25.14) over the photon angles and the frequency is nothing but the difference between the work done in the two cases by the braking forces during the time t, between zero and infinity.

It was assumed in Eq. (25.14) that the path of the particle in the medium (L') is infinite, and therefore, the individual terms contribute infinite intensities resulting from the usual bremsstrahlung. These terms cancel. Furthermore, the terms also simplify when L' is finite, and the overall result of Eq. (25.14) is independent of L', if L' is much greater than the length of coherent interaction. Due to the symmetry of the "distribution function" relative to t = 0, the following equality holds:

$$\int_{-\infty}^{0} \widetilde{\mathbf{A}} \, dt = -\int_{0}^{\infty} \widetilde{\mathbf{A}} \, dt. \qquad (25.15)$$

In the sixth and seventh term of Eq. (25.14), the integral-cofactors define the radiation field generated by the particle along various sections of its trajectory (relative to the particle coordinates at the time t = 0), and the expressions under the integral depend upon the unique determination of **r** and **v** at t = 0. This implies that the averaging over the integrals is independent. This independence and Eq. (25.15) lead to the following result:

$$W_{n\omega} = c \left| \int_{-\infty}^{0} \mathbf{A}\,dt + \overline{\int_{0}^{\infty} \widetilde{\mathbf{A}}\,dt} \right|^2 . \qquad (25.16)$$

The integrals between the vertical bars determine the amplitudes of the radiation field when the particle is slowed down at the boundary and ejected into the other medium. Thus, the additional radiation intensity resulting from the boundary is proportional to the integral of the square of the absolute value of the sum of the corresponding amplitudes (integration over the solid angle).

For the following discussion we have to take the average of Eq. (25.12) over all possible particle trajectories. To do this, we use Gol'dman's result [80]; we showed that in the small-angle approximation for relativistic particle radiation, it suffices to assume*

$$\int w(\mathbf{r}, \vartheta, t) \exp[i(\mathbf{kr} - \omega t)]\,d\mathbf{r} = \exp(\alpha_1\theta^2 + \alpha_2\theta\theta_0 + \alpha_3), \qquad (25.17)$$

$$\alpha_1 = -\left(\frac{i\omega}{8q}\right)^{1/2} \mathrm{cth}\,(2i\omega q)^{1/2}t, \quad \alpha_2 = \left(\frac{i\omega}{8q}\right)^{1/2} \mathrm{sh}^{-1}\,(2i\omega q)^{1/2}t,$$

$$\alpha_3 = -\frac{i\omega\gamma t}{2g^2} - \ln[\mathrm{sh}\,(2i\omega q)^{1/2}t] - \theta_0^2 \left(\frac{i\omega}{8q}\right)^{1/2} \mathrm{cth}\,(2i\omega q)^{1/2}t + \ln\left(\frac{i\omega}{8\pi^2 q}\right)^{1/2},$$

$$\vartheta = \theta - \theta_0, \quad \omega \gg \omega_0,$$

where $w(\mathbf{r}, \vartheta, t)$ denotes a distribution function satisfying the usual kinetic equation in the Fokker-Planck approximation with the initial condition $w(\mathbf{r}, \vartheta, 0) = \delta(\mathbf{r})\delta(\theta - \theta_0)$:

$$g = E/mc^2, \qquad q = \overset{\bullet}{\omega}_{\mathrm{cr}}/4g^4, \qquad \gamma = 1 + (\omega_{\mathrm{cr}}^2/\omega^2), \qquad (25.18)$$

and θ and θ_0 denote the angles between the wave vector $\mathbf{k}$ of the photons emitted and $\mathbf{v}$ and $\mathbf{v}_0$ respectively.

We obtain with Eq. (25.17)

$$\widetilde{\mathbf{A}} = \frac{e\omega}{2\pi c} \int \theta w(\mathbf{r}, \vartheta, t)\, e^{i(\mathbf{kr} - \omega t)}\,d\mathbf{r}\,d\theta = \frac{\theta_0}{\mathrm{ch}^2\,(2i\omega q)^{1/2}t} \exp\left[-\theta_0^2\left(\frac{i\omega}{8q}\right)^{1/2} \mathrm{th}\,(2i\omega q)^{1/2}t - \frac{i\omega\gamma t}{2g^2}\right]. \qquad (25.19)$$

When we switch in Eq. (25.16) to integration over the dimensionless quantity x, which is related to t by the relation x = $(2i\omega q)^{\frac{1}{2}}$t, and when we use Eq. (25.19) and make some simple transformations, we obtain the following result:

$$W_{n\omega} = \frac{e^2 g^2}{\pi^2 cz} \left| \frac{1}{1+z} - \sigma\gamma \int_{0}^{\infty} \exp(-\sigma z\,\mathrm{th}\,x - \sigma\gamma x)\,dx \right|^2, \qquad (25.20)$$

$$z = \theta_0^2 g^2, \qquad g = E/mc^2, \qquad \sigma = (i\omega/2\overset{\bullet}{\omega}_{\mathrm{cr}})^{1/2},$$

$$\gamma = 1 + \omega_{\mathrm{cr}}^2/\omega^2, \qquad \omega_{\mathrm{cr}} = \omega_0 g, \qquad \overset{\bullet}{\omega}_{\mathrm{cr}} = E_s^2 g^2/m^2 c^3 L.$$

With the results of Ternovskii [81], we obtain after some analogous calculations and transformations a quantum-mechanical generalization of Eq. (25.10):

$$W_{n\omega} = \frac{e^2 g^2}{2\pi^2 p_0^2 \{k^2 + [p_0^2 + (p_0 - k)^2]\,z\}} \left| \frac{p_0^2 + (p_0 - k)^2 - k^2}{1+z} + \right.$$

$$\left. + k^2 \sigma \int_{0}^{\infty} \frac{dx}{\mathrm{ch}\,x}\, e^{-\sigma z\,\mathrm{th}\,x - \sigma\gamma x} - \sigma\gamma\,[p_0^2 + (p_0 - k)^2] \int_{0}^{\infty} e^{-\sigma z\,\mathrm{th}\,x - \sigma\gamma x}\,dx \right|^2, \qquad (25.21)$$

---

*The polarization of the medium has been taken into account and the misprints of [80] corrected.

$$g = \frac{2p_0 - k}{2}, \qquad \sigma = \sqrt{ik/8qg^2 p_0(p_0 - k)}, \qquad \gamma = \frac{1 + p_0(p_0 - k)k_0^2}{k^2},$$

$$k_0^2 = 4\pi n Z e^2, \qquad qg^2 = 2\pi n Z^2 e^4 \ln(\theta_2/\theta_1), \qquad m = \hbar = c = 1,$$

where k denotes the momentum of the emitted photon and p_0 the momentum of the particle at the time $t = 0$.

Generalization of the result of Eq. (25.21) to the case of two media leads to

$$W_{n\omega} = \frac{e^2 g^2}{2\pi^2 p_0^2 \{k^2 + [p_0^2 + (p_0 - k)^2] \cdot z\}} \left| k^2 \sigma_1 \int_0^\infty \frac{dx}{\mathrm{ch}\, x} e^{-\sigma_1 z\, \mathrm{th}\, x - \sigma_1 \gamma_1 x} - \right.$$

$$- \sigma_1 \gamma_1 [p_0^2 + (p_0 - k)^2] \int_0^\infty e^{-\sigma_1 z\, \mathrm{th}\, x - \sigma_1 \gamma_1 x} dx -$$

$$\left. - k^2 \sigma_2 \int_0^\infty \frac{dx}{\mathrm{ch}\, x} e^{-\sigma_2 z\, \mathrm{th}\, x - \sigma_1 \gamma_2 x} + \sigma_2 \gamma_2 [p_0^2 + (p_0 - k)^2] \int_0^\infty e^{-\sigma_2 z\, \mathrm{th}\, x - \sigma_1 \gamma_2 x} dx \right|^2, \qquad (25.22)$$

where the subscripts 1 and 2 denote that the quantities refer to the first and second medium, respectively.

The spectral density and the total energy of the transition radiation are

$$W_\omega = \frac{\pi}{g^2} \int_0^\infty W_{n\omega} dz, \qquad W = \pi \int_0^{p_0} \frac{dk}{g^2} \int_0^\infty W_{n\omega} dz. \qquad (25.23)$$

Let us dwell on the transition radiation at frequencies $\omega \ll E/\hbar$. When a charged particle moves through a boundary separating a vacuum from a medium, the calculation of the spectral density of the transition-radiation energy results in the following expressions:

$$W_\omega = \frac{e^2}{\pi c} \left[\left(1 + 2\frac{\omega^2}{\omega_{cr}^2} \right) \ln\left(1 + \frac{\omega_{cr}^2}{\omega^2} \right) - 2 \right], \qquad \omega \lesssim \omega_{cr}, \qquad (25.24)$$

$$W_\omega = \frac{e^2}{\pi c} \left[\frac{1}{6} \left(\frac{\omega_{cr}}{\omega} \right)^4 + \frac{8}{21} \left(\frac{\omega_{cr}^*}{\omega} \right)^2 \right], \qquad \omega \gg \omega_{cr}, \qquad (25.25)$$

if $\omega_{cr}^* \ll \omega_{cr}$. We obtain for $\omega_{cr}^* \gg \omega_{cr}$

$$W_\omega = \frac{e^2}{\pi c} \ln \frac{\omega_{cr}^2}{\omega^2}, \qquad \omega < \omega_{cr}^{4/3} \omega_{cr}^{*-1/3}, \qquad (25.26)$$

$$W_\omega = \frac{e^2}{\pi c} \ln \frac{2}{3} \left(\frac{\omega_{cr}^*}{\omega} \right)^{1/3}, \qquad \omega_{cr}^{4/3} \omega_{cr}^{*-1/3} < \omega \ll \omega_{cr}^*, \qquad (25.27)$$

$$W_\omega = \frac{e^2}{\pi c} \frac{8}{21} \left(\frac{\omega_{cr}^*}{\omega} \right)^2, \qquad \omega \gg \omega_{cr}^* \qquad (25.28)$$

In order to determine the spectral density of the energy at frequencies $\omega \sim \omega_{cr}^*$ for $\omega_{cr}^* \gg \omega_{cr}$, a numerical integration was performed. The results are shown in Fig. 22, where the function $\pi c e^{-2} (\omega/\omega_{cr}^*)^{\frac{1}{2}} W_\omega$ is plotted to the ordinate axis, and the quantity $(\omega/\omega_{cr}^*)^{\frac{1}{2}}$ to the abscissa axis. The solid curve corresponds to our results and the dashed curve to the results of some preceding publications [80-82].*

The spectral density of the transition-radiation energy can be approximated by the formula

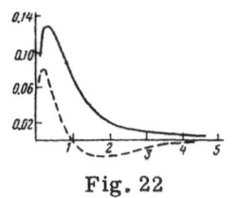

Fig. 22

$$W_\omega = \frac{e^2}{\pi c}\left(\ln\frac{a_1\sqrt{\omega}+\sqrt{\omega+a_2\omega_{cr}^*}}{a_3\sqrt{\omega}}+\frac{a_3\sqrt{\omega}}{a_1\sqrt{\omega}+\sqrt{\omega+a_2\omega_{cr}^*}}-1\right),\qquad(25.29)$$

$$a_1=6\sqrt{3/7}-1,\qquad a_2={}^{48}/_7,\qquad a_3=6\sqrt{3/7},$$

$$\omega_{cr}^{1/2}\omega_{cr}^{*-1/2}<\omega\ll E/\hbar;$$

the maximum deviations of the approximation are less than several per cent.

Equation (25.29) holds if, along the path $s\sim(E/mc^2)^2\,c/\omega$ along which the transition radiation is generated, no electrons are absorbed ($s\ll L$), but many collisions between electrons and the atoms of the medium occur ($s\gg n^{-1/3}$). This leads us to the applicability range of the result defined by Eq. (25.29):

$$(mc^2)^2/E_s^2\ll\omega/\omega_{cr}^*\ll n^{1/3}L\,(mc^2)^2/E_s^2.\qquad(25.30)$$

In the case of electrons, the quantity at the left is of the order 10^{-3} and that at the right is of the order 10^5; thus, relation (25.30) refers to large ($\omega\gg\omega_{cr}^*$), as well as to relatively small ($\omega\ll\omega_{cr}^*$) frequencies for which Eq. (25.29) results in Eq. (25.28) and Eq. (25.27), respectively.

By integrating Eq. (25.29) over the frequency, we obtain the energy of the transition radiation

$$W=\frac{e^2\omega_{cr}^*}{\pi c}\frac{a_2}{a_3\,(a_3-2)}\left(\frac{a_3}{a_3-2}\ln\frac{a_3}{2}-1\right)\approx\frac{0.34e^2\,E_s^2E^2}{\pi L\,(mc^2)^4}.\qquad(25.31)$$

This energy is proportional to the square of the particle's energy. Equation (25.31) holds in the interval

$$\omega_0L(mc^2)^3/cE_s^2<E\ll L(mc^2)^4/E_s^2\hbar c,\qquad(25.32)$$

outside this interval, the dependence of the transition-radiation energy upon the particle energy is given either by the polarization of the medium or quantum effects, and the energy of the transition radiation is proportional to the energy of the particle.

The frequency dependence of the spectral energy density of the transition radiation generated by multiple scattering is displayed in Fig. 23. The function W_ω is plotted to the ordinate axis in units $e^2/\pi c$ (the function was obtained from Eq. (25.29); the dimensionless quantity ω/ω_{cr}^* is plotted to the abscissa axis. The area defined by the curve and the coordinate axis represents the radiation energy. The dashed lines in the figure indicate a rectangle which has the same area. Figure 24 displays for comparison the frequency dependence of the spectral energy density of the transition radiation resulting from the polarization of the medium. The function defined by Eq. (25.24) is plotted to the ordinate axis in the same units as before and to the same scale; the ratio ω/ω_{cr} is plotted to the abscissa axis, with $\omega_{cl}=\omega_0/\sqrt{1-\beta^2}$. The area of the rectangle indicated by the dashed lines represents the energy of the transition radiation. We see from a comparison of the curves that the spectral density of the transition radiation generated at frequencies $\omega\sim\omega_{cr}^*$ (for $\omega_{cr}^*\gg\omega_{cr}$) by multiple scattering drops off,

*Comparison of the curves of Fig. 22 reveals that the results obtained by the authors of [80-82] describe the total boundary effect for the radiation at frequencies $\omega\ll\omega_{cr}^*$; the curves coincide for $\omega\ll\omega_{cr}^*$.

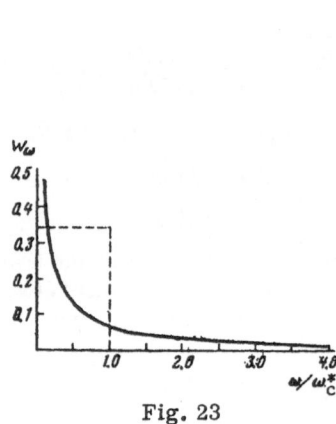

Fig. 23

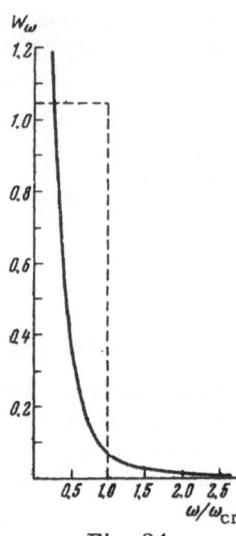

Fig. 24

with increasing ω, much more slowly than the transition radiation generated by the polarization of the medium at frequencies $\omega \sim \omega_{cr}$ (for $\omega_{cr}^* \ll \omega_{cr}$).

It is worthwhile to compare our results with those for the bremsstrahlung in condensed matter [83-89]. It follows from this comparison that in frequency regions in which the intensity of the transition radiation is determined by the polarization of the medium or by multiple scattering and in which the transition-radiation intensity is proportional to $\ln(\omega_{cr}^2/\omega^2)$ or $\ln(2/3)(\omega_{cr}^*/\omega)^{\frac{1}{2}}$, respectively, the bremsstrahlung is weaker than in a discrete medium; the intensity reduction is ω^2/ω_{cr}^2 and $(3/2)(\omega/\omega_{cr}^*)^{\frac{1}{2}}$ times, respectively. The fact that the corresponding frequency intervals are identical indicates that the polarization of the medium and multiple scattering have the same consequence, namely an increase in the intensity of the transition radiation and a decrease in the intensity of the bremsstrahlung. The reason is that the path of coherent interaction between the particle and the electromagnetic wave of the particular frequency is reduced. A differential effect occurs in the case of the transition radiation. In the case of bremsstrahlung, the result is proportional to the relative change in path length.

§26. Bremsstrahlung at Frequencies

Exceeding Optical Frequencies

The intensity of the bremsstrahlung of a relativistic particle in condensed matter has been discussed extensively in several previous publications [83-89].

When the initial conditions are put into a more definite form, both the degree of polarization and the angular distribution of the bremsstrahlung play a role. We will consider the emission of quanta with energies much smaller than the particle energy. This makes it possible to employ the classical theory.

We assume that a charged relativistic particle has the velocity $\mathbf{v}_0$ at the initial time $t = 0$ and moves in the medium for some finite time T. When averaged over all possible trajectories [85], the spectral density of the radiation energy per unit solid angle is given by

$$W_{n\omega} = \frac{e^2\omega^2}{2\pi^2 c^3} \operatorname{Re} \int_0^T dt \int [\mathbf{n}\mathbf{v}] w_1(\mathbf{v}, \mathbf{r}, t) \, d\mathbf{v} d\mathbf{r} \int_0^{T-t} d\tau \int [\mathbf{n}\mathbf{v}'] w_2(\mathbf{v}, \mathbf{r}; \mathbf{v}', \mathbf{r}'; \tau) \, e^{(i\mathbf{k}\boldsymbol{\rho} - \omega\tau)} \, d\mathbf{v} d\boldsymbol{\rho}, \qquad (26.1)$$

where k denotes the wave vector of the waves emitted; $n = k/k$; $\mathbf{v}$ and $\mathbf{r}$ denote the velocity and the particle coordinates at the time t, respectively; $\mathbf{v'}$ and $\mathbf{r'}$ denote the velocity and the particle coordinates at a later time $t + \tau$; $\boldsymbol{\rho} = \mathbf{r'} - \mathbf{r}$; $w_1(\mathbf{v}, \mathbf{r}, t)$ is the distribution function for the initial condition $w_1(\mathbf{v}, \mathbf{r}, 0) = \delta(\mathbf{v_0} - \mathbf{v})\delta(\mathbf{r})$; $w_2(\mathbf{v}, \mathbf{r}; \mathbf{v'}, \mathbf{r'}; \tau)$ denotes the probability of the values $\mathbf{v'}$ and $\mathbf{r'}$ under the condition that $\mathbf{v'}$ and $\mathbf{r'}$ assume the values $\mathbf{v}$ and $\mathbf{r}$ at the time t, respectively. The functions w_1 and w_2 satisfy the usual kinetic equation.

Since the angles between the vectors $\mathbf{v_0}$, $\mathbf{v}$, $\mathbf{v'}$, and $\mathbf{n}$ are small and the velocity remains unchanged in the scattering process, we can switch to the following angular vectors:

$$\boldsymbol{\vartheta} = \frac{\mathbf{v} - \mathbf{v_0}}{v_0}, \quad \boldsymbol{\vartheta}' = \frac{\mathbf{v'} - \mathbf{v_0}}{v_0}, \quad \boldsymbol{\theta_0} = \mathbf{n} - \frac{\mathbf{v_0}}{v_0}, \quad \boldsymbol{\theta} = \mathbf{n} - \frac{\mathbf{v}}{v_0}, \quad \boldsymbol{\theta}' = \mathbf{n} - \frac{\mathbf{v'}}{v_0}, \tag{26.2}$$

$$W_{n\omega} = \frac{e^2\omega^2}{2\pi^2 c} \operatorname{Re} \int_0^T dt \int \boldsymbol{\theta} w_1(\boldsymbol{\vartheta}, \mathbf{r}, t) \, d\boldsymbol{\vartheta} d\mathbf{r} \int_0^{T-t} d\tau \int \boldsymbol{\theta}' w_2(\boldsymbol{\vartheta}, \mathbf{r}; \boldsymbol{\vartheta}', \mathbf{r}'; \tau) e^{i(\mathbf{k}\rho - \omega\tau)} d\boldsymbol{\vartheta}' d\rho. \tag{26.3}$$

The problem is to calculate the integral. According to the result of Gol'dman (Eq. (15) of [80]), when the polarization of the medium is taken into account (Eq. (6) in [16]) at frequencies exceeding optical frequencies ($\omega \gg \omega_0$, $\omega_0^2 = 4\pi n Z e^2/m$), we obtain in the Fokker-Planck approximation:

$$\int w_2(\boldsymbol{\vartheta}, \mathbf{r}; \boldsymbol{\vartheta}', \mathbf{r}'; \tau) e^{i(\mathbf{k}\rho - \omega\tau)} d\rho = \exp[\alpha_0(\tau) + (\theta^2 + \theta'^2)\alpha_1(\tau) + \theta\theta'\alpha_2(\tau)], \tag{26.4}$$

where

$$\alpha_0(\tau) = -i\omega\gamma\tau/2g^2 - \ln \operatorname{sh}(2i\omega q)^{1/2}\tau,$$

$$\alpha_1(\tau) = -(i\omega/8q)^{1/2} \operatorname{cth}(2i\omega q)^{1/2}\tau, \quad \alpha_2(\tau) = (i\omega/2q)^{1/2} \operatorname{sh}^{-1}(2i\omega q)^{1/2}\tau. \tag{26.5}$$

We used the following notations:

$$g = \frac{E}{mc^2}, \qquad q = \frac{\overset{*}{\omega}_{cr}}{4g^4}, \qquad \gamma = 1 + \frac{\omega_{cr}^2}{\omega^2}, \qquad \omega_{cr} = \omega_0 \frac{E}{mc^2},$$

$$\overset{*}{\omega}_{cr} = \frac{E_s^2 E^2}{(mc^2)^4} \frac{c}{L},$$

$$\omega_0^2 = 4\pi n Z e^2/m, \qquad E_s = \sqrt{4\pi \cdot 137}\, mc^2, \qquad L^{-1} = \frac{4nZ}{137}\left(\frac{e^2}{mc^2}\right)^2 \ln 191\, Z^{-1/3}. \tag{26.6}$$

With the well-known relation

$$\int w_1(\boldsymbol{\vartheta}, \mathbf{r}, t)\, d\mathbf{r} = (4\pi q t)^{-1} \exp(-\vartheta^2/4qt), \tag{26.7}$$

we obtain after integration of Eq. (26.3) over ρ and $\mathbf{r}$:

$$W_{n\omega} = \frac{e^2\omega^2}{8\pi^3 cq} \operatorname{Re} \int_0^T \frac{dt}{t} \int \boldsymbol{\theta} \exp\left(-\frac{\theta^2}{4qt}\right) d\boldsymbol{\vartheta} \int_0^{T-t} d\tau \int \boldsymbol{\theta}' \exp[\alpha_0(\tau) +$$

$$+ (\theta^2 + \theta'^2)\alpha_1(\tau) + \theta\theta'\alpha_2(\tau)]\, d\boldsymbol{\vartheta}'. \tag{26.8}$$

For the purpose of calculating the degree of polarization of the bremsstrahlung, both the vectors $\boldsymbol{\theta}$ and $\boldsymbol{\theta}'$ must be represented as the sum of two vectors, one of which is parallel to the vector $\boldsymbol{\theta_0}$ and the other perpendicular to it. We denote these vectors by the symbols $\parallel$ and $\perp$, respectively:

$$\theta = \theta_\perp + \theta_\parallel, \qquad \theta' = \theta'_\perp + \theta'_\parallel,$$

$$\theta_\perp = \theta - \frac{\theta_0\theta}{\theta_0^2}\theta_0, \qquad \theta_\parallel = \frac{\theta_0\theta}{\theta_0^2}\theta_0, \tag{26.9}$$

$$\theta'_\perp = \theta' - \frac{\theta_0\theta'}{\theta_0^2}\theta_0, \qquad \theta'_\parallel = \frac{\theta_0\theta'}{\theta_0^2}\theta_0.$$

The scalar product $\theta\theta'$ of Eq. (26.8) splits into the sum of the following scalar products:

$$\theta\theta' = \theta_\perp\theta'_\perp + \theta_\parallel\theta'_\parallel. \tag{26.10}$$

The part of Eq. (26.8) which, after inserting Eqs. (26.9), contains $\theta_\parallel\theta'_\parallel$ describes the angular distribution of the radiation intensity of the waves polarized in the plane of the wave vector and the initial direction of motion of the particle. We denote this intensity by $W_{n\,\omega\,\parallel}$. The other part of Eq. (26.8) corresponds to waves polarized perpendicular to the plane. We denote the second part by $W_{n\,\omega\,\perp}$. After integration over ϑ', we obtain

$$W_{n\omega\perp} = \frac{e^2\omega^2}{16\pi^3cq\,\theta_0^2}\,\mathrm{Re}\int_0^T\frac{dt}{t}\int_0^{T-t}\frac{\alpha_2}{\alpha_1^2}\,e^{\alpha_0}d\tau\int[\theta_0\theta]^2\exp\left[\theta^2\left(\alpha_1-\frac{\alpha_2^2}{4\alpha_1}\right)-\frac{\vartheta^2}{4qt}\right]d\vartheta,$$

$$\tag{26.11}$$

$$W_{n\omega\,\parallel} = \frac{e^2\omega^2}{16\pi^2cq\,\theta_0^2}\,\mathrm{Re}\int_0^T\frac{dt}{t}\int_0^{T-t}\frac{\alpha_2}{\alpha_1^2}\,e^{\alpha_0}d\tau\int(\theta_0\theta)^2\exp\left[\theta^2\left(\alpha_1-\frac{\alpha_2^2}{4\alpha_1}\right)-\frac{\vartheta^2}{4qt}\right]d\vartheta.$$

When we now integrate over ϑ, we find

$$W_{n\omega\perp} = \frac{e^2\omega^2}{32\pi cq}\,\mathrm{Re}\int_0^T\frac{dt}{t}\int_0^{T-t}\frac{\alpha_1}{\alpha_1^2 p^2}\exp\left(\alpha_0-\frac{\theta_0^2}{4qt}+\frac{\theta_0^2}{16\cdot pq^2t^2}\right)d\tau,$$

$$\tag{26.12}$$

$$W_{n\omega\,\parallel} = \frac{e^2\omega^2}{32\pi cq}\,\mathrm{Re}\int_0^T\frac{dt}{t}\int_0^{T-t}\frac{\alpha_2}{\alpha_1^2 p^2}\left(1+\frac{\theta_0^2}{8pq^2t^2}\right)\exp\left(\alpha_0-\frac{\theta_0^2}{4qt}+\frac{\theta_0^2}{16pq^2t^2}\right)d\tau,$$

where

$$p = \frac{\alpha_2^2}{4\alpha_1}+\frac{1}{4qt}-\alpha_1.$$

By inserting the functions α_0, α_1, and α_2 given by Eqs. (26.5) into Eqs. (26.12), we can rewrite the results in the following way:

$$W_{n\omega\perp} = \frac{e^2\omega^2 q}{\pi^2 c}\,\mathrm{Re}\int_0^T tdt\int_0^{T-t}\frac{\exp\left(-\frac{\theta_0^2}{4qt}\frac{\sqrt{2i\omega q}\,t\,\mathrm{th}\,\sqrt{2i\omega q}\,\tau}{1+\sqrt{2i\omega q}\,t\,\mathrm{th}\,\sqrt{2i\omega q}\,\tau}-\frac{i\omega\gamma\tau}{2g^2}\right)}{(1+\sqrt{2i\omega q}\,t\,\mathrm{th}\,\sqrt{2i\omega q}\,\tau)^2\,\mathrm{ch}^2\,\sqrt{2i\omega q}\,\tau}\,d\tau,$$

$$W_{n\omega\,\parallel} = \frac{e^2\omega^2 q}{\pi^2 c}\,\mathrm{Re}\int_0^T tdt\int_0^{T-t}\left[1+\frac{\theta_0^2}{2qt\,(1+\sqrt{2i\omega q}\,t\,\mathrm{th}\,\sqrt{2i\omega q}\,\tau)}\right]\times$$

$$\times\frac{\exp\left(-\frac{\theta_0^2}{4qt}\frac{\sqrt{2i\omega q}\,t\,\mathrm{th}\,\sqrt{2i\omega q}\,\tau}{1+\sqrt{2i\omega q}\,t\,\mathrm{th}\,\sqrt{2i\omega q}\,\tau}-\frac{i\omega\gamma\tau}{2g^2}\right)}{(1+\sqrt{2i\omega q}\,t\,\mathrm{th}\,\sqrt{2i\omega q}\,\tau)^2\,\mathrm{ch}^2\,\sqrt{2i\omega q}\,\tau}\,d\tau. \tag{26.13}$$

In deriving Eqs. (26.13), the assumption was made that the energy losses are relatively small. It is generally accepted that a charged particle loses a small portion of its energy during the passage through a layer of condensed matter, provided that the thickness of the layer is much smaller than the unit radiation length (cT ≪ L). Beyond the layer, the particle trajectory is linear. One must bear in mind that in certain frequency ranges and angular regions the radiation intensity described by Eqs. (26.13) can be smaller than the radiation intensity resulting from a sudden arrest of the particle at the time t = 0, or from the ejection of a particle into vacuum at the time t = T. In these ranges, the contribution resulting from the linear trajectory section is important for the radiation. This contribution is given by the expression

$$
W'_{n\omega} = \frac{e^2}{\pi^2 c} \frac{\theta_0^2}{(1-\beta^2+\theta_0^2)^2} + \frac{e^2\omega}{\pi^2 c} \frac{\theta_0}{1-\beta^2+\theta_0^2} \operatorname{Im} \int_0^T dt \int \theta \exp\left[\alpha_0(t)+(\theta_0^2+\theta^2)\alpha_1(t)+\theta_0\theta\alpha_2(t)\right] d\vartheta +
$$

$$
+ \frac{e^2\omega}{\pi^2 c} \frac{\theta_0}{1-\beta^2+\theta_0^2} \operatorname{Im} \int_0^\infty dt \int \theta \exp\left[\alpha_0(T)+(\theta_0^2+\theta^2)\alpha_1(T)+\theta_0\theta\alpha_2(T) - \right.
$$

$$
\left. -\frac{i\omega t}{2}(1-\beta^2+\theta^2)\right] d\vartheta - \frac{e^2\omega^2}{8\pi^3 cqT} \operatorname{Re} \int_0^T dt \int_0^\infty dt' \int \theta \exp\left[\frac{i\omega t'}{2}(1-\beta^2+\theta^2) - \right.
$$

$$
\left. -\frac{\theta^2}{4qT}\right] d\vartheta \int \theta' \exp\left[\alpha_0(t)+(\theta^2+\theta'^2)\alpha_1(t)+\theta\theta'\alpha_2(t)\right] d\vartheta' +
$$

$$
+ \frac{e^2\omega^2}{16\pi^3 cqT} \operatorname{Re} \int_0^\infty dt \int_0^\infty dt' \int \theta^2 \exp\left[-\frac{i\omega t}{2}(1-\beta^2+\theta^2)+\frac{i\omega t'}{2}(1-\beta^2+\theta^2)-\frac{\theta^2}{4qT}\right] d\vartheta. \tag{26.14}
$$

The first term corresponds to the radiation resulting from the sudden arrest of the particle at the time t = 0. The second and third terms describe the interference of the field generated along the path in vacuum before the entry into the medium, with the field generated along the particle's path in the medium and the vacuum, respectively, after the particle has passed through the layer of matter; the fourth term denotes the interference of the field generated along the path in the medium with the field resulting from the linear section of the trajectory after the exit of the particle from the medium; finally, the fifth term describes the radiation resulting from the ejection of the particle into vacuum, after the particle passed through the layer of matter. In order to simplify the notations, we modified the time origin in the last three terms so that t = 0 corresponds to the time at which the particle passes from the medium into vacuum.

The ϑ and ϑ' integration in Eq. (26.14) is done in the same fashion as in the derivation of Eq. (26.13), i.e., the formula is divided into two sections corresponding to the radiation of waves of various polarizations. After integrating and after adding the results to Eq. (26.13), we obtain the angular distribution of the spectral energy density of the bremsstrahlung generated when a charged particle passes through a layer of condensed matter. We write the results for the radiation of waves of the various polarization states in the following form:

$$
W_{n\omega\perp} = \frac{e^2 g^2}{2\pi^3 c} \operatorname{Re} \sigma \left\{ -2 \int_0^{x_0} x\,dx \int_0^{x_0-x} \frac{\exp\left[-\sigma\left(\gamma y + \frac{z\,\mathrm{th}\,y}{1+x\,\mathrm{th}\,y}\right)\right]}{(1+x\,\mathrm{th}\,y)^2\,\mathrm{ch}^2 y}\,dy + \right.
$$

$$
+ 2x_0 \int_0^{x_0} dx \int_0^\infty \frac{\exp\left\{-\sigma\left[\gamma x - y + \frac{z(\mathrm{th}\,x-y)}{1+x(\mathrm{th}\,x-y)}\right]\right\}}{[1+x_0(\mathrm{th}\,x-y)]^3\,\mathrm{ch}^2 x}\,dy - x_0 \int_0^\infty dx \int_0^\infty \frac{\exp\left\{-\sigma(x-y)\left[1+\frac{z}{1+x_0(x-y)}\right]\right\}}{[1+x_0(x-y)]^3}\,dy \right\},
$$

$$
\tag{26.15}
$$

$$W_{n\omega\parallel} = \frac{e^2 g^2}{\pi^2 c} \operatorname{Re} \left\{ -2\sigma \int_0^{x_0} dx \int_0^{x-x} \frac{\exp\left[-\sigma\left(\gamma y + \frac{z\,\mathrm{th}\,y}{1+x\,\mathrm{th}\,y}\right)\right]}{(1+x\,\mathrm{th}\,y)^2\,\mathrm{ch}^2 y}\left(\frac{x}{2}+\frac{\sigma z}{1+x\,\mathrm{th}\,y}\right) dy + \right.$$

$$+ \frac{z}{(1+z)^2} - \frac{2\sigma z}{1+z}\int_0^{x_0}\frac{\exp\left[-\sigma(\gamma x + z\,\mathrm{th}\,x)\right]}{\mathrm{ch}^2 x}\,dx - \frac{2\sigma z}{1+z}\int_0^{\infty}\frac{\exp\left\{-\sigma\left[\gamma x_0 + x + \frac{z(\mathrm{th}\,x_0 + x)}{1+x\,\mathrm{th}\,x_0}\right]\right\}}{(1+x\,\mathrm{th}\,x_0)^2\,\mathrm{ch}^2 x_0}\,dx +$$

$$+ 2\sigma \int_0^{x_0} dx \int_0^{\infty}\frac{\exp\left\{-\sigma\left[\gamma x - y + \frac{z(\mathrm{th}\,x - y)}{1+x_0(\mathrm{th}\,x - y)}\right]\right\}}{[1+x_0(\mathrm{th}\,x - y)]^2\,\mathrm{ch}^2 x}\left[\frac{x_0}{2}+\frac{\sigma z}{1+x_0(\mathrm{th}\,x - y)}\right] dy -$$

$$\left. - \sigma \int_0^{\infty} dx \int_0^{\infty}\frac{\exp\left\{-\sigma(x-y)\left[1+\frac{z}{1+x_0(x-y)}\right]\right\}}{[1+x_0(x-y)]^2}\left[\frac{x_0}{2}+\frac{\sigma z}{1+x_0(x-y)}\right] dy\right\}, \qquad (26.16)$$

where x and y denote new dimensionless integration variables which are related to time by the factor $\sqrt{2i\omega q}$, we have introduced the notations

$$x_0 = \sqrt{2i\omega q}\,T, \quad z = \theta_0^2 g^2, \quad \sigma = \sqrt{i\omega/2\omega_{\mathrm{cr}}^{\bullet}} \qquad (26.17)$$

and used the notations stated in Eqs. (26.6).

The degree of polarization of the bremsstrahlung is given by the formula

$$p = \frac{W_{n\omega\parallel} - W_{n\omega\perp}}{W_{n\omega\parallel} + W_{n\omega\perp}}. \qquad (26.18)$$

Of particular interest is the radiation at small layer thicknesses. Let us assume that $\omega \gg \omega_{\mathrm{cr}}$. The parameter γ is frequently close to unity in this region. When we assume $\gamma = 1$ and expand the results in a power series in T, we obtain the first nonvanishing terms of the expansion for the angular distribution and the spectral density of the radiation energy in the form

$$W_{n\omega} = \frac{4e^2 qTg^4}{\pi^2 c}\frac{1+z^2}{(1+z)^4}, \quad W_{\omega} = \frac{8e^2 qTg^2}{3\pi c}. \qquad (26.19)$$

These terms are much greater than the following expansion terms which are proportional to the square of T, provided that $x_0/\sigma \ll 1$, i.e., if during the time T multiple scattering does not lead the particle beyond an angle of about $\sqrt{1-\beta^2}$. In this case, Eqs. (26.15), (26.16), and (26.18) lead to the respective results

$$W_{n\omega\perp} = \frac{2e^2 qTg^4}{\pi^2 c(1+z)^2}, \quad W_{n\omega\parallel} = \frac{2e^2 qTg^4}{\pi^2 c}\frac{(1-z)^2}{(1+z)^4}, \quad P = -\frac{2z}{1+z^2}. \qquad (26.20)$$

The results of a calculation of the corresponding quantities for a single scattering process involving an angle determined from multiple-scattering theory for the layer under consideration agree with the results stated in Eqs. (26.19) and (26.20). This agreement between the results is a consequence of the interference of radiations generated in individual collision processes.

When the layer is thick, the first terms in Eqs. (26.15) and (26.16) outweigh all the other terms and the error which is made when the upper limit of the y integration is replaced by infinity is not greater than the error resulting from neglecting the other terms, whereas the formulas become simpler. When we simplify the equations in this fashion and use th y as a new integration variable, the results which are valid for a large layer thickness can be written in the form*

$$W_{n\omega\perp} = \frac{e^2\omega}{2\pi^2 c} \, \text{Im} \int\limits_0^T \eta t dt \int\limits_0^1 \frac{\exp\left(-\frac{\sigma z x}{1+\eta t x}\right)}{(1+\eta t x)^2} \left(\frac{1-x}{1+x}\right)^{\frac{\sigma\gamma}{2}} dx \quad (\eta = \sqrt{2i\omega q}),$$ (26.21)

$$W_{n\omega\parallel} = \frac{e^2\omega}{\pi^2 c} \, \text{Im} \int\limits_0^T dt \int\limits_0^1 \frac{\exp\left(-\frac{\sigma z x}{1+\eta t x}\right)}{(1+\eta t x)^2} \left(\frac{1-x}{1+x}\right)^{\frac{\sigma\gamma}{2}} \left(\frac{\eta t}{2} + \frac{\sigma z}{1+\eta t x}\right) dx.$$ (26.22)

The results are complicated functions of the layer thickness and of the parameters σ and γ, which characterize the influence of multiple scattering and the polarization of the medium upon the radiation, respectively. These functions can be examined in detail by numerical integration using computers.

§ 27. Optical Bremsstrahlung in an Absorbing Medium

In this paragraph we will calculate the angular distribution and the degree of polarization of the radiation generated by a charged particle moving from vacuum into a medium along the normal to the separating boundary; the multiple scattering in the medium will be taken into account.

Equations (4.13)-(4.18) will be used as formulas to start with. These equations describe the radiation field by means of the law of motion of the charge. It follows from these equations that in the case of normal incidence ($\beta_x = \beta_y = 0$, $\beta_z = -\beta$, $x = y = 0$, $z = -vt$ for $t < 0$), the radiation field in vacuum is determined by the sum of the following Hertz vectors:

$$\Pi_{\omega 2\parallel} = -\frac{evk\exp\left(i\frac{\omega}{c}R\right)}{\pi\omega^2 R} \frac{(\varepsilon - \beta\sqrt{\varepsilon - \sin^2\vartheta_z})\cos\vartheta_z}{(1-\beta^2\cos^2\vartheta_z)(\varepsilon\cos\vartheta_z + \sqrt{\varepsilon - \sin^2\vartheta_z})},$$ (27.1)

$$\Pi_{\omega 1\parallel} = \frac{iek\exp\left(i\frac{\omega}{c}R\right)\cos\vartheta_z}{\pi\omega R(\varepsilon\cos\vartheta_z + \sqrt{\varepsilon - \sin^2\vartheta_z})} \int\limits_0^\infty [v_z - (v_x\cos\vartheta_x +$$

$$+ v_y\cos\vartheta_y)\frac{\sqrt{\varepsilon - \sin^2\vartheta_z}}{\sin^2\vartheta_z}\right] \exp\left[i\omega t - i\frac{\omega}{c}(x\cos\vartheta_x + y\cos\vartheta_y + z\sqrt{\varepsilon - \sin^2\vartheta_z})\right]dt,$$ (27.2)

$$\Pi_{\omega 1\perp} = \frac{ie\exp\left(i\frac{\omega}{c}R\right)\cos\vartheta_z}{\pi\omega R(\cos\vartheta_z + \sqrt{\varepsilon - \sin^2\vartheta_z})} \int\limits_0^\infty [iv_x + jv_y + k(v_x\cos\vartheta_x +$$

$$+ v_y\cos\vartheta_y)\frac{\cos\vartheta_z}{\sin^2\vartheta_z}\right] \exp\left[i\omega t - i\frac{\omega}{c}(x\cos\vartheta_x + y\cos\vartheta_y + z\sqrt{\varepsilon - \sin^2\vartheta_z})\right]dt.$$ (27.3)

The problem is to average the radiation intensity over all possible particle trajectories. It follows from Eqs. (27.1)-(27.3) and (4.21) that in the case of a particle which moves uniformly and linearly in vacuum along the normal to the separating boundary, while waves polarized in the plane of incidence propagate in the medium, the spectral density of the radiation energy per unit solid angle is given by:

—————————
*These results were obtained some time ago and outlined by the author in [17]. In the corresponding formulas (22) and (21) of [17], errors were made: the factors before the integrals must be two times smaller.

$$W_{n\omega\parallel} = \frac{e^2\omega^2}{\pi^2 c^3} \frac{\sin^2\vartheta_z \cos^2\vartheta_z}{|\varepsilon\cos\vartheta_z + \sqrt{\varepsilon - \sin^2\vartheta_z}|^2} \left\{ \frac{\dot{v}^2|\varepsilon - \beta\sqrt{\varepsilon - \sin^2\vartheta_z}|^2}{\omega^2(1 - \beta^2\cos^2\vartheta_z)^2} + \right.$$

$$+ \frac{2v}{\omega(1 - \beta^2\cos^2\vartheta_z)} \operatorname{Im}(\varepsilon - \beta\sqrt{\varepsilon - \sin^2\vartheta_z})^* \int_0^T [v_z - (v_x\cos\vartheta_x +$$

$$+ v_y\cos\vartheta_y)\frac{\sqrt{\varepsilon - \sin^2\vartheta_z}}{\sin^2\vartheta_z}] \exp\left[i\omega t - i\frac{\omega}{c}(x\cos\vartheta_x + y\cos\vartheta_y +\right.$$

$$+ z\sqrt{\varepsilon - \sin^2\vartheta_z})\Big] dt + 2\operatorname{Re}\int_0^T \Big[v_z - (v_x\cos\vartheta_x +$$

$$+ v_y\cos\vartheta_y)\frac{\sqrt{\varepsilon - \sin^2\vartheta_z}^*}{\sin^2\vartheta_z}\Big] \exp\left(2\frac{\omega}{c}z\operatorname{Im}\sqrt{\varepsilon - \sin^2\vartheta_z}\right) dt \int_0^{T-t} \Big[v_z' -$$

$$- (v_x'\cos\vartheta_x + v_y'\cos\vartheta_y)\frac{\sqrt{\varepsilon - \sin^2\vartheta_z}}{\sin^2\vartheta_z}\Big] \exp\left[i\omega\tau - i\frac{\omega}{c}(\chi\cos\vartheta_x + \eta\cos\vartheta_y + \zeta\sqrt{\varepsilon - \sin^2\vartheta_z})\right] d\tau\}, \qquad (27.4)$$

the radiation intensity of the waves which are polarized in a plane perpendicular to the plane of incidence is given by

$$W_{n\omega\perp} = \frac{2e^2\omega^2}{\pi^2 c^3} \frac{\cos^2\vartheta_z}{|\cos\vartheta_z + \sqrt{\varepsilon - \sin^2\vartheta_z}|^2\sin^2\vartheta_z} \operatorname{Re}\int_0^T (v_x\cos\vartheta_y -$$

$$- v_y\cos\vartheta_x) \exp\left(2\frac{\omega}{c}z\operatorname{Im}\sqrt{\varepsilon - \sin^2\vartheta_z}\right) dt \int_0^{T-t} (v_x'\cos\vartheta_y -$$

$$- v_y'\cos\vartheta_x) \exp\left[i\omega\tau - i\frac{\omega}{c}(\chi\cos\vartheta_x + \eta\cos\vartheta_y + \zeta\sqrt{\varepsilon - \sin^2\vartheta_z})\right] d\tau \qquad (27.5)$$

where x, y, z denote the particle coordinates at the time t; χ, η, and ζ are the particle coordinates at the time $t + \tau$ ($\tau > 0$); v and v' denote the particle velocity at the time t and the later time $t + \tau$, respectively; and T denotes the time during which the particle moves in the medium.

Under the assumption that the deviation of the particle from the initial direction of motion is small, we can switch to the angular vectors $\boldsymbol{\theta}$ and $\boldsymbol{\theta}'$:

$$\boldsymbol{\theta}v_0 = \mathbf{v} - \mathbf{v}_0, \quad \boldsymbol{\theta}'v_0 = \mathbf{v}' - \mathbf{v}_0, \qquad (27.6)$$

then, Eqs. (27.4) and (27.5) assume the following form:

$$W_{n\omega\parallel} = \frac{e^2\omega^2\beta^2}{\pi^2 c} \frac{\sin^2\vartheta_z\cos^2\vartheta_z}{|\varepsilon\cos\vartheta_z + \sqrt{\varepsilon - \sin^2\vartheta_z}|^2} \left\{ \frac{|\varepsilon - \beta\sqrt{\varepsilon - \sin^2\vartheta_z}|^2}{\omega^2(1 - \beta^2\cos^2\vartheta_z)^2} - \right.$$

$$- \frac{2}{\omega(1 - \beta^2\cos^2\vartheta_z)} \operatorname{Im}(\varepsilon - \beta\sqrt{\varepsilon - \sin^2\vartheta_z})^* \int_0^T [1 + (\theta_x\cos\vartheta_x +$$

$$+ \theta_y\cos\vartheta_y)\frac{\sqrt{\varepsilon - \sin^2\vartheta_z}}{\sin^2\vartheta_z}] \exp\left[i\omega t - i\frac{\omega}{c}(x\cos\vartheta_x + y\cos\vartheta_y + z\sqrt{\varepsilon - \sin^2\vartheta_z})\right] dt + 2\operatorname{Re}\int_0^T [1 + (\theta_x\cos\vartheta_x +$$

$$+ \theta_y\cos_y)\frac{\sqrt{\varepsilon - \sin^2\vartheta_z}^*}{\sin^2\vartheta_z} + (\theta_x'\cos\vartheta_x + \theta_y'\cos\vartheta_y)\frac{\sqrt{\varepsilon - \sin^2\vartheta_z}}{\sin^2\vartheta_z} + (\theta_x\cos\vartheta_x + \theta_y\cos\vartheta_y)(\theta_x'\cos\vartheta_x +$$

$$+ \theta_y'\cos\vartheta_y)\frac{|\varepsilon - \sin^2\vartheta_z|}{\sin^4\vartheta_z}] \exp\left(2\frac{\omega}{c}z\operatorname{Im}\sqrt{\varepsilon - \sin^2\vartheta_z}\right) dt \times$$

$$\times \int_0^{T-} \exp\left[i\omega\tau - i\frac{\omega}{c}(\chi\cos\vartheta_x + \eta\cos\vartheta_y + \zeta\sqrt{\varepsilon - \sin^2\vartheta_z})\right] dt\}, \qquad (27.7)$$

$$W_{n\omega\perp} = \frac{2e^2\omega^2\beta^2}{\pi^2 c} \frac{\cos^2\vartheta_z}{|\cos\vartheta_z + \sqrt{\varepsilon - \sin^2\vartheta_z}|^2 \sin^2\vartheta_z} \operatorname{Re} \int_0^T (\theta_x \cos\vartheta_y -$$

$$- \theta_y \cos\vartheta_x) \exp\left(2\frac{\omega}{c} z \operatorname{Im}\sqrt{\varepsilon - \sin^2\vartheta_z}\right) dt \int_0^{T-t} (\theta'_x \cos\vartheta_y -$$

$$- \theta'_y \cos\vartheta_x) \exp\left[i\omega\tau - i\frac{\omega}{c}(\chi\cos\vartheta_x + \eta\cos\vartheta_y + \zeta\sqrt{\varepsilon - \sin^2\vartheta_z})\right] d\tau. \tag{27.8}$$

Averaging over all possible trajectories is done in the following way:

$$W_{n\omega\parallel} = \frac{e^2\omega^2\beta^2}{\pi^2 c} \frac{\sin^2\vartheta_z \cos^2\vartheta_z}{|\varepsilon\cos\vartheta_z + \sqrt{\varepsilon - \sin^2\vartheta_z}|^2} \left\{ \frac{|\varepsilon - \beta\sqrt{\varepsilon - \sin^2\vartheta_z}|^2}{\omega^2(1 - \beta^2\cos^2\vartheta_z)^2} - \right.$$

$$- \frac{2}{\omega(1 - \beta^2\cos^2\vartheta_z)} \operatorname{Im}\left(\varepsilon - \beta\sqrt{\varepsilon - \sin^2\vartheta_z}\right)^* \int_0^T dt\, w_1(\mathbf{r}, \theta, t)\left[1 + \right.$$

$$+ (\theta_x\cos\vartheta_x + \theta_y\cos\vartheta_y)\frac{\sqrt{\varepsilon - \sin^2\vartheta_z}}{\sin^2\vartheta_z}\right] \exp\left[i\omega t - i\frac{\omega}{c}(x\cos\vartheta_x + \right.$$

$$+ y\cos\vartheta_y + z\sqrt{\varepsilon - \sin^2\vartheta_z})\right] d\mathbf{r}d\theta + 2\operatorname{Re}\int_0^T dt \int w_1(\mathbf{r}, \theta, t)[1 +$$

$$+ (\theta_x\cos\vartheta_x + \theta_y\cos\vartheta_y)\frac{\sqrt{\varepsilon - \sin^2\vartheta_z}^*}{\sin^2\vartheta_z} + (\theta'_x\cos\vartheta_x +$$

$$+ \theta'_y\cos\vartheta_y)\frac{\sqrt{\varepsilon - \sin^2\vartheta_z}}{\sin^2\vartheta_z} + (\theta_x\cos\vartheta_x + \theta_y\cos\vartheta_y)(\theta'_x\cos\vartheta_x + \theta'_y\cos\vartheta_y)\frac{|\varepsilon - \sin^2\vartheta_z|}{\sin^4\vartheta_z}\right] \times$$

$$\times \exp\left(2\frac{\omega}{c}z\operatorname{Im}\sqrt{\varepsilon - \sin^2\vartheta_z}\right) \int_0^{T-t} w_2(\rho, \theta, \theta', \tau)\exp\left[i\omega\tau - \right.$$

$$- i\frac{\omega}{c}\left(\chi\cos\vartheta_x + \eta\cos\vartheta_y + \zeta\sqrt{\varepsilon - \sin^2\vartheta_z}\right)\right] d\tau d\mathbf{r}d\theta d\rho d\theta'\Big\}, \tag{27.9}$$

$$W_{n\omega\perp} = \frac{2e^2\omega^2\beta^2}{\pi^2 c} \frac{\cos^2\vartheta_z}{|\cos\vartheta_z + \sqrt{\varepsilon - \sin^2\vartheta_z}|^2 \sin^2\vartheta_z} \operatorname{Re} \int_0^T dt \int_0^{T-t} d\tau \int w_1(\mathbf{r}, \theta, t) \times$$

$$\times w_2(\rho, \theta, \theta', \tau)(\theta_x\cos\vartheta_y - \theta_y\cos\vartheta_x)\exp \times$$

$$\times \left(2\frac{\omega}{c}z\operatorname{Im}\sqrt{\varepsilon - \sin^2\vartheta_z}\right)(\theta'_x\cos\vartheta_y - \theta'_y\cos\vartheta_x)\exp\left[i\omega\tau - \right.$$

$$- i\frac{\omega}{c}\left(\chi\cos\vartheta_x + \eta\cos\vartheta_y + \zeta\sqrt{\varepsilon - \sin^2\vartheta_z}\right)\right] d\mathbf{r}d\theta d\rho d\theta', \tag{27.10}$$

where

$$\theta = i\theta_x + j\theta_y, \quad \theta' = i\theta'_x + j\theta'_y, \quad \mathbf{r} = ix + jy, \quad \rho = i\chi + j\eta, \tag{27.11}$$

and $w_1(\mathbf{r}, \theta, t)$ and $w_2(\rho, \theta, \theta', \tau)$ denote the distribution functions with the initial conditions $w_1(\mathbf{r}, \theta, 0) = \delta(\mathbf{r})\delta(\theta)$ and $w_2(\rho, \theta, \theta', 0) = \delta(\rho)\delta(\theta - \theta')$; these functions satisfy the usual kinetic equation. In the Fokker-Planck approximation, the kinetic equation has the form

$$\frac{\partial w}{\partial \tau} + \mathbf{v}\frac{\partial w}{\partial \rho} = q\Delta_\theta w. \tag{27.12}$$

Integration over ρ in Eqs. (27.9) and (27.10) reduces to the determination of the function

$$u_2(\theta, \theta', \tau) = \int w_2 e^{i(\omega\tau - k_1\rho_1)}d\rho, \tag{27.13}$$

where

$$\rho_1 = i\chi + j\eta + k\zeta, \qquad \zeta = \int_0^\tau v_\zeta(\tau)\,d\tau,$$

$$k_1 = \frac{\omega}{c}\,(i\cos\vartheta_x + j\cos\vartheta_y + k\sqrt{\varepsilon - \sin^2\vartheta_z}). \tag{27.14}$$

We multiply both sides of Eq. (27.12) by $\exp(i\omega\tau - ik_1\rho_1)$ and integrate over ρ. When we recall that

$$\int \frac{\partial w_2}{\partial\tau}\,e^{i(\omega\tau - k_1\rho_1)}\,d\rho = \frac{\partial u_2}{\partial\tau} - i\omega u_2 + i\frac{\omega}{c}\,v_\zeta\sqrt{\varepsilon - \sin^2\vartheta_z},$$

$$\int \frac{\partial w_2}{\partial\rho}\,e^{i(\omega\tau - k_1\rho_1)}\,d\rho = i\,(i\cos\vartheta_x + j\cos\vartheta_y)\,\frac{\omega}{c}\,u_2, \tag{27.15}$$

$$\int \Delta_\theta w_2 e^{i(\omega\tau - k_1\rho_1)}\,d\rho = \Delta_\theta u_2,$$

we obtain the following differential equations:

$$\frac{\partial u_2}{\partial\tau} - i\,(\omega - vk_1)\,u_2 = q\Delta_\theta u_2. \tag{27.16}$$

The calculation of the integral of Eq. (27.13) reduces to solving this equation. After inserting into this equation the vector k_1 from Eq. (27.14) and v, we obtain

$$\frac{\partial u_2}{\partial\tau} - i\omega\left[1 - \beta\,(\theta_x\cos\vartheta_x + \theta_y\cos\vartheta_y) + \beta\sqrt{\varepsilon - \sin^2\vartheta_z}\left(1 - \frac{\theta^2}{2}\right)\right]u_2 - q\left(\frac{\partial^2 u_2}{\partial\theta_x^2} + \frac{\partial^2 u_2}{\partial\theta_y^2}\right) = 0. \tag{27.17}$$

We search for a solution in the form

$$u_2(\theta,\,\theta',\,\tau) = \exp(\gamma_1\theta^2 + \gamma\theta + \gamma_0)\quad(\gamma = i\gamma_x + j\gamma_y), \tag{27.18}$$

where γ_1, $\boldsymbol{\gamma}$, and γ_0 are functions of time. Differentiation leads to

$$\frac{\partial u_2}{d\tau} = (\dot\gamma_1\theta^2 + \dot\gamma\theta + \dot\gamma_0)\,u_2, \tag{27.19}$$

$$\frac{\partial^2 u_2}{\partial\theta_x^2} + \frac{\partial^2 u_2}{\partial\theta_y^2} = (4\gamma_1 + 4\gamma_1^2\theta^2 + \gamma^2 + 4\gamma_1\gamma\theta)\,u_2.$$

After substituting Eqs. (27.19) into Eqs. (27.17) and equating the sum of terms equal to zero (i.e., we equate the terms proportional to the first, second, and zeroth power of the angles θ_x and θ_y to zero), we obtain the following equation system:

$$\dot\gamma_1 + \frac{i\omega\beta}{2}\sqrt{\varepsilon - \sin^2\vartheta_z} = 4q\gamma_1^2,$$

$$\dot\gamma_x + i\omega\beta\cos\vartheta_x = 4q\gamma_1\gamma_x,$$

$$\dot\gamma_y + i\omega\beta\cos\vartheta_y = 4q\gamma_1\gamma_y, \tag{27.20}$$

$$\dot\gamma_0 - i\omega\,(1 + \beta\sqrt{\varepsilon - \sin^2\vartheta_z}) = q\,(4\gamma_1 + \gamma^2).$$

The solution of this system has the form

$$\gamma_1 = -\frac{\eta}{4q}\,\operatorname{cth}\eta\tau, \qquad \gamma_x = -\frac{i\omega\beta}{\eta}\left(\operatorname{cth}\eta\tau - \frac{C_x}{\operatorname{sh}\eta\tau}\right)\cos\vartheta_x. \tag{27.21}$$

$$\gamma_y = -\frac{i\omega\beta}{\eta}\left(\operatorname{cth}\eta\tau - \frac{C_y}{\operatorname{sh}\eta\tau}\right)\cos\vartheta_y,$$

$$\gamma_0 = -\ln\operatorname{sh}\eta\tau + i\omega\tau\left(1 + \beta\sqrt{\varepsilon - \sin^2\vartheta_z}\right) + \ln\frac{\eta}{4\pi q} -$$

$$-\omega^2\beta^2\frac{q}{\eta^3}\left[\left(\eta\tau - \operatorname{cth}\eta\tau - C_x^2\operatorname{cth}\eta\tau + \frac{2C_x}{\operatorname{sh}\eta\tau}\right)\cos^2\vartheta_x + \left(\eta\tau - \operatorname{cth}\eta\tau - C_y^2\operatorname{cth}\eta\tau + \frac{2C_y}{\operatorname{sh}\eta\tau}\right)\cos^2\vartheta_y\right], \quad (27.21)$$

where

$$\eta = \sqrt{2i\omega\beta q\sqrt{\varepsilon - \sin^2\vartheta_z}}. \quad (27.22)$$

Since the distribution function $w_2(\rho, \theta, \theta', \tau)$ satisfies the initial condition $w_2(\rho, \theta, \theta', 0) = \delta(\rho)\,\delta(\theta - \theta')$, we obtain $u_2(\theta, \theta', 0) = \delta(\theta - \theta')$ from the definition given in Eq. (27.13). The constants C_x and C_y are obtained from this condition. These constants are

$$C_x = \theta_x'\frac{\sqrt{\varepsilon - \sin^2\vartheta_z}}{\cos\vartheta_x} + 1, \qquad C_y = \theta_y'\frac{\sqrt{\varepsilon - \sin^2\vartheta_z}}{\cos\vartheta_y} + 1. \quad (27.23)$$

After inserting Eqs. (27.23) into Eqs. (27.21), and then Eqs. (27.21) into Eq. (27.18), we find

$$u_2(\theta, \theta', \tau) = \exp\left[-\frac{\eta(\theta^2 + \theta'^2)}{4q}\operatorname{cth}\eta\tau + \frac{\eta\theta\theta'}{2q\operatorname{sh}\eta\tau} + \frac{\eta(1 - \operatorname{ch}\eta\tau)}{2q\sqrt{\varepsilon - \sin^2\vartheta_z}\operatorname{sh}\eta\tau}\left(\theta_x\cos\vartheta_x + \theta_y\cos\vartheta_y + \frac{\sin^2\vartheta_z}{\sqrt{\varepsilon - \sin^2\vartheta_z}} + \right.\right.$$

$$\left.+ \theta_x'\cos\vartheta_x + \theta_y'\cos\vartheta_y\right) + i\omega\tau\left(1 + \beta\sqrt{\varepsilon - \sin^2\vartheta_z}\right) + \frac{i\omega\beta\tau}{2}\frac{\sin^2\vartheta_z}{\sqrt{\varepsilon - \sin^2\vartheta_z}} - \ln\operatorname{sh}\eta\tau + \ln\frac{\eta}{4\pi q}\bigg]. \quad (27.24)$$

Equation (27.9) contains the integral

$$u_1(\theta, t) = \int w_1 e^{i(\omega t - \mathbf{k_1 r})}d\mathbf{r}, \quad (27.25)$$

and here we have $u_1(\theta, 0) = \delta(\theta)$. The function $u_2(\theta, \theta', \tau)$ satisfies this initial condition if we set in the function $\theta' = 0$. It follows that $u_1(\theta, t) = u_2(\theta, 0, t)$. Hence

$$u_1(\theta, t) = \exp\left[-\frac{\eta\theta^2}{4q}\operatorname{cth}\eta t + \frac{\eta(1 - \operatorname{ch}\eta t)}{2q\sqrt{\varepsilon - \sin^2\vartheta_z}\operatorname{sh}\eta t}\left(\theta_x\cos\vartheta_x + \theta_y\cos\vartheta_y + \right.\right.$$

$$\left.+ \frac{\sin^2\vartheta_z}{\sqrt{\varepsilon - \sin^2\vartheta_z}}\right) + \frac{i\omega\beta t\sin^2\vartheta_z}{2\sqrt{\varepsilon - \sin^2\vartheta_z}} + i\omega t\left(1 + \beta\sqrt{\varepsilon - \sin^2\vartheta_z}\right) - \ln\operatorname{sh}\eta t + \ln\frac{\eta}{4\pi q}\bigg]. \quad (27.26)$$

A similar calculation results in

$$u_0(\theta, t) = \int w_1\exp\left(2\frac{\omega}{c}z\operatorname{Im}\sqrt{\varepsilon - \sin^2\vartheta_z}\right)d\mathbf{r}. \quad (27.27)$$

It follows from the kinetic equation that the function $u_0(\theta, t)$ satisfies equation

$$\frac{\partial u_0}{dt} = q\left(\frac{\partial^2 u_0}{\partial\theta_x^2} + \frac{\partial^2 u_0}{\partial\theta_y^2}\right) - 2\omega\beta\left(1 - \frac{\theta^2}{2}\right)\operatorname{Im}\sqrt{\varepsilon - \sin^2\vartheta_z}. \quad (27.28)$$

We look for a solution of this equation in the form

$$u_0(\theta, t) = \exp\left(\xi_1\theta^2 + \xi_2\right). \quad (27.29)$$

We recall that

$$\frac{\partial u_0}{\partial t} = \left(\dot{\xi}_1\theta^2 + \dot{\xi}_2\right)u_0, \qquad \frac{\partial^2 u_0}{\partial\theta_x^2} + \frac{\partial^2 u_0}{\partial\theta_y^2} = 4\xi_1\left(1 + \xi_1\theta^2\right)u_0. \quad (27.30)$$

After inserting Eq. (27.30) into Eq. (27.28), we obtain

$$\dot{\xi}_1 = \omega\beta\operatorname{Im}\sqrt{\varepsilon - \sin^2\vartheta_z} + 4q\xi_1^2, \qquad \dot{\xi}_2 + 2\omega\beta\operatorname{Im}\sqrt{\varepsilon - \sin^2\vartheta_z} = 4q\xi_1. \quad (27.31)$$

The initial condition for the function $u_0(\theta, t)$ is $u_0(\theta, 0) = \delta(\theta)$. The corresponding solution of the equation system (27.31) is given by the formulas

$$\xi_1 = -\frac{\eta_1}{4q}\,\text{ctg}\,\eta_1 t, \qquad \xi_2 = -\frac{\eta_1^2 t}{2q} - \ln\sin\eta_1 t + \ln\frac{\eta_1}{4\pi q}, \tag{27.32}$$

where

$$\eta_1 = \sqrt{4\omega\beta q\,\text{Im}\,\sqrt{\varepsilon - \sin^2\vartheta_z}}. \tag{27.33}$$

Thus, we obtain

$$u_0(\theta, t) = \exp\left(-\frac{\eta_1\theta^2}{4q}\,\text{ctg}\,\eta_1 t - \frac{\eta_1^2 t}{2q} - \ln\sin\eta_1 t + \ln\frac{\eta_1}{4\pi q}\right). \tag{27.34}$$

We use the formula

$$\int \exp(-p\theta^2 + \mathbf{k}\theta + s\theta\theta' + \mathbf{k}'\theta' - p'\theta'^2)\,d\theta\,d\theta' = \frac{4\pi^2}{4pp' - s^2}\exp\left(\frac{p\mathbf{k}'^2 + s\mathbf{k}\mathbf{k}' + p'\mathbf{k}^2}{4pp' - s^2}\right) \tag{27.35}$$

in the integration over θ and θ'; in this integration we make use of the consequences resulting from this formula after differentiation with respect to the components of the vectors $\mathbf{k}$ and $\mathbf{k}'$. We note that for

$$p' = \frac{\eta}{4q}\,\text{cth}\,\eta\tau, \qquad p = \frac{1}{4q}(\eta\,\text{cth}\,\eta\tau + \eta_1\,\text{ctg}\,\eta_1 t),$$

$$s = \frac{\eta}{2q}\frac{1}{\text{sh}\,\eta\tau}, \qquad \mathbf{k} = \mathbf{k}' = \frac{\eta(1 - \text{ch}\,\eta\tau)(\mathbf{i}\cos\vartheta_x + \mathbf{j}\cos\vartheta_y)}{2q\sqrt{\varepsilon - \sin^2\vartheta_z}\,\text{sh}\,\eta\tau} \tag{27.36}$$

the expression under the integral in Eq. (27.35) agrees with $u_0(\theta, t)u_2(\theta, \theta', \tau)$, except for a factor which is independent of the integration variables. The results have the form

$$J = \int u_0(\theta, t)\,u_2(\theta, \theta', \tau)\,d\theta\,d\theta' = \frac{\eta_1}{(\eta\,\text{th}\,\eta\tau + \eta_1\,\text{ctg}\,\eta_1 t)\,\text{ch}\,\eta\tau\sin\eta_1 t}\exp\left[\frac{\eta\sin^2\vartheta_z}{4q(\varepsilon - \sin^2\vartheta_z)}\left(\eta\tau - \right.\right.$$

$$\left.\left. -\frac{\eta_1\,\text{ctg}\,\eta_1 t\,\text{th}\,\eta\tau}{\eta\,\text{th}\,\eta\tau + \eta_1\text{ctg}\,\eta_1 t}\right) + i\omega\tau(1 + \beta\sqrt{\varepsilon - \sin^2\vartheta_z}) - \frac{\eta_1^2 t}{2q}\right], \tag{27.37}$$

$$\int(\theta'_x\cos\vartheta_x + \theta'_y\cos\vartheta_y)\,u_0 u_2\,d\theta\,d\theta' = J\frac{(1 - \text{ch}\,\eta\tau)\eta_1\text{ctg}\,\eta_1 t - \eta\,\text{sh}\,\eta\tau}{(\eta\,\text{th}\,\eta\tau + \eta_1\,\text{ctg}\,\eta_1 t)\,\text{ch}\,\eta\tau}\frac{\sin^2\vartheta_z}{\sqrt{\varepsilon - \sin^2\vartheta_z}}, \tag{27.38}$$

$$\int(\theta_x\cos\vartheta_x + \theta_y\cos\vartheta_y)\,u_0 u_2\,d\theta\,d\theta' = -\frac{J\eta\,\text{th}\,\eta\tau}{\eta\,\text{th}\,\eta\tau + \eta_1\,\text{ctg}\,\eta_1 t}\frac{\sin^2\vartheta_z}{\sqrt{\varepsilon - \sin^2\vartheta_z}}, \tag{27.39}$$

$$\int(\theta_x\cos\vartheta_x + \theta_y\cos\vartheta_y)(\theta'_x\cos\vartheta_x + \theta'_y\cos\vartheta_y)\,u_0 u_2\,d\theta d\theta' =$$

$$= \frac{J2q\sin^2\vartheta_z}{(\eta\,\text{th}\,\eta\tau + \eta_1\,\text{ctg}\,\eta_1 t)\,\text{ch}\,\eta\tau} + J\frac{\eta[\eta\,\text{sh}\,\eta\tau - (1 - \text{ch}\,\eta\tau)\eta_1\,\text{ctg}\,\eta_1 t]\,\text{sh}\,\eta\tau}{(\eta\,\text{th}\,\eta\tau + \eta_1\,\text{ctg}\,\eta_1 t)^2\,\text{ch}^2\,\eta\tau}\frac{\sin^4\vartheta_z}{\varepsilon - \sin^2\vartheta_z}, \tag{27.40}$$

$$\int(\theta_x\cos\vartheta_y - \theta_y\cos\vartheta_x)(\theta'_x\cos\vartheta_y - \theta'_y\cos\vartheta_x)\,u_0 u_2\,d\theta d\theta' = \frac{J2q\sin^2\vartheta_z}{(\eta\,\text{th}\,\eta\tau + \eta_1\,\text{ctg}\,\eta_1 t)\,\text{ch}\,\eta\tau}, \tag{27.41}$$

$$\int u_1(\theta, t)\,d\theta = \frac{1}{\text{ch}\,\eta_1 t}\exp\left[\frac{\eta}{4q}\frac{\sin^2\vartheta_z}{\varepsilon - \sin^2\vartheta_z}(\eta t - \text{th}\,\eta t) + i\omega t(1 + \beta\sqrt{\varepsilon - \sin^2\vartheta_z})\right], \tag{27.42}$$

$$\int(\theta_x\cos\vartheta_x + \theta_y\cos\vartheta_y)\,u_1(\theta, t)\,d\theta = \frac{1 - \text{ch}\,\eta t}{\text{ch}^2\,\eta t}\exp\left[\frac{\eta}{4q}\frac{\sin^2\vartheta_z}{\varepsilon - \sin^2\vartheta_z}(\eta t - \text{th}\,\eta t) + \right.$$

$$\left. + i\omega t(1 + \beta\sqrt{\varepsilon - \sin^2\vartheta_z})\right]\frac{\sin^2\vartheta_z}{\sqrt{\varepsilon - \sin^2\vartheta_z}}. \tag{27.43}$$

With these formulas, we obtain the angular distribution of the spectral energy density of the radiation of waves polarized in the plane of incidence and perpendicular to this plane:

$$W_{n\omega\parallel} = \frac{e^2\omega^2\beta^2}{\pi^2 c} \frac{\sin^2\vartheta\cos^2\vartheta}{|\varepsilon\cos\vartheta + \sqrt{\varepsilon-\sin^2\vartheta}|^2} \left\{ \frac{|\varepsilon-\beta\sqrt{\varepsilon-\sin^2\vartheta}|^2}{\omega^2(1-\beta^2\cos^2\vartheta)^2} - \right.$$

$$- \frac{2}{\omega(1-\beta^2\cos^2\vartheta)} \operatorname{Im}\left(\varepsilon-\beta\sqrt{\varepsilon-\sin^2\vartheta}\right)^* \int_0^T \frac{dt}{\operatorname{ch}^2\eta t} \exp\left[\frac{\eta}{4q}\frac{\sin^2\vartheta}{\varepsilon-\sin^2\vartheta}\times\right.$$

$$\times\left.(\eta t - \operatorname{th}\eta t) + i\omega t\left(1+\beta\sqrt{\varepsilon-\sin^2\vartheta}\right)\right] + 2\operatorname{Re}\int_0^T dt \int_0^{T-t} d\tau \times$$

$$\times \frac{\eta_1^2\cos\eta_1 t}{p^2\operatorname{ch}^2\eta\tau\sin^2\eta_1 t}\left(2i\frac{\operatorname{Im}\sqrt{\varepsilon-\sin^2\vartheta}}{\sqrt{\varepsilon-\sin^2\vartheta}} + \frac{\eta_1\operatorname{ctg}\eta_1 t}{p}\frac{|\varepsilon-\sin^2\vartheta|}{\varepsilon-\sin^2\vartheta} + \frac{2q}{\eta_1}\frac{|\varepsilon-\sin^2\vartheta|}{\sin^2\vartheta}\operatorname{tg}\eta_1 t\right)\times$$

$$\times \left.\exp\left[\frac{\eta}{4q}\frac{\sin^2\vartheta}{\varepsilon-\sin^2\vartheta}\left(\eta\tau - \frac{\eta_1}{p}\operatorname{ctg}\eta_1 t\operatorname{th}\eta\tau\right) + i\omega\tau\left(1+\beta\sqrt{\varepsilon-\sin^2\vartheta}\right) - \frac{\eta_1^2 t}{2q}\right]\right\}, \qquad (27.44)$$

$$W_{n\omega\perp} = \frac{4e^2\omega^2\beta^2 q\cos^2\vartheta}{\pi^2 c\,|\cos\vartheta + \sqrt{\varepsilon-\sin^2\vartheta}|^2}\operatorname{Re}\int_0^T dt \int_0^{T-t} d\tau \frac{\eta_1}{p^2\operatorname{ch}^2\eta\tau\sin\eta_1 t}\times$$

$$\times \exp\left[\frac{\eta}{4q}\frac{\sin^2\vartheta}{\varepsilon-\sin^2\vartheta}\left(\eta\tau - \frac{\eta_1}{p}\operatorname{ctg}\eta_1 t\operatorname{th}\eta\tau\right) + i\omega\tau\left(1+\beta\sqrt{\varepsilon-\sin^2\vartheta}\right) - \frac{\eta_1^2 t}{2q}\right], \qquad (27.45)$$

where

$$p = \eta\operatorname{th}\eta\tau + \eta_1\operatorname{ctg}\eta_1 t, \qquad \eta_1 = \sqrt{4\omega\beta q\operatorname{Im}\sqrt{\varepsilon-\sin^2\vartheta}},$$

$$\eta = \sqrt{2i\omega\beta q\sqrt{\varepsilon-\sin^2\vartheta}}. \qquad (27.46)$$

In the case of wave absorption, the expressions under the integrals of Eqs. (27.44) and (27.45) are exponentially attenuated in time. In the interference term (i.e., the second term) of Eq. (27.44), the corresponding exponential function is $\exp\left(-\omega\beta t\operatorname{Im}\sqrt{\varepsilon-\sin^2\vartheta}\right)$; in Eq. (27.45) and in the third term of Eq. (27.44) the term

$$\exp\left(-\omega\beta\tau\operatorname{Im}\sqrt{\varepsilon-\sin^2\vartheta} - \frac{\eta_1^2 t}{2q}\right) \equiv \exp\left[-\omega\beta(\tau+2t)\operatorname{Im}\sqrt{\varepsilon-\sin^2\vartheta}\right]$$

appears. During the time t_p, during which the particle passes through the wave-absorption region

$$t_p \sim \left(\omega\beta\operatorname{Im}\sqrt{\varepsilon-\sin^2\vartheta}\right)^{-1} \qquad (27.47)$$

the exponent of the exponential function decreases from a quantity of the order of zero to minus unity. This means that the basic contribution to the result stems from the integration from zero to t_p.

Let us assume that the time T of particle motion in the medium is much greater than t_p, and expand Eqs. (27.44) and (27.45) in power series in q. We restrict ourselves to the linear terms of the expansion. The result is

$$W_{n\omega\parallel} = \frac{e^2\beta^2}{\pi^2 c}\frac{\sin^2\vartheta\cos^2\vartheta}{|\varepsilon\cos\vartheta+\sqrt{\varepsilon-\sin^2\vartheta}|^2}\left\{\frac{|(\varepsilon-1)(1-\beta^2+\beta\sqrt{\varepsilon-\sin^2\vartheta})|^2}{(1-\beta^2\cos^2\vartheta)^2|1+\beta\sqrt{\varepsilon-\sin^2\vartheta}|^2} + \right.$$

$$+ \frac{4\beta q|\varepsilon-\beta\sqrt{\varepsilon-\sin^2\vartheta}|^2}{\omega(1-\beta^2\cos^2\vartheta)}\operatorname{Im}\frac{2\sqrt{\varepsilon-\sin^2\vartheta}(1+\beta\sqrt{\varepsilon-\sin^2\vartheta})+\beta\sin^2\vartheta}{(\varepsilon-\beta\sqrt{\varepsilon-\sin^2\vartheta})(1+\beta\sqrt{\varepsilon-\sin^2\vartheta})^4} + $$

$$+ \frac{q}{\omega J P^4}\left[P^3\left(2+\frac{J^2+R^2}{\beta^2\sin^2\vartheta}\right) - 2P^2(1+R-2J^2) - 16PJ^2(1+R) + (P^2+12PJ^2-16J^4)\beta^2\sin^2\vartheta\right]\right\}, \qquad (27.48)$$

$$J = \beta \, Im \, \sqrt{\varepsilon - \sin^2 \vartheta}, \quad R = \beta \, Re \, \sqrt{\varepsilon - \sin^2 \vartheta},$$

$$P = \beta^2 (Im \, \sqrt{\varepsilon - \sin^2 \vartheta})^2 + (1 + \beta \, Re \, \sqrt{\varepsilon - \sin^2 \vartheta})^2,$$

$$W_{n\omega\perp} = \frac{e^2 \beta q \cos^2 \vartheta}{\pi^2 \omega c \, |\cos \vartheta + \sqrt{\varepsilon - \sin^2 \vartheta}|^2 \, |1 + \beta \sqrt{\varepsilon - \sin^2 \vartheta}|^2 Im \, \sqrt{\varepsilon - \sin^2 \vartheta}}. \tag{27.49}$$

Terms proportional to q^2 can be ignored if

$$q \ll \omega \beta \, Im \, \sqrt{\varepsilon - \sin^2 \vartheta}. \tag{27.50}$$

This inequality determines the applicability range of the results stated in Eqs. (27.48) and (27.49).

In the case of a nonrelativistic particle, Eqs. (27.48) and (27.49) simplify and assume the following form:

$$W_{n\omega\parallel} = \frac{e^2}{\pi^2 c} \frac{\cos^2 \vartheta}{|\varepsilon \cos \vartheta + \sqrt{\varepsilon - \sin^2 \vartheta}|^2} \left[\beta^2 |\varepsilon - 1|^2 \sin^2 \vartheta + \right.$$

$$\left. + 8\beta^3 \frac{q |\varepsilon|^2}{\omega} \sin^2 \vartheta \, Im \frac{\sqrt{\varepsilon - \sin^2 \vartheta}}{\varepsilon} + \beta \frac{q}{\omega} \frac{|\varepsilon - \sin^2 \vartheta|}{Im \, \sqrt{\varepsilon - \sin^2 \vartheta}} \right], \tag{27.51}$$

$$W_{n\omega\perp} = \frac{e^2 \beta q \cos^2 \vartheta}{\pi^2 \omega c \, |\cos \vartheta + \sqrt{\varepsilon - \sin^2 \vartheta}|^2 \, Im \, \sqrt{\varepsilon - \sin^2 \vartheta}}. \tag{27.52}$$

It follows from the kinetic equation (27.12) that the coefficient q is equal to one quarter of the mean square angle of multiple scattering per unit time

$$4q = v \langle \theta^2 \rangle, \tag{27.53}$$

where $\langle \theta^2 \rangle$ denotes the mean square angle of multiple scattering per unit path.

According to the results of the theory given in [90], we have

$$\langle \theta^2 \rangle = \frac{N 8\pi e^4 Z (Z + 1)(1 - \beta^2)}{m^2 c^4 \beta^4} \left[\ln \frac{241\beta}{Z^{1/3}(1 - \beta^2)} - 1 - \frac{\beta^2}{4} \right], \tag{27.54}$$

where Z denotes the specific atomic number of the element and N the number of atoms per cm^3. Hence, we have

$$q = \frac{N 2\pi e^4 Z (Z + 1)(1 - \beta^2)}{m^2 c^3 \beta^3} \left[\ln \frac{241\beta}{Z^{1/3}(1 - \beta^2)} - 1 - \frac{\beta^2}{4} \right]. \tag{27.55}$$

The radiation which is generated by a charged particle during its passage through the boundary of an absorbing medium was previously discussed in [91]. There, the authors disregarded the scattering-induced relative phase shift of the waves emitted from the various sections of the path. This explains the discrepancies of the results stated in Eq. (27.48) and formula (8) of [91]. In the case of waves polarized in a plane perpendicular to the plane of incidence, and, in the nonrelativistic approximation, also for waves which are polarized in the plane of incidence, inclusion of the phase shift in the calculations does not cause strong modifications and the results of the radiation-intensity calculations coincide, except for a relatively small term.

It is worthwhile to compare the theory with the recent experiment [41]. Both the angular and spectral distributions of the intensity of the radiation generated by 30 keV ($\beta = 0.33$) electrons incident on a metal target were measured. By changing the accelerating potential, the authors found that the observed radiation consists of two parts. The intensity of one part is proportional to the electron energy and can be adequately explained by transition-radiation theory. The radiation intensity of the other part is inversely proportional to the electron energy. It was observed that when a silver target is bombarded with electrons, the latter intensity has a sharp peak at $\lambda = 3250$ Å, a wavelength at which the absorption of silver is relatively low. When the spectral and angular distributions of this part of the radiation intensity were compared with the results of [91] regarding the bremsstrahlung, it turned out that the experimental results are quite well represented by the theory. The intensity of the radiation was 5 times greater than the theoretical intensity calculated with data obtained from measurements of the optical properties of silver [92]; the mean square angle of multiple scattering was determined with Eq. (27.54). The latter equation agrees quantitatively with the empirical formula for the most probable angle of multiple scattering (see Bothe, [93]):

$$\theta_\lambda = \frac{8}{V} \frac{V + 511}{V + 1022} Z \left(\frac{\rho x}{A} \right)^{1/2}, \tag{27.56}$$

where V denotes the kinetic energy of the electron (keV); x is the thickness of the layer (μ); Z is the atomic number of the element; A denotes the atomic weight; and ρ the density (g/cm^3). In the case of 30 keV electrons, the theoretical formula (27.54) renders $\langle \theta^2 \rangle = 4.8$ rad/μ for the mean square of the multiple-scattering angle. It follows from the empirical formula (27.56) that the mean square of the multiple-scattering angle is 1.7 times greater than the result of the theory outlined in [90]: $\langle \theta^2 \rangle = 2\theta_\lambda = 8$ rad/μ. Thus, if we base our considerations on the experimental value of the mean-square multiple scattering angle, the discrepancy is reduced: the measured radiation intensity is three times greater than that calculated with the results of [91]. The calculation with Eq. (27.48) reveals that the intensity of the bremsstrahlung of waves polarized in the plane of incidence is 1.72 times greater than that stated in [91]. The total intensity is 1.4 times greater. Thus, the experimentally determined intensity is twice the theoretical intensity. When other measurements of the optical properties of silver are employed [94], the discrepancy with the experimental results amounts to a factor of 4. Let us consider the possible reasons for this discrepancy.

An electromagnetic wave with a wavelength $\lambda = 3250$ Å is absorbed in silver along a path of about one tenth of a micron. Along this path the mean-square multiple-scattering angle for 30 keV electrons amounts to 0.8 rad, according to the empirical equation (27.56). This value is at the limit of the applicability range of Eqs. (27.48) and (27.49). Therefore, one can hardly expect a good quantitative description of the results. On the other hand, it is hardly likely that a correct inclusion of the effect of large deviations (excluding scattering at obtuse angles) might explain the large radiation intensity. It seems that the scattering at obtuse angles plays an important role. At a depth of several tenths of a micron, electrons with an energy of 30 keV are scattered at large angles and the probability that electrons return to the metal surface is not small. The electrons lose energy and consequently, the intensity of the bremsstrahlung increases so that the contribution to the observed radiation can be relatively large. In order to exclude electron scattering at obtuse angles, the thickness l of the target must be smaller than the inverse mean square of the multiple scattering angle per unit path length, i.e., the mean square of the multiple-scattering angle over the thickness of the target must be smaller than unity:

$$l \langle \theta^2 \rangle < 1. \tag{27.57}$$

In the case of 30 keV electrons, the silver target must not be thicker than 0.15 μ. Since in [41] a 5-μ-thick silver target was used, the contribution of electrons scattered at obtuse angles to the radiation is evidently not small. New experiments with thinner targets are of definite interest.

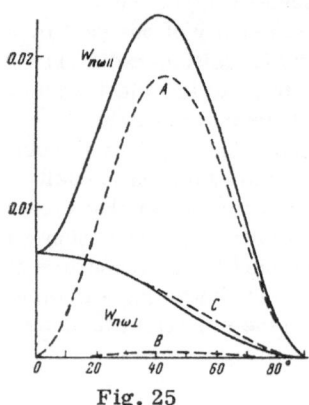

Fig. 25

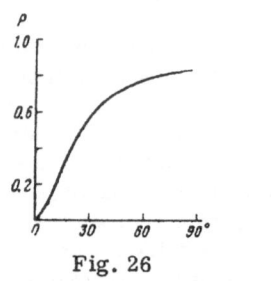

Fig. 26

A comparison of the optical data from various sources indicates that the data differ strongly. This seems to result from the various conditions of sample preparation, which imply varying microstructures of the surface layer. Another possibility is that the target is contaminated by admixtures. Hence, when one compares the results obtained with Eqs. (27.48) and (27.49) with the experimental values, one must employ independent measurements of the optical properties of the targets.

As an example we will outline the results of calculations using Eqs. (27.48) and (27.49) in the case of a silver target ($\lambda = 3290$ Å, $\beta = 0.3$). The optical data for this wavelength was taken from [95]. The values for the real and imaginary parts of the dielectric constant are $\varepsilon' = -0.0453$ and $\varepsilon'' = 0.5802$, respectively in [95]. In this case, the attenuation is not small and the condition (27.50) for the applicability of Eqs. (27.48) and (27.49) is fulfilled. The empirical value of the mean-square multiple-scattering angle was used [Eq. (27.56)]. The results of the calculations are represented by Figs. 25 and 26. Figure 25 shows the angular distribution of the radiation intensity. The dashed curves A, B, and C correspond to the first, second, and third terms in Eq. (27.48), respectively; A refers to the transition radiation, B to the interference term taken with a minus sign (this term is negative), and C refers to the angular distribution of the intensity of the bremsstrahlung of waves polarized in the plane of incidence. The solid curves indicate the angular distribution of the total radiation intensity of waves polarized in the plane of incidence and the angular distribution for waves polarized in a plane perpendicular to the plane of incidence (this part of the intensity results entirely from bremsstrahlung). All functions are plotted to the ordinate axis in units $e^2/\pi^2 c$. We see from Fig. 25 that the bremsstrahlung contributes substantially in the entire angular range from 0 to 90°. The bremsstrahlung intensity assumes its maximum in a direction parallel to the normal to the boundary ($\vartheta = 0°$) and decreases monotonically with increasing ϑ; the intensity tends to zero for $\vartheta \to 90°$. Figure 26 shows the relation between the degree of polarization of the radiation and the angle of observation (i.e., the angle between the direction of quanta emission and the normal to the boundary). The degree of polarization is a monotonic function of ϑ. At $\vartheta = 0$ the radiation is not polarized (only bremsstrahlung quanta are emitted in this direction). At $\vartheta = 90°$ the degree of polarization reaches its maximum value and amounts to about 0.84.

When the magnetic permeability of the medium is not equal to unity, calculations similar to those of this graph lead to the following results [18][*]:

---

[*]The following misprints are found in the work [18] of the author: in the square brackets of the factor before the exponential function in Eq. (3) cos ϑ_z must appear instead of $(\varepsilon\mu - \sin^2\vartheta_z)^{\frac{1}{2}}$; in the first term in parentheses, before the double integral sign in Eq. (24), ch $\eta\tau$ must be omitted; in the square brackets in Eqs. (30) and (37), there must appear $-2P^2 \times (1 + R - 2J^2) - 16PJ^2(1 + R)$ instead of $-2P^2(1 + R) + 16PJ^2R(1 + R)$.

$$W_{n\omega\parallel} = \frac{e^2\omega^2\beta^2}{\pi^2 c} \frac{\sin^2\vartheta\cos^2\vartheta}{|\varepsilon\cos\vartheta + \sqrt{\varepsilon\mu - \sin^2\vartheta}|^2} \left\{ \frac{|\varepsilon - \beta\sqrt{\varepsilon\mu - \sin^2\vartheta}|^2}{\omega^2(1 - \beta^2\cos^2\vartheta)^2} - \right.$$

$$- 2\,\mathrm{Im}\, \frac{(\varepsilon - \beta\sqrt{\varepsilon\mu - \sin^2\vartheta})^*}{\omega(1 - \beta^2\cos^2\vartheta)} \int_0^T \frac{dt}{\mathrm{ch}^2\eta t} \exp\left[\frac{\eta}{4q} \frac{\sin^2\vartheta}{\varepsilon\mu - \sin^2\vartheta}(\eta t - \mathrm{th}\,\eta t) + i\omega t\left(1 + \right.\right.$$

$$\left. + \beta\sqrt{\varepsilon\mu - \sin^2\vartheta}\right) \Big] + 2\,\mathrm{Re}\int_0^T dt \int_0^{T-t} d\tau \frac{\eta_1^2\cos\eta_1 t}{p^2\,\mathrm{ch}^2\eta\tau\sin^2\eta_1 t} \cdot \left(2i\frac{\mathrm{Im}\sqrt{\varepsilon\mu - \sin^2\vartheta}}{\sqrt{\varepsilon\mu - \sin^2\vartheta}} + \right.$$

$$\left. + \frac{\eta_1\,\mathrm{ctg}\,\eta_1 t}{p} \frac{|\varepsilon\mu - \sin^2\vartheta|}{\varepsilon\mu - \sin^2\vartheta} + \frac{2q}{\eta_1} \frac{|\varepsilon\mu - \sin^2\vartheta|}{\sin^2\vartheta} \,\mathrm{tg}\,\eta_1 t\right) \cdot \exp\left[\frac{\eta}{4q} \frac{\sin^2\vartheta}{\varepsilon\mu - \sin^2\vartheta}(\eta\tau - \right.$$

$$\left.\left. - \frac{\eta_1}{p}\,\mathrm{ctg}\,\eta_1 t\,\mathrm{th}\,\eta\tau) + i\omega\tau(1 + \beta\sqrt{\varepsilon\mu - \sin^2\vartheta}) - \frac{\eta_1^2 t}{2q}\right]\right\}, \qquad (27.58)$$

$$W_{n\omega\perp} = \frac{4e^2\omega^2\beta^2 q\,|\mu|^2\cos^2\vartheta}{\pi^2 c\,|\mu\cos\vartheta + \sqrt{\varepsilon\mu - \sin^2\vartheta}|^2}\,\mathrm{Re}\int_0^T dt \int_0^{T-t} d\tau \frac{\eta_1}{p^2\,\mathrm{ch}^2\eta\tau\sin\eta_1 t} \times$$

$$\times \exp\left[\frac{\eta}{4q} \frac{\sin^2\vartheta}{\varepsilon\mu - \sin^2\vartheta}\left(\eta\tau - \frac{\eta_1}{p}\,\mathrm{ctg}\,\eta_1 t\,\mathrm{th}\,\eta\tau\right) + i\omega\tau(1 + \beta\sqrt{\varepsilon\mu - \sin^2\vartheta}) - \frac{\eta_1^2 t}{2q}\right], \qquad (27.59)$$

where

$$p = \eta\,\mathrm{th}\,\eta\tau + \eta_1\,\mathrm{ctg}\,\eta_1 t,$$

$$\eta_1 = \sqrt{4\omega\beta q\,\mathrm{Im}\sqrt{\varepsilon\mu - \sin^2\vartheta}}, \qquad \eta = \sqrt{2i\omega\beta q\sqrt{\varepsilon\mu - \sin^2\vartheta}} \qquad (27.60)$$

and it was assumed that

$$\mathrm{Im}\,\varepsilon \geqslant 0, \qquad \mathrm{Im}\,\mu \geqslant 0, \qquad \mathrm{Im}\sqrt{\varepsilon\mu - \sin^2\vartheta} \geqslant 0. \qquad (27.61)$$

Let us compare the consequences of Eqs. (27.58) and (27.59) with the results of Eqs. (25.22) and (25.21), which describe the angular distribution of the bremsstrahlung intensity of waves of both polarizations in directions which include small angles with the initial direction of motion of a relativistic particle at frequencies greatly exceeding optical frequencies. In this case, we set in Eqs. (27.58) and (27.59) $\mu = 1$, $1 - \varepsilon = \omega_0^2/\omega^2 \ll 1$, $\vartheta \ll 1$, and $1 - \beta^2 \ll 1$.

We recall that ϑ denotes the angle between the wave vector and the positive z axis ($\vartheta < \pi/2$). The particle moves in the direction of the negative z axis. Thus, Eqs. (27.58) and (27.59) refer to the emission of quanta under obtuse angles relative to the initial direction of particle motion. Hence, the comparison must be made after reversing the direction of particle motion, i.e., after replacing β by $-\beta$. Then, the angle between the direction of quanta emission and the direction of particle motion will be an acute angle, as in Eqs. (27.22) and (27.21). The reversal of the sign of the velocity in Eqs. (27.58) and (27.59) is equivalent to a transition to intensity averaging over the actual trajectories in reversed direction relative to the actual direction of particle motion in vacuum, when the particle passes into the medium. The conditions defined by Eqs. (27.11) are now final conditions. We will ignore the transition radiation, extend the τ integration to infinity, and omit the first two terms in Eq. (27.58). Then we obtain results which coincide with Eqs. (25.22) and (25.21). In other words, the averaging over the actual direction of particle motion in vacuum leads to the same results as the averaging over the initial direction of particle motion when the particle enters into the medium — a result which we could anticipate.

Equations (27.58) and (27.59) may be of interest insofar as the group velocity in a ferrodielectric can be negative. Let us consider the radiation conditions which might arise in this frequency region.

We expand the formulas of Eqs. (27.58) and (27.59) in a power series in q. Restricting ourselves to the linear terms of the expansion (which appear under the integral sign) and integrating over time, we obtain

$$W_{n\omega\parallel} = \frac{e^2\beta^2}{\pi^2 c} \frac{\sin^2\vartheta\cos^2\vartheta}{|\varepsilon\cos\vartheta + \sqrt{\varepsilon\mu - \sin^2\vartheta}|^2} \left\{ \frac{||\varepsilon - 1)(1 + \beta\sqrt{\varepsilon\mu - \sin^2\vartheta}) - \beta^2(\varepsilon\mu - 1)|^2}{(1 - \beta^2\cos^2\vartheta)^2|1 + \beta\sqrt{\varepsilon\mu - \sin^2\vartheta}|^2} + \right.$$

$$+ \frac{4\beta q|\varepsilon - \beta\sqrt{\varepsilon\mu - \sin^2\vartheta}|^2}{\omega(1 - \beta^2\cos^2\vartheta)} \operatorname{Im} \frac{2\sqrt{\varepsilon\mu - \sin^2\vartheta}(1 + \beta\sqrt{\varepsilon\mu - \sin^2\vartheta}) + \beta\sin^2\vartheta}{(\varepsilon - \beta\sqrt{\varepsilon\mu - \sin^2\vartheta})(1 + \beta\sqrt{\varepsilon\mu - \sin^2\vartheta})^4} +$$

$$+ \frac{q}{\omega J P^4}\left| P^3\left(2 + \frac{|\varepsilon\mu - \sin^2\vartheta|}{\sin^2\vartheta}\right) - 2P^2(1 + R - 2J^2) - 16PJ^2(1 + R) + (P^2 + \right.$$

$$\left.\left. + 12PJ^2 - 16J^4)\beta^2\sin^2\vartheta\right]\right\}, \tag{27.62}$$

$$J = \beta\operatorname{Im}\sqrt{\varepsilon\mu - \sin^2\vartheta}, \quad R = \beta\operatorname{Re}\sqrt{\varepsilon\mu - \sin^2\vartheta}, \quad P = J^2 + (1 + R)^2,$$

$$W_{n\omega\perp} = \frac{e^2\beta q|\mu|^2\cos^2\vartheta}{\pi^2\omega c|\mu\cos\vartheta + \sqrt{\varepsilon\mu - \sin^2\vartheta}|^2|1 + \beta\sqrt{\varepsilon\mu - \sin^2\vartheta}|^2\operatorname{Im}\sqrt{\varepsilon\mu - \sin^2\vartheta}}. \tag{27.63}$$

Except for a small angular region near the diffraction angle of the Cerenkov radiation, i.e., near the angle

$$\vartheta_r = \arcsin\frac{|\sqrt{\varepsilon\mu\beta^2 - 1}|}{\beta} \tag{27.64}$$

(for $\beta^2|\varepsilon\mu| > 1$, $0 < |\sqrt{\varepsilon\mu\beta^2 - 1}| < \beta$, $\operatorname{Im}\sqrt{\varepsilon\mu - \sin^2\vartheta_r} \ll |\operatorname{Re}\sqrt{\varepsilon\mu - \sin^2\vartheta_r}|$), the following terms of the expansion of Eqs. (27.58) and (27.59) in powers of q can be ignored at frequencies involving negative group velocities ($\operatorname{Im}\varepsilon\mu < 0$) under the condition

$$4qT_0 \ll 1, \quad T_0 = 1/\omega\beta\operatorname{Im}\sqrt{\varepsilon\mu - \sin^2\vartheta}. \tag{27.65}$$

In order to obtain the angular distribution and the degree of polarization of the radiation near the above angle, numerical integrations over time must be performed in Eqs. (27.58) and (27.59). This results in a graphical representation of the smearing-out (by multiple scattering) of the intensity peak of the Cerenkov radiation generated in this frequency range under the obtuse angle relative to the velocity vector.

In the nonrelativistic approximation, Eqs. (27.62) and (27.63) simplify considerably and assume the following form:

$$W_{n\omega\parallel} = \frac{e^3}{\pi^2\omega c} \frac{\cos^2\vartheta}{|\varepsilon\cos\vartheta + \sqrt{\varepsilon\mu - \sin^2\vartheta}|^2}\left[\omega\beta^2|\varepsilon - 1|^2\sin^2\vartheta + \right.$$

$$\left. + 8\beta^3 q|\varepsilon|^2\sin^2\vartheta\operatorname{Im}\frac{\sqrt{\varepsilon\mu - \sin^2\vartheta}}{\varepsilon} + \frac{\beta q|\varepsilon\mu - \sin^2\vartheta|}{\operatorname{Im}\sqrt{\varepsilon\mu - \sin^2\vartheta}}\right], \tag{27.66}$$

$$W_{n\omega\perp} = \frac{e^2\beta q|\mu|^2\cos^2\vartheta}{\pi^2\omega c|\mu\cos\vartheta + \sqrt{\varepsilon\mu - \sin^2\vartheta}|^2\operatorname{Im}\sqrt{\varepsilon\mu - \sin^2\vartheta}}. \tag{27.67}$$

The first terms in Eqs. (27.48), (27.62), and (27.66) correspond to the transition radiation; the second terms describe the interference of the field (generated along the path in the vacuum) with the radiation field generated along the path in the medium — a radiation field which is disturbed by the multiple scattering. The latter terms describe the bremsstrahlung of waves which are polarized in the plane of incidence.

We note that the coefficient q which appears in the kinetic equation (27.12) and in the results is equal to one quarter of the mean square of the multiple-scattering angle per unit time; in the case of a nonrelativistic particle, this coefficient is inversely proportional to the cube of the particle velocity. Hence, the second term in Eq. (27.66) is independent of the particle velocity, whereas the latter term is inversely proportional to the square of the particle velocity.

CHAPTER 6

Excitation of Surface Waves

As is known from [96], along a boundary separating two nonmagnetic media with positive and negative dielectric constants, surface H waves with a field exponentially decreasing inside the two media can propagate. The excitation of surface waves by a charged particle moving uniformly in one direction has been discussed in several publications [97-101]. Dissipation was disregarded in all these publications.

In this chapter we calculate the field of surface waves generated by an arbitrarily moving charged particle at the boundary between vacuum and a nonmagnetic absorbing medium. The case of uniform linear motion under arbitrary angles relative to the boundary has been thoroughly discussed in [21].

§ 28. Field of a Surface Wave Generated

by a Moving Particle

We determined the field of an arbitrarily moving charged particle in Chapter 1 for the case of a boundary separating two media. The results of Chapter I [Eqs. (3.8), (3.9), and (3.11)-(3.14)] form the basis for the calculation of the field of surface waves. The field is given by Hertz vectors in which the expressions under the integral sign have poles. In nonmagnetic media ($\mu_j = \mu_s = 1$) which are considered in the following discussion, the poles appear only in Eqs. (3.13) and (3.8) describing the reflected and diffracted field polarized in the plane of propagation. This means that the surface waves are polarized in the plane of propagation, i.e., their electric vector is parallel to a plane which passes through the wave vector and the normal to the separating boundary; the magnetic field vector is parallel to the boundary.

When we set in Eqs. (3.13) and (3.8) $\varepsilon_j = 1$, $\varepsilon_s = \varepsilon$, $\mu_j = \mu_s = 1$ and integrate over the trajectory, we obtain Hertz vectors of the electromagnetic field generated in vacuum by a moving particle:

$$\Pi_{\omega j}^{i} = -\frac{ek}{4\pi^2\omega} \iiint \frac{e^{i\omega t}\,(\varepsilon k_{jz} - k_{sz})}{\beta \varkappa^2 k_{jz}\,(\varepsilon k_{jz} + k_{sz})} [\beta_z \varkappa^2 + (\beta_x k_x + \beta_y k_y)\,k_{jz}] \cdot \exp\,[ik_x\,(x - x_\zeta) +$$
$$+ ik_y\,(y - y_\zeta) + ik_{jz}\,(z + z_\zeta)]\,dk_x dk_y d\zeta \quad (z > 0,\, z_\zeta > 0), \qquad (28.1)$$

$$\Pi_{\omega s}^{j} = -\frac{ek}{2\pi^2\omega} \iiint \frac{e^{i\omega t}}{\beta \varkappa^2\,(\varepsilon k_{jz} + k_{sz})} [\beta_z \varkappa^2 - (\beta_x k_x + \beta_y k_y)\,k_{sz}] \cdot \exp\,[ik_x\,(x - x_\zeta) +$$
$$+ ik_y\,(y - y_\zeta) + ik_{jz}z - ik_{sz}z_\zeta]\,dk_x dk_y d\zeta \quad (z > 0,\, z_\zeta < 0). \qquad (28.2)$$

After changing the signs of the vectors of Eqs. (3.13) and (3.8), exchanging the positions of the subscripts j and s, changing the signs before the components k_{jz} and k_{sz}, assuming $\varepsilon_j = 1$, $\varepsilon_s = \varepsilon$, $\mu_j = \mu_s = 1$ and integrating over the trajectory, we obtain both the reflected and diffracted fields in the medium (z < 0):

$$\Pi_{\omega s}^{s} = \frac{ek}{4\pi^2\omega} \iiint \frac{e^{i\omega t}\,(\varepsilon k_{jz} - k_{sz})}{\beta \varkappa^2 k_{sz}\,(\varepsilon k_{jz} + k_{sz})} [\beta_z \varkappa^2 - (\beta_x k_x + \beta_y k_y)\,k_{sz}] \cdot \exp\,[ik_x\,(x - x_\zeta) +$$
$$+ ik\,(y - y_\zeta) - ik_{sz}\,(z + z_\zeta)]\,dk_x dk_y d\zeta \quad (z < 0,\, z_\zeta < 0), \qquad (28.3)$$

$$\Pi_{\omega j}^{s} = -\frac{eke}{2\pi^2\omega} \iiint \frac{e^{i\omega t}}{\beta \varkappa^2 (ek_{jz}+k_{sz})} [\beta_z \varkappa^2 + (\beta_x k_x + \beta_y k_y) k_{jz}] \cdot \exp[ik_x(x-x_\zeta) +$$
$$+ ik_y(y-y_\zeta) - ik_{sz}z + ik_{jz}z_\zeta] dk_x dk_y d\zeta \quad (z < 0, \; z_\zeta > 0). \quad (28.4)$$

In Eqs. (28.1)–(28.4) we used the notations

$$\varkappa = \sqrt{k_x^2 + k_y^2}, \qquad k_{jz} = \sqrt{\frac{\omega^2}{c^2} - \varkappa^2}, \qquad k_{sz} = \sqrt{\varepsilon \frac{\omega^2}{c^2} - \varkappa^2}. \quad (28.5)$$

it follows from the condition that the field is finite at infinity, that the imaginary parts of $\varkappa$, k_{jz} and k_{sz} are positive or vanish.

As before, we use a Cartesian coordinate system with the z axis pointing into the medium j (which is vacuum in the case considered) and the xy plane coinciding with the boundary. We omit the symbol ∥ which indicates that the waves under consideration are polarized in the plane of propagation. The Hertz vectors receive the superscripts j and s, which indicate that the Hertz vector describes the field in vacuum (z > 0) and in the medium (z < 0), respectively. As before, the subscripts j and s indicate the region of particle motion: j denotes that the particle moves in vacuum ($z_\zeta > 0$) and s indicates particle motion in the medium ($z_\zeta < 0$). The particle coordinates at the time t are denoted by x_ζ, y_ζ, z_ζ in Eqs. (28.1)–(28.4).

The Fourier components of the electric and magnetic fields are given by the Hertz vectors of Eqs. (28.1)–(28.4):

$$\mathbf{E}_\omega^j = \operatorname{grad} \operatorname{div} \mathbf{\Pi}_\omega^j + \frac{\omega^2}{c^2} \mathbf{\Pi}_\omega^j,$$

$$\mathbf{E}_\omega^s = \frac{1}{\varepsilon} \operatorname{grad} \operatorname{div} \mathbf{\Pi}_\omega^s + \frac{\omega^\varkappa}{c^2} \mathbf{\Pi}_\omega^s, \quad (28.6)$$

$$\mathbf{H}_\omega^{j,s} = -i\frac{\omega}{c} \operatorname{rot} \mathbf{\Pi}_\omega^{j,s}.$$

These are the formulas which are required for calculating the surface waves generated by a charged particle on the boundary between vacuum and a nonmagnetic medium.

An integration over k_x and k_y must be performed in Eqs. (28.1)–(28.4) in order to determine the field of the surface waves. First, we switch to the new variables φ and $\varkappa$, which are related to k_x and k_y by the equations $k_x = \varkappa \cos \varphi$, $k_y = \varkappa \sin \varphi$. Integration over φ results in Bessel functions. When we replace the Bessel functions by a half-sum of Hankel functions, the integration over $\varkappa$ can be extended from minus infinity to plus infinity and the integration path can be closed in the complex $\varkappa$ plane by a semicircle of infinite radius. The expressions under the integral have a pole at $\varkappa = \frac{\omega}{c}\sqrt{\frac{\varepsilon}{1+\varepsilon}}$ and branch points at $\varkappa = \omega/c$ and $\varkappa = \omega/c\sqrt{\varepsilon}$.

The pole gives us a surface wave, whereas the integration over the rest of the contour is more complicated since, in addition to the surface field, we obtain contributions from both the transition and the Cerenkov radiation. We restrict our considerations to the contribution by the pole and outline the range in which the result is valid.

At the pole, the quantities k_{jz} and k_{sz} are

$$k_{jz} = -\frac{\omega}{c}\frac{1}{\sqrt{1+\varepsilon}}, \quad k_{sz} = \frac{\omega}{c}\frac{\varepsilon}{\sqrt{1+\varepsilon}}.$$

When absorption takes place, the imaginary parts of $\varkappa$, k_{jz}, and k_{sz} must be positive. For $\omega > 0$ this requirement is satisfied by the following roots:

$$\sqrt{1+\varepsilon} = \frac{1}{\sqrt{2}}\sqrt{|1+\varepsilon| + 1 + \varepsilon'} + \frac{i}{\sqrt{2}}\sqrt{|1+\varepsilon| - 1 - \varepsilon'}, \tag{28.7}$$

$$\sqrt{\varepsilon} = \frac{1}{\sqrt{2}}\sqrt{|\varepsilon| + \varepsilon'} + \frac{i}{\sqrt{2}}\sqrt{|\varepsilon| - \varepsilon'},$$

where ε' denotes the real part of the dielectric constant: $\varepsilon = \varepsilon' + i\varepsilon''$ ($\varepsilon'' > 0$ for $\omega > 0$). The complex conjugated roots are used for $\omega < 0$.

When we take the residues in Eqs. (28.1)–(28.4) we obtain the field of a surface wave which is generated by a moving charged particle:

$$\Pi_{\omega j}^{j} = -\frac{ek\varepsilon\sqrt{\varepsilon}}{c\,(\varepsilon^2-1)\,\sqrt{1+\varepsilon}} \int \frac{1}{\beta}\left[(\beta_x \cos\varphi_\zeta + \beta_y \sin\varphi_\zeta) H_1^{(1)}\left(\frac{\omega}{c}\sqrt{\frac{\varepsilon}{1+\varepsilon}}\,\rho_\zeta\right) + \right.$$

$$\left. + i\beta_z\sqrt{\varepsilon}\,H_0^{(1)}\left(\frac{\omega}{c}\sqrt{\frac{\varepsilon}{1+\varepsilon}}\,\rho_\zeta\right)\right] \exp\left(i\omega t - i\frac{\omega}{c}\frac{z+z_\zeta}{\sqrt{1+\varepsilon}}\right) d\zeta, \tag{28.8}$$

$$\Pi_{\omega s}^{j} = \Pi_{\omega j}^{j}\left(\beta_z \to \frac{\beta_z}{\varepsilon},\ z_\zeta \to \varepsilon z_\zeta\right), \qquad \Pi_{\omega j}^{s} = \Pi_{\omega j}^{j}(z \to \varepsilon z),$$

$$\Pi_{\omega s}^{s} = \Pi_{\omega j}^{j}\left(\beta_z \to \frac{\beta_z}{\varepsilon},\ z \to \varepsilon z,\ z_\zeta \to \varepsilon z_\zeta\right).$$

Here, as well as in the following discussion, the arrows in the parentheses indicate the changes which must be made in the expressions in the parentheses in order to obtain the result defined by the left side of the corresponding equation. The following notation has been introduced:

$$\rho_\zeta = \sqrt{(x-x_\zeta)^2 + (y-y_\zeta)^2}, \qquad \varphi_\zeta = \text{arctg}\,\frac{y-y_\zeta}{x-x_\zeta}. \tag{28.9}$$

In the wave zone

$$\rho_\zeta \frac{\omega}{c}\left|\sqrt{\frac{\varepsilon}{1+\varepsilon}}\right| \gg 1, \tag{28.10}$$

the results simplify and the first of Eqs. (28.8) assumes the following form:

$$\Pi_{\omega j}^{j} = -\frac{iek\sqrt{-2i\omega}\,\varepsilon}{\sqrt{\pi c}\,\omega\,(\varepsilon^2-1)}\left(\frac{\varepsilon}{1+\varepsilon}\right)^{1/4} \int \frac{1}{\beta}\frac{1}{\sqrt{\rho_\zeta}}\,(\beta_x \cos\varphi_\zeta + \beta_y\sin\varphi_\zeta - \beta_z\sqrt{\varepsilon}) \times$$

$$\times \exp\left(i\omega t + i\frac{\omega}{c}\sqrt{\frac{\varepsilon}{1+\varepsilon}}\rho_\zeta - i\frac{\omega}{c}\frac{z+z_\zeta}{\sqrt{1+\varepsilon}}\right) d\zeta. \tag{28.11}$$

In order to obtain the other Hertz vectors it suffices to make the replacements indicated in Eqs. (28.8).

It follows from Eqs. (28.6) that the electromagnetic field in the wave zone is given by the following formulas which include the Hertz vectors calculated above:

$$\mathbf{E}_{\omega}^{j} = \frac{\omega^2}{c^2}\frac{1}{1+\varepsilon}\,\Pi_{\omega}^{j}\,[\varepsilon\mathbf{k} + \sqrt{\varepsilon}\,(\mathbf{i}\cos\varphi_\zeta + \mathbf{j}\sin\varphi_\zeta)],$$

$$\mathbf{E}_{\omega}^{s} = \frac{\omega^2}{c^2}\frac{1}{1+\varepsilon}\,\Pi_{\omega}^{s}\,[\mathbf{k} + \sqrt{\varepsilon}\,(\mathbf{i}\cos\varphi_\zeta + \mathbf{j}\sin\varphi_\zeta)], \tag{28.12}$$

$$\mathbf{H}_{\omega}^{j,\,s} = \frac{\omega^2}{c^2}\sqrt{\frac{\varepsilon}{1+\varepsilon}}\,\Pi_{\omega}^{j,\,s}\,(\mathbf{i}\sin\varphi_\zeta - \mathbf{j}\cos\varphi_\zeta).$$

We digress and disregard the absolute contributions from the contours and the pole term. However, we wish to note that under certain conditions both the pole and the branching can lead to the same field configuration and that a complete or partial compensation of the pole term is possible. We restrict our considerations to conditions under which such a compensation does not occur and consequently, the contribution of the pole can be considered separately. Note-worthy enough, at the contour sections, the imaginary part of $\varkappa$ can assume any positive values, whereas the real part of $\varkappa$ is restricted to the intervals

$$0 < \operatorname{Re} \varkappa < \frac{\omega}{c}, \qquad 0 < \operatorname{Re} \varkappa < \frac{\omega}{c} \operatorname{Re} \sqrt{\varepsilon}.$$

These intervals determine the integration interval over the contour sections. One can show that (at least in the wave zone) the results do not cancel if the pole is situated beyond the integration contour, i.e., when the following inequalities hold:

$$\operatorname{Re} \sqrt{\frac{\varepsilon}{1+\varepsilon}} > 1, \qquad \operatorname{Re} \sqrt{\frac{\varepsilon}{1+\varepsilon}} > \operatorname{Re} \sqrt{\varepsilon}. \tag{28.13}$$

In this case, the contribution of the pole corresponds to small wavelengths which are not in-cluded in the integrals over the contour sections. A detailed discussion of the applicability of the pole contribution is obtained when the integrals along the contour sections are calculated. This problem requires additional studies.

§ 29. Surface Waves in the Case of Normal Incidence

Let us consider the simplest case of a uniform, linear motion perpendicular to the bound-ary. When we set in Eqs. (28.8) $\beta_x = \beta_y = 0$, $\beta_z = -\beta$, $x_\zeta = 0$, $y_\zeta = 0$ and integrate over the trajectory, we obtain the field of the surface waves generated when the particle passes from the vacuum into the medium (in the direction of the negative z axis):

$$\Pi_{\omega j}^{j} = \frac{eke^2\beta}{\omega(\varepsilon^2-1)(\sqrt{1+\varepsilon}+\beta)} H_0^{(1)}\left(\frac{\omega}{c}\sqrt{\frac{\varepsilon}{1+\varepsilon}}\rho\right) \exp\left(-i\frac{\omega}{c}\frac{z}{\sqrt{1+\varepsilon}}\right),$$

$$\Pi_{\omega j}^{s} = \Pi_{\omega j}^{j}(z \to \varepsilon z),$$

$$\Pi_{\omega s}^{j} = -\frac{eke\beta}{\omega(\varepsilon^2-1)(\sqrt{1+\varepsilon}+\beta\varepsilon)} H_0^{(1)}\left(\frac{\omega}{c}\sqrt{\frac{\varepsilon}{1+\varepsilon}}\rho\right) \exp\left(-i\frac{\omega}{c}\frac{z}{\sqrt{1+\varepsilon}}\right), \tag{29.1}$$

$$\Pi_{\omega s}^{s} = \Pi_{\omega s}^{j}(z \to \varepsilon z).$$

When we use these results along with Eqs. (28.6), we obtain the electromagnetic field of the surface waves:

$$\mathbf{E}_{\omega j}^{j} = \frac{e\varepsilon^2\sqrt{\varepsilon}\,\omega\beta}{c^2(\varepsilon^2-1)(1+\varepsilon)(\sqrt{1+\varepsilon}+\beta)} \left[i\,(\mathbf{i}\cos\varphi + \mathbf{j}\sin\varphi) H_1^{(1)}\left(\frac{\omega}{c}\sqrt{\frac{\varepsilon}{1+\varepsilon}}\rho\right) + \right.$$

$$\left. + \mathbf{k}\sqrt{\varepsilon}\,H_0^{(1)}\left(\frac{\omega}{c}\sqrt{\frac{\varepsilon}{1+\varepsilon}}\rho\right) \right] \exp\left(-i\frac{\omega}{c}\frac{z}{\sqrt{1+\varepsilon}}\right), \ \ \mathbf{E}_{\omega j}^{s} = \mathbf{E}_{\omega j}^{j}(z \to \varepsilon z),$$

$$\mathbf{E}_{\omega s}^{j} = -\frac{e\varepsilon\sqrt{\varepsilon}\,\omega\beta}{c^2(\varepsilon^2-1)(1+\varepsilon)(\sqrt{1+\varepsilon}+\beta\varepsilon)} \left[i\,(\mathbf{i}\cos\varphi + \mathbf{j}\sin\varphi) H_1^{(1)}\left(\frac{\omega}{c}\sqrt{\frac{\varepsilon}{1+\varepsilon}}\rho\right) + \right.$$

$$\left. + \mathbf{k}\sqrt{\varepsilon}\,H_0^{(1)}\left(\frac{\omega}{c}\sqrt{\frac{\varepsilon}{1+\varepsilon}}\rho\right) \right] \exp\left(-i\frac{\omega}{c}\frac{z}{\sqrt{1+\varepsilon}}\right), \ \ \mathbf{E}_{\omega s}^{s} = \mathbf{E}_{\omega s}^{j}(z \to \varepsilon z), \tag{29.2}$$

$$\mathbf{H}_{\omega j}^{j} = \frac{iee^2\sqrt{\varepsilon}\,\omega\beta}{c^2(\varepsilon^2-1)\sqrt{1+\varepsilon}(\sqrt{1+\varepsilon}+\beta)} (\mathbf{i}\sin\varphi - \mathbf{j}\cos\varphi) H_1^{(1)}\left(\frac{\omega}{c}\sqrt{\frac{\varepsilon}{1+\varepsilon}}\rho\right) \exp\left(-i\frac{\omega}{c}\frac{z}{\sqrt{1+\varepsilon}}\right),$$

$$\mathbf{H}_{\omega s}^{j} = -\frac{iee\sqrt{\varepsilon}\,\omega\beta}{c^2(\varepsilon^2-1)\sqrt{1+\varepsilon}(\sqrt{1+\varepsilon}+\beta\varepsilon)} (\mathbf{i}\sin\varphi - \mathbf{j}\cos\varphi) H_1^{(1)}\left(\frac{\omega}{c}\sqrt{\frac{\varepsilon}{1+\varepsilon}}\rho\right) \exp\left(-i\frac{\omega}{c}\frac{z}{\sqrt{1+\varepsilon}}\right),$$

$$\mathbf{H}_{\omega j}^{s} = \mathbf{H}_{\omega j}^{j}(z \to \varepsilon z), \quad \mathbf{H}_{\omega s}^{s} = \mathbf{H}_{\omega s}^{j}(z \to \varepsilon z).$$

In the wave zone, we have

$$H_0^{(1)}\left(\frac{\omega}{c}\sqrt{\frac{\varepsilon}{1+\varepsilon}}\rho\right) \simeq iH_1^{(1)}\left(\frac{\omega}{c}\sqrt{\frac{\varepsilon}{1+\varepsilon}}\rho\right) \simeq \sqrt{\frac{-2ic\sqrt{1+\varepsilon}}{\pi\rho\omega\sqrt{\varepsilon}}}\exp\left(i\frac{\omega}{c}\sqrt{\frac{\varepsilon}{1+\varepsilon}}\rho\right), \qquad (29.3)$$

and Eq. (29.1) assumes the form

$$\Pi_{\omega j}^{j} = \frac{ek\beta\varepsilon\sqrt{2c\varepsilon\sqrt{\varepsilon(1+\varepsilon)}}}{\sqrt{i\pi\rho\omega^3}(\varepsilon^2-1)(\sqrt{1+\varepsilon}+\beta)}\exp\left(i\frac{\omega}{c}\frac{\sqrt{\varepsilon}\rho-z}{\sqrt{1+\varepsilon}}\right), \qquad \Pi_{\omega j}^{s} = \Pi_{\omega j}^{j}(z \to \varepsilon z),$$

$$\Pi_{\omega s}^{j} = -\frac{ek\beta\sqrt{2c\varepsilon\sqrt{\varepsilon(1+\varepsilon)}}}{\sqrt{i\pi\rho\omega^3}(\varepsilon^2-1)(\sqrt{1+\varepsilon}+\beta\varepsilon)}\exp\left(i\frac{\omega}{c}\frac{\sqrt{\varepsilon}\rho-z}{\sqrt{1+\varepsilon}}\right), \quad \Pi_{\omega s}^{s} = \Pi_{\omega s}^{j}(z \to \varepsilon z). \tag{29.4}$$

The vectors $\Pi_{\omega j}^{j}$ and $\Pi_{\omega j}^{s}$ indicate the field of the surface wave when the particle is suddenly stopped at the boundary. The vectors $\Pi_{\omega s}^{j}$ and $\Pi_{\omega s}^{s}$ indicate the field of the surface wave resulting from the ejection of the particle into the medium. The vectors $\Pi_{\omega j}^{j}$ and $\Pi_{\omega s}^{j}$ describe the field of the surface wave in vacuum (z > 0), whereas the vectors $\Pi_{\omega j}^{s}$ and $\Pi_{\omega s}^{s}$ refer to the medium (z < 0). When the particle motion is uniform and linear and the particle passes through the boundary, the field of the surface wave in vacuum is given by the sum $\Pi_{\omega}^{j} = \Pi_{\omega j}^{j} + \Pi_{\omega s}^{j}$, and the field of the surface wave in the medium by the sum $\Pi_{\omega}^{s} = \Pi_{\omega j}^{s} + \Pi_{\omega s}^{s}$:

$$\Pi_{\omega}^{j} = \frac{ek\beta\sqrt{-2i\omega c\varepsilon\sqrt{\varepsilon}(1+\beta\sqrt{1+\varepsilon})}}{\omega^2\sqrt{\pi\rho\sqrt{1+\varepsilon}}(\sqrt{1+\varepsilon}+\beta)(\sqrt{1+\varepsilon}+\beta\varepsilon)}\exp\left(i\frac{\omega}{c}\frac{\sqrt{\varepsilon}\rho-z}{\sqrt{1+\varepsilon}}\right),$$

$$\Pi_{\omega}^{s} = \Pi_{\omega}^{j}(z \to \varepsilon z). \tag{29.5}$$

The spectral density of the energy flux through the unit area in vacuum and in the medium can be calculated with the above Hertz vectors in the wave zone, according to the following equations:

$$S_{\omega}^{j} = \frac{\omega^4}{c^3}|\Pi_{\omega}^{j}|^2\left|\frac{\varepsilon}{1+\varepsilon}\right|\mathrm{Re}\left(\sqrt{\frac{\varepsilon}{1+\varepsilon}}\frac{\rho}{\rho}-\frac{k}{\sqrt{1+\varepsilon}}\right),$$

$$S_{\omega}^{s} = \frac{\omega^4}{c^3}|\Pi_{\omega}^{s}|^2\left|\frac{\varepsilon}{1+\varepsilon}\right|\mathrm{Re}\left(\frac{1}{\sqrt{\varepsilon(1+\varepsilon)}}\frac{\rho}{\rho}-\frac{k}{\sqrt{1+\varepsilon}}\right). \tag{29.6}$$

Since $\mathrm{Re}\sqrt{\frac{\varepsilon}{1+\varepsilon}}>0$, i.e., since the factor at ρ/ρ in S_{ω}^{j} is positive, the energy flux S_{ω}^{j} in vacuum is parallel to the trajectory. In the region defined by the relations (28.13), we have $\mathrm{Re}\sqrt{\varepsilon(1+\varepsilon)}<0$, and therefore, the energy flux in the medium is parallel to the particle trajectory. When no absorption occurs, the energy flux in the vacuum is antiparallel to that in the medium. With absorption, we have $\mathrm{Re}\sqrt{1+\varepsilon}>0$ and the projections of the Poynting vectors onto the z axis are negative, i.e., the energy of the surface wave flows through the boundary, from the vacuum into the medium.

When we insert into Eq. (29.6) the Hertz vectors from Eqs. (29.4) and (29.5), integrate S_{ω}^{j} over z from zero to infinite, and S_{ω}^{s} from minus infinity to zero, and multiply the results by $2\pi\rho$, we obtain the spectral density of the energy flux through a cylinder of radius ρ for the vacuum and for the medium:

Case in which the particle is arrested at the boundary:

$$W^j_{\omega j} = \frac{2e^2\beta^2|\varepsilon|^2}{c\,|\,(\sqrt{1+\varepsilon}+\beta)\,(\varepsilon-1)\,|^2}\left|\frac{\varepsilon}{1+\varepsilon}\right|^{3/2}\frac{\mathrm{Re}\,\sqrt{\varepsilon\,(1+\varepsilon^*)}}{\mathrm{Im}\,\sqrt{1+\varepsilon}}\exp\!\left(-2\rho\,\frac{\omega}{c}\,\mathrm{Im}\sqrt{\frac{\varepsilon}{1+\varepsilon}}\right),\qquad (29.7)$$

$$W^s_{\omega j} = -\frac{2e^2\beta^2|\varepsilon|}{c\,|\,(\sqrt{1+\varepsilon}+\beta)\,(\varepsilon-1)\,|^2}\left|\frac{\varepsilon}{1+\varepsilon}\right|^{3/2}\frac{\mathrm{Re}\,\sqrt{\varepsilon\,(1+\varepsilon)}}{\mathrm{Im}\,(\varepsilon^*\,\sqrt{1+\varepsilon})}\exp\!\left(-2\rho\,\frac{\omega}{c}\,\mathrm{Im}\sqrt{\frac{\varepsilon}{1+\varepsilon}}\right);\qquad (29.8)$$

Case in which the particle is ejected into the medium

$$W^j_{\omega s} = \frac{2e^2\beta^2}{c\,|\,(\sqrt{1+\varepsilon}+\beta\varepsilon)\,(\varepsilon-1)\,|^2}\left|\frac{\varepsilon}{1+\varepsilon}\right|^{3/2}\frac{\mathrm{Re}\,\sqrt{\varepsilon\,(1+\varepsilon^*)}}{\mathrm{Im}\,\sqrt{1+\varepsilon}}\exp\!\left(-2\rho\,\frac{\omega}{c}\,\mathrm{Im}\sqrt{\frac{\varepsilon}{1+\varepsilon}}\right),\qquad (29.9)$$

$$W^s_{\omega s} = -\frac{2e^2\beta^2}{c\,|\,(\sqrt{1+\varepsilon}+\beta\varepsilon)\,(\varepsilon-1)\,|^2}\left|\frac{\varepsilon}{1+\varepsilon}\right|^{3/2}\frac{\mathrm{Re}\,\sqrt{\varepsilon\,(1+\varepsilon)}}{|\varepsilon|\,\mathrm{Im}\,(\varepsilon^*\,\sqrt{1+\varepsilon})}\exp\!\left(-2\rho\,\frac{\omega}{c}\,\mathrm{Im}\sqrt{\frac{\varepsilon}{1+\varepsilon}}\right);\qquad (29.10)$$

Case in which the particle motion is uniform and linear from the vacuum into the medium:

$$W^j_{\omega} = \frac{2e^2\beta^2|1+\beta\sqrt{1+\varepsilon}|^2|\varepsilon|}{c\,|\,(\sqrt{1+\varepsilon}+\beta)\,(\sqrt{1+\varepsilon}+\beta\varepsilon)\,|^2}\left|\frac{\varepsilon}{1+\varepsilon}\right|^{3/2}\frac{\mathrm{Re}\,\sqrt{\varepsilon\,(1+\varepsilon^*)}}{\mathrm{Im}\,\sqrt{1+\varepsilon}}\exp\!\left(-2\rho\,\frac{\omega}{c}\,\mathrm{Im}\sqrt{\frac{\varepsilon}{1+\varepsilon}}\right),\qquad (29.11)$$

$$W^s_{\omega} = -\frac{2e^2\beta^2|1+\beta\sqrt{1+\varepsilon}|^2}{c\,|\,(\sqrt{1+\varepsilon}+\beta)\,(\sqrt{1+\varepsilon}+\beta\varepsilon)\,|^2}\left|\frac{\varepsilon}{1+\varepsilon}\right|^{3/2}\frac{\mathrm{Re}\,\sqrt{\varepsilon\,(1+\varepsilon)}}{\mathrm{Im}\,(\varepsilon^*\,\sqrt{1+\varepsilon})}\exp\!\left(-2\rho\,\frac{\omega}{c}\,\mathrm{Im}\sqrt{\frac{\varepsilon}{1+\varepsilon}}\right).\qquad (29.12)$$

When no absorption occurs, the results simplify considerably and assume the following form:

$$W^j_{\omega j} = \frac{2e^2\beta^2\varepsilon_1^5}{c\,(\beta^2+\varepsilon_1-1)\,(\varepsilon_1^2-1)^2\,\sqrt{\varepsilon_1-1}}\,,\qquad W^s_{\omega j} = -W^j_{\omega j}/\varepsilon_1^2,$$

$$W^j_{\omega s} = \frac{2e^2\beta^2\varepsilon_1^3}{c\,(\beta^2\varepsilon_1^2+\varepsilon_1-1)\,(\varepsilon_1^2-1)^2\,\sqrt{\varepsilon_1-1}}\,,\qquad W^s_{\omega s} = -W^j_{\omega s}/\varepsilon_1^2,\qquad (29.13)$$

$$W^j_{\omega} = \frac{2e^2\beta^2\,(\beta^2\varepsilon_1-\beta^2+1)}{c\,(\beta^2+\varepsilon_1-1)\,(\beta^2\varepsilon_1^2+\varepsilon_1-1)\,(\varepsilon_1-1)^{3/2}}\,,\qquad W^s_{\omega} = -W^j_{\omega}/\varepsilon_1^2,$$

where $\varepsilon_1 = |\varepsilon| = -\varepsilon > 1$.

The spectral density of the excitation energy of surface waves is given by the sum of the fluxes in vacuum and the medium. When the particle moves through the medium, the spectral density is

$$W_{\omega} = \frac{2e^2\beta^2\,(\beta^2\varepsilon_1-\beta^2+1)\,(\varepsilon_1+1)\,\varepsilon_1}{c\,(\beta^2+\varepsilon_1-1)\,(\beta^2\varepsilon_1^2+\varepsilon_1-1)\,\sqrt{\varepsilon_1-1}}\,.\qquad (29.14)$$

Surface-wave excitation (without absorption) by a uniformly, linearly moving particle (particle motion perpendicular to the boundary between vacuum and the medium) was also discussed in [99]. Our results for W^j_{ω}, W^s_{ω}, and W_{ω} are twice as large as the results obtained in [99] (the corresponding, rather complex formulas (4) and (5) of [99] for the energy flux of the surface wave in the medium and for the total flux in both the medium and vacuum can be simplified, and one can verify that the expressions under the integral agree with our results, except for a factor of 2). The discrepancy results from the fact that in [99] the considerations were restricted to the range in which the frequency change involves positive values.

§ 30. Coherent Generation of a Surface Wave

We will now consider the excitation of surface waves by a charged particle which moves uniformly and linearly parallel to the boundary. We choose the plane defined by the normal to the surface and the direction of motion as xz plane. We assume that a particle moves in vacuum at a distance z_0 from the boundary in the positive x direction. In this particular case $(\beta_x = \beta, \beta_y = \beta_z = 0, y_\zeta = 0, z_\zeta = z_0, t = x_\zeta/v)$, the initial formula of Eq. (28.11) assumes the form

$$\Pi_{\omega j}^j = \frac{iek\sqrt{-2i\omega}\,\varepsilon}{\sqrt{\pi c}\,\omega(\varepsilon^2-1)} \left(\frac{\varepsilon}{1+\varepsilon}\right)^{1/4} \int \frac{\cos\varphi_\zeta}{\sqrt{\rho_\zeta}} \exp\left(i\frac{\omega}{v}x_\zeta + i\frac{\omega}{c}\sqrt{\frac{\varepsilon}{1+\varepsilon}}\rho_\zeta - i\frac{\omega}{c}\frac{z+z_0}{\sqrt{1+\varepsilon}}\right)dx_\zeta, \qquad (30.1)$$

where

$$\rho_\zeta = \sqrt{(x-x_\zeta)^2 + y^2}, \quad \varphi_\zeta = \text{arc tg}\frac{y}{x-x_\zeta}. \qquad (30.2)$$

Equation (30.1) defines the field of the surface wave in the wave zone in vacuum (z > 0). The Hertz vector of the electromagnetic field in the medium is $\Pi_{\omega j}^* = \Pi_{\omega j}^j(z \to \varepsilon z)$.

Assume that the particle is ejected from the point x $= -l$ and suddenly stopped at the point x = 0. In this case, the integration in Eq. (30.1) must be extended from $-l$ to 0. At sufficiently large distances from the particle trajectory, $(\rho_\zeta)^{\frac{1}{2}}$ in the factor before the exponential function and φ_ζ vary but slightly in the region of integration and they can be considered constants; the exponent of the exponential function can be expanded in powers of x_ζ and only the linear term of the expansion need be used. The result is

$$\Pi_{\omega j}^j = \frac{ekv\varepsilon\sqrt{-2i\omega}\sqrt{\varepsilon(1+\varepsilon)}}{\sqrt{\pi\rho c}\,\omega^2(\varepsilon^2-1)} \frac{\exp\left(i\frac{\omega}{c}\frac{\sqrt{\varepsilon}\,\rho - z - z_0}{\sqrt{1+\varepsilon}}\right)}{\sqrt{1+\varepsilon} - \beta\sqrt{\varepsilon}\cos\varphi} \times$$

$$\times \left\{1 - \exp\left[-il\frac{\omega}{v}\left(1 - \beta\sqrt{\frac{\varepsilon}{1+\varepsilon}}\cos\varphi\right)\right]\right\}\cos\varphi, \qquad (30.3)$$

where

$$\rho = \sqrt{x^2 + y^2}, \quad \varphi = \text{arc tg}\frac{y}{x}. \qquad (30.4)$$

Without absorption, Eq. (30.3) assumes the form

$$\Pi_{\omega j}^j = -\frac{ekv\varepsilon_1\sqrt{-2i\omega}\sqrt{\varepsilon_1(\varepsilon_1-1)}}{\sqrt{\pi\rho c}\,\omega^2(\varepsilon_1^2-1)} \cdot \frac{\exp\left(\frac{\omega}{c}\cdot\frac{i\sqrt{\varepsilon_1}\,\rho - z - z_0}{\sqrt{\varepsilon_1-1}}\right)}{\sqrt{\varepsilon_1-1} - \beta\sqrt{\varepsilon_1}\cos\varphi} \times$$

$$\times \left\{1 - \exp\left[-il\frac{\omega}{v}\left(1 - \beta\sqrt{\frac{\varepsilon_1}{\varepsilon_1-1}}\cos\varphi\right)\right]\right\}\cos\varphi. \qquad (30.5)$$

For $\varphi = \varphi_r$, where

$$\varphi_r = \text{arc cos}\frac{\sqrt{\varepsilon_1-1}}{\beta\sqrt{\varepsilon_1}}, \qquad (30.6)$$

the denominator in the equation tends to zero. The projection of the particle velocity vector on this direction is equal to the phase velocity of the surface wave:

$$V = c\sqrt{\frac{\varepsilon_1-1}{\varepsilon_1}}. \qquad (30.7)$$

Coherent interaction takes place between the particle and the surface wave which propagates at an angle $\varphi = \varphi_r$ relative to the velocity vector of the particle. Cerenkov surface waves are generated. We note that the factor in the braces of Eq. (30.5) vanishes for $\varphi = \varphi_r$. An expansion reveals that the amplitude of the surface wave generated in these directions is finite and proportional to the particle's path length l.

After inserting Eq. (30.5) into Eq. (29.6), integrating over z, and multiplying the result by ρ, we obtain the spectral density of the energy flux of the surface wave in vacuum in the interval $(\varphi, \varphi + d\varphi)$:

$$\frac{dW^j_{\omega j}}{d\varphi} = \frac{4e^2\beta^2\varepsilon_1^4 \sin^2\left[\frac{\omega l}{2v}\left(1 - \beta\sqrt{\frac{\varepsilon_1}{\varepsilon_1-1}}\cos\varphi\right)\right]}{\pi c\,(\varepsilon_1^2-1)^2\,\sqrt{\varepsilon_1-1}\,(\sqrt{\varepsilon_1-1}-\beta\sqrt{\varepsilon_1}\cos\varphi)^2}\exp\left(-2\frac{\omega}{c}\frac{z_0}{\sqrt{\varepsilon_1-1}}\right)\cos^2\varphi. \qquad (30.8)$$

In the direction perpendicular to the velocity vector of the particle, the intensity of the surface-wave excitation vanishes.

When the projection of the particle's path length on the direction $\varphi = \pm|\varphi_r|$ is large relative to the wavelength of the surface wave, the angular distribution of the excitation intensity [Eq. (30.8)] has sharp maxima at $\varphi = \pm|\varphi_r|$, and the values of the maxima are proportional to the square of the path length:

$$\left.\frac{dW^j_{\omega j}}{d\varphi}\right|_{\max} = \frac{e^2\varepsilon_1^3\omega^2 l^2 \exp\left(-2\frac{\omega}{c}\frac{z_0}{\sqrt{\varepsilon_1-1}}\right)}{\pi v^2 c\,(\varepsilon_1^2-1)^2\,\sqrt{\varepsilon_1-1}}. \qquad (30.9)$$

The width of the maxima is inversely proportional to the path length:

$$\Delta\varphi \approx \frac{2\pi v\,\sqrt{\varepsilon_1-1}}{\omega l\,\sqrt{\beta^2\varepsilon_1-\varepsilon_1+1}}. \qquad (30.10)$$

By integrating Eq. (30.8) over the angles φ we obtain the spectral density of the energy flux in vacuum:

$$W^j_{\omega j} = \frac{4e^2\varepsilon_1^3\omega l \exp\left(-2\frac{\omega}{c}\frac{z_0}{\sqrt{\varepsilon_1-1}}\right)}{vc\,(\varepsilon_1^2-1)^2\,\sqrt{\beta^2\varepsilon_1-\varepsilon_1+1}}. \qquad (30.11)$$

The spectral density of the energy flux in vacuum is proportional to the path length of the particle.

When the particle's path length is much smaller than the wavelength of the surface wave, the angular distribution of the intensity of surface wave excitation is proportional to $\cos^2\varphi$. Moreover, when also the time of motion of the particle is small relative to the period of the oscillation, the intensity is proportional to the square of the particle's path length.

The calculation of the angular distribution of the spectral density of the energy flux in the medium leads to a negative expression whose absolute value is ε_1^2 times smaller than in the vacuum case of Eq. (30.8).

For $l \gg \frac{\omega}{c}\mathrm{Re}\sqrt{\frac{\varepsilon}{1+\varepsilon}}$ it becomes necessary to include the square term of the expansion of the function in the exponent of the exponential function of Eq. (30.1) whenever the distance between the point of observation and the particle trajectory decreases. The function

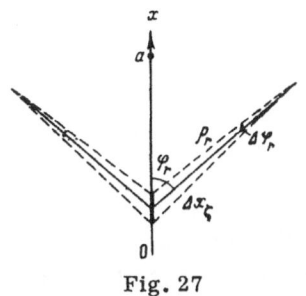

Fig. 27

in the exponent has an extremum. We use this extremum in the stationary phase integration technique. After separating terms which depend upon the integration variable, we expand the function in a Taylor series around some point x_r. When we go only to the quadratic term in the expansion, we obtain

$$i\frac{\omega}{v}x_\zeta + i\frac{\omega}{c}\sqrt{\frac{\varepsilon}{1+\varepsilon}}\rho_\zeta = i\frac{\omega}{v}x_r + i\frac{\omega}{c}\sqrt{\frac{\varepsilon}{1+\varepsilon}}\rho_r +$$

$$+ i\frac{\omega}{v}\left(1-\beta\sqrt{\frac{\varepsilon}{1+\varepsilon}}\cos\varphi_r\right)(x_\zeta - x_r) + i\frac{\omega}{c}\sqrt{\frac{\varepsilon}{1+\varepsilon}}\frac{\sin^2\varphi_r}{2\rho_r}(x_\zeta - x_r)^2,$$

$$(30.12)$$

where ρ_r and φ_r denote the values of the functions ρ_ζ and φ_ζ at the point $x_\zeta = x_r$.

For a point of stationary phase to exist, the following inequality must be satisfied:

$$\beta\,\mathrm{Re}\sqrt{\frac{\varepsilon}{1+\varepsilon}} > 1. \qquad (30.13)$$

When x_r and the point of observation are selected so that the corresponding φ_r value is equal to

$$\varphi_r = \mathrm{arc\,cos\,Re}\frac{\sqrt{1+\varepsilon}}{\beta\sqrt{\varepsilon}}, \qquad (30.14)$$

the imaginary part of the linear term of the expansion in Eq. (30.12) vanishes, i.e., the phase is stationary. We assume that φ_r is defined by Eq. (30.14), i.e., that the expansion of Eq. (30.12) is made near the point of stationary phase. We assume that $-l < x_r < 0$, i.e., that the point of observation is situated between lines drawn from the points $x = -l$ and $x = 0$ (at identical angles φ_r) to the velocity vector of the particle. When this is done, the greatest contribution to the integral results from the region Δx_r, which is of the order (see Fig. 27):

$$\Delta x_r \sim \sqrt{\frac{c\rho_r}{\omega}}\left|\left(\frac{1+\varepsilon}{\varepsilon}\right)^{1/4}\frac{1}{\sin\varphi_r}\right| \qquad (30.15)$$

has its center at the point $x_\zeta = x_r$ and is characterized by the angular dimensions

$$\Delta\varphi_r \sim \sqrt{\frac{c}{\omega\rho_r}}\left|\frac{1+\varepsilon}{\varepsilon}\right|^{1/4}. \qquad (30.16)$$

In the wave zone, $\Delta\varphi_r \ll 1$ holds and therefore $\sqrt{\rho_\zeta}$ and $\cos\varphi_\zeta$ vary only insignificantly in the region Δx_r; in the factor before the exponential function in Eq. (30.1), these quantities can be replaced by the constants $\sqrt{\rho_r}$ and $\cos\varphi_r$ and taken before the integral region. If the entire important part of the integration interval (Δx_r) is inside the interval $-(l, 0)$, and the real part of the linear term of the expansion (30.12) can be ignored, i.e., when the attenuation of surface waves along a path of the order $\Delta x_r \cos\varphi_r$ is small,

$$\sqrt{\frac{\omega\rho_r}{c}}\left|\left(\frac{1+\varepsilon}{\varepsilon}\right)^{1/4}\mathrm{ctg}\,\varphi_r\left|\mathrm{Im}\sqrt{\frac{\varepsilon}{1+\varepsilon}}\right| \ll 1, \qquad (30.17)$$

then the integration in Eq. (30.1) reduces to the multiplication of the expression under the integral sign by

$$\sqrt{\frac{2i\rho c\rho_r}{\omega \sin^2 \varphi_r}} \sqrt{\frac{1+\varepsilon}{\varepsilon}} \tag{30.18}$$

and the replacement of x_ζ by x_r:

$$\Pi_{\omega j}^j = -\frac{2ieke}{\omega(\varepsilon^2-1)}\,|\,\mathrm{ctg}\,\varphi_r\,|\exp\left(i\frac{\omega}{v}x_r + i\frac{\omega}{c}\frac{\sqrt{\varepsilon}\rho_r - z - z_0}{\sqrt{1+\varepsilon}}\right). \tag{30.19}$$

This result is the analog of the spatial cylindrical wave of Cerenkov radiation.

In calculating the field defined by Eq. (30.19) it was assumed that the points at which the particle is ejected ($x_\zeta = -l$) and suddenly stopped ($x_\zeta = 0$) are situated beyond the region Δx_r, which is essential for the integration. If this is not the case, the integration does not result in Eq. (30.19). The means that the field generated by the surface wave does not have the form stated in Eq. (30.19) in the entire region between the straight lines drawn under angles φ_r from the points at which the particles were ejected and suddenly stopped. This condition defines the region in which the diffraction of the surface wave of Eq. (30.19) can be ignored. The angular dimensions of this region are of the order of the $\Delta\varphi_r$ defined by Eq. (30.16).

At sufficiently large distances, for which

$$\Delta x_r \gg l, \tag{30.20}$$

the essential integration interval includes the point of particle ejection as well as the point of sudden particle arrest, and therefore, the result of the integration differs from Eq. (30.19). In other words, the space region in which the wave defined by Eq. (30.19) can exist is limited by the following additional inequality:

$$\sqrt{\frac{c\rho_r}{\omega}}\left|\left(\frac{1+\varepsilon}{\varepsilon}\right)^{1/4}\frac{1}{\sin\varphi_r}\right| \ll l; \tag{30.21}$$

when this inequality holds, the diffraction areas do not overlap. When the inverse inequality holds, the square term of the expansion given in Eq. (30.12) for the function in the exponent of the exponential function in Eq. (30.1) can be neglected. Then, the result of the integration is given by Eq. (30.3). In the intermediate region, the field defined by Eq. (30.19) transforms into a wave of the type defined by Eq. (30.3).

Without absorption, Eq. (30.19) simplifies and assumes the form

$$\Pi_{\omega j}^j = \frac{2ieke_1}{\omega(\varepsilon_1^2-1)}\frac{\sqrt{\varepsilon_1-1}}{\sqrt{\beta^2\varepsilon_1-\varepsilon_1+1}}\exp\left(i\frac{\omega}{v}x_r + \frac{\omega}{c}\frac{i\sqrt{\varepsilon_1}\rho_r - z - z_0}{\sqrt{\varepsilon_1-1}}\right), \tag{30.22}$$

where $\varepsilon_1 = |\varepsilon| = -\varepsilon > 1$, $\beta^2\varepsilon_1 > \varepsilon_1 - 1$.

We recall that the Hertz vector of the surface-wave field in the medium is $\Pi_{\omega j}^{\bullet} = \Pi_{\omega j}^j(z \to ez)$. Thus, without absorption, we have $\Pi_{\omega j}^{\bullet} = \Pi_{\omega j}^j(z \to -\varepsilon_1 z)$. The calculation of the spectral density of the energy flux per unit length of the particle trajectory in vacuum leads to the following result (no absorption assumed):

$$\frac{dW_{\omega j}^j}{dx} = \frac{4\omega e^2 e_1^3 \exp\left(-2\frac{\omega}{c}\frac{z_0}{\sqrt{\varepsilon_1-1}}\right)}{vc\,(\varepsilon_1^2-1)^2\,\sqrt{\beta^2\varepsilon_1-\varepsilon_1+1}}. \tag{30.23}$$

We obtain a negative expression for the spectral density of the energy flux of the surface wave in the medium. This negative expression is related to the spectral density of the flux in vacuum [Eq. (30.23)] by the following simple relation:

$$\frac{dW_{\omega j}^{s}}{dx} = -\frac{dW_{\omega j}^{j}}{dx}\frac{1}{\varepsilon_1^2}. \tag{30.24}$$

The spectral density of the excitation energy of the surface wave per unit particle path length is the sum of the above results. This sum is

$$\frac{dW_{\omega j}}{dx} = \frac{4\omega e^2 \varepsilon_1 \exp\left(-2\frac{\omega}{c}\frac{z_0}{\sqrt{\varepsilon_1-1}}\right)}{vc\,(\varepsilon_1^2-1)\,\sqrt{\beta^2\varepsilon_1-\varepsilon_1+1}}. \tag{30.25}$$

The coherent generation of a surface wave at the boundary separating vacuum from a non-absorbing medium was discussed by Barsukov and Naryshkina in [101]. Their results differ from our results only by a numerical coefficient: the spectral density of the energy flux in vacuum, in the medium, and the total flux are twice the corresponding values defined by Eqs. (19) and (22) in [101] (in [101] the assumption is made that the integration is extended over both positive and negative frequencies; in the second formula of Eqs. (19) in [101], an error was made: instead of the first power of the expression in parentheses, i.e., instead of $(|\varepsilon|-1)$, in the denominator of the expression under the integral the square of the expression in parentheses must be inserted). This difference results because in the calculation of the residue at the pole, a factor of 2 was lost in [101].*

§ 31. Excitation of Surface Waves in the Case of Inclined Particle Incidence

We will now discuss the generation of surface waves by a charged particle which is incident on the boundary under an acute angle. It is assumed that the particle motion is uniform and linear. We take the plane defined by the normal to the boundary and the direction of particle motion as the xz plane and set $y_\zeta = 0$, $x_\zeta = v_x t$, $z_\zeta = v_z t$, $\beta_y = 0$, $\zeta = vt$. The ζ integration in Eqs. (28.8) for $\Pi_{\omega j}^{j}$ and $\Pi_{\omega j}^{s}$ must be performed from minus infinity to zero, and for $\Pi_{\omega s}^{j}$ and $\Pi_{\omega s}^{s}$ from zero to infinity.

In the wave zone, we obtain from Eqs. (28.8) at sufficiently large distances from the point at which the particle intersects the boundary (where ρ_ζ in the factors before the exponential function and φ_ζ may be considered constants, and only the linear terms of the expansion of the functions in the exponent of the exponential function need be employed)

$$\Pi_{\omega j}^{j} = \frac{eke\sqrt{2c\sqrt{\varepsilon\,(1+\varepsilon)}}}{\sqrt{i\pi\rho\omega^3}\,(\varepsilon^2-1)}\frac{(\beta_x \cos\varphi - \beta_z\sqrt{\varepsilon})\exp\left(i\frac{\omega}{c}\frac{\sqrt{\varepsilon}\,\rho - z}{\sqrt{1+\varepsilon}}\right)}{\sqrt{1+\varepsilon}-\beta_x\sqrt{\varepsilon}\cos\varphi - \beta_z},$$

$$\Pi_{\omega s}^{j} = -\frac{eke\sqrt{2c\sqrt{\varepsilon\,(1+\varepsilon)}}}{\sqrt{i\pi\rho\omega^3}\,(\varepsilon^2-1)}\frac{\left(\beta_x \cos\varphi - \dfrac{\beta_z}{\sqrt{\varepsilon}}\right)\exp\left(i\frac{\omega}{c}\frac{\sqrt{\varepsilon}\,\rho - z}{\sqrt{1+\varepsilon}}\right)}{\sqrt{1+\varepsilon}-\beta_x\sqrt{\varepsilon}\cos\varphi - \beta_z\varepsilon}, \tag{31.1}$$

$$\Pi_{\omega j}^{s} = \Pi_{\omega j}^{j}\,(z \to \varepsilon z), \qquad \Pi_{\omega s}^{s} = \Pi_{\omega s}^{j}\,(z \to \varepsilon z).$$

---

*In [101], in the initial equations (15) and (17) for the components of the particle field incorrect expressions were introduced (see, e.g., [11]).

The vectors $\Pi^j_{\omega j}$ and $\Pi^s_{\omega s}$ describe the surface wave resulting from an instantaneous arrest of the particle at the boundary (particle moving through vacuum). The vectors $\Pi^j_{\omega s}$ and $\Pi^s_{\omega s}$ refer to the ejection of the particle into the medium (both cases imply the same direction and the same velocity). When the particle motion through the boundary is uniform and linear, the field of the surface wave in vacuum ($z > 0$) is given by the sum of the vectors $\Pi^j_{\omega j}$ and $\Pi^j_{\omega s}$. In this case, the sum of the vectors $\Pi^s_{\omega j}$ and $\Pi^s_{\omega s}$ defines the surface wave in the medium ($z < 0$).

The spectral density of the energy flux per unit area in vacuum and in the medium are expressed by the corresponding Hertz vectors of the surface waves:

$$S^j_\omega = \frac{\omega^4}{c^3}\,|\Pi^j_\omega|^2\,\left|\frac{\varepsilon}{1+\varepsilon}\right|\,\mathrm{Re}\left[\sqrt{\frac{\varepsilon}{1+\varepsilon}}\,(i\cos\varphi + j\sin\varphi) - \frac{k}{\sqrt{1+\varepsilon}}\right],$$

$$S^s_\omega = \frac{\omega^4}{c^3}\,|\Pi^s_\omega|^2\,\left|\frac{\varepsilon}{1+\varepsilon}\right|\,\mathrm{Re}\left[\frac{1}{\sqrt{\varepsilon(1+\varepsilon)}}\,(i\cos\varphi + j\sin\varphi) - \frac{k}{\sqrt{1+\varepsilon}}\right].$$

(31.2)

When the angle of grazing incidence decreases to the limit $\beta_z \to 0$, the quantities $\Pi^j_{\omega j} + \Pi^j_{\omega s}$ and $\Pi^s_{\omega j} + \Pi^s_{\omega s}$ tend to zero. This means that the surface waves generated in this limit case do not result in the corresponding fields. When no absorption occurs near the angle of coherent generation of a surface wave (i.e., near the angle $\varphi = \arccos\dfrac{\sqrt{1+\varepsilon}}{\beta_x\sqrt{\varepsilon}}$), which is proportional to $|\beta_z|$), the above quantities are inversely proportional to β_z for arbitrarily small, but finite $|\beta_z|$. This region provides the principal contribution to the radiation intensity, which is inversely proportional to $|\beta|$. We note that in the case of grazing incidence it might become necessary to take into account the scattering of the particle in the medium. When large scattering angles are involved over small distances, the effect of the scattering upon the generation of a surface wave can be estimated by retaining only the portion of the surface-wave field generated along the path in vacuum.

Let us now consider the field of the surface wave at smaller distances from the particle trajectory — distances for which the nonlinear dependence of the function ρ_ζ upon ζ is essential. First, we perform the integration in Eq. (28.11). We rewrite Eq. (28.11) under the assumption that the motion of the particle is uniform and linear in the xz plane:

$$\Pi^j_{\omega j} = \frac{iek\sqrt{-2i\omega}\,\varepsilon}{\sqrt{\pi c}\,\omega\,(\varepsilon^2 - 1)}\left(\frac{\varepsilon}{1+\varepsilon}\right)^{1/4}\int_{-\infty}^{0}\frac{\beta_x\cos\varphi_\zeta - \beta_z\sqrt{\varepsilon}}{\beta\sqrt{\rho_\zeta}}\exp\left(i\,\frac{\omega}{c}\,\frac{z}{\sqrt{1+\varepsilon}} + if_\zeta\right)d\zeta,$$

(31.3)

where

$$f_\zeta = \frac{\omega}{v}\left(\zeta + \beta\sqrt{\frac{\varepsilon}{1+\varepsilon}}\,\rho_\zeta - \beta\,\frac{z_\zeta}{\sqrt{1+\varepsilon}}\right).$$

(31.4)

In the case of grazing incidence of the particle on the boundary and of weak absorption of the surface wave $\left(\mathrm{Re}\sqrt{\dfrac{\varepsilon}{1+\varepsilon}} \gg \mathrm{Im}\sqrt{\dfrac{\varepsilon}{1+\varepsilon}}\right)$ the exponential factor under the integral sign may vary strongly when the integration variable changes but slightly and this factor can have a point of stationary phase if

$$|1+\varepsilon| - \mathrm{Re}(\beta_z\sqrt{1+\varepsilon^*}) < \beta_x\,\mathrm{Re}\sqrt{\varepsilon(1+\varepsilon^*)}$$

(31.5)

(here, and in the following discussion, $\beta_x > 0$ is assumed in order to be more specific). We use this expression and employ the stationary phase technique.

The function f_ζ defined by Eq. (31.4) contains all terms of the exponent of the exponential in the expression under the integral sign of Eq. (31.3), terms which depend upon the integration variable. We expand the function f_ζ in a Taylor series near the point ζ_r of stationary phase. At this point, the values of the functions ρ_ζ and φ_ζ will be denoted by ρ_r and φ_r, as before. When we restrict ourselves to the square terms of the expansion, we obtain

$$f_\zeta = \frac{\omega}{v}\left(1 - \frac{\beta_z}{\sqrt{1+\varepsilon}}\right)\zeta_r + \frac{\omega}{c}\sqrt{\frac{\varepsilon}{1+\varepsilon}}\rho_r + \frac{\omega}{v}\left(1 - \frac{\beta_z}{\sqrt{1+\varepsilon}}\right.$$

$$\left. - \beta_x\sqrt{\frac{\varepsilon}{1+\varepsilon}}\cos\varphi_r\right)(\zeta - \zeta_r) + \frac{\omega}{c}\sqrt{\frac{\varepsilon}{1+\varepsilon}}\frac{\beta_x^2}{\beta^2}\frac{\sin^2\varphi_r}{2\rho_r}(\zeta - \zeta_r)^2. \tag{31.6}$$

In the extremum of the function f_ζ, the value of φ_ζ is

$$\varphi_r = \arccos\frac{|1+\varepsilon| - \mathrm{Re}\,(\beta_z\sqrt{1+\varepsilon^*})}{\beta_x\,\mathrm{Re}\,\sqrt{\varepsilon(1+\varepsilon^*)}}. \tag{31.7}$$

The most essential contribution to the integral stems from a region of the order

$$\Delta\zeta_r \sim \frac{\beta}{\beta_x}\sqrt{\frac{c\rho_r}{\omega\sin^2\varphi_r}}\left|\frac{1+\varepsilon}{\varepsilon}\right|^{1/4} \tag{31.8}$$

which has its center at the point ζ_r. The angular dimensions of the projection of the region upon the boundary are

$$\Delta\varphi_r \sim \sqrt{\frac{c}{\omega\rho_r}}\left|\frac{1+\varepsilon}{\varepsilon}\right|^{1/4}. \tag{31.9}$$

When, in the region defined by the relation (31.8), the imaginary part of the linear term of the expansion of the function f_ζ in a Taylor series around the point of stationary phase has an absolute value much smaller than unity, i.e., if

$$\left|\mathrm{Im}\frac{\beta_z + \beta_x\sqrt{\varepsilon}\cos\varphi_r}{\beta_x\sqrt{1+\varepsilon}}\right|\sqrt{\frac{\omega\rho_r}{c\sin^2\varphi_r}}\left|\frac{1+\varepsilon}{\varepsilon}\right|^{1/4} \ll 1 \tag{31.10}$$

and when the point of observation is such that the corresponding ζ_r value relates to a greater distance from the coordinate origin than $\Delta\zeta_r$, the integration in Eq. (31.3) leads to the following result:

$$\Pi_{\omega j}^j = \frac{2ieke}{\omega(\varepsilon^2 - 1)}\frac{\beta_x\cos\varphi_r - \beta_z\sqrt{\varepsilon}}{\beta_x|\sin\varphi_r|}\exp\left[i\frac{\omega}{c}\left(\frac{z_r}{\beta_z} + \sqrt{\frac{\varepsilon}{1+\varepsilon}}\rho_r - \frac{z+z_r}{\sqrt{1+\varepsilon}}\right)\right], \tag{31.11}$$

where z_r denotes the coordinate of the stationary phase point and ρ_r is the projection of the distance between the point of observation and the corresponding stationary phase point on the boundary.

Inequality (31.10) defines the spatial region in which the wave of Eq. (31.11) exists; this wave results from the absorption and the inclined trajectory of the particle relative to the boundary. The condition that the stationary phase point be at greater distances from the coordinate origin than $\Delta\zeta_r$ determines the surface region in which the diffraction of the propagating wave of Eq. (31.11) can be ignored. It follows from the existence of a stationary phase point on the integration path that the region in which the result of Eq. (31.11) is valid is situated beyond the angle formed by the rays which can be drawn from the coordinate origin to the projections of the particle's velocity vector onto the surface (drawn under angles $\pm |\varphi_r|$).

According to Eq. (28.8) we have $\Pi^s_{\omega j} = \Pi^j_{\omega j}$ $(z \to \varepsilon z)$, and therefore, the calculation of the vector $\Pi^s_{\omega j}$ resembles the calculation made before; the result is given by Eq. (31.11); however, z must be replaced by εz. In similar calculations of the field of the surface wave generated by the particle motion in the medium, formulas are derived which differ from Eqs. (31.5)-(31.10) insofar as $\varepsilon \beta_z$ appears instead of β_z. The final result is $\Pi^j_{\omega s} = \Pi^j_{\omega j}$ $(\varphi_r \to \varphi'_r,\ \beta_z \to \frac{\beta_z}{\varepsilon},\ z_r \to \varepsilon z_r)$, where $\varphi^!_r = \varphi_r$ $(\beta_z \to \varepsilon \beta_z)$; according to Eq. (28.8), we have $\Pi^s_{\omega s} = \Pi^j_{\omega s}$ $(z \to \varepsilon z)$. In contrast to the equations for $\Pi^j_{\omega j}$ and $\Pi^s_{\omega j}$, these results are valid within the angle formed by rays drawn from the point at which the particle intersects the boundary under angles $\pm |\varphi_r|$ (except for the diffracting region).

With these results we are in a position to calculate the spectral flux density of the Poynting vector through the planes $y = \rho_r |\sin \varphi_r|$ and $y = -\rho_r |\sin \varphi_r|$ (and, correspondingly, $y = \rho_r |\sin \varphi^!_r|$ and $y = -\rho_r |\sin \varphi_r|$); we refer to the Poynting vector for the surface wave generated by the particle motion in vacuum and in the medium:

$$W_{\omega j} = \frac{2e^2}{c} \left| \frac{\varepsilon}{1-\varepsilon^2} \right|^2 \left| \frac{\varepsilon}{1+\varepsilon} \right| \frac{|\beta_x \cos \varphi_r - \beta_z \sqrt{\varepsilon}|^2}{|\beta_x \beta_z \sin \varphi_r|} \left(\mathrm{Re} \sqrt{\frac{\varepsilon}{1+\varepsilon}} \, \mathrm{Im} \sqrt{1+\varepsilon} \; + \right.$$
$$\left. + \, \mathrm{Re} \, \frac{1}{\sqrt{\varepsilon(1+\varepsilon)}} \, \mathrm{Im} \, \frac{\sqrt{1+\varepsilon}}{\varepsilon} \right) \mathrm{Im} \sqrt{1+\varepsilon} \, \exp\left(-2 \frac{\omega}{c} \rho_r \, \mathrm{Im} \sqrt{\frac{\varepsilon}{1+\varepsilon}} \right), \qquad (31.12)$$

$$W_{\omega s} = -\frac{2e^2}{c} \left| \frac{\varepsilon}{1-\varepsilon^2} \right|^2 \frac{|\beta_x \sqrt{\varepsilon} \cos \varphi'_r - \beta_z|^2}{|(1+\varepsilon)\beta_x \beta_z \sin \varphi'_r|} \left(\mathrm{Re} \sqrt{\frac{\varepsilon}{1+\varepsilon}} \, \mathrm{Im} \sqrt{1+\varepsilon} \; + \right.$$
$$\left. + \, \mathrm{Re} \, \frac{1}{\sqrt{\varepsilon(1+\varepsilon)}} \, \mathrm{Im} \, \frac{\sqrt{1+\varepsilon}}{\varepsilon} \right) \mathrm{Im} \, \frac{\sqrt{1+\varepsilon}}{\varepsilon} \, \exp\left(-2 \frac{\omega}{c} \rho_r \, \mathrm{Im} \sqrt{\frac{\varepsilon}{1+\varepsilon}} \right). \qquad (31.13)$$

When no absorption takes place, the results simplify considerably and assume the form

$$W_{\omega j} = \frac{2e^2 \varepsilon_1 (\varepsilon_1 - 1 + \beta_z^2 \varepsilon_1^2)}{c |\beta_z| \sqrt{\beta_x^2 \varepsilon_1 - \varepsilon_1 + 1} \sqrt{\varepsilon_1 - 1} (\varepsilon_1^2 - 1)}, \qquad (31.14)$$

$$W_{\omega s} = \frac{2e^2 (\varepsilon_1 - 1 + \beta_z^2)}{c |\beta_z| \sqrt{\beta_x^2 \varepsilon_1 - \varepsilon_1 + 1} \sqrt{\varepsilon_1 - 1} (\varepsilon_1^2 - 1)}, \qquad (31.15)$$

where $\varepsilon_1 = -\varepsilon > 1$, $\beta_x^2 \varepsilon_1 > \varepsilon_1 - 1$. In the case of uniform and linear particle motion through the boundary, the total energy flux is given by the sum $W_{\omega j} + W_{\omega s}$.

The spectral density of the excitation energy of surface waves is inversely proportional to the grazing angle. In the particular case of a nonrelativistic particle, this dependence was derived in [97] in qualitative form. The infinite intensity which results in the limit $\beta_z \to 0$ stems from the assumption that the motion of the particle is uniform and linear and that the interaction between the particle and the surface wave is coherent over the entire infinite path (which interaction leads to the Cerenkov effect for surface waves).

We note in conclusion of this section that if the inequality which is the inverse of that stated in (31.10) holds, and when also the inverse inequality which results from (31.10) by replacing β by $\varepsilon \beta$ is fulfilled, the results of the calculation of the field of the surface wave are given by formulas (31.1).

§32.   Work of the Braking Force

The energy losses of a charged particle can be defined as the work done by the moving charge against the reactive force of the field. We are interested in the portion of the work which is related to the generation of surface waves and which is determined by the contribution of the pole. We restrict ourselves to uniform and linear motion of a particle, with the particle incidence inclined relative to the boundary.

First, let us calculate the work done by the charge in exciting surface waves upon sudden arrest of the particle at the boundary. It suffices to use Eq. (28.1) in this case. We use the xz plane as the plane of particle motion and set $y_\zeta = 0$, $y = 0$, $\beta_y = 0$, $x_\zeta = v_x t$, $z_\zeta = v_z t$, $t = \zeta/v$, $x = v_x t'$, $z = v_z t'$, $t' = \zeta'/v$. The formula for the work done by the charge along the entire path of particle motion in vacuum has the form

$$A_j = \frac{e^2}{4\pi^2} \int\limits_{-\infty}^{+\infty} \frac{d\omega}{\omega} \int\limits_{-\infty}^{0} d\zeta \int\limits_{-\infty}^{0} d\zeta' \int\limits_{0}^{\infty} d\varkappa \int\limits_{-\pi}^{\pi} d\varphi \, \frac{\varkappa}{k_{jz} v^2} \left(\frac{\varepsilon k_{jz} - k_{sz}}{\varepsilon k_{jz} + k_{sz}} \right) (\varkappa^2 v_z^2 -$$

$$- k_{jz}^2 v_x^2 \cos^2 \varphi) \exp \left\{ \frac{i}{v} [(\omega - \varkappa v_x \cos \varphi)(\zeta - \zeta') + k_{jz} v_z (\zeta + \zeta')] \right\}. \qquad (32.1)$$

First, we integrate over ζ and ζ'. In order to avoid diverging expressions when the order of integration is changed, we separate from the integration interval a small region near the coordinate origin (at the end of the calculation, we have this region shrink to zero). The result is

$$A_j = -\frac{e^2}{4\pi^2} \iiint \frac{(\varepsilon k_{jz} - k_{sz})(\varkappa^2 v_z^2 - k_{jz}^2 v_x^2 \cos^2 \varphi) \exp(2ik_{jz} v_z \zeta_0/v)}{\omega k_{jz} (\varepsilon k_{jz} + k_{sz}) [k_{jz}^2 v_z^2 - (\omega - \varkappa v_x \cos \varphi)^2]} \varkappa \, d\omega \, d\varkappa \, d\varphi. \qquad (32.2)$$

Then, we integrate over φ and use the following formulas:

$$\frac{1}{k_{jz}^2 v_z^2 - (\omega - \varkappa v_x \cos \varphi)^2} = \frac{1}{2k_{jz} v_z} \left(\frac{1}{\omega + k_{jz} v_z - \varkappa v_x \cos \varphi} - \frac{1}{\omega - k_{jz} v_z - \varkappa v_x \cos \varphi} \right), \qquad (32.3)$$

$$\int\limits_{-\pi}^{\pi} \frac{d\varphi}{a + b \cos \varphi} = \frac{2\pi}{\sqrt{a^2 - b^2}}, \qquad \int\limits_{-\pi}^{\pi} \frac{\cos^2 \varphi \, d\varphi}{a + b \cos \varphi} = 2\pi \frac{a}{b^2} \left(\frac{a}{\sqrt{a^2 - b^2}} - 1 \right), \qquad (32.4)$$

where the real parts of the roots remaining after taking out a are positive. The result of the φ integration can be written in the form

$$A_j = -\frac{e^2}{4\pi v_z} \iint \frac{(\varepsilon k_{jz} - k_{sz})}{\omega k_{jz}^2 (\varepsilon k_{jz} + k_{sz})} \left[\varkappa^3 v_z^2 \left(\frac{1}{\sqrt{a_1^2 - b^2}} - \frac{1}{\sqrt{a_2^2 - b^2}} \right) - \right.$$

$$\left. - \frac{k_{jz}^2}{\varkappa} \left(\frac{a_1^2}{\sqrt{a_1^2 - b^2}} - \frac{a_2^2}{\sqrt{a_2^2 - b^2}} - a_1 + a_2 \right) \right] \exp(2ik_{jz} v_z \zeta_0/v) \, d\varkappa \, d\omega, \qquad (32.5)$$

where $a_1 = \omega + k_{jz} v_z$, $a_2 = \omega - k_{jz} v_z$, $b = \varkappa v_x$.

In the ensuing integration over the frequencies one can switch to the interval extending from zero to infinity for the real part of the integral in Eq. (32.5); the integration over $\varkappa$ can be extended over the entire real axis. When we close the contour for the $\varkappa$ integration in the upper complex half-plane (Im $\varkappa > 0$), limit ourselves to the calculation of the pole contribution (the calculation reduces to multiplying the expression under the integral by $\frac{2\pi i \sqrt{\varepsilon}}{\varepsilon^2 - 1} (\varepsilon k_{jz} + k_{sz})$ and to replace $\varkappa$ by $\frac{\omega}{c} \sqrt{\frac{\varepsilon}{1 + \varepsilon}}$), let the integration interval $(\zeta_0, 0)$ shrink to zero, and omit the frequency integration, we obtain the spectral density of the excitation energy of surface waves in the form

$$W_{\omega j} = \frac{e^2}{\beta_z c} \operatorname{Im} \frac{\varepsilon}{(\varepsilon^2-1)\sqrt{1+\varepsilon}} \left[\frac{(\sqrt{1+\varepsilon}-\beta_z)^2-\beta_z^2\varepsilon^2}{\sqrt{(\sqrt{1+\varepsilon}-\beta_z)^2-\beta_x^2\varepsilon}} - \frac{(\sqrt{1+\varepsilon}+\beta_z)^2-\beta_z^2\varepsilon^2}{\sqrt{(\sqrt{1+\varepsilon}+\beta_z)^2-\beta_x^2\varepsilon}} + 2\beta_z \right]. \quad (32.6)$$

We can calculate in similar fashion both the work done against the reactive force of the field and the work done in the ejection of the particle from the boundary into the medium. To do this, we have to use Eq. (28.3). The corresponding expression for the spectral density of the excitation enegy of surface waves, $W_{\omega s}$, has the form

$$W_{\omega s} = \frac{e^2}{\beta_z c} \operatorname{Im} \frac{1}{(\varepsilon^2-1)\sqrt{1+\varepsilon}} \left[\frac{(\sqrt{1+\varepsilon}-\beta_z\varepsilon)^2-\beta_z^2}{\sqrt{(\sqrt{1+\varepsilon}-\beta_z\varepsilon)^2-\beta_x^2\varepsilon}} - \frac{(\sqrt{1+\varepsilon}+\beta_z\varepsilon)^2-\beta_z^2}{\sqrt{(\sqrt{1+\varepsilon}+\beta_z\varepsilon)^2-\beta_x^2\varepsilon}} + 2\beta_z\varepsilon \right]. \quad (32.7)$$

When the particle motion through the boundary is uniform and linear, the calculations must employ all equations (28.1)-(28.4). In this case we obtain the following additional expressions along the path in the vacuum and in the medium:

$$W'_{\omega j} = -\frac{2e^2}{\beta_z c} \operatorname{Im} \frac{\varepsilon}{(\varepsilon^2-1)(1+\varepsilon)^{3/2}} \left[\frac{(\sqrt{1+\varepsilon}-\beta_z)^2-\beta_z^2\varepsilon}{\sqrt{(\sqrt{1+\varepsilon}-\beta_z)^2-\beta_x^2\varepsilon}} - \frac{(\sqrt{1+\varepsilon}+\beta_z\varepsilon)^2-\beta_z^2\varepsilon}{\sqrt{(\sqrt{1+\varepsilon}+\beta_z\varepsilon)^2-\beta_x^2\varepsilon}} + \beta_z(1+\varepsilon) \right], \quad (32.8)$$

$$W'_{\omega s} = -\frac{2e^2}{\beta_z c} \operatorname{Im} \frac{\varepsilon}{(\varepsilon^2-1)(1+\varepsilon)^{3/2}} \left[\frac{(\sqrt{1+\varepsilon}-\beta_z\varepsilon)^2-\beta_z^2\varepsilon}{\sqrt{(\sqrt{1+\varepsilon}-\beta_z\varepsilon)^2-\beta_x^2\varepsilon}} - \frac{(\sqrt{1+\varepsilon}+\beta_z)^2-\beta_z^2\varepsilon}{\sqrt{(\sqrt{1+\varepsilon}+\beta_z)^2-\beta_x^2\varepsilon}} + \beta_z(1+\varepsilon) \right]. \quad (32.9)$$

The spectral density of the excitation energy of surface waves is given by the sum of the expressions (32.6)-(32.9) when the particle motion through the boundary is uniform and linear. This sum is equal to

$$W_\omega = \frac{e^2}{\beta_z c} \operatorname{Im} \frac{1}{(1+\varepsilon)^2\sqrt{1+\varepsilon}} \left[\varepsilon \frac{(\sqrt{1+\varepsilon}-\beta_z)^2-\beta_z^2\varepsilon(2+\varepsilon)}{\sqrt{(\sqrt{1+\varepsilon}-\beta_z)^2-\beta_x^2\varepsilon}} - \right.$$

$$\left. -\varepsilon \frac{(\sqrt{1+\varepsilon}+\beta_z)^2-\beta_z^2\varepsilon(2+\varepsilon)}{\sqrt{(\sqrt{1+\varepsilon}+\beta_z)^2-\beta_x^2\varepsilon}} - \frac{(\sqrt{1+\varepsilon}-\beta_z\varepsilon)^2-\beta_z^2(1+2\varepsilon)}{\sqrt{(\sqrt{1+\varepsilon}-\beta_z\varepsilon)^2-\beta_x^2\varepsilon}} + \frac{(\sqrt{1+\varepsilon}+\beta_z\varepsilon)^2-\beta_z^2(1+2\varepsilon)}{\sqrt{(\sqrt{1+\varepsilon}+\beta_z\varepsilon)^2-\beta_x^2\varepsilon}} \right]. \quad (32.10)$$

It was assumed in Eqs. (32.6)-(32.10) that the real parts of the roots in the denominators of the terms in the square brackets remain positive (terms which remain after taking out the factors $(\sqrt{1+\varepsilon}\pm\beta_z)$ and $(\sqrt{1+\varepsilon}\pm\beta_z\varepsilon)$).

When no absorption takes place, Eqs. (32.6), (32.7), and (32.10) simplify and assume the form

$$W_{\omega j} = \frac{2e^2}{\beta_z c} \frac{\varepsilon_1}{(\varepsilon_1^2-1)\sqrt{\varepsilon_1-1}} \left[\beta_z - \operatorname{Re} \frac{(i\sqrt{\varepsilon_1-1}+\beta_z)^2-\beta_z^2\varepsilon_1^2}{\sqrt{(i\sqrt{\varepsilon_1-1}+\beta_z)^2+\beta_x^2\varepsilon_1}} \right], \quad (32.11)$$

$$W_{\omega s} = \frac{2e^2}{\beta_z c} \frac{1}{(\varepsilon_1^2-1)\sqrt{\varepsilon_1-1}} \left[\beta_z\varepsilon_1 - \operatorname{Re} \frac{(i\sqrt{\varepsilon_1-1}+\beta_z\varepsilon_1)^2-\beta_z^2}{\sqrt{(i\sqrt{\varepsilon_1-1}+\beta_z\varepsilon_1)^2+\beta_x^2\varepsilon_1}} \right], \quad (32.12)$$

$$W_\omega = \frac{2e^2}{\beta_z c} \frac{1}{(\varepsilon_1-1)^2\sqrt{\varepsilon_1-1}} \operatorname{Re} \left[\frac{(i\sqrt{\varepsilon_1-1}+\beta_z\varepsilon_1)^2-\beta_z^2(1-2\varepsilon_1)}{\sqrt{(i\sqrt{\varepsilon_1-1}+\beta_z\varepsilon_1)^2+\beta_x^2\varepsilon_1}} - \varepsilon_1 \frac{(i\sqrt{\varepsilon_1-1}+\beta_z)^2+\beta_z^2\varepsilon_1(2-\varepsilon_1)}{\sqrt{(i\sqrt{\varepsilon_1-1}+\beta_z)^2+\beta_x^2\varepsilon_1}} \right], \quad (32.13)$$

where $\varepsilon_1 = |\varepsilon| = -\varepsilon > 1$, and the imaginary parts of the roots in the denominators in the terms in square brackets are positive.

Recently, in experiments with fast electrons (energies of several tens of keV) in grazing incidence on a silver target, a radiation was observed [41], the intensity of which exceeds both the transition-radiation and the bremsstrahlung intensities (at normal incidence) by about two orders of magnitude. This radiation has an intensity maximum whose position does not coincide with the transmission band of silver, but is slightly shifted into the region of greater wavelengths. The position of the intensity maximum corresponds to the position of the intensity maximum for surface-wave excitation [41, 102]. But the surface waves did not result in spherical waves and therefore, they could not be directly observed as radiation at great distances. These surface waves may strongly contribute to the radiation only if a mechanism exists which efficiently transforms the waves into spatial waves, for example, scattering at surface irregularities.

In order to compare the theory with experiments, one must undertake a careful, quantitative analysis of the results (including absorption) which were obtained under the assumption that the boundary is completely smooth. Such an analysis can be made with digital computers.

References

1. V. L. Ginzburg and I. M. Frank, Zh. Eksp. Teor. Fiz., 16:15 (1946).
2. F. G. Bass and V. M. Yakovenko, Usp. Fiz. Nauk, 86:189 (1965).
3. I. M. Frank, Usp. Fiz. Nauk, 87:189 (1965).
4. I. M. Frank, Izv. AN SSSR, Ser. Fiz., 6:3 (1942).
5. V. E. Pafomov, FIAN Preprint, No. 31 (1966).
6. V. E. Pafomov, Izv. VUZOV Radiofizika, 10:240 (1967).
7. V. E. Pafomov, Zh. Eksp. Teor. Fiz., 33:1074 (1957).
8. V. E. Pafomov, Zh. Eksp. Teor. Fiz., 36:1853 (1959).
9. V. E. Pafomov, Zh. Eksp. Teor. Fiz., 39:134 (1960).
10. V. E. Pafomov, Dokl. Akad. Nauk SSSR, 133:1315 (1960).
11. V. E. Pafomov, Trudy FIAN, 16:94 (1961).
12. V. E. Pafomov, Izv. VUZOV Radiofizika, 5:484 (1962).
13. V. E. Pafomov, Izv. VUZOV Radiofizika, 5:1072 (1962).
14. V. E. Pafomov, Zh. Tekh. Fiz., 33:557 (1963).
15. V. E. Pafomov, FIAN Preprint, No. A-72 (1964).
16. V. E. Pafomov, Zh. Eksp. Teor. Fiz., 47:530 (1964).
17. V. E. Pafomov, Zh. Eksp. Teor. Fiz., 49:1222 (1965).
18. V. E. Pafomov, Zh. Eksp. Teor. Fiz., 52:208 (1967).
19. V. E. Pafomov and I. M. Frank, FIAN Preprint, No. A-76 (1965).
20. V. E. Pafomov and I. M. Frank, Yadern. Fiz., 5:631 (1967).
21. V. E. Pafomov and E. P. Fetisov, Zh. Eksp. Teor. Fiz., 53:965 (1967).
22. F. Frank and R. Mieses, Differential and Integral Equations of Theoretical Physics [Russian translation], ONTI (1937).
23. I. E. Tamm and I. M. Frank, Dokl. Akad. Nauk SSSR, 14:109 (1937).
24. I. E. Tamm, J. Phys., 1:439 (1939).
25. I. M. Frank, Usp. Fiz. Nauk, 58:111 (1956).
26. B. M. Bolotovskii, Usp. Fiz. Nauk, 65:295 (1961).
27. V. G. Veselago, Fiz. Tverd. Tela, 8:3571 (1966); V. G. Veselago and E. G. Rudashevskii, FIAN Preprint, No. 35 (1966); Fiz. Tverd. Tela, 8:2862 (1966).
28. V. E. Pafomov, Zh. Eksp. Teor. Fiz., 32:366 (1957).
29. I. M. Frank, The Role of the Group Velocity of Light in the Radiation in a Refractive Medium, OIYaI, Dubna (1958); Zh. Eksp. Teor. Fiz., 36:823 (1959); Usp. Fiz. Nauk, 68:397 (1959).

30.   K. A. Barsukov, Zh. Eksp. Teor. Fiz., 36:1485 (1959).
31.   K. A. Barsukov and A. A. Kolomenskii, Zh. Tekh. Fiz., 29:954 (1959).
32.   V. L. Ginzburg, Zh. Eksp. Teor. Fiz., 33:1403 (1957).
33.   G. M. Garibyan, Zh. Eksp. Teor. Fiz., 33:1403 (1957).
34.   A. Rezanov, Zh. Eksp. Teor. Fiz., 16:878 (1946).
35.   G. Beck, Phys. Rev., 47:795 (1948); Problems of Modern Physics, No. 7, p. 116 (1953).
36.   P. Goldsmith and J. V. Jelley, Phil. Mag., 4:836 (1959).
37.   S. Mikhalyak, Dissertation Moscow State University (1961); Yadern. Fiz., 3:89 (1966).
38.   H. Boersch, C. Radeloff, and G. Sauerbrey, Phys. Rev. Lett., 7:52 (1961); Z. Phys.,
      165:464 (1961).
39.   S. Tanaka and J. Katayama, J. Phys. Soc. Japan, 19:40 (1964).
40.   P. von Blanckenhagen, H. Boersch, D. Fritsch, H. G. Seifert, and G. Sauerbrey, Phys.
      Lett., 11:296 (1964).
41.   H. Boersch, P. Dobberstein, D. Fritzsche, and G. Sauerbrey, Z. Phys., 187:97 (1965).
42.   N. A. Korkhmazyan, Izv. Akad. Nauk Arm. SSR, 11(6):87 (1958).
43.   G. M. Garibyan, Izv. Akad. Nauk Arm. SSR, 11(4):7 (1958).
44.   N. A. Korkhmazyan, Izv. Akad. Nauk Arm. SSR, 13(2):139 (1960).
45.   G. M. Garibyan, Zh. Eksp. Teor. Fiz., 38:1814 (1960).
46.   N. A. Korkhmazyan, Uzv. Akad. Nauk Arm. SSR, 15(1):115 (1962).
47.   V. A. Engibaryan and B. V. Khachatryan, Izv. Akad. Nauk Arm. SSR, Ser. Fiz., 1:11
      (1966).
48.   R. A. Ferrell, Phys. Rev., 111:1214 (1958).
49.   G. M. Garibyan and G. A. Chalikyan, Zh. Eksp. Teor. Fiz., 35:1282 (1958).
50.   G. M. Garibyan and G. A. Chalikyan, Izv. Akad. Nauk Arm.SSR, 12(3):49 (1959).
51.   K. A. Barsukov, Zh. Tekh. Fiz., 30:1337 (1960).
52.   R. A. Ferrell and E. A. Stern, J. Am. Phys., 30:810 (1962).
53.   E. A. Stern, Phys. Rev. Lett., 8:7 (1962).
54.   B. A. Lippmann, Plasma Oscillation and Transition Radiation, UCRL–6774, Livermore
      (1962).
55.   R. H. Ritchie and H. B. Eldridge, Phys. Rev., 126:1935 (1962).
56.   V. P. Silin and E. P. Fetisov, Phys. Rev. Lett., 7:374 (1961); Zh. Eksp. Teor. Fiz.,
      45:1572 (1963).
57.   W. Steinmann, Phys. Rev. Lett., 5:470 (1960).
58.   W. Steinmann, Z. Phys., 163:92 (1961).
59.   R. W. Brown, P. Wessel, and E. P. Trounson, Phys. Rev. Lett., 5:472 (1960).
60.   A. L. Frank, E. T. Arakawa, and R. D. Birkhoff, Phys. Rev., 126:1947 (1962).
61.   E. T. Arakawa, L. C. Emerson, D. C. Hammer, and R. D. Birkhoff, Phys. Rev., 131:719
      (1963).
62.   L. C. Emerson, E. T. Arakawa, R. H. Ritchie, and R. D. Birkhoff, Emission Spectra of
      Electron–Irradiated Metal Foils, ORNL–3540 (1963).
63.   D. K. Aitken, R. E. Jennings, A. S. L. Parsons, and R. N. F. Walker, Proc. Phys. Soc.,
      82:710 (1963).
64.   E. T. Arakawa, N. O. Davis, and R. D. Birkhoff, Phys. Rev., 135:A224 (1964).
65.   F. R. Arutyunyan, Zh. V. Petrosyan, and R. A. Oganesyan, Pis'ma Zh. Eksp. Teor. Fiz.,
      3:193 (1966); Zh. Eksp. Teor. Fiz., 51:760 (1966); Optika i Spektr., 21:399 (1966).
66.   G. M. Garibyan, Dokl. Akad. Nauk Arm. SSR, 40:21 (1965).
67.   G. M. Garibyan and M. P. Lorikyan, Dokl. Akad. Nauk Arm. SSR, 40:21 (1965).
68.   Ya. B. Fainberg and N. A. Khizhnyak, Zh. Eksp. Teor. Fiz., 32:883 (1957).
69.   P. V. Bliokh, Izv. VUZOV Radiofizika, 2:63 (1959).
70.   I. M. Frank, Usp. Fiz. Nauk, 68:397 (1959).
71.   M. L. Ter-Mikaelyan, Dokl. Akad. Nauk SSSR, 134:318 (1960); Izv. Akad. Nauk Arm. SSR,
      Ser. Fiz.-Matem. Nauk, 14(2):103 (1961).

72.   M. L. Ter-Mikaelyan and A. D. Gazazyan, Zh. Eksp. Teor. Fiz., 39:1993 (1960).
73.   G. M. Garibyan and I. I. Gol'dman, Dokl. Akad. Nauk Arm. SSR, 31(4):219 (1960).
74.   A. Ts. Amatuni and N. A. Korkhmazyan, Izv. Akad. Nauk Arm. SSR, Ser. Fiz.-Matem. Nauk, 13(5):55 (1960).
75.   K. A. Barsukov and B. M. Bolotovskii, Zh. Eksp. Teor. Fiz., 45:303 (1963).
76.   K. A. Barsukov and B. M. Bolotovskii, Izv. VUZOV Radiofizika, 7:291 (1964).
77.   G. M. Garibyan, Zh. Eksp. Teor. Fiz., 37:527 (1959); Dokl. Akad. Nauk Arm. SSR, Vol. 36, No. 2 (1963).
78.   K. A. Barsukov, Zh. Eksp. Teor. Fiz., 37:1106 (1959).
79.   G. M. Garibyan and I. Ya. Pomeranchuk, Zh. Eksp. Teor. Fiz., 37:1828 (1959).
80.   I. I. Gol'dman, Zh. Eksp. Teor. Fiz., 38:1866 (1960).
81.   G. M. Garibyan, Zh. Eksp. Teor. Fiz., 39:332 (1960).
82.   F. F. Ternovskii, Zh. Eksp. Teor. Fiz., 39:171 (1960).
83.   L. D. Landau and I. Ya. Pomeranchuk, Dokl. Akad. Nauk SSSR, 92:535 (1953).
84.   L. D. Landau and I. Ya. Pomeranchuk, Dokl. Akad. Nauk SSSR, 92:735 (1953).
85.   A. B. Migdal, Dokl. Akad. Nauk SSSR, 96:49 (1954).
86.   M. L. Ter-Mikaelyan, Dokl. Akad. Nauk SSSR, 96:1033 (1954).
87.   M. L. Ter-Mikaelyan, Izv. Akad. Nauk SSSR, Ser. Fiz., 19:657 (1955).
88.   E. L. Feinberg, Usp. Fiz. Nauk, 58:193 (1956).
89.   A. B. Migdal, Zh. Eksp. Teor. Fiz., 32:633 (1957).
90.   B. P. Nigam, M. K. Sundaresan, and E. T. Wu, Phys. Rev., 115:491 (1959).
91.   R. H. Ritchie, J. C. Ashley, and L. C. Emerson, Phys. Rev., 135:A759 (1964).
92.   E. A. Taft, and H. R. Philipp, Phys. Rev., 121:1100 (1961).
93.   R. H. Huebner, E. T. Arakawa, R. A. MacRae, and R. N. Hamm, J. Opt. Soc. Am., 54:1434 (1964).
94.   W. Bothe, Handbuch der Physik, Vol. 24, Berlin (1927).
95.   H. Landolt and R. Börnstein, Zahlenwerte und Funktionen aus Physik (Constants and Functions of Physics), Berlin (1955).
96.   L. D. Landau and E. M. Lifshits, Electrodynamics of Continuous Media, Gostekhizdat, Moscow (1957), p. 364.
97.   R. H. Ritchie, Phys. Rev., 106:874 (1957).
98.   E. A. Stern and R. A. Ferrell, Phys. Rev., 120:130 (1960).
99.   V. Ya. Eidman, Izv. VUZOV Radiofizika, 8:188 (1965).
100. K. A. Barsukov and L. G. Naryshkina, Zh. Tekh. Fiz., 36:225 (1966).
101. K. A. Barsukov and L. G. Naryshkina, Zh. Tekh. Fiz., 36:800 (1966).
102. G. E. Jones, L. S. Cram, and E. A. Arakawa, Phys. Rev., 147:515 (1966).

INTRODUCTION OF HEAVY AND LIGHT RADIOACTIVE NUCLEI INTO NUCLEAR EMULSIONS*

T. A. Romanova

Introduction

Particle track photography with thick photographic emulsions has obtained great importance for the development of nuclear physics and elementary particle physics. One of the advantages of this technique for the study of nuclear reactions and elementary particles stems from the possibility of observing individual interactions after introduction of target nuclei into the emulsion. When this is done, the photographic plate plays the role of both detector and target and contains the nuclei of the elements contributing to the emulsion components, as well as the nuclei of elements introduced (loaded) into the emulsion in some way, depending upon the particular problem under consideration. Depending upon the purpose of the study, in order to increase the efficiency of an experiment one must try to increase the concentration of the nuclei of elements which are anyhow contained in the emulsion (in order to increase the scanning efficiency), or one must introduce as many as possible heavy (or light) nuclei originally not contained in the emulsion. This is a difficult problem because one must maintain the principal properties of the photographic emulsion which is composed of certain elements in a specific way. Particular difficulties are encountered when the emulsion is provided for recording particles at the ionization minimum.

Long storability of emulsions loaded with the desired nuclei is usually desirable for experimental purposes because the irradiation time may vary within wide limits. This means that one must avoid chemical reactions between the elements of the photographic emulsion and the newly introduced elements which may have an adverse effect upon the emulsion properties.

It is often impossible to introduce nuclei of the desired elements to any large extent during the preparation of the photographic emulsion because these elements may have a detrimental effect upon several stages of the complicated emulsion-preparation process. Another method is more suitable: the additional elements are introduced in the form of stable compounds into emulsions of standard photographic plates. Plates with various types of emulsions manufactured in our country and abroad can be stored for a long time, and, as could be shown by experiments, are stable with respect to operations in which they are loaded with the desired

---

*The systematic names of the developers used in this work are employed interchangeably with their trade names: Elon = Metol = monomethyl-p-aminophenolsulfate; Amidol = 2,4-diaminophenol; Rodinal = p-aminophenol — Translator.

nuclei. We therefore use ready-made emulsion layers for loading with nuclei which are introduced in the form of stable compounds. The element introduced and the chemical compound in which the element is introduced can retard or enhance the latent image formation and may even change the physical and chemical properties of the emulsions.

After examining the processes which occur in a photographic emulsion when compounds of uranium, lithium, and deuterium, as well as nuclei of elements originally present in the emulsions are introduced, we developed and employed reliable methods for introducing these elements whereby the maximum concentration of nuclei introduced is retained, the nuclei are uniformly distributed in the emulsion, and a good discrimination of the recorded particles is obtained.

Special methods and processing techniques referring to sensitization, development, and fixing were developed in order to obtain adequate nuclear particle tracks in the loaded emulsions and to improve the physical and chemical properties of the loaded emulsions.

A number of research studies in physics could be done owing to these developments and techniques. The majority of these studies were done with direct participation of the author. In the physics experiments, the author introduced the substances to be examined into the emulsions for nuclear research and processed the emulsions under conditions which were optimal for each particular experiment. The technological developments and the results derived from this work were reported at several All-Union and international conferences [3, 8, 10, 14, 46, 47, 61].

The work proceeded along the following principal directions:

1. Introduction of heavy radioactive and light elements into various nuclear emulsions.

2. Testing and improvement of processing methods for loaded emulsion layers.

3. Methods for increasing the discrimination between the recorded particle tracks.

4. Study of the effect which a dilution with gelatin has upon emulsions sensitive to particles with minimum ionization.

The first section of this article is a review of the work concerning the introduction of uranium into various emulsions. The second part recites the results of research on the introduction of hydrogen and deuterium into emulsions sensitive to particles with minimum ionization. The third section describes methods for introducing lithium isotopes into emulsions. The fourth section describes results of research on increasing the gelatin concentration in nuclear emulsions sensitive to particles with minimum ionization. Finally, the fifth section gives a short review of the investigation methods used in the physics experiments.

The bibliography includes a complete list of the work of the author [1-25].

I. Introduction of Uranium into Emulsions

for Nuclear Research

The following questions were considered in [1, 2, 7-11, 13-16, 18, 19]:

1. Loading and development of nuclear emulsions sensitive to protons with energies of up to 50 MeV.

2. Introduction of uranium into nuclear pellicles with a thickness of 250-300 μ, sensitive at the ionization minimum:

   a) loading NIKFI-R emulsion sheets with a thickness of 250-300 μ with uranyl acetate;

b) making visible (by development) relativistic particles in emulsion layers loaded with uranyl acetate;

c) checking the number of uranium nuclei introduced into emulsion layers;

d) recommendations.

3. Desensitizing effect of uranium salts.

4. Reducing effect of triethanol amine.

5. Study of the tanning effect of uranium.

1. Loading and Development of Nuclear Emulsions Sensitive

to Protons with Energies of up to 50 MeV

In work with thick emulsion sheets (for increasing the efficiency of fission-process observations) it is necessary to introduce the greatest possible amount of uranium nuclei into the emulsion layer and to distribute these nuclei as uniformly as possible over the entire emulsion layer.

In the articles [25, 51, 59], which preceded our work, uranyl nitrate was used in small concentrations as a "loading" material because at high concentrations the emulsions were desensitized and the number of grains along the tracks was strongly reduced, occasionally to the extent that the tracks disappeared completely. When uranyl nitrate was used as a loading substance it was observed that the introduction of a high concentration of uranium nuclei by means of a concentrated bath solution is limited by the pH of the solution; the pH value drops sharply when the uranyl nitrate concentration is increased and becomes as low as pH = 2 at 10% solutions. This affects the adsorption, the formation of the latent image, and the development. Furthermore, the free nitric acid resulting from hydrolysis of uranyl nitrate affects both the gelatin and the silver and this leads to a dissolution of crystallization centers and the latent image centers, as well as to a destruction of the gelatin.

Our subsequent attempts to increase the uranium concentration introduced into finished 100μ plates sensitive to 50 MeV protons led us to use uranyl acetate $UO_2(C_2H_3O_2) \cdot 2H_2O$ as "loading" material, since this compound has excellent buffering properties. The use of uranyl acetate makes it possible to increase the uranium-nucleus concentration introduced into the emulsion by treating the emulsion layers in a highly concentrated solution without fissure development in the emulsion or displacement of the emulsion layers, which were observed when uranyl nitrate was used for the loading.

Several studies [1, 2, 15-19], which had the goal of increasing the efficiency of fission-process observations, required the development of a technique for loading photographic layers with uranium salts [16]. It became possible to introduce large amounts of uranium and to distribute it uniformly over the entire 100-μ-thick emulsion layer. The amount of uranium introduced into the photographic emulsion was determined by α-particle counts in the emulsion which had been developed a certain time after loading, and by the number of radioactive decay pulses recorded with a BFL ring counter. Our experiments revealed that the pellicles must be kept in water prior to loading them with the uranyl compounds. A fourfold increase in the uranium concentration is obtained when an emulsion is loaded which prior to loading was soaked and allowed to swell. According to our result, the uniformity of the loading in photographic layers with a thickness of 100μ or more is guaranteed only when the emulsions are kept in water before loading. The uranium concentration in the layer and the uranium distribution over the layer depend upon the soaking time, the temperature of the loading bath, and the uranium concentration in the loading solution. The maximum concentration of loaded nuclei is obtained with a 7% uranyl acetate bath.

Our technique for loading 100-μ-thick emulsions with uranyl acetate comprised the following steps: (1) soaking the emulsion in water in order to obtain a swelling of the emulsion layer to 2.5-2.6 times the original thickness; (2) loading the uranium while the emulsion is kept in a 2.5-10% uranyl acetate bath for 30 min at a temperature of 26°C; (3) drying after soaking, in an air current with a temperature of 20-22°C.

We found that an extended washing of the layers before the development does not wash out the uranium to any noticeable extent. The presence of uranium in a photographic layer reduces the sensitivity, accelerates the regression of the latent image created by charged particles, and makes the ensuing development and processing more difficult because the emulsion is darkened and its pH value is lowered. The introduction of large uranium quantities made it necessary to develop special processing techniques.

Several experiments which we had made for the purpose of establishing adequate development conditions for emulsions loaded in uranyl acetate baths of high concentrations (from 5-10%) led to the use of a paraaminophenol developer, with the developing split into two stages. In the first stage, the emulsion layers are soaked with an alkali solution (3%); in the second stage, the plates are immersed in the developer solution containing all developer substances except for the alkali. This method is advantageous for obtaining a uniform development over the entire emulsion volume and high discernibility of the particle tracks. A preliminary soaking in alkali solutions levels the pH value of the uranium-containing emulsion layer and increases the swelling. This in turn facilitates the penetration of the developer into the emulsion layer and improves the uniformity of the development in the depth of the emulsion layer. A certain exhaustion of the developer during its penetration into the depth of the emulsion layer is compensated by the high alkali concentration of the emulsion itself.

We preferred paraaminophenol to Metol as a developer because Metol has a strong tendency to create fog, particularly at elevated temperatures. The sensitivity to bromide is greater in the case of paraaminophenol than with Metol, and this permits one to use bromide in the development of thick emulsion layers for the purpose of blocking the development process in the cold development stage.

We obtained good results in experiments using a hot development stage for 100-μ-thick plates which had been previously treated with an alkali solution, and also when we used a development at room temperature after preliminary treatment with an alkali solution. In both cases, the emulsions had been loaded in a 5-7% uranyl acetate bath. One obtains (1) full development over the entire depth of the emulsion layer; and (2) uniform development in the depth, indicated by a uniform grain density on α-particle tracks located at various depths in the emulsion layer.

These are the formulas for our developer for developing 100- to 200-μ-thick emulsions having high uranium concentrations:

Solution 1		Solution 2	
Na_2CO_3	50 g	Paraaminophenol	4 g
Water	fill to 1000 ml	Anhydrous sulfite	25 g
		Water	fill to 1000 ml

Discrimination of Particle Tracks in Plates Recording Protons with Energies of up to 50 MeV. It is very important to record various particle types with plates of the same type in certain physics experiments. Our work proved that it is possible to use plates sensitive to protons having energies up to 50 MeV for obtaining excellent discrimination between α-particle tracks and uranium-fission fragments, to develop only the tracks of the fission fragments, and to identify the tracks of the various particles (fission fragments, α particles, and protons) with a minimum of background fog. In order to obtain these results, we used the method of incomplete development, in which the development process is interrupted before all exposed grains begin to become visible in the development. In discrimination experiments, a two-solution developer was employed (alkali → developing agent).

The use of a particular alkali compound and its concentration were established from the conditions of the problem under consideration. Various methods of incomplete development were employed in the discrimination of various particle tracks in emulsions loaded with uranyl acetate: development at reduced temperatures, development with a retarding agent, development at reduced pH value, and reduction of the development time. A long search ended with a technique for the selective development of particle tracks in C_2 Ilford emulsions loaded in a 5% uranyl acetate solution.

Development of All Tracks at Minimum Background. A temperature-cycle development with the following processing conditions was employed: (1) the plates were kept in a 5% anhydrous sodium carbonate solution for 20 min at 18°C and then for 10 min at 3–4°C; (2) cold soaking in the paraaminophenol developer for 30 min at 3–4°C; (3) dry development for 30 min at 22°C.

Discrimination of Fission-Fragment and α-Particle Tracks in the Presence of Proton Tracks. (1) Soaking in a 5% Na_2HPO_4 solution for 30 min. Paraaminophenol developer of the usual composition. (2) 30 min development at 18°C.

We could obtain the same results in a temperature cycle in which the emulsion underwent a preliminary treatment in an alkali solution, with the development stopped in the cold stage: (1) 5% (anhydrous) Na_2CO_3 solution as alkali, applied for 30 min; (2) cold development at 3–4°C for 20 min.

Selective Development of Fission Fragments. Soaking in a 5% Na_2HPO_4 solution for 30 min. (a) Development in the paraaminophenol developer at 17°C using potassium bromide (0.1 g per 100 cm³ of developer) for 40 min. (b) A similar discrimination could be obtained by a low-temperature development at 3–4°C with preliminary soaking in a 5% Na_2HPO_4 solution for 30 min.

The following results were obtained in a series of tests on the introduction of uranium into 100-μ-thick emulsion sheets sensitive to 30-40 MeV protons.

1. The uranium nuclei concentration could be increased to $6 \cdot 10^{18}$ per cm² of emulsion layer of 100 μ thickness, with the uranium uniformly distributed over the entire layer thickness.

2. A complete, uniform development over the entire depth of the 100-μ-thick emulsion layers loaded with 7% uranyl acetate solution could be obtained under adequate processing conditions.

3. It proved possible to obtain an excellent discrimination between α particles and fission fragments and to develop the fission-fragment tracks by employing the incomplete development technique in plates sensitive to 30-40 MeV mesons.

2. Introduction of Uranium into 250- to 300-μ-thick Nuclear Emulsions

Sensitive to Particles at the Ionization Minimum

Uranium-loaded emulsions sensitive at the ionization minimum had to be employed in studies of the fission mechanism in uranium nuclei bombarded by neutrons, γ rays of various energies, and π and μ mesons, and, more important, in studies of rare events like the emission of electrons in the fission of nuclei by neutrons of various energies. The greatest possible uranium concentrations had to be introduced into very thick emulsion layers in order to increase the scanning efficiency.

We could show in this series of experiments [8, 9, 11, 13, 14] that it is possible to observe relativistic particles resulting from the fission of uranium nuclei in 250-μ-thick emulsions which were sensitive to particles at the ionization minimum even after loading with uranium to the greatest possible concentration.

In the work which had been reported earlier or more recently, uranium nuclei had been introduced only into emulsions which were not sensitive to particles at the ionization minimum. Studies were made with plates carrying a photographic layer not thicker than 50 μ. Several researchers [60] have declared that it is impossible to load emulsion layers thicker than 50 μ with uranium, even in the case of layers which are sensitive only to particles of lower energies. An exception is the work of [26-29] in which P_8 and Ilford C_3 emulsions of 100 μ thickness were loaded with uranium. Standard plates, capable of recording protons with energies of up to 50 MeV and having a thickness of 100 μ, were used by [1, 2, 15-19]; these plates contained a uniform uranium "load" and permitted good discrimination between fission fragments, α particles, and protons.

Our work required the observation of particles at the ionization minimum, which are emitted from fission fragments resulting from the interaction with certain types of particles. Uranium is a source of α, β, and γ radiation; moreover, when a particle flux is incident on such a loaded plate (neutrons, protons, etc.), γ rays are emitted which create a strong background fog and reduce the recognizability of minimum-ionization tracks. Occasionally, the background fog can be so strong that the observation of minimum-ionization tracks is completely impossible.

In order to obtain a high detection efficiency, it was necessary to introduce the greatest possible amount of uranium into 200- to 300-μ-thick NIKFI-R pellicles which are sensitive to particles at the ionization minimum and which contain almost twice as much gelatin as emulsions for the recording of strongly ionizing particles. As has been shown in [29], uranium is adsorbed only on the gelatin. To date, no systematic research on the effect of uranium and its compounds on emulsion layers has been reported. Physicists using photographic emulsions are interested in the properties of the uranium nuclei, but not in the chemical properties of uranium or its compounds in the emulsions.

But almost all papers on the loading of emulsions with uranium include hints that uranium retards the formation of the latent image and desensitizes emulsions. It was also noted that the track definition depends upon the uranium concentration in the solution and that the development becomes impossible once high concentrations are employed. On the other hand, the latent image created by particles in nuclear emulsions fades and this fading proceeds rather rapidly.

With these facts in mind, the majority of the researchers use only thin layers which can be loaded by soaking them with aqueous solutions of uranyl salts at low uranium concentrations. This is done in order to avoid a desensitization; if one is willing to put up with the desensitizing effect of high-concentration uranyl solutions, the concentration is increased in order to single out fission-fragment tracks alone. There are no reports on loading emulsions sensitive to minimum-ionization particles in the case of emulsions with up to 300 μ thickness.

Conditions which are most favorable for the introduction of uranium into these emulsions had to be established in the development of the loading technique. We had to consider carefully the properties of this element, and become familiar with the effect of uranyl compounds upon the physical and photochemical properties of the emulsions.

Loading NIKFI-R Emulsions of 250-300 μ Thickness with Uranyl Acetate.   In developing a method for optimal loading of thick layers which are sensitive to particles at the ionization minimum, we had to consider the principal influences upon loading. The emulsion layers were loaded with uranyl acetate by soaking the emulsions in aqueous solutions of uranyl acetate salts. In these tests, the variables were: the swelling prior to loading, the concentration of the uranyl bath, the pH value of the uranyl bath, the duration of the soaking, the temperature, and the drying.

Preliminary swelling of the emulsion is particularly important for the soaking of thick emulsion layers which are sensitive to particles at the ionization minimum. As we have mentioned above, the gelatin content of these emulsions is half that of the usual nuclear emulsions. When no preliminary bath is employed, a hardened surface zone develops which prevents the penetration of the bath solution to the depth of the emulsion. Then, the result is irregular loading and a low uranium concentration in the layer. It could be shown that preliminary swelling makes it possible to increase the uranium concentration in the layer by almost an order of magnitude.

Initial swelling, which facilitates the transfer of uranyl ions to the depth of the emulsion, reduces the time during which the emulsions must be kept in the loading bath, to a minimum. This is very important in view of the detrimental effect which uranium has upon the photographic layer (dissolution of the silver centers by the uranyl bath and the hardening effect of an uranyl acetate bath, causing an undesirable shrinking of the emulsion).

The concentration of uranium nuclei introduced increases when the concentration of the loading bath is increased, but the rate of increase drops with increasing bath concentrations. In the case of NIKFI-R emulsions, the maximum concentration of uranium nuclei is obtained when a 5% uranyl acetate bath is used.

The effect which the pH of the uranyl acetate bath has upon the adsorption was examined. In agreement with the results of other authors [52, 58], the tests indicated that the absorption increases with increasing pH of the loading solution. We found during these tests that a correspondence between the uranium-nucleus concentration in the photographic layer and the uranium concentration in the loading bath does not exist (see Table 1). The concentration of the nuclei in the emulsion layer considerably exceeds the uranium concentration in the solution and this can probably be explained by the increased volume of the emulsion layer due to swelling of the gelatin, or, by adsorption of uranium on the gelatin. This means that the adsorption of the gelatin causes the concentration in the loading bath to drop.

Furthermore, the loading tests established that uranium concentration in the layer and the distribution of the uranium over the depth of the layer depend upon the time of soaking. The minimum time during which the emulsion must be kept in the loading bath in order to obtain a uniform uranium distribution over the layer was established.

Development of Relativistic Particles in 250- to 300-μ-thick Layers Loaded with Uranyl Acetate.   The above-described loading procedure for NIKFI-R emulsions required special photographic processing methods in order to obtain the desired development of nuclear particle tracks.

In our studies on the most suitable development of loaded pellicles, we employed several variations, using paraaminophenol, 2,4-diaminophenol, and Elon (monomethyl-p-aminophenyl sulfate)/hydroquinone developers with and without hypersensitization. The two-solution paraaminophenol development which we had suggested for the development of emulsion layers loaded with uranium nuclei could not be employed. It turned out that the paraaminophenol developer is not capable of developing tracks of relativistic particles, not even in unloaded emul-

TABLE 1

Concentration of the uranium salt in the solution, %	Uranium concentration	
	in the solution, g/cm^3	in the loaded plates, g/cm^3
2.5	$1.4 \cdot 10^{-2}$	$7.95 \cdot 10^{-2}$
2.5	$1.4 \cdot 10^{-2}$	$7.94 \cdot 10^{-2}$
2.5	$1.4 \cdot 10^{-3}$	$7.66 \cdot 10^{-2}$
2.5	$1.4 \cdot 10^{-2}$	$6.94 \cdot 10^{-2}$
2.5	$1.4 \cdot 10^{-2}$	$7.84 \cdot 10^{-2}$
2.5	$1.4 \cdot 10^{-2}$	$6.91 \cdot 10^{-2}$
2.5	$1.4 \cdot 10^{-2}$	$7.4 \cdot 10^{-2}$
5.0	$2.8 \cdot 10^{-2}$	$8.91 \cdot 10^{-2}$
5.0	$2.8 \cdot 10^{-2}$	$10.3 \cdot 10^{-2}$
5.0	$2.8 \cdot 10^{-2}$	$10.7 \cdot 10^{-2}$
5.0	$2.8 \cdot 10^{-2}$	$10.0 \cdot 10^{-2}$

sions. We found that an increase in the developer efficiency and an increase in the development time (increase in the pH value of the alkali solution, increase in the temperature, and an extended hot stage) cause a two- to threefold growth of the developed grain and of the background fog, without increasing the number of grains along the track. Two-solution paraaminophenol development makes visible the tracks of mesons having energies of 25–30 MeV, when the maximum uranium loading ($7 \cdot 10^{18}$ per cm^2 of a 250-μ-thick emulsion layer) is introduced [7, 9].

When we increased the sensitization of the loaded emulsions by treating them with triethanolamine solutions after the loading, the tracks of relativistic particles could be made visible, but the number of grains was 15–18 per 100 μ track section. The scanning of emulsions of this type is very difficult because the optical density of α-particle tracks is high and the background formed by individual grains is excessive. Neither could acceptable results be obtained with various types of Elon/hydroquinone developers. It was not possible to to obtain a uniform development over the entire emulsion layer and good discrimination between relativistic particles and the background in 200- to 250-μ-thick emulsion layers.

2,4-Diaminophenol developers must be used in order to develop tracks resulting from particles at the ionization minimum. Preliminary tests revealed that uranium-loaded plates cannot be developed under the usual development conditions. Difficulties seem to result from retarded diffusion of the developing solutions, an effect which must be ascribed to the hardening of the gelatin (as will be explained below, the development itself takes place while the full amount of uranyl salts is present). Other difficulties stem from the desensitizing effect which uranium has upon the centers of the latent image as well as upon the development process per se.

When the development efficiency was increased by increasing the pH of the developer solution, and when the diffusion time was doubled relative to that used in the development of unloaded plates, the tracks of relativistic particles could be developed in pellicles with a thickness not exceeding 100 μ. The sensitivity of the control plates could not be obtained (22 instead of 28 grains on the track). Typically, the development of thick layers which had been loaded with a double salt at pH = 5 did not improve the appearance of minimum-ionization tracks. Optimum results were obtained with a 2,4-diaminophenol developer when the wet loading solutions were hypersensitized with triethanolamine and rapidly dried afterwards. When this was done, the sensitivity of the solutions was not only maintained, but doubled. A 2,4-diaminophenol developer was used in a temperature cycle (pH = 6.6).

Fixing. We found that the fixing process plays an important role in the development of minimum ionization tracks, as far as the definition of these tracks in uranium-loaded emulsions is concerned.

The time of fixing uranium-loaded plates is usually increased 3–4 times or more relative to unloaded plates, since the fixing occurs at a large uranium concentration, which has a hardening effect on the emulsion layer. Furthermore, the fixing of loaded plates can obviously be considered as a fixing of undeveloped emulsions in which the silver grains are very small and can be rapidly destroyed by the dissolving effect of a fixing solution containing sodium thiosulfate.

Our tests regarding the use of small quantities (0.125–0.150 g/liter) of sodium sulfate in fixing solutions for the processing of conventional unloaded emulsions gave good results insofar as they increased the transparency of the layers, accelerated the fixing, and prevented the oxidation of silver grains. We therefore tried to use sodium sulfite for fixing loaded emulsion plates. The result was that it is not advantageous to use small concentrations of sodium sulfite. Addition of 10 g anhydrous sodium sulfite to the fixing solution halves the fixing time and a sharp development of tracks from particles at the ionization minimum is obtained. 250- to 300-μ-thick emulsions become transparent. The sulfite played a decisive role for some emulsion batches, because it was not possible to obtain adequate development of the tracks without the sulfite.

Checking the Number of Uranium Nuclei Introduced in an Emulsion Layer. It was mentioned above that the number of uranium nuclei introduced into an emulsion was determined by counting the number of α-particle tracks generated by the natural radioactivity of uranium. For accurate determinations of the uranium concentration, one must know, first, the time during which the particles formed in the emulsion layer are recorded; second, one must be sure that during this time no α-particle tracks disappeared, i.e., one must know the exposure dependence of the number of α particles in the emulsions under inspection (in order to obtain accurate results, the exposure time must be as large as possible); and third, the number of α-particle tracks formed in the wet stage must be negligibly small relative to the number of α-particle tracks formed during the dry processing stages.

Work on the determination of the uranium concentration in photographic emulsions during the various processing stages in two developers (paraaminophenol developer and 2,4-diaminophenol developer) revealed that a two-solution paraaminophenol development must be employed for determining the number of uranium nuclei, because after the first stage (involving a treatment with a 3% sodium acetate solution), in the photographic layer a small amount of uranium remains, whereas the development itself in the 2,4-diaminophenol developer takes place at the full uranium load [9]. According to our results, a preliminary soaking does not wash out the uranium. Moreover, when the development with paraaminophenol is employed, the α-particle tracks appear more clearly and are heavier than in the development with 2,4-diaminophenol. The number of α-particle tracks from naturally radioactive uranium is proportional to the time, provided that the two-solution paraaminophenol developer is used; the proportionality holds even in the case of plates which were loaded by immersing them for 8 h in uranyl acetate solutions. During this time, no fading of the latent image of the α particles is observed, whereas the exposure time for counting α particles in plates sensitive to lower energies is limited by the regression of the α-particle tracks. This means that when the two-solution paraaminophenol is used with plates sensitive to high-energy particles, the amount of uranium introduced into the layer can be determined with a higher degree of accuracy than in the case of the 2,4-diaminophenol developer.

In view of the varying extent of the hardening effect of uranium upon various emulsion batches, the different storages, and the exhaustion of the uranyl bath, the number of uranium nuclei must be redetermined whenever a new experiment is started. To do this, one or two unexposed plates are scanned.

Recommendations. Based on detailed research on the loading and the development, we could recommend a method for loading and subsequent processing of uranium-loaded 250-μ-thick emulsion layers, in which an excellent development of relativistic particles with a small background is obtained.

In order to develop tracks of particles at the ionization minimum it is advisable to use highly concentrated (2–4%) uranyl acetate baths and to introduce maximum swelling (2.5–2.7

TABLE 2

Experiment No.	Uranyl acetate solution, %	Initial optical density before application of uranyl acetate bath	2 h		4 h		24 h		48 h	
			without fixing bath	with treatment in fixing bath	without fixing bath	with treatment in fixing bath	without fixing bath	with treatment in fixing bath	without fixing bath	with treatment in fixing bath
1	2	1.78	1.56	1.44	1.52	1.20	1.46	1.08	—	0.46
2	2	1.54	1.42	1.16	1.16	1.01	—	—	—	—
3	2	1.63	1.59	1.46	1.56	1.26	—	1.00	—	0.42
4	5	2.52	—	2.36	—	—	—	1.70	—	—

times) of the emulsion before the uranyl acetate loading. This preliminary swelling helps to increase the uranium concentration in the emulsion by almost an order of magnitude and guarantees a uniform uranium distribution over the entire emulsion layer.

Immediately after the loading with uranium, triethanolamine is introduced into the emulsion. Triethanolamine plays a double role: it acts as plasticizer after the hardening and as a hypersensitizing agent. The efficiency of triethanolamine is increased when the drying is done rapidly. The drying is done in an air current at a temperature of 25-26°C. The emulsions must be developed immediately after their exposure. A 2,4-diaminophenol developer at pH = 6.6 is to be used in a temperature cycle development.

In view of the above-mentioned hardening effect of uranium, and the fact that uranium is not washed out before the developing stage, the time for all diffusion processes must be approximately doubled. The fixing is done in a 40% sodium thiosulfate solution to which 1% sodium sulfate must be added. The washing must be extended relative to unloaded emulsions of the same type and the same thickness. A 3% glycerin bath must be employed before the drying.

The number of uranium nuclei must be checked in each new experiment; to do this, one or two unexposed plates which are loaded simultaneously with the plates designed for the experiment are singled out, stored for 8-10 hours in order to accumulate a number of α-particle tracks, and then developed in a two-bath paraaminophenol developer.

The following conditions must be observed in order to obtain a reproducible recording of relativistic particles in loaded emulsion plates: (1) the same emulsion number must be used; (2) chemically pure triethanolamine must be used for the hypersensitization; (3) the plates must be dried rapidly after immersion in the hypersensitizing bath; (4) the time period from the end of the drying to the beginning of the processing must not exceed 2-2.5 hours.

3. Desensitizing Effect of Uranium Salts

In work with uranium-loaded plates, several researchers [1, 2, 9, 11, 13, 14, 51, 52, 58] and we, ourselves, noted a desensitizing effect of uranium salts. Systematic research on the effect of uranium upon photographic emulsions has not been reported previously in the literature.

Piggiotto [52] examined the effect of the pH value upon the stability of the latent image and could show that uranium does not have any specific effect upon photographic processes. The development is affected only by the pH value which the emulsion assumes after immersion in the uranyl bath; the pH value of the emulsion has an influence upon the developer's pH value, which is of basic importance for the development process.

TABLE 3

Experiment No.	Control plates	Treatment with uranyl bath	
		before exposure	after exposure
	Overall optical density D		
1	2.46	1.54	2.40
	1.04	0.68	1.06
	0.70	0.48	0.68
	0.54	0.31	0.52
2	2.38	1.48	2.36
	1.94	1.22	1.90
	1.46	1.04	1.42
	1.22	0.84	1.22
3	2.04	1.42	2.00
	1.64	1.02	1.60
	1.04	0.63	1.02
	0.58	0.32	0.60

In [58], the desensitizing effect of uranium was explained by the reduced pH value of concentrated uranium solutions. But, the pH value which emulsions assume after treatment in uranyl salt solutions must affect not only the development, but also the formation of the latent image and its regression. The acidity rises with increasing concentration of the uranyl bath solutions and H^+ ions retard the formation of the latent image.

We could show that the desensitizing effect of uranium can be explained, apart from the pH effect of the uranium solutions, by silver oxidation which in turn can depend upon the pH of the solution. The oxidizing effect of uranium was established from changes in the overall optical density of developed emulsions. To this end, 25-μ-thick NIKFI emulsions which had been exposed and developed to a certain optical density D, were immersed into solutions with various uranyl acetate and uranium salt concentrations for various lengths of time. After that, some of the plates were treated with a slightly diluted fixing bath and their optical density was determined after washing and drying. The rest of the plates were only washed and dried. After that, the optical density of these plates was measured with a densitometer.

As a result, a decrease in the optical density was observed after the plates had been kept for two hours in the uranyl bath. The data listed in Table 2 indicates that the optical density is greatly reduced in plates which had been treated with a fixing bath. This can probably be explained by the fact that in the dissolution of the silver a complex salt is formed which is slightly soluble in water (an analogy to potassium ferricyanate which reduces the image and is only slightly soluble in water without a fixative). In other words, our experiments proved the dissolving and, hence, the oxidizing effect of uranyl ions upon colloidal silver.

Thus, we have reason to assume that the desensitizing effect of uranium results not only from the pH of the uranyl solution, but that the uranyl ions can definitely affect crystallization centers as well as latent image centers and reduce or completely eliminate them.

In research on the effect of desensitizing agents upon the x-ray sensitivity of photographic emulsions, Bogomolov [30] could show that the desensitizing effect depends upon the type of radiation as well as upon the sequence in which desensitization and irradiation follow each other. We therefore undertook several experiments for the purpose of establishing the desensitizing properties of uranyl compounds for various types of radiations like light, relativistic particles, and γ rays. 2.5% uranyl acetate baths were employed before and after irradiation. In order to eliminate the hardening effect of the uranyl compounds, to facilitate the sensitometric measurements, and to save time, we used 10- to 15-μ-thick NIKFI-R nuclear emulsion plates for this work.

A large number of tests rendered the following results.

1. A clear decrease in the sensitivity to light results from uranyl ions introduced before the exposure; no sensitivity changes are observed when the uranyl ions are added after the exposure (Table 3).

TABLE 4

Source of γ rays	Control plates	Treatment in uranyl bath	
		before irradia-tion	after irradia-tion
	Total optical density		
RaTh	1.02	0.90	0.42
	0.96	0.78	0.33
RaBe	3.00	2.56	1.58

TABLE 5

Test No.	Control plates	Treatment in uranyl bath	
		before ex-posure	after ex-posure
		Number of grains on a 100-μ-long track section	
1	28	15-16	24-26
2	32	10-12	22
3	31	12-13	20

TABLE 6

Loading	2.5% uranyl acetate solution			Introduction of triethanolamine			
				before uranyl acetate loading		after uranyl acetate loading	
Development conditions	normal	normal	developed twice normal time	normal	developed twice normal time	normal	developed twice normal time
Sensitivity, i.e., number of grains along a 100-μ track	25-26	None	Low-energy electrons	None	20-22	32-33	50-51
Number of background grains in a cell (237 μ^2)	15.6	15.0	23.3	17.1	34.4	17.1	24

2. The sensitivity to γ rays (Table 4) depends only insignificantly upon the time during which the emulsion layer is kept in an uranyl bath before the exposure. But, a marked sensitivity decrease is observed when the treatment sequence is changed, i.e., when the emulsion is treated after exposure. The results can probably be explained by oxidizing desensitization, a theory according to which the size of the oxidized particle determines the extent of the oxidizing effect. The extent of the oxidizing effect increases with increasing size of the oxidized particle [33].

The development centers for x-ray radiation and hence, for γ radiation (since they are of the same type) are highly dispersed and, consequently, are oxidized more rapidly than the development centers for visible light, which are much larger than they.

In the case of relativistic particles, a decrease in the sensitivity upon introduction of uranium into emulsion is observed for introduction, before and after exposure. The decrease in the sensitivity is more pronounced when the emulsion is treated in an uranium bath before exposure than in a treatment after exposure (Table 5). To date, the underlying mechanism is unknown.

4. Reducing Effect of Triethanolamine

It is a well-known fact that the sensitivity of nuclear emulsions and other emulsions can be increased by triethanolamine (TEA). There is disagreement among the researchers whether triethanolamine is a reducing agent and whether the effect of triethanolamine differs from that of other alkaline substances. Nikolae [31] and Trukhin [32] measured the reduction potential of TEA and other alkaline substances and helped to solve this problem. Their research has shown that TEA sensitization is not related to a direct pH dependence of the emulsion sensitivity, but results from the effect of the pH value upon the reducing capability of triethanolamine itself.

We wished to find out how the hypersensitizing effect of triethanolamine can aid in the development of minimum-ionization tracks in loaded plates coated with emulsions containing uranium in amounts of $4-5 \cdot 10^{18}$ per cm^2 of a 200-μ-thick emulsion layer. A test series was made, the results of which revealed the hypersensitizing effect of triethanolamine (Table 6).

1. When introduced into the loading bath, triethanolamine does not have a noticeable effect.

2. Triethanolamine treatment before the loading has no effect. It was not possible to restore the initial sensitivity even when the development time and the temperature of the hot stage were increased. When this is done, only an increased background is observed.

3. When the triethanolamine treatment is applied after the loading (at up to $4 \cdot 10^{18}$ nuclei/cm^2) of 250-μ-thick plates (with the previously established optimum development time), the sensitivity could be restored and even increased without increasing the background; the sensitivity was doubled relative to that of unloaded emulsions of the same thickness.

4. It was established that triethanolamine hypersensitization is effective only when the uranium-nucleus concentrations in emulsions are less than $5 \cdot 10^{18}$ per cm^2 of 250-μ-thick layers, provided that 2-2.5% uranyl acetate solutions are employed. When the bath concentrations are higher, increasing triethanolamine concentrations prevent the development of minimum ionization tracks and, frequently also of α-particle and proton tracks. This may result from the increased oxidizing effect of concentrated uranium salt solutions, as well as from the hardening effect of concentrated uranyl baths.

The replacement of triethanolamine by other alkalis did not render any advantages (an increase in the grain number per unit track length was accompanied by an increase in the number of emulsion-background grains and of γ-ray-induced background grains). This means that triethanolamine, which is adsorbed by emulsions and cannot be washed out, protects the emulsions from the oxidizing effect of the uranium during all stages of the processing, before and after the exposure. A particularly strong "antioxidizing" effect on minimum-ionization tracks is observed when the triethanolamine is introduced before the exposure.

Based on our tests on the application of triethanolamine before and after the exposure, one can conclude that triethanolamine affects loaded, as well as unloaded emulsions only insignificantly, as far as the effect of the exposure is concerned; the resulting effect is comparable to that of alkalis. The greater effect of triethanolamine applied after loading relative to the application before loading seems to point to the oxidizing effect of uranium, since according to Kartuzhanskii's results [33], after oxidation (CrO_3), triethanolamine always leads to a higher sensitivity than an oxidizing agent employed after triethanolamine. Thus, the efficiency of triethanolamine after loading indicates that uranium is an oxidizing agent.

These results make it understandable that 2,4-diaminophenol development of uranium-loaded plates treated with triethanolamine before their exposure is a highly efficient processing method. In this case, the presence of uranium in the layer plays a favorable role insofar as it aids in the development of minimum-ionization tracks. The uranium helps to develop clearly tracks of relativistic particles, but restrains the development of the background caused by γ radiation, which is always present during natural radioactive decay processes and nuclear reactions during the exposure of nuclear emulsions. The latent image generated by γ rays seems,

TABLE 7

Uranium salt	Concentra-tion of ura-nium salt in solution, %	Concentra-tion of ura-nium in solution, %	pH	Melting point, °C	
				Test I	Test II
$UO_2(NO_3)_2 \cdot 6H_2O$	4.75	2.25	2.6	59	70
$UO_2(CH_3COO)_2 \cdot 2H_2O$	4.00	2.25	3.9	63	84
$NaUO_2(CH_3COO)_3$	4.40	2.25	4.8	54	74
H_2O	—	—	6.5	37	36

TABLE 8

Uranium salt	Concentra-tion of ura-nium salt in solution, %	Concentra-tion of ura-nium in solution, %	pH	Millimeters spent for coagulation	
				Test I	Test II
$UO_2(NO_3)_2 \cdot 6H_2O$	4.75	2.25	2.6	4.0	5.3
$UO_2(CH_3COO)_2 \cdot 2H_2O$	4.00	2.25	3.9	3.7	
$NaUO_2(CH_3COO)_3$	4.40	2.25	4.8	4.4	6.6
H_2O	—	—	6.5	—	—

to be more highly dispersed than the latent image of relativistic particles and therefore, the oxidation by uranium is accelerated, while the appearance of the latent image in the development is retarded.

5. The Hardening Effect of Uranium

In studies of the effect of uranium salts upon the physical and chemical properties of the emulsion, particular attention was paid to the hardening of the gelatin by the ions of the UO_2^{++} solution. This hardening seems to have a great bearing on 250- to 300-μ-thick "R" plates in which the gelatin concentration is half that of other nuclear emulsion plates.

In studies of the hardening effect, we used solutions of various uranyl salts: $UO_2(C_2H_3O_2) \cdot 2H_2O$; $UO_2(NO_3)_2 \cdot 6H_2O$; $UO_2Na(C_2H_3O_2)_3$. We varied both the concentration and the pH value of the solutions in which 250-μ-thick NIKFI-R emulsion layers were treated.

The hardening of the emulsion layers is confirmed by the following:

1. Increase in the melting point of loaded emulsions (Table 7).

2. Shrinking of the emulsion in the loading bath and reduction of the swelling of loaded emulsions during their soaking in water (Fig. 1).

3. Need for increasing the times of all processes involving diffusion into emulsions loaded with uranium. (For example, the duration of the fixing was increased 3-4 times relative to that of unloaded emulsions.)

4. The hardening can be assessed from the amount of uranyl salts spent in the coagulation of gelatin with a subsequent separation of a dispersed phase from the dispersed medium. Measurements similar to determinations of the "chromium number" involved solutions of uranium acetate, uranium nitrate, and the binary salt with equal amounts (by weight) of uranium, and pH determinations. The coagulation was determined from treadlike formations in a 5% gelatin sol (Table 8).

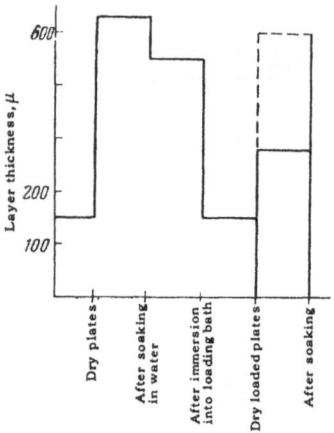

Fig. 1. Thickness of the emulsion layer during uranium loading and during soaking after irradiation. The solid line denotes soaking in water, the dashed line soaking in a 3% sodium carbonate solution.

It was found that the hardening is strongly dependent upon the concentration of the solution, the conditions of soaking, and the pH value of the solution. In order to obtain a uniform loading, the emulsions must undergo a preliminary soaking in water because, otherwise, a strong surface hardening occurs and hence, the diffusion of the uranyl salts into the emulsion layer during the bath period, as well as diffusion processes during the processing of the plates are hampered. The result is a poor uranium concentration and a nonuniform uranium distribution over the layer, which poses problems in the interpretation of the physical events observed in the plates.

It is noteworthy that the hardening effect decreases in proportion to the uranium removal from the layer due to washing operations in the photographic processing. This can be assessed from the slight thickness increase in the layer. But the thickness increase in loaded emulsions is more uniform than that in unloaded (control) emulsions. This is particularly the case in the transition stage between fixing and washing. In unloaded emulsions, the osmotic pressure causes a strong swelling of the emulsion layer. The removal of heavy salts from loaded emulsions is not accompanied by a sharp increase in the emulsion thickness and therefore, we can assume that the bending of particle tracks in loaded emulsions is insignificant.

In finished, uranium-salt-containing plates, a thickness smaller than that of control plates is measured, and the melting point of emulsions loaded in uranyl baths is higher than than that of control emulsions. This remanent hardening can be explained not only by the small amount of uranium remaining after the processing (5-7%), but also by other physical and chemical properties (irreversible hardening). The hardening leads to a substantial increase in the brittleness of the emulsion layers and 250-μ-thick layers crumble and peel from the glass before, as well as after development. The layer must be plasticized in order to reduce its brittleness. Triethanolamine proved to be a highly efficient plasticizer for our purposes. Triethanolamine must be introduced immediately after treating the plates in the uranyl acetate bath.

The uranyl-induced hardening effect is proportional to time, as can be inferred from the time-dependent increase in the melting point of the plates. The effect of triethanolamine stabilizes the hardening. For example, in the case of 250-μ-thick plates, the melting point remained constant, even up to 6 hours after loading. We see that the presence of uranium in the layer is one of the principal obstacles for the processing of loaded plates.

In order to establish adequate development conditions the uranium concentration in the photographic layer was checked during the various stages of the development process; the two developing agents paraaminophenol and 2,4-diaminophenol were employed. The thickness of the emulsion was measured in each stage. Figure 2 illustrates the results of this work. With the paraaminophenol developer, the uranyl salt is rapidly removed from the layer, when the latter is treated with a 3% sodium carbonate bath. During the development, the concentration of the residual uranium is low. The fact that the uranium is rapidly washed out in a sodium solution can be explained by a softening of the gelatin and the formation of a more readily soluble

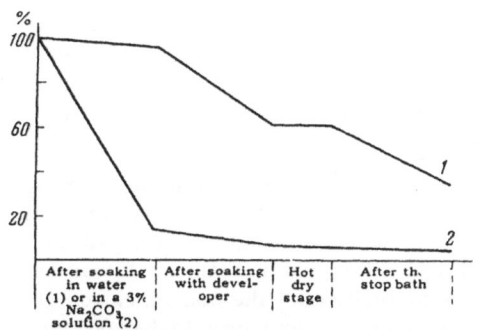

Fig. 2. Relative uranium concentration in emulsion layers during various processing stages. 1) 2,4-Diaminophenol development; 2) paraaminophenol development.

salt. An alkali bath leads to a swelling of the emulsion layer, which facilitates the penetration of the developing substance.

The fixing takes place at the low residual uranium concentration. Development with the 2,4-diaminophenol developer does not involve a noticeable removal of the uranium salt when the layer is soaked in water for an extended period of time. Consequently, the development itself takes place at the full uranium concentration in the unswelled emulsion. The uranium concentration during fixing is high.

BASIC CONCLUSIONS OF SECTION I

1. It was shown that the uranyl salts in loaded nuclear emulsions are adsorbed.

2. The effect of the pH value of the uranyl bath upon the adsorption and the appearance of particle tracks was established. It could be shown that an increased adsorption and an improved development of the tracks are obtained at increased pH values of the uranyl-loading bath.

3. The advantages of uranyl acetate for the loading of emulsions relative to the previously used uranium salts were determined.

4. Development methods for emulsions loaded with uranyl salts were investigated. A two-stage development technique, in which paraaminophenol is used as developing agent, was suggested. It was shown that all particles can be made visible in emulsions sensitive to protons with energies of up to 50 MeV; optimum development conditions with this developer were established for 300-μ-thick NIKFI-R emulsions in order to achieve high particle discrimination (relativistic particle tracks are not developed). Temperature-cycle development with 2,4-diaminophenol and triethanolamine for hypersensitizing wet, loaded emulsions can be recommended in order to make appear minimum-ionization tracks in NIKFI-R emulsions.

5. The fixing methods were examined in order to establish optimum development of tracks of minimum-ionization particles in NIKFI-R emulsions. The optimum conditions for fixing with sodium sulfite were determined. This fixing technique makes it possible to half the fixing time and to obtain transparent emulsion layers.

6. Methods for increasing particle discrimination were examined. It could be shown that with emulsions sensitive to protons up to 50 MeV, incomplete development provides good discrimination between α particles and fission fragments and helps to develop only fission fragments so that cracks of various particles are easily identified.

7. The desensitizing effect of uranyl salts upon various emulsions for nuclear research was investigated. Apart from the reduction of the pH value of concentrated uranium salt solutions, the desensitizing effect of uranium salts can be explained by silver oxidation which, in turn, depends upon the pH value of the solution.

8. The desensitizing effect of uranyl baths was examined in relation to the type of radiation used (light, γ rays, and relativistic particles) and the sequence in which the desensitization was introduced. It could be shown that, in the case of irradiation with light, the sensitivity is reduced when uranyl salts are introduced before exposure, whereas no changes are

observed when these salts are introduced after exposure; in the case of γ-ray irradiation, a substantial decrease in the sensitivity was observed in the opposite sequence, i.e., when the salts were introduced after irradiation. The sensitivity reduction is more pronounced in the case of relativistic particles when the emulsions are treated in the uranyl bath before their exposure. The results can be explained by oxidation leading to desensitization.

9. The hardening effect of UO_2^{++} was examined. It turned out that hardening is a strong obstacle to both the loading and the development of particle tracks. Hardening is particularly noticeable in the case of 250- to 300-μ-thick NIKFI-R emulsions which contain a low gelatin concentration. The great thickness of these emulsions implies long loading times. It was established that the hardening of the emulsion layer depends strongly upon the concentration of the uranyl solution, its pH, and the treatment before the loading. It was shown that maximum swelling of the emulsion before loading is very desirable.

10. The effect of triethanolamine upon the sensitivity of loaded emulsions was tested. It could be shown that triethanolamine, which is adsorbed by the emulsion and not washed out, compensates the oxidizing effect of uranyl in all states of the emulsion processing before, as well as after, exposure. The "antioxidizing" effect which triethanolamine has upon the development of minimum-ionization tracks is particularly strong when the triethanolamine is introduced before the exposure. Triethanolamine is more efficient when it is introduced after the loading than when introduced before the loading, and this is another proof of the oxidizing effect which uranyl has upon silver.

11. These studies and systematic quantitative checks of the uranium concentration in each stage of the emulsion processing led to the use of 2,4-diaminophenol developers for loaded emulsions which had been treated with triethanolamine before their exposure. The presence of large quantities of uranyl during the actual development is favorable for the development of relativistic particle tracks and for the reduction of the γ-ray induced background.

12. Based on a detailed study of the uranium properties and the effect of various factors on both loading and development, a method for optimum uranium enrichment of 250- to 300-μ-thick emulsions sensitive to relativistic particles could be recommended. A method for the subsequent treatment of these layers could be suggested so that optimum development of relativistic particle tracks and low background result. Practical recommendations for checking the uranium-nucleus concentration in the emulsions could be given.

II. Introduction of Hydrogen and Deuterium into Emulsions Sensitive to Particles at the Ionization Minimum [3-5]

In research on various physical problems with the nuclear emulsion technique, particularly in studies of elementary particle interactions, one must increase the hydrogen concentration in the emulsions and introduce deuterium in very large amounts. Usually a soaking in normal or heavy water is employed, and the emulsions are exposed while still wet [35, 54]. Tests with q_5-I and NIKFI-R emulsions revealed several important shortcomings of this method.

1. Introduction of a large quantity of water into layers which are sensitive at the ionization minimum is limited by the irregularity of the swelling, the finite extent of swelling of the layer, and the drop in sensitivity. Our experiments have shown that a reduction of the grain density along the track and a rapid regression of the latent image occur when NIKFI-R plates are slightly wetted with water. When the hydrogen concentration per cm^3 of wet emulsion assumes its maximum value (1.5-1.6 times the hydrogen concentration of the dry emulsion), the grain density decreases from 30 grains per 100 μ track length to 12-13 grains. The drying of

the plates sets a limit for the irradiation time and poses difficulties for an exact control of the uranium-nucleus concentration.

The "loading" technique requires thick emulsions with a minimum background-grain concentration (special emulsions) because otherwise the development of very weak tracks remains inadequate. It is convenient to introduce solid substances containing hydrogen isotopes — substances which make it possible to obtain high-grain density tracks without a reduction of the number of nuclei introduced per cm^3 of the emulsion.

The solution of this problem, particularly the selection of a deuterated agent, is difficult insofar as the substance introduced must be stable, not decompose the emulsion, and not reduce the sensitivity. Lithium acetate, $CH_3COOLi \cdot 2H_2O$, appeared promising. But while we conducted our experiments, the necessary deuterated compounds were not available. Numerous experimental results indicate that when hydrogen is combined with oxygen, trivalent nitrogen, divalent sulfur, and halogens, hydrogen is exchanged against deuterium from water or alcohol. On the other hand, when hydrogen is bonded to carbon, the exchange is slow or does not take place at all, with the exchange-reaction rate depending strongly upon the structure of the substance, the temperature, and the presence of catalysts.

A systematic study of exchange reactions between hydrogen and various carboxylic acids revealed that hydrogen atoms bound to carbon in radicals of the acids are only slowly exchanged for deuterium of heavy water, even when the temperatures are high and when acidic or basic catalysts are present. According to Bonhofer, Geib, and Reitz, cited in [35], the reaction mechanism in alkaline solutions involves the ionization of the hydrogen in C—H by activation through the carboxylic group. The affinity of the hydrogen in the C—H bond of carbonic acid salts depends upon the chemical properties of the metals built into those salts.

No information is available on the exchange of hydrogen by deuterium in lithium acetate. Our experiments have shown that in the deuteration of lithium acetate, only 15-20% of the hydrogen can be replaced in the radical. We obtained in subsequent studies deuterated lithium acetate from malonic acid [36]. In malonic acid, the hydrogen is rapidly replaced by deuterium at the α position in the carboxyl group; this replacement results from the activating effect of the two neighboring substitutes.

Deuterated compounds are easily obtained from the following reaction: malonic acid + $D_2O \rightarrow$ heavy malonic acid $\rightarrow$ (pyrolysis) $\rightarrow$ heavy acetic acid [37]:

$$CH_2(COOH)_2 + 4D_2O \rightarrow CD_2(COOD)_2 + 4HDO,$$

$$CD_2(COOD)_2 \xrightarrow{140-143°} CD_3COOD + CO_2.$$

The deuterium concentration in the resulting salt was determined by combustion of the salt over copper oxide and determination of the density of the combustion residue by floating. Deuterated malonic acid with a concentration of 97-98% was prepared upon our request in the Physico-Chemical Institute of Kiev (this work was supervised by Miklukhin). Apart from this, we prepared lithium hydroxide LiOD in an inert gas atmosphere or in vacuum in order to prevent the formation of lithium nitride (Li_3N). Deuterated lithium acetate can be obtained by treating deuterated acetic acid with deuterated lithium hydroxide:

$$CD_3COOD + LiOD \rightarrow CD_3COOLi \cdot 2D_2O.$$

300- to 320-μ-thick NIKFI R emulsion plates were used for developing a loading technique employing lithium acetate. The effect of various factors was examined, e.g., soaking of the solutions before the loading process, temperature of the soaking bath, concentration of the loading bath, optimum time for keeping the emulsions in the soaking bath, drying of the loaded emulsions, and subsequent processing of the loaded emulsions (development, fixing, and washing).

The following loading technique can be recommended from our experiments. Hydrogen and deuterium in the form of the stable salt of lithium acetate $CH_3COOLi \cdot 2H_2O$ or $CD_3COOLi \cdot 2D_2O$ are introduced into emulsions which are sensitive at the ionization minimum. A 30% aqueous solution of lithium acetate (pH = 7.4) is required for the loading. When deuterium is introduced, heavy water serves as the solvent (D_2O).

In order to increase the amount of salt introduced and to obtain an increased loading uniformity, an initial soaking in water at temperatures between 27 and 28°C is employed for 120 min (in the case of deuterium, heavy water is used for the soaking). An initial soaking can increase the salt concentration by 2.5 times and guarantees a uniform salt distribution after the loading. The soaking in the 30% lithium acetate solution lasts 120 min at 26°C. Both the loading and the preliminary soaking are performed in stainless steel trays with tightly fitting lids. The plates are dried in a special drying cabinet in which a gentle, uniform air current is maintained. The air is humidified by natural water or by heavy water; to do this, the air in the cabinet is first dried with silica gel, then humidified over a water surface. When plates loaded with deuterium are dried, the air is humidified with heavy water in order to prevent an isotope exchange. The drying takes 7-8 hours and the air for drying is slightly heated. The amount of material introduced into each loaded lot is checked by spectral analysis of the lithium. In the case of deuterium loading one must take into account that the hydrogen content in the water contained in the emulsion (which has a certain humidity) is completely replaced by deuterium. The uniformity of the loading over the depth of the emulsion is checked by scanning for (α, H_1^3) stars which result from irradiation with slow neutrons.

This technique makes it possible to increase the hydrogen concentration up to $0.6 \cdot 10^{23}$ nuclei/cm^3 or, to introduce $0.3 \cdot 10^{23}$ deuterium nuclei into 1 cm^3 of dry emulsion (standard emulsions contain $0.3 \cdot 10^{23}$ hydrogen nuclei per cm^3). Thus, one observes an increase in the concentration of hydrogen and deuterium nuclei which exceeds by 40-50% the hydrogen concentration obtained by water soaking. Moreover, the introduction of hydrogen and deuterium in the form of stable compounds implies several other advantages. For example, the grain density on tracks of particles at the ionization minimum is not decreased (a decrease was observed after soaking in water), but rather rises by 25-30%. In some cases, in emulsions with a definite, low sensitivity, a twofold increase in the number of grains on the track was observed (see Table 9, which lists data on the number of grains per 100 μ track length of minimum ionization tracks). The swelling coefficient K_s of the plates equals 1.3, whereas in the case of water soaking this coefficient amounts to 2-2.5. The reduction of swelling decreases distortions. The loaded emulsions can be kept for three days. In the majority of NIKFI-R plates, the number of background grains in loaded emulsions is not increased when optimum processing techniques are employed. No regression of minimum ionization tracks occurs when the loaded plates are stored for two days. The presence of lithium makes it possible to check the concentration of the material introduced, by means of spectral analysis or chemical analysis. The scanning for (α, H_1^3) stars which result from the irradiation by slow neutrons makes it possible to check the uniformity of the loading in the layer (see Table 10, in which the distribution of (α, H_1^3) stars over identical plate areas irradiated with slow neutrons is compiled).

The development is done with the usual temperature cycle method involving a dry, hot development in 2,2-diaminophenol developer (pH = 6.7). Since the pH of the soaked emulsion slightly exceeds the pH of the unloaded emulsion (the pH increases due to the presence of lithium acetate), the temperature of the hot stage of loaded emulsions must be somewhat lower than that of standard emulsions. The fixing is done in a 40% sodium thiosulfate solution at low temperatures with sodium sulfate added in order to decrease the distortions. The washing involves decantation in distilled water to which sodium sulfate has been added (2°C). A slow drying is used. The plates are immersed in a glycerin bath before the drying.

TABLE 9

Emulsion	Without loading, %	Loaded with a 30% lithium acetate solution, %	Loaded with a 20% lithium acetate solution, %	Remarks
Laboratory emulsion	26±4	34±4	31.2±4	Loaded and un-
» »	28.4±5	35.5±5	32.3±5	loaded plates
NIKFI 2724	20±3	43±3	42.6±3	processed simul-
2720	32±3	44±3	—	taneously and under
2756	30±4	35±5	—	identical conditions
2726	32±5	34.5±5	30.2±4	
Laboratory emulsion 1	28±5	35±4	34.6±4	
» » 2	26±5	36±5	—	
» » 3	29±5	33±5	—	NIKFI emulsions
» » 4	28±4	35±4	—	were used for the
» » 5	30±4	35±4	—	laboratory tests
» » 6	32±4	36±4	—	
» » 7	24±4	32±4	—	

TABLE 10

Depth in developed 400-μ emulsion	$\alpha + H_1^3$ average in field of view	Depth in developed 400-μ emulsion	$\alpha + H_1^3$ average in field of view
25	3.6±0.4	10	5.3±0.3
50	3.45±0.4	30	5.2±0.3
100	3.3±0.4	50	5.2±0.3
150	3.4±0.4	70	5.3±0.3
175	3.3±0.4	90	5.5±0.3
		110	5.2±0.3

Determination of the Number of Hydrogen and Deuterium Nuclei.
In determinations of the number of hydrogen nuclei, the elementary composition of the emulsion is of importance. We used the results of a quantitative analysis of NIKFI-R emulsions. According to the NIKFI data for R emulsions, there is 0.0492 ± 0.00402 g hydrogen in 1 cm^3 of emulsion with a residual humidity of 3%, i.e., there are $2.7 \cdot 10^{22}$ hydrogen nuclei per cm^3 of emulsion.

The amount of lithium acetate introduced was determined by spectral analysis in each lot of emulsion plates. The accuracy of this method amounts to 1%. The concentration of the introduced lithium varied from lot to lot, from 0.03 to 0.035 g/cm^3 of loaded emulsion, which corresponds to 1.7-$2.2 \cdot 10^{22}$ hydrogen nuclei per cm^3 of loaded emulsion. Variations result from the instability of the physical and chemical properties of this type of emulsion, with the properties varying from one lot to another. One must also consider the additional humidity which is inherent to emulsions loaded with lithium acetate. Since the emulsions are hygroscopic, the moisture content depends upon the content and changes from lot to lot. When this moisture content is taken into account, the hydrogen content increases to $0.8 ± 0.3 \cdot 10^{22}$ hydrogen nuclei per cm^3.

The total number of hydrogen nuclei is calculated as follows: NH is obtained from the element composition; to this amount one must add the additional moisture and the hydrogen content obtained from the spectral check for lithium.

The hydrogen concentration can be determined with an accuracy of 7%. In determinations of the number of deuterium nuclei, the amount of loaded material is found in the same fashion as in the case of hydrogen. However, one must take into account that during the immersion of the plates in heavy water, a part of the hydrogen in the NH and NH_2 groups of gelatin can be replaced by deuterium. We performed an analysis of emulsions which had been treated with heavy water and found that about 20% of the hydrogen in the gelatin is replaced by deuterium during the loading process. Furthermore, one must take into account the incomplete replacement of the hydrogen in lithium acetate by deuterium. In the material at our disposition, the replacement was as high as 98%.

When we take into account the additional humidity which results in the form of heavy water from the hygroscopicity of heavy water, we obtain $0.25-0.3 \cdot 10^{23}$ hydrogen nuclei per cm^3 as the deuterium concentration in the various plates loaded with lithium acetate. The calculation is done in the following form: the ND amount of loaded deuterium (this amount is obtained by analysis) + additional moisture in the form of heavy water + amount of hydrogen in the gelatin, replaced by deuterium − incomplete replacement of the hydrogen in the substance by deuterium. The error in the determination of the deuterium nuclei amounts to about 15%.

III. Introduction of Lithium into Emulsion Plates

[11, 16, 17, 21, 22]

Studies of the interaction of 14-MeV neutrons with Li^6 and Li^7 nuclei encounter great experimental difficulties because the reaction products (tritons, deuterons, and protons) have very similar energies and consequently, are hard to discriminate in experiments. A strong background of recoil protons created by the neutrons occurs at the same time. This complicated process was studied by introducing the lithium isotopes Li^6 and Li^7 into nuclear emulsions.

The goal of our work was to develop a technique for loading certain 100- and 200-μ-thick nuclear emulsions with Li^6 and Li^7 and for processing the loaded emulsions. We considered the following problems.

1. Selection of a stable lithium-containing compound which does not affect the photographic and physicochemical properties of the emulsion.

2. Uniform distribution of the substance over the entire depth of the emulsion.

3. Development of a method for determining the amount of lithium adsorbed on the emulsion.

4. Determination of the optimum development in order to obtain the best possible discrimination between protons and deuterons, and tritium nuclei and α particles resulting from the interaction of 14 MeV neutrons with Li^6 and Li^7 nuclei.

Darling et al. [56] used 100-μ-thick C_2-J emulsions loaded with lithium. However, they did not outline their loading and processing technique. It seems that their lithium compound was unsuccessfully tried, because the distribution of the events recorded depended strongly upon the depth in the emulsion.

Half of the total number of events occurred in the upper 35 μ of the emulsion. At a depth of 45 μ, the concentration of loaded nuclei is one third of the concentration at the surface and, consequently, not the entire thickness of the emulsion was used to full advantage. An irregular distribution of the loaded substance over the depth of the emulsion makes it difficult to introduce corrections for tracks leaving the emulsion. Titterton et al. [55] loaded plates with lithium sulfate, which has a detrimental effect on the emulsion and hampers the development process.

After a series of tests, we used for the loading lithium acetate−a stable substance which can be introduced without strong swelling of the emulsion. Lithium acetate is easily adsorbed

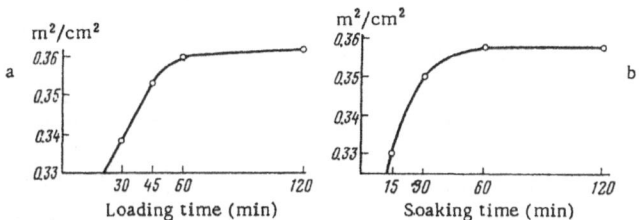

Fig. 3. a) Loading curve; b) soaking curve.

on the gelatin, soluble in water (which is very important for introducing the substance into an emulsion layer), and the pH of the aqueous solution is very close to the pH of the emulsion itself (7.2). An increased pH value of the solution is known to reduce the regression — a fact which is of great importance. The hygroscopicity of lithium acetate makes it possible to dry loaded emulsions uniformly. Lithium acetate containing the Li^6 isotope was prepared from a small amount of an imperfect isotope substance which contained 92% Li^6. A 10% lithium acetate solution was used for the loading because, during storage, a high lithium acetate concentration could cause lithium acetate crystallization on the surface of the loaded emulsion. This crystallization leads to a deformation of the layer and disturbs the uniformity of the lithium acetate distribution in the emulsion.

The lithium was loaded in the following fashion. A 200-μ-thick E^1-J nuclear emulsion was first soaked for one hour with distilled water of 22°C; then the emulsion was kept for 45 min in a 10% lithium acetate bath at a temperature of 20°C.

The number of lithium nuclei was determined as follows. The adsorbed lithium acetate was extracted from the emulsion by repeated washing in a certain volume of distilled water at room temperature (the washing lasted several days). The lithium concentration in the solution was determined by spectral analysis.

The lithium concentration could be determined with an accuracy of 3% when an average of about 2.9-2.8 · 10^{-4} g/cm^2 was introduced into a 200-μ-thick emulsion. Analysis revealed that the loading of the emulsions of various series is reproducible with an accuracy of about 10%.

According to our experiments (Fig. 3), the diffusion of lithium acetate into the emulsion proceeds rapidly. The lithium loading of emulsions reaches saturation within 45 min; on the other hand, 96% of the lithium can be washed out in 45 min. The removal of lithium by washing makes it possible to determine the lithium concentration introduced into each plate used for experiments; to this end, the solution in which a plate is immersed before the developing is spectroscopically analyzed. However, the number of nuclei obtained from these analyses must be increased by 4% in order to allow for the incomplete removal of lithium by a 45-min immersion.

It is important to note that appropriate selection of the loading substance and a preliminary swelling of the emulsion before the loading guarantee uniform lithium-nucleus distribution over the entire depth of the emulsion. This can be reliably confirmed in experiments in which loaded emulsions are irradiated with thermal neutrons and the depth distribution of the events observed is established. The number of events observed as a function of the depth in the developed emulsion layer was constant, within the limits of statistical error in the depth interval ranging from 0 to 70-80 μ.

The uniformity of the loading is an important advantage over the work of Darling and made it possible to take full advantage of thick emulsions. The thickness increase in the loaded emulsions amounted to 10-15%, whereas the stopping power of the loaded emulsions was 5-7% smaller than that of unloaded emulsions.

Emulsions loaded with lithium were developed with a temperature-cycle method in which a hot dry stage with a 2,4-diaminophenol developer of the following composition was used:

$$
\begin{array}{ll}
\text{2,4-Diaminophenol} \dots\dots\dots\dots\dots & \text{2.5 g} \\
\text{Sulfite (crystalline)} \dots\dots\dots\dots & \text{20 g} \\
\text{KBr} \dots\dots\dots\dots\dots\dots\dots\dots & \text{0.5 g} \\
\text{H}_2\text{O} \dots\dots\dots\dots\dots\dots\dots\dots & \text{fill to 1 liter}
\end{array}
$$

Since the soaking before the development does not completely remove the lithium from the emulsion and since the pH of the loaded emulsions is slightly higher than the pH of normal emulsions, the temperature of the hot stage was slightly reduced relative to the normal hot-stage temperature.

In order to obtain reproducible results one must maintain constant-temperatures during the cold stage of the development and, more important, during the hot stage (deviations must not be greater than $\pm 0.1^{\circ}$C); the pH value must remain constant (the deviations must not exceed 0.1 pH units).

We could obtain satisfactory determinations of the point from which α and H_1^3 tracks start when we used this technique. In the case of emulsions which had been irradiated with thermal neutrons, the point at which the tracks parted could be established in 92-95% of the total number of events.

The particle-track density permitted a reliable determination of the tracks resulting from the (n, α) reaction at neutron energies of 14 MeV. Discrimination was possible even when the angle between the outgoing α particle and the neutron beam amounted to 90°.

The spread in the number of events observed as a function of the depth agreed with the statistical error in the depth interval from 0 to 70-80 μ. This attests to the uniformity of the development and to an excellent track discrimination.

IV. Dilution of NIKFI-R Emulsions with Gelatin

The modern NIKFI-R emulsions, as well as the emulsions of other companies (G_5, NT_4, NTB_3) contain 86-87 wt.% silver halide. Emulsions with such a high volume concentration of silver halide have the following shortcomings for certain applications: 1) high stopping power; 2) high shrinkage coefficient ($K_s = 2.9$); and 3) low efficiency of the development (an additional background results from the high degree of grain packing) [57], etc.

In our work, we diluted NIKFI-R emulsions with gelatin in order to employ the diluted emulsions for recording relativistic particles and low-energy particles. Thick-layer plates with an emulsion thickness of 300 μ were prepared with various gelatin concentrations. The gelatin was introduced as a solution into liquid NIKFI-R emulsions. Glucose was used as an additional plasticizer because 200-μ-thick emulsions peel from the glass if no plasticizer is used.

The plates were exposed to relativistic mesons and 14-MeV neutrons. The volume concentration of the silver halogenide had been diluted relative to the gelatin volume 1.5, 2, 4, 5, and 8 times (Table 11 lists the dilution values). The plates were developed with the temperature-cycle method.

In the majority of diluted "R" emulsions, the reduction of the grain density along the track is proportional to the reduction of the silver halogenide volume per cm^3 of the emulsion (Fig. 4). There exists a range of volume ratios V_g / V_{AgBr} (with V_g denoting the gelatin volume) in which tracks of particles with low ionization can be observed. This interval comprises the

TABLE 11

Emulsion No.	Dilution	AgBr vol.: total emulsion volume	AgBr vol.: gelatine vol.	Emulsion density	Shrinkage coefficient	AgBr Concentration (by weight) in dry emulsion
I	1	0.51	1.05	3.9	2.9	87
II	1.5	0.34	0.69	2.84	2.2	80
III	2	0.29	0.50	2.64	1.9	74
IV	4	0.22	0.25	2.1	1.63	62
V	6	0.16	0.18	1.9	1.5	57.6
VI	8	0,13	0.14	1.75	1.3	48

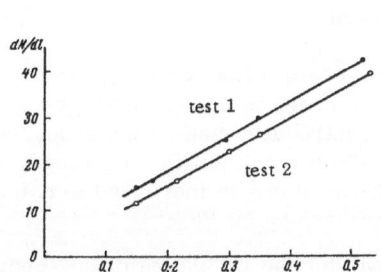

Fig. 4. Relation between the grain density on the track (dN/dl) and the AgBr volume per cm^3 of emulsion (E = 300 MeV).

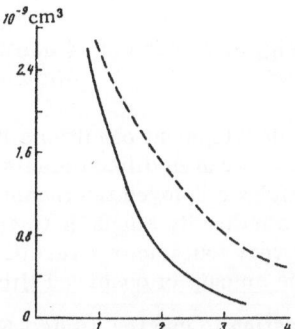

Fig. 5. Relation between the number of background grains in the emulsion volume and the degree of dilution. The solid line refers to test 1 with an "R" 2978 emulsion; the dashed line refers to test 2 with an "R" 2993 emulsion.

volume ratios 1.5–2.5. The number of grains per 100 μ track length varies from 24 to 18 grains. At higher dilution ratios, relativistic particle tracks can no longer be observed in the majority of the emulsions.

The background is strongly reduced by the dilution with gelatin, particularly in the dilution interval ranging from 1 to 2.5–3 (Fig. 5). This effect seems to be related to the fact that in undiluted emulsions the high degree of halide packing and the low efficiency of the development cause an additional background. In the work of Dodd et al. [57] this effect was not observed in G$_5$-J emulsions. The strong reduction of the background makes it possible to partially compensate for the reduced grain density in diluted emulsions by employing a stronger development. In diluted emulsions, increased pH values, extended cold-stage times, and increased temperatures in the hot stage do not lead to an increased background, as is the case in the majority of undiluted emulsions which we examined.

By increasing the hot-stage temperature to 27°C and prolonging the cold-stage development, we could raise the number of grains per 100 μ track length from 20–21 to 25–28 in diluted emulsions III (diluted by a factor of 2). The background was 50% lower than in the control emulsion I which had been developed at 24°C. The reduction in the background lets us believe

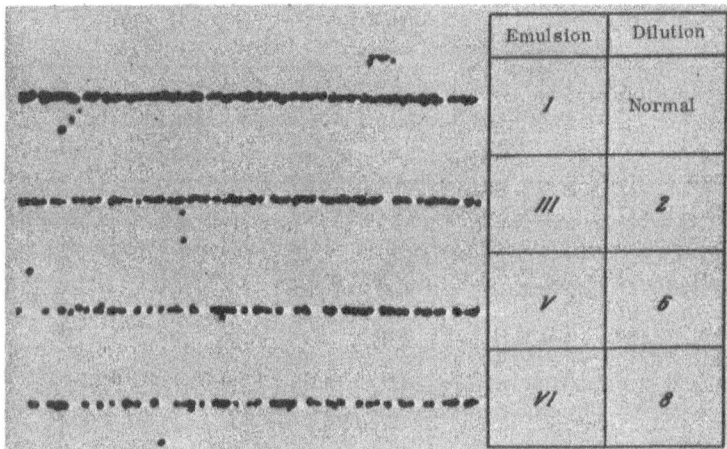

Fig. 6. Tracks of 14–MeV protons in emulsions which had been
diluted to various degrees.

that by adequate development conditions it will possible to obtain tracks whose grain density
is close to that of tracks in diluted emulsions. Recently, in order to increase the grain density
we have successfully employed triethanolamine which was introduced before the exposure. In
this case, the grain density amounts to 40-45 grains per 100 μ track length. Loading of diluted
emulsions gives very satisfactory results. When the gelatin volume is increased to twice the
AgBr volume, the amount of observed lithium acetate increases by 50-70%.

It is interesting to use the diluted NIKFI-R emulsions for the recording of low–energy
particles, particularly for 14-MeV protons. To do this, diluted emulsions were used in which
the volume ratio of silver halide to gelatin had been reduced from 4.5 to 8 times. Figure 6
displays tracks of 14 MeV protons obtained at various degrees of dilution. When the gelatin
volume was increased 8 times, the range of the protons increased 60-70%. The shrinkage coefficient of the emulsion dropped to 1.2-1.3. Both the end and the beginning of the track could
be clearly distinguished.

The fact that no track regression occurred during 10 days was one of the big advantages
of diluted emulsions used for the recording of low–energy particles (Table 12). It seems that
an increase in the grain size and in the sensitivity of the grains implies a latent image of increased stability.

It appeared promising to use diluted emulsions for recording heavy charged particles with
E = 0.3 MeV since both the grain density and the length of the track are increased. Diluted
emulsions can be very valuable in studies of nuclear interactions involving high–energy particles and light nuclei.

Thus, it is possible to employ diluted "R" emulsions for recording particles with relativistic energies when the degree of dilution is as high as 2-2.5. For recording low–energy
particles, one can successfully use "R" emulsions in which the volume concentration of the
silver halide has been reduced 8 times or more, depending upon the energy of the particles
to be recorded.

V. Using the Results of the Methodological Studies

in Physics Experiments

The results of our studies and the techniques developed were successfully employed for
several series of physical studies. The majority of these studies were performed with direct
participation of the author.

TABLE 12

Storage time, h	Track length, μ	Number of grains per 100μ track length at a distance of 140μ from the end of the track, %
0	1516±20	73±2
48	1510±20	75±2
·72	1496±20	73±2
240	1522±20	78±2

1. Fission of uranium nuclei by slow π^- mesons, fast neutrons with energies up to 460 MeV, and γ rays with energies of up to 250 MeV [1, 2] (Figs. 7-10).

Fission induced by π^- mesons was observed first. The investigations helped to establish the physics involved in the fission process of the uranium nucleus at high excitation energies, and to determine the fission probability obtained with various types of radiation. The results of this work were communicated during the International Conference on the Physics of Fission in 1956 in Canada [61].

New techniques for introducing uranium salts into the emulsion layer were required for this type of work. The new techniques made it possible to increase the uranium concentration in the photographic emulsions almost 30 times relative to the previously obtained concentrations. A uniform distribution of the uranium over the entire photographic layer was accomplished for the first time. This is very important for the interpretation of the processes considered. Since charged particles with low ionization power had to be observed (50-MeV protons and π mesons with energies of up to 5 MeV), a special development technique was needed. The author developed a two-stage development in which paraaminophenol was employed as the developing agent.

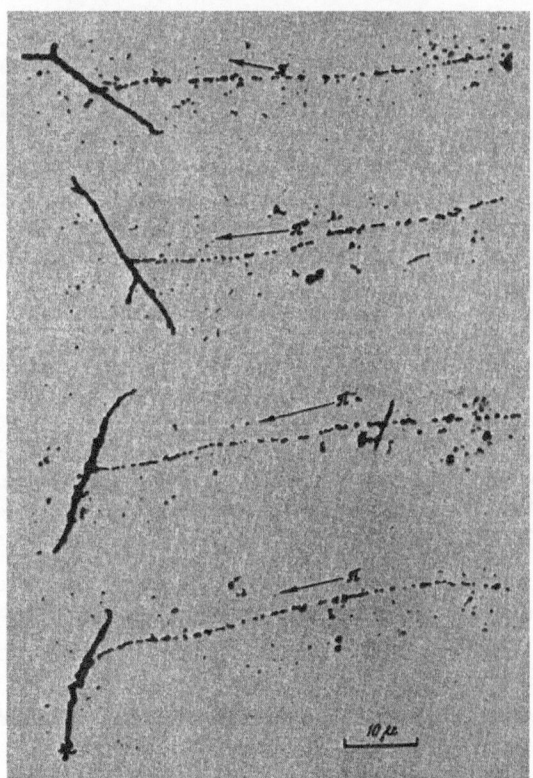

Fig. 7. Photomicrographs of the π^--meson-induced fission of uranium nuclei into fragments (two-solution development with paraaminophenol).

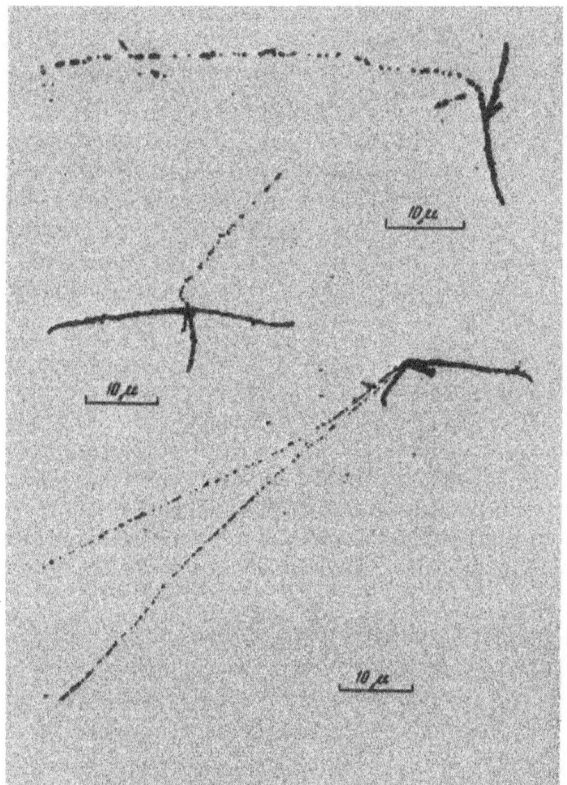

Fig. 8. Photomicrographs of the π^--meson-induced
fission of uranium nuclei into three and four fission
fragments (two-solution development with paraaminophenol).

The division of the development process into two stages, the successful selection of the developing agent, and the determination of optimal processing conditions made it possible to use in physics experiments emulsions containing high uranium concentrations and to obtain complete and uniform development over the entire thickness of the photoemulsions. The method which had been worked out for selective development of emulsions sensitive to protons with energies of up to 50 MeV helped to obtain a good discrimination between α particles and fission fragments and made it possible to make appear the fission fragments alone.

Both the loading technique and the processing suggested by the author were widely used in studies of the angular distribution of fission fragments resulting from photo-induced fission [45, 46, 48, 49]. The results of the work were reported in the International Conference in Geneva [46].

2. Determination of the reaction cross sections of the reactions $Li^6(nT)\alpha$; $Li^6(n, D)He$; $Li^6(n, n')\alpha + D$; $Li^6(n, n')\alpha + p$; $Li^7(nT)He^5$ involving lithium isotopes [12, 17, 21, 22, 24, 40]. This set of experiments required the development of a method for introducing large quantities of lithium into a photographic emulsion, the accurate determination of the lithium concentration in the layer, and a development technique which permits one to distinguish between particles of similar masses (protons, deuterons, tritons) and energies.

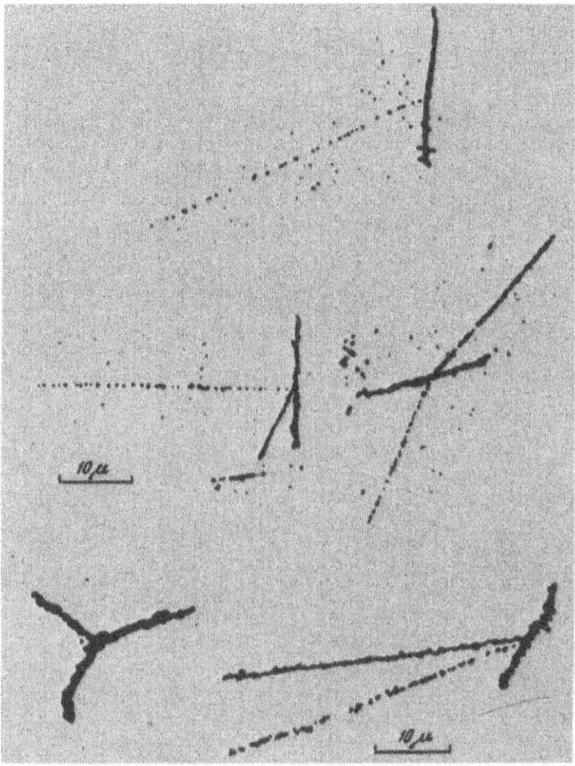

Fig. 9. Photomicrographs of the fission of uranium
nuclei by γ rays having energies of up to 250 MeV;
emission of fast charged particles.

After laborious methodological studies, the problems of selection of appropriate loading
substances, uniform distribution of the lithium-isotope nuclei over the thickness of the
emulsion layer, and accurate determination of the lithium-atom concentration in the emulsion
were solved. Experimental difficulties related to the identification of tritons, deuterons, and
protons of similar energies were overcome. The points at which α particles and H_1^3 are ejected
could be clearly determined. These are important improvements on the work done by several
foreign researchers [55, 56] and helped research on complicated reactions involving interactions of neutrons with lithium isotopes.

3. Studies of the interaction between $\pi^{\pm}$ mesons with energies of 300 MeV on the one
hand, and hydrogen, deuterium, and the nuclei of the photoemulsion on the other (Figs. 11
and 12). New results concerning the various processes and the cross section of the interaction between $\pi^{\pm}$ mesons and nuclei were obtained. The enrichment of domestic NIKFI-R
emulsions with hydrogen isotopes in the form of stable compounds had to be accomplished
for this work since then it became possible to double the hydrogen-atom concentration and to
introduce large quantities of deuterium without reducing the sensitivity to relativistic particles.
A method was developed for determining the number of hydrogen and deuterium atoms, with
the isotope exchange being taken into account. Difficulties resulting from the isotope exchange
during the actual experiment were overcome. Thus, a series of intersecting, important ex-

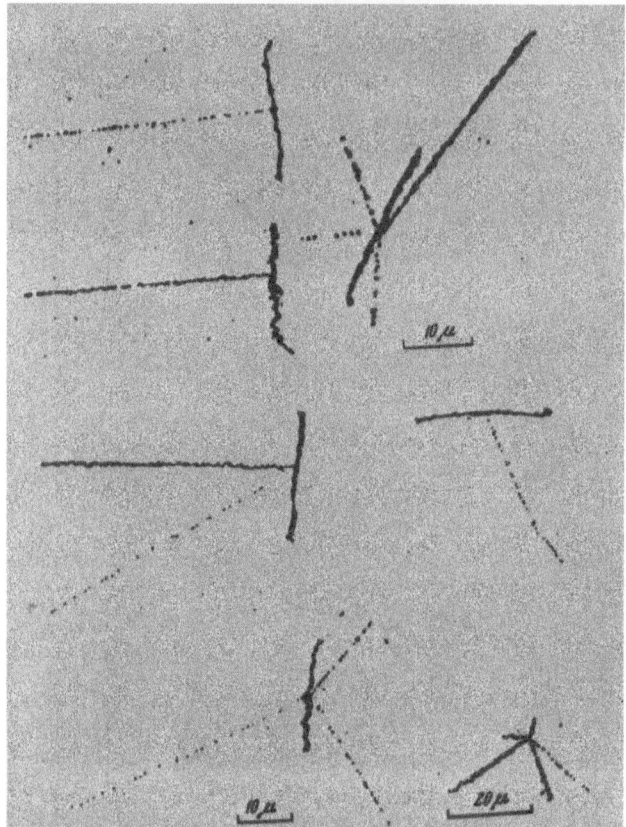

Fig. 10. Photomicrographs of the fission of uranium
nuclei by neutrons having energies of 150 and 380 MeV;
emission of fast charged particles.

periments could be performed for which the previously mentioned research work formed a
methodological basis [3, 4, 23, 25, 47, 50, 50a]. One must also mention a series of experiments
in which both the total and differential cross section of the 300-MeV π^- meson scattering on
hydrogen and deuterium were determined [41, 50]. The work on the gelatin dilution of NIKFI-R
emulsions relates to this research. The application of this method in physics experiments is
promising insofar as the concentration of light nuclei in diluted emulsions is increased, the
path length of low-energy particles is raised, and consequently, the accuracy of the track measurements is improved.

    4. The possibility of catalytic uranium–nucleus fission by slow μ mesons was examined
for the first time (Figs. 13 and 14) [7, 9]. The work was reported at the International Conference on High-Energy Physics in Kiev, 1959 [47]. Electrons emitted from uranium nuclei
during the fission induced by neutrons of various energies (Figs. 15 and 16) [44] were observed
for the first time; the goal was to gather new information on the fission mechanism in nuclei.
The results of a series of studies and the methodological developments of the author formed
the technical basis for this series of important and interesting experiments.

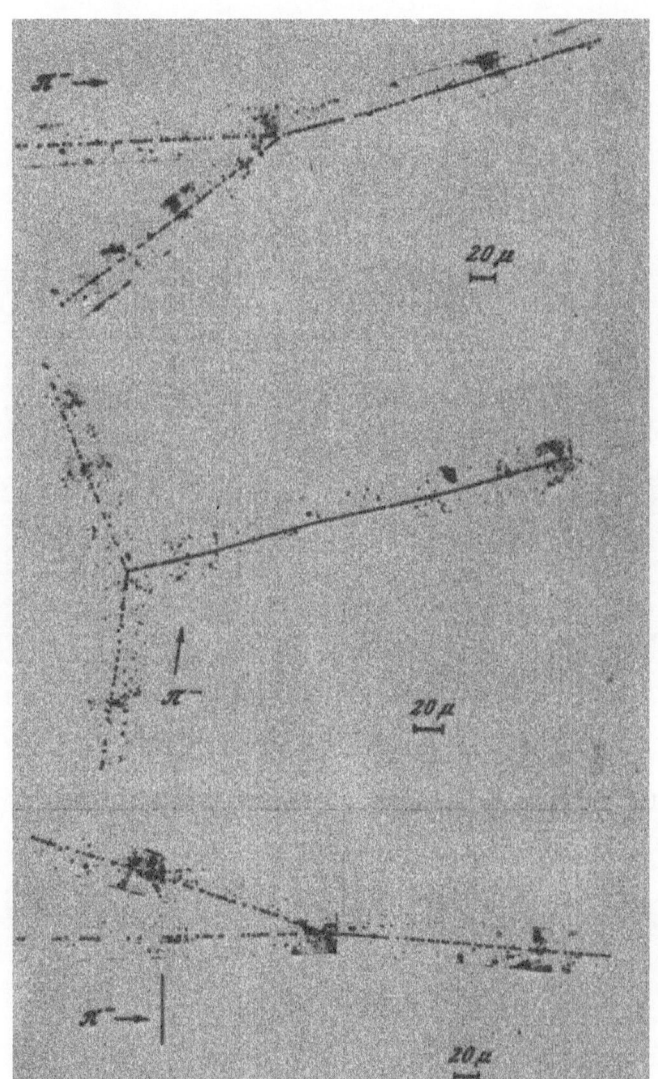

Fig. 11. Photomicrographs of the scattering of π^- mesons
on hydrogen.

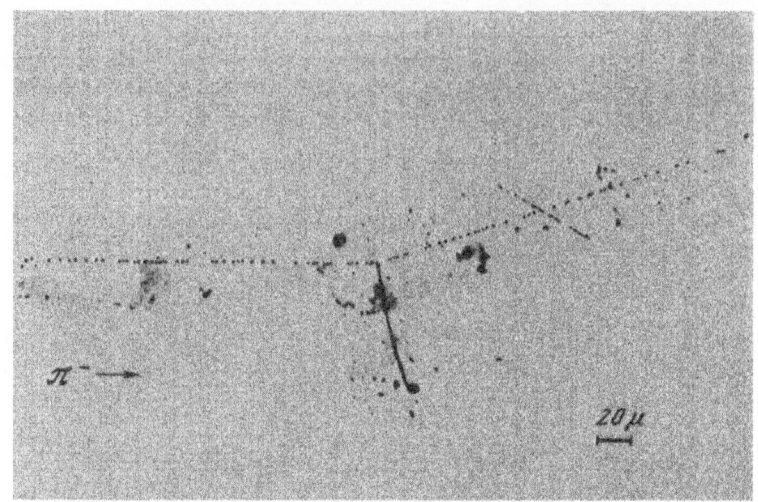

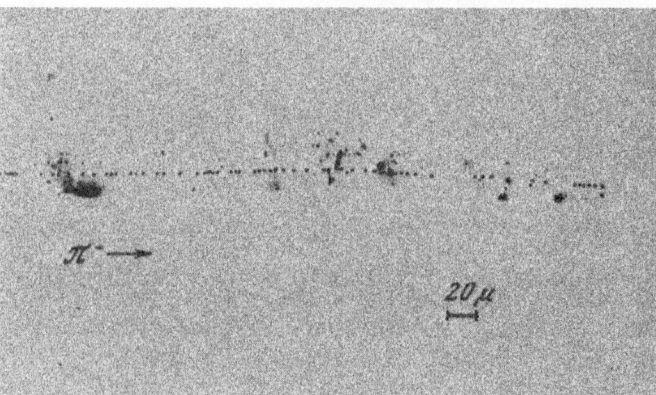

Fig. 12. Photomicrographs of the scattering of π^- mesons on deuterons.

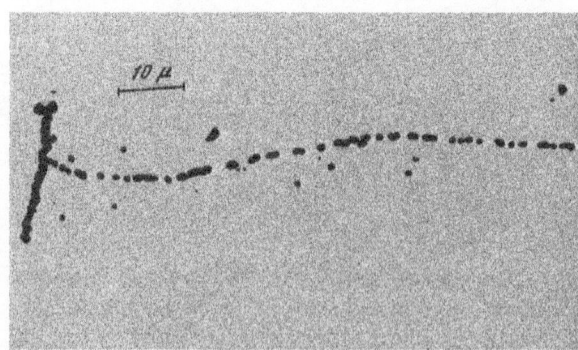

Fig. 13. Photomicrograph of the uranium fission induced by slow mesons (paraaminophenol development without a hot stage).

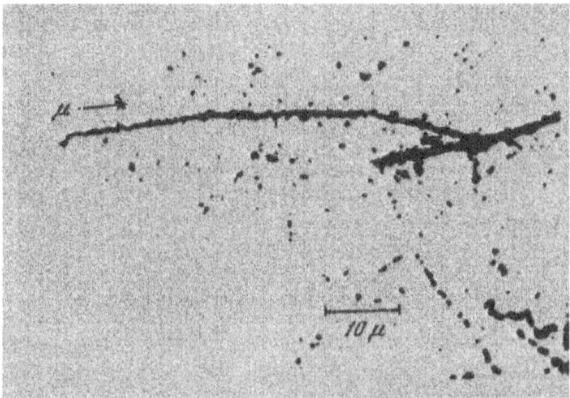

Fig. 14. Photomicrograph of the uranium fission
induced by slow mesons (paraaminophenol development with a temperature cycle).

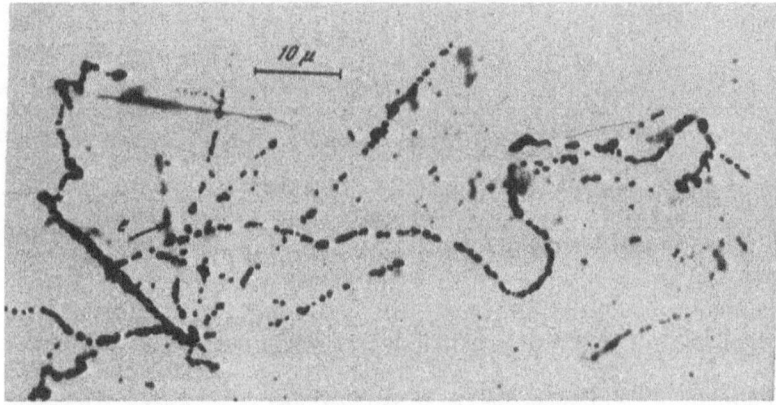

Fig. 15. Photomicrograph of the fission of uranium nuclei by thermal
neutrons; emission of electrons with energies of 150 keV.

This research became possible since relativistic electrons ejected from fragments
in the fission process could be recorded in the presence of a strong background resulting from
radioactivity and the γ rays generated by the particle flux. It was necessary to overcome the
hardening and desensitizing effect which uranyl salts exert upon emulsions sensitive to particles with minimum ionization (NIKFI-R emulsions). After laborious work, it could be shown
for the first time that it is possible to observe the tracks of relativistic electrons at grain
densities which were increased 2-2.5 times in photographic layers with high uranium concentrations. This had been considered impossible before, due to the desensitizing effect of
the uranium. Effects which were observed for the first time could be reliably evaluated since
it had become possible to obtain a clear development of relativistic particle tracks over the
entire depth of the emulsion layer; at the same time the background was poorly developed and
the uranium distribution in 200- to 300-μ-thick emulsions was uniform.

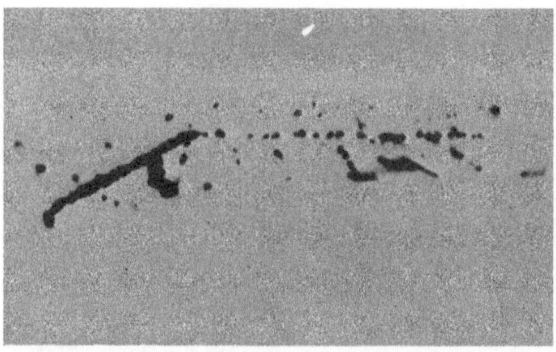

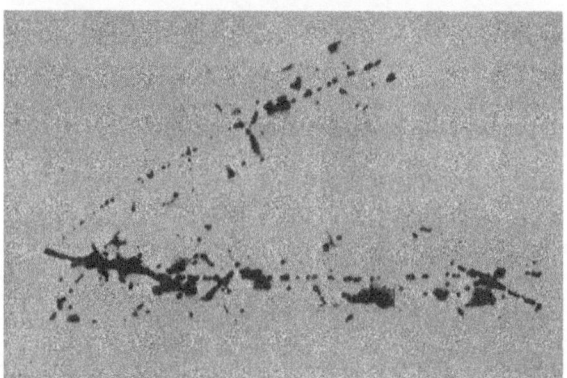

Fig. 16. Photomicrographs of the uranium fission
induced by neutrons with energies of 14 MeV; emission of fast electrons.

References

1. G. A. Belovitskii, T. A. Romanova, L. V. Sukhov, and I. M. Frank, Zh. Eksp. Teor. Fiz.,
    28(6):729 (1955).
2. G. A. Belovitskii, T. A. Romanova, L. V. Sukhov, and I. M. Frank, Zh. Eksp. Teor. Fiz.,
    29(5):537 (1956).
3. L. S. Dul'kova, T. A. Romanova, I. B. Sokolova, and K. D. Tolstov, All-Union Conference
    on High-Energy Particle Physics, Moscow (1956).
4. L. S. Dul'kova, T. A. Romanova, I. B. Sokolova, K. D. Tolstov, et al., Dokl. Akad. Nauk
    SSSR, Vol. 107, No. 1 (1956).
5. T. A. Romanova, Pribory i Tekhn. Eksp., No. 2, p. 21 (1958).
6. T. A. Romanova, Pribory i Tekhn. Eksp., No. 4, p. 93 (1958).
7. G. E. Belovitskii, T. A. Romanova, K. T. Kashchukeev, et al., OIYaI, Preprint No. 388
    (1959).
8. T. A. Romanova and L. D. Chikil'dina, Third International Conference on Nuclear Par-
    ticle-Track Photography, Moscow (1960).
9. G. E. Belovitskii, K. T. Kashchukeev, A. Mikhul, M. G. Petrashku, and T. A. Romanova,
    Zh. Eksp. Teor. Fiz., 38(2):404 (1960).
10. N. M. Kulikova, L. É. Lazareva, N. V. Nikitina, and T. A. Romanova, Third International
    Conference on Nuclear Particle-Track Photography, Moscow (1960).

11.   T. A. Romanova and L. D. Chikil'dina, FIAN Scientific Report A-209 (1961).
12.   K. M. Mikhailina, N. A. Nomofilov, T. A. Romanova, V. A. Sveridov, and K. D. Tolstov, in: Neutron Physics, p. 249 (1961).
13.   T. A. Romanova and L. D. Chikil'dina, Pribory i Tekhn. Eksp., No. 5, p. 88 (1962).
14.   T. A. Romanova, Reports of the Conference on the Thick-Layer Photoemulsion Technique, Vol. 1, pp. 39-42, OIYaI (1947).
15.   G. A. Belovitskii, T. A. Romanova, L. V. Sukhov, and I. M. Frank, FIAN Scientific Report (April 1950).
16.   T. A. Romanova and G. E. Belovitskii, FIAN Scientific Report A-127 (1951).
17.   T. A. Romanova, G. E. Belovitskii, D. V. Karumyan, A. D. Kuznetsov, and K. D. Tolstov, FIAN Scientific Report (1950).
18.   G. E. Belovitskii, T. A. Romanova, and L. V. Sukhov, FIAN Scientific Report A-172 (1950).
19.   G. E. Belovitskii, T. A. Romanova, L. V. Sukhov, and I. M. Frank, FIAN Scientific Report A-42.
20.   T. A. Romanova, V. P. Perelygin, F. A. Tikhomirov, and K. D. Tolstov, FIAN Scientific Report (1953).
21.   G. E. Belovitskii, K. M. Mikhailina, V. P. Perelygin, and T. A. Romanova, FIAN Scientific Report (1953).
22.   T. A. Romanova, V. P. Perelygin, and F. A. Tikhomirov, Scientific Report (1953).
23.   G. E. Belovitskii, T. A. Romanova, L. S. Dul'kova, I. B. Sokolova, and L. V. Sukhov, FIAN Scientific Report A-681 (1953).
24.   K. M. Mikhailina, A. A. Nomofilov, T. A. Romanova, V. A. Sveridov, and K. D. Tolstov, FIAN Scientific Report A-507 (1955).
25.   L. S. Dul'kova, T. A. Romanova, and I. B. Sokolova, Scientific Report (1955).
26.   N. A. Perfilov, FIAN Report (1948).
27.   N. A. Perfilov et al., Trudy Radievogo Inst., 7(2):3 (1956).
28.   N. A. Perfilov et al., Trudy Radievogo Inst., 7(2):2 (1956).
29.   O. V. Lozhkin and V. P. Shamov, Zh. Eksp. Teor. Fiz., 28:739 (1955).
30.   K. S. Bogomolov, Zh. Prikl. Khim., 22(8):831 (1949).
31.   M. Nikolae, Nuclear Particle-Track Photography (1962), p. 182.
32.   M. I. Trukhin, OIYaI r.r. 1749 (1964).
33.   A. L. Kartuzhanskii, L. V. Shur, et al., Zh. Prikl. Nauch. Fotog. Kinemat., Vol. 8, No. 4.
34.   M. I. Adamovich, G. B. Kuz'michev, V. G. Larionova, and S. P. Kharlamov, Dokl. Akad. Nauk SSSR, 20:715 (1955).
35.   G. P. Miklukhin, Isotopes and Organic Chemistry (1961), p. 238.
36.   G. P. Miklukhin, Isotopes and Organic Chemistry (1961), p. 240.
37.   I. P. Gragorev, M. F. Rekasheva, et al., Zh. Org. Khim., 31(4):1113 (1961).
38.   A. Mikhul and M. G. Petrashku, G-2233, OIYaI, Dubna (1958).
39.   L. S. Dul'kova, I. B. Sokolova, and M. G. Shafranova, Dokl. Akad. Nauk SSSR, Vol. 111 (1956).
40.   V. P. Perelygin and K. D. Tolstov, Atomnaya Energiya, Vol. 9, No. 6 (1960).
41.   L. S. Dul'kova, I. B. Sokolova, and M. G. Shafranova, Zh. Eksp. Teor. Fiz., Vol. 35, No. 2(8) (1958).
42.   G. E. Belovitskii, Zh. Eksp. Teor. Fiz., 38(1):658 (1961).
43.   G. E. Belovitskii, Zh. Eksp. Teor. Fiz., 38(2):404 (1960).
44.   G. E. Belovitskii, Yadern. Fiz., Vol. 3, No. 3 (1966).
45.   B. I. Bannik, N. M. Kulikova, N. V. Nikitina, L. E. Lazareva, and V. A. Yakovlev, Zh. Eksp. Teor. Fiz., Vol. 33, No. 1(7) (1957).
46.   A. I. Baz', N. M. Kulikova, L. E. Lazareva, N. V. Nikitina, et al., Reports of the Second Geneva Conference 15 (P), 2037.

47. K. D. Tolstov, Report of the 9th International Conference on High-Energy Physics, p. 196 (1959).

48. N. M. Kulikova, N. V. Nikitina, and N. B. Popov, in: Photomeric and Photonuclear Processes, Consultants Bureau, New York (1967), p. 163.

49. L. E. Lazareva and N. V. Nikitina, Supplement No. 1 to Atomnaya Energiya.

50. L. S. Dul'kova, I. B. Sokolova, and M. G. Shafranova, II-189, OIYaI, Dubna (1958).

50a. G. E. Belovitskii, Zh. Eksp. Teor. Fiz., Vol. 35, No. 4(10) (1958).

51. P. Demers, Canad. J. Res., A25(4):223 (1947).

52. E. Piggiotto, Compt. Rend., 228(2):173 (1949).

53. E. Piggiotto, Compt. Rend., 228(3):247 (1949).

54. G. Goldhaber, et al., Nucl. Sci. Abst., 7(23):6559 (1953).

55. Titterton et al., Radiol., 23(272):465 (1950).

56. Darling, Roberts, et al., Phys. Rev., No. 12 (1953).

57. E. C. Dodd and Waller, Phot. Sens. Sympos., pp. 266-271 (1951).

58. H. Faraggi, Ann. Phys., 6:325 (1951).

59. Tsien, Chastel et al., Phys. Rev., 72:13 (1947).

60. Johansson, et al., Nucl. Phys., No. 2, p. 136.

61. Reports of the International Conference on the Physics of Nuclear Fission, Canada, 1956.

NEUTRON-TRANSFER THEORY
FOR INHOMOGENEOUS MEDIA

A. V. Stepanov

§1. Introduction

The transition of neutrons through a homogeneous moderator has been rather thoroughly studied in both theory and experiment. In several cases of practical interest, we have to consider the neutron transfer in media with extended inhomogeneities in space. Problems of this type are encountered when the methods of neutron physics are used in nuclear geophysics or when the neutron diffusion in boiling liquids is examined. The parameters of the medium change irregularly from point to point and this leads to fluctuations of the neutron density. Both the average neutron density and the mean square density are of particular interest in practice. An averaging over the fluctuations in the medium is performed. The experimental results concerning the neutron density are usually compared with the calculated results obtained from a solution of the kinetic equation with averaged parameters. Under certain conditions, the calculated values may differ considerably from the average neutron density and errors in the interpretation of the experimental values are the consequence. A moderator with a periodic structure, for example, a lattice of a heterogeneous reactor, is an important particular case of an inhomogeneous medium. A large number of publications deals with the transition of neutrons through media of this type.* However, an analytic solution to this problem could be given only when very restrictive conditions for the moderator properties were assumed: either one ignores the neutron absorption, or one limits the discussion to a medium with a simple geometry and uses the diffusion approximation.

In the statistical approach suggested in the present article for describing the transition of neutrons in an inhomogeneous medium, a medium with a periodic structure is a special case of the medium with random inhomogeneities. The representatives of the statistical ensemble of inhomogeneous media differ by their displacement in space ("phase shift"). An exact expression for the average neutron density cannot be determined and one must consider the inhomogeneity of the medium as a disturbance and use the well-known principles of perturbation theory. Our approximations differ from the standard approximation, and the statistical approach outlined below can serve as an addition to well-known methods in the theory of heterogeneous reactors.

---

*A bibliography can be found in the monographs [1, 2].

As has been mentioned above, the averaging over the neutron density involves the distribution of the fluctuations in the medium. This means that first of all we must find the neutron density in the medium characterized by a fixed (yet arbitrary) dependence of the macroscopical cross sections $\Sigma(\mathbf{r})$ upon the coordinates $N(\mathbf{r}, \Sigma(\mathbf{r}))$. After that, we must calculate the functional integral

$$\langle N(\mathbf{r}) \rangle = \int N(\mathbf{r}, \Sigma(\mathbf{r})) P(\Sigma) d\Sigma, \tag{1.1}$$

where $P(\Sigma)$ denotes the distribution law of the cross sections Σ. Naturally, these calculations can be completely performed only in a small number of cases, whenever the structure of the medium is relatively simple.

One might think that perturbation theory is a universal method. We might try to use the expression for the neutron density in a homogeneous medium as a zeroth approximation and consider the inhomogeneity as a perturbation. However, the perturbation-theory series diverges at large distances from the neutron sources, as we will show in concrete examples of neutron-diffusion theory. The region of large distances is of greatest interest in practice. It turns out that one can use a modified perturbation theory which is known as perturbation theory for the "mass operator" [3-6] in quantum-field theory and in multiparticle quantum theory. This modification of perturbation theory was also used for describing the transition of electromagnetic radiation through fluctuating media [7-9]. In § 2 we derive an equation for the average Green function of the kinetic equation of [10, 11] in a first, nonvanishing approximation of the perturbation theory for the "mass operator." By neglecting in the zeroth approximation the inhomogeneity of the medium, we include in adequate form only the average values of the cross sections Σ_0, which are assumed to be constant. The real distribution law $P(\Sigma)$ is replaced by the approximation

$$P_0(\Sigma) = \delta(\Sigma - \Sigma_0). \tag{1.2}$$

In the first nonvanishing approximation of perturbation theory, we include the second moments of the distribution law $P(\Sigma)$: we calculate the correlation functions $\langle \Sigma_s(\mathbf{r}) \Sigma_s(\mathbf{r}') \rangle$, $\langle \Sigma_a(\mathbf{r}) \Sigma_a(\mathbf{r}') \rangle$, etc. Here Σ_s and Σ_a denote the macroscopic scattering and absorption cross sections. In the case of an ensemble of homogeneous diffusion media (with cross sections changing from one place of the medium to another), we can use an expression similar to Eq. (1.2):

$$P(\Sigma) = \frac{1}{2} [\delta(\Sigma - \Sigma_0 + \Delta) + \delta(\Sigma - \Sigma_0 - \Delta)], \tag{1.3}$$

where

$$\Delta^2 = \langle \Sigma^2 \rangle - \langle \Sigma \rangle^2. \tag{1.4}$$

When the ensuing terms of the perturbation-theory series are taken into account, one must calculate higher moments of $P(\Sigma)$. In the event that the fluctuations in the medium obey a normal (Gaussian) distribution law, the Feynman graph method [3-7] is useful for developing the perturbation-theory series (see also § 2). The correction functions for various models of diffusion media are considered in § 3 [12]. An approximated average Green's function which describes the diffusion of thermal neutrons in an inhomogeneous infinite medium is derived in § 4. In the particular case of a plane lattice, we show that the constants for the attenuation of the neutron flux agree with the results obtained by another method [13].* In § 5, the relaxation length of the average neutron flux is derived; this average flux satisfies the single-velocity stationary kinetic equation involving isotropic scattering [14].

---

*In the solution of concrete problems, it is suitable to use the neutron flux instead of the neutron density.

Nonstationary diffusion of thermal neutrons is discussed in § 6. The energy dependence of the flux of diffusing neutrons is taken into account in §7.

A first approximation to the perturbation theory for the "mass operator" was sufficient in the case of small-scale fluctuations. The characteristic size l of an inhomogeneity must be small relative to L_0, the length of the neutron diffusion path in a homogeneous medium. Estimates of the subsequent terms of the series of perturbation theory are calculated in § 8, and an expression for the average neutron flux in a medium with large-scale fluctuations ($l \gg L_0$) and with a Gaussian distribution law is derived.

In Appendix A we derive an equation in functional derivatives which is satisfied by the average neutron flux in any type of medium with fluctuating parameters. An approximate equation for the mean square of the neutron flux is given in Appendix B. The energy distribution of moderated neutrons is considered in Appendix C; the distribution function is averaged over the fluctuations of the absorption cross section.

§ 2. Derivation of General Formulas

1. The homogeneous kinetic equation which describes the neutron transfer in a medium without sources can be written in symbolic form as

$$\hat{A}\psi(x) = 0, \tag{2.1}$$

where $\hat{A}$ denotes a linear operator acting upon x; x is a set of variables upon which the neutron-distribution function ψ depends. The particular form of the operator $\hat{A}$ is determined by the conditions of the particular problem under consideration. For example, in the case of stationary diffusion of thermal neutrons in a homogeneous medium, we have

$$\hat{A} = D\nabla^2 - \frac{1}{T}, \tag{2.2}$$

where D denotes the diffusion coefficient and T the lifetime of the thermal neutrons until their absorption. For example, when the scattering properties of the medium change irregularly from point to point and in the course of time, then the operator $\hat{A}$ includes components which have an irregular dependence on the spatial coordinates r and the time t. We separate the fluctuating part from the operator $\hat{A}$:

$$\hat{A}(x) = \hat{B}(x) - \hat{\mu}(x), \tag{2.3}$$

where $\hat{B}(x)$ denotes the regular part of the operator $\hat{A}(x)$:

$$\hat{B}(x) = \langle \hat{A}(x) \rangle, \tag{2.4}$$

$$\langle \hat{\mu}(x) \rangle = 0. \tag{2.5}$$

The angle brackets denote the calculation of the average over an ensemble of inhomogeneous media (mathematical expectation). In the case of a reactor, this reduces to an averaging over the volume of an elementary cell.*

We denote the Green's function of Eq. (2.1) by $G(x|y)$. We have the well-known relation

$$\hat{A}(x)\, G(x|y) = -\delta(x - y), \tag{2.6}$$

where $\delta(x - y)$ denotes the product of the δ functions with the various variables as arguments. With Eq. (2.3), we can rewrite Eq. (2.6) in integral form:

---

*The averaging over the neutron flux must be performed at a fixed distance between the source and the detector.

$$G\left(x\,|\,y\right) = G_0\left(x\,|\,y\right) - \int dx' G_0\left(x\,|\,x'\right)\hat{\mu}\left(x'\right)G\left(x'\,|\,y\right). \tag{2.7}$$

The operators included in $\hat{\mu}$ operate, as usual, upon the expressions to the right of $\hat{\mu}$. The function $G_0(x|y)$, which satisfies the equation

$$\hat{B}\left(x\right)G_0\left(x\,|\,y\right) = -\delta\left(x - y\right), \tag{2.8}$$

is the Green's function for the kinetic equation

$$\hat{B}\left(x\right)\psi_0\left(x\right) = 0, \tag{2.9}$$

which describes the neutron transfer in a medium with regular nonfluctuating properties. We search for a solution to Eq. (2.7) and use the following successive approximation method:

$$G\left(x\,|\,y\right) = G_0\left(x\,|\,y\right) - \int dx' G_0\left(x\,|\,x'\right)\hat{\mu}\left(x'\right)G_0\left(x'\,|\,y\right) +$$
$$+ \iint dx'\,dx'' G_0\left(x\,|\,x'\right)\hat{\mu}\left(x'\right)G_0\left(x'\,|\,x''\right)\hat{\mu}\left(x''\right)G_0\left(x''\,|\,y\right) - \dots \tag{2.10}$$

By averaging both sides of Eq. (2.10) over the distribution law $\hat{\mu}$, we obtain the following representation of the function $\langle G(x\,|\,y)\rangle$ as a perturbation-theory series:

$$\langle G\left(x\,|\,y\right)\rangle = G_0\left(x\,|\,y\right) + \iint dx'\,dx'' G_0\left(x\,|\,x'\right)\langle\hat{\mu}\left(x'\right)G_0\left(x'\,|\,x''\right)\hat{\mu}\left(x''\right)\rangle G_0\left(x''\,|\,y\right) -$$
$$- \iiint dx'\,dx''\,dx''' G_0\left(x\,|\,x'\right)\langle\hat{\mu}\left(x'\right)G_0\left(x'\,|\,x''\right)\hat{\mu}\left(x''\right)G_0\left(x''\,|\,x'''\right)\hat{\mu}\left(x'''\right)\rangle G_0\left(x'''\,|\,y\right) +$$
$$+ \int\dots\int dx'\dots dx^{\text{IV}} G_0^{-}\left(x\,|\,x'\right)\langle\hat{\mu}\left(x'\right)G_0\left(x'\,|\,x''\right)\hat{\mu}\left(x''\right)G_0\left(x''\,|\,x'''\right)\hat{\mu}\left(x'''\right)\times$$
$$\times G_0\left(x'''\,|\,x^{\text{IV}}\right)\hat{\mu}\left(x^{\text{IV}}\right)\rangle G_0\left(x^{\text{IV}}\,|\,y\right) - \dots \tag{2.11}$$

However, an analysis of real problems reveals that this expansion cannot be used for describing the asymptotic behavior of the neutron density at large distances from the source and in several other cases of practical importance. As a matter of fact, far from a stationary plane source of monoenergetic neutrons, the neutron density decreases exponentially like $e^{-\varkappa\,|\,z-z_0|}$, where z_0 denotes the plane on which the source is situated. In a homogeneous medium, the neutron density decreases like $e^{-\varkappa_0\,|\,z-z_0|}$. It is easy to verify that the correction to $G_0(z|z_0)$ defined by Eq. (2.11) is proportional to powers of $|z-z_0|$, i.e., for large $|z-z_0|$, the series of Eq. (2.11) converges poorly, or not at all. Thus, we have to look for another method. We average the equation for the function $G(x|y)$ defined by Eq. (2.7). We obtain

$$\langle G\left(x\,|\,y\right)\rangle = G_0\left(x\,|\,y\right) - \int dx' G_0\left(x\,|\,x'\right)\langle\hat{\mu}\left(x'\right)G\left(x'\,|\,y\right)\rangle. \tag{2.12}$$

By introducing a new, unknown operator $M(x'|x'')$ we can remove the correlation function $\langle\hat{\mu}(x')G(x'\,|\,y)\rangle$ and obtain an analytical equation for the function $\langle G(x\,|\,y)\rangle$:

$$\langle G\left(x\,|\,y\right)\rangle = G_0\left(x\,|\,y\right) + \int dx' \int dx'' G_0\left(x\,|\,x'\right)M\left(x'\,|\,x''\right)\langle G\left(x''\,|\,y\right)\rangle. \tag{2.13}$$

When we use Eq. (2.8), the last equation can be rewritten in the form (see Appendix A):

$$\hat{B}(x)\langle G\left(x\,|\,y\right)\rangle = -\delta\left(x-y\right) - \int dx' M\left(x\,|\,x'\right)\langle G\left(x'\,|\,y\right)\rangle. \tag{2.14}$$

The operator $M(x|x')$ must be expressed by the functions $G_0(x|x')$ and $\langle G(x|x')\rangle$. To do this, we multiply both sides of Eq. (2.13) from the left by $G_0^{-1}(s|x)$ and from the right by $\langle G(y|v)\rangle^{-1}$ and integrate over x and y:

$$\int dx \int dy\, G_0^{-1}(s\,|\,x)\,\langle G(x\,|\,y)\,\langle G(y\,|\,v)\rangle^{-1} = \iint dx\,dy\, G_0^{-1}(s|x)\,G_0(x|y)\,\langle G(y|v)\rangle^{-1} +$$

$$+ \int dx \int dy \int dx' \int dx''\, G_0^{-1}(s\,|\,x)\,G_0(x\,|\,x')\,M(x''\,|\,x')\,\langle G(x''\,|\,y)\rangle\,\langle G(y\,|\,v)\rangle^{-1}. \tag{2.15}$$

With the condition

$$\int dx'\, G(x\,|\,x')\,G^{-1}(x'\,|\,y) = \delta(x-y), \tag{2.16}$$

we obtain from Eq. (2.15) the following relation for the operator $M(x|x')$:*

$$M(x\,|\,x') = G_0^{-1}(x\,|\,x') - \langle G(x\,|\,x')\rangle^{-1}. \tag{2.17}$$

We convert the expansion of the function $\langle G(x\,|\,x')\rangle$ into the perturbation theory series of Eq. (2.11) and set

$$\langle G(x\,|\,y)\rangle^{-1} = G_0^{-1}(x\,|\,y) - \langle \hat{\mu}(x)\,G_0(x\,|\,y)\,\hat{\mu}(y)\rangle + \int dx'\,\langle \hat{\mu}(x)\,G_0(x\,|\,x')\,\hat{\mu}(x')\,G_0(x'\,|\,y)\,\hat{\mu}(y)\rangle -$$

$$- \iint dx'\,dx''\,\langle \hat{\mu}(x)\,G_0(x\,|\,x')\,\hat{\mu}(x')\,G_0(x'\,|\,x'')\,\hat{\mu}(x'')\,G_0(x''\,|\,y)\,\hat{\mu}(y)\rangle +$$

$$+ \int dx'\,\langle \hat{\mu}(x)\,G_0(x\,|\,x')\,\hat{\mu}(x')\rangle \int dx''\,G_0(x'\,|\,x'')\,\langle \hat{\mu}(x'')\,G_0(x''\,|\,y)\,\hat{\mu}(y)\rangle - \dots \tag{2.18}$$

and accordingly, we obtain from Eq. (2.17)

$$M(x\,|\,y) = \langle \hat{\mu}(x)\,G_0(x\,|\,y)\,\hat{\mu}(y)\rangle - \int dx'\,\langle \hat{\mu}(x)\,G_0(x\,|\,x')\,\hat{\mu}(x')\,G_0(x'|y)\,\hat{\mu}(y)\rangle +$$

$$+ \iint dx'\,dx''\,\langle \hat{\mu}(x)\,G_0(x\,|\,x')\,\hat{\mu}(x')\,G_0(x'\,|\,x'')\,\hat{\mu}(x'')\,G_0(x''\,|\,y)\,\hat{\mu}(y)\rangle -$$

$$- \iint dx'\,dx''\,\langle \hat{\mu}(x)\,G_0(x\,|\,x')\,\hat{\mu}(x')\rangle\,G_0(x'\,|\,x'')\,\langle \hat{\mu}(x'')\,G_0(x''\,|\,y)\,\hat{\mu}(y)\rangle + \dots \tag{2.19}$$

Obviously, Eq. (2.19) is an expansion of the operator $M(x|y)$ in moments of the function $\hat{\mu}(x)$.[†] The general expression for the k-th term of the series of Eq. (2.19) can be found in the work of Finkel'berg [8].

In what follows we will restrict ourselves to the first term of the expansion in Eq. (2.19):

$$M_1(x\,|\,y) = \langle \hat{\mu}(x)\,G_0(x\,|\,y)\,\hat{\mu}(y)\rangle. \tag{2.20}$$

The applicability range of this approximation will be outlined below.

By inserting $M_1(x|y)$ in Eq. (2.14) we obtain the following equations for the average Green's function:

$$\hat{B}(x)\,\langle G(x\,|\,y)\rangle = -\,\delta(x-y) - \int dx'\,\langle \hat{\mu}(x)\,G_0(x\,|\,x')\,\hat{\mu}(x')\rangle\,\langle G(x\,|\,y)\rangle. \tag{2.21}$$

This equation will be the subject of the following discussion. The corresponding homogeneous equation

$$\hat{B}(x)\,\langle \psi(x)\rangle = -\int dx'\,\langle \hat{\mu}(x)\,G_0(x\,|\,x')\,\hat{\mu}(x')\rangle\,\langle \psi(x')\rangle \tag{2.22}$$

helps to formulate the eigenvalue problem.

---

*The operator $M(x|x')$ is a nonlocal operator in the general case.

† The expansion in Eq. (2.19) is the analog of the perturbation-theory series for the "mass operator" in quantum-field theory [3, 4] and statistical physics [5, 6].

In the following section we consider a special form of a perturbation of $\hat{\mu}(x)$ (a random function of Gaussian form), which makes it possible to employ a graph technique for solving the equation for the average Green's function.

2. We assume that the random function $\hat{\mu}(x)$ has a Gaussian distribution. This assumption means that all correlation functions $\langle \hat{\mu}(x_1)\hat{\mu}(x_2)\ldots\hat{\mu}(x_s)\rangle$ comprising an odd number of $\hat{\mu}$ vanish (and not only $\langle\hat{\mu}\rangle$), and that all correlation functions with an even number of $\hat{\mu}$ are expressed by paired correlation functions $\langle\hat{\mu}(x_1)\hat{\mu}(x_2)\rangle$. This latter point makes it possible to use efficiently a graphical representation of the perturbation-theory series.

Let us return to Eq. (2.10). We represent the functions $G_0(x|y)$ by a straight horizontal line joining points x and y, and the functions $G(x|y)$ and $\langle G(x|y)\rangle$, respectively, by a double horizontal line and a thick horizontal line, drawn between points x and y (Fig. 1).

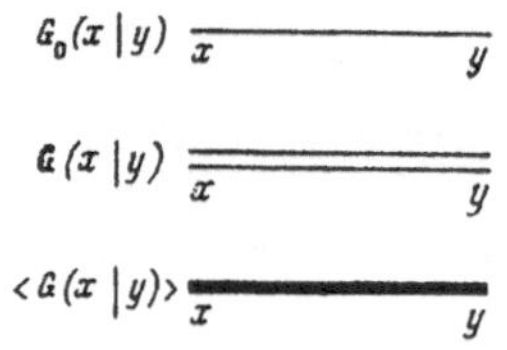

$$
\begin{aligned}
&G_0(x\,|\,y) \quad \underline{\qquad\qquad} \\
&\quad x \qquad\qquad y \\
&G(x\,|\,y) \quad =\!\!=\!\!=\!\!=\!\!= \\
&\quad x \qquad\qquad y \\
&\langle G(x\,|\,y)\rangle \; \blacksquare\!\!\blacksquare\!\!\blacksquare \\
&\; x \qquad\quad y
\end{aligned}
$$

Fig. 1

We denote the operator $\hat{\mu}(x)$ by a dashed vertical line running upward from point x. The integration is performed over the variables of each intersection point from which a dashed line starts. Obviously, a graphical representation of Eq. (2.10) has the form of Fig. 2.

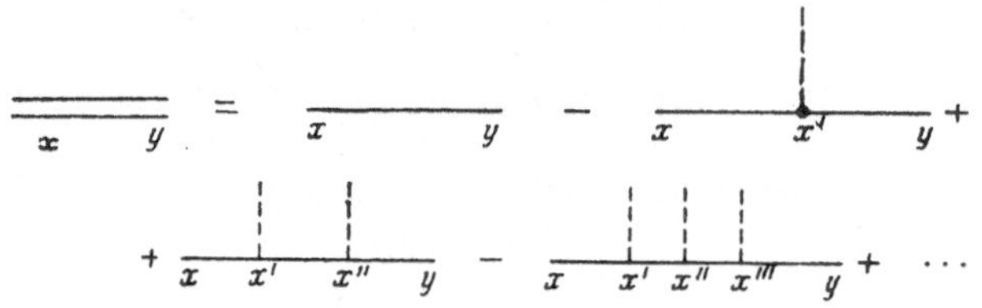

$$ \tag{2.23} $$

Fig. 2

With the above assumptions regarding the statistical properties of $\hat{\mu}$, we average Eq. (2.23) (represented by Fig. 2). The contribution of the sections containing an odd number of dashed lines vanishes, and the averaging over the individual components is reduced to a pairwise connection of dashed lines. We obtain the graphical representation of $\langle G(x|y)\rangle$ shown in Fig. 3 [see Eq. (2.11)]:

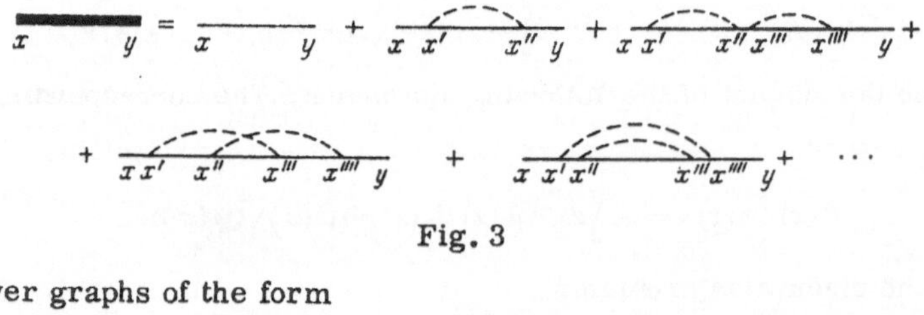

$$ \tag{2.24} $$

Fig. 3

By summation over graphs of the form

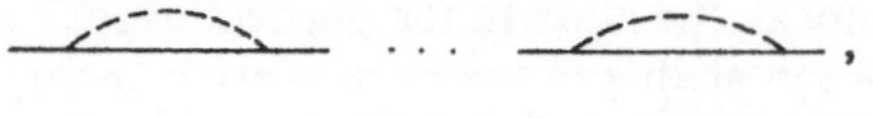

$$ \tag{2.25} $$

Fig. 4

we obtain the following equation for the function $\langle G(x'|y)\rangle$:

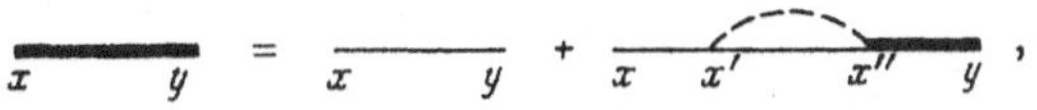

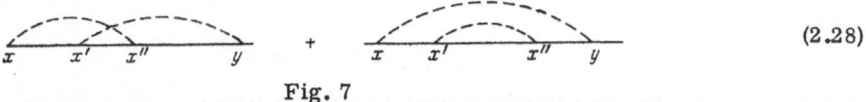

$$\tag{2.26}$$

Fig. 5

which is equivalent to Eq. (2.21) above.

The graphical representation of the operator $M_1(x|x')$ has the form

$$\tag{2.27}$$

Fig. 6

and the principal terms among the omitted terms of the perturbation-theory series for the operator M are given by

$$\tag{2.28}$$

Fig. 7

3. Thus, in the M_1 approximation [Eqs. (2.21) and (2.26)] the information on the inhomogeneity of the medium (information which is required for determining the average Green's function $\langle G(x|y)\rangle$) appears in a correlator pair $\langle \hat{\mu}(x)G_0(x|x')\hat{\mu}(x')\rangle$. In what follows in this article, we will consider diffusion media whose parameters have only spatial fluctuations; the perturbation $\hat{\mu}(x) \equiv \hat{\mu}(\mathbf{r}, \xi)$ will be written in the form

$$\hat{\mu}(\mathbf{r}, \xi) = \sum_{i=1}^{N} \mu_i(\mathbf{r})\hat{\alpha}_i(\mathbf{r}, \xi), \tag{2.29}$$

where $\mu_i(\mathbf{r})$ denotes the random amplitudes ($\langle \mu_i \rangle = 0$) and $\hat{\alpha}_i(\mathbf{r}, \xi)$ operators acting upon the variables $x = \{\mathbf{r}, \xi\}$. By substituting Eq.(2.29) into the expression for the mass operator M_1 of Eq. (2.20) we find

$$M_1(\mathbf{r}, \xi|\mathbf{r}', \xi') = \sum_{i=1}^{N} \sum_{j=1}^{N} K_{ij}(\mathbf{r}|\mathbf{r}')\hat{\alpha}_i(\mathbf{r}, \xi)G_0(\mathbf{r}, \xi|\mathbf{r}', \xi')\hat{\alpha}_j(\mathbf{r}', \xi'). \tag{2.30}$$

We used the notation

$$K_{ij}(\mathbf{r}|\mathbf{r}') \equiv \langle \mu_i(\mathbf{r})\mu_j(\mathbf{r}')\rangle. \tag{2.31}$$

In cases of practical interest, the fluctuating media are homogeneous in space (i.e., homogeneous on the average):

$$K_{ij}(\mathbf{r}|\mathbf{r}') = K_{ij}(\mathbf{r} - \mathbf{r}'). \tag{2.32}$$

When the arguments are identical, we obtain

$$K_{ij}(\mathbf{r}|\mathbf{r}') \equiv K_{ij}(0) = \langle \mu_i(\mathbf{r})\mu_j(\mathbf{r})\rangle = \varepsilon_{ij}. \tag{2.33}$$

For the following discussion we will need the Fourier representation of the correlation function $K_{ij}(\mathbf{r} - \mathbf{r}')$:

$$k_{ij}(\mathbf{p}) = \int d\mathbf{r}\, e^{i\mathbf{p}\mathbf{r}}\, K_{ij}(\mathbf{r}). \tag{2.34}$$

Accordingly, we find

$$K_{ij}(\mathbf{r}) = \frac{1}{(2\pi)^3} \int d\mathbf{p} e^{-i\mathbf{p}\mathbf{r}} k_{ij}(\mathbf{p}). \tag{2.35}$$

It follows from Eqs. (2.33) and (2.35) that

$$\varepsilon_{ij} = \frac{1}{(2\pi)^3} \int d\mathbf{p} k_{ij}(\mathbf{p}). \tag{2.36}$$

For solving Eq. (2.21) we must approximate the correlation function K_{ij} by rather simple analytical expressions. The following paragraph analyzes functions K_{ij} for various models of an inhomogeneous medium.

§3. Correlation Functions $K_{ij}(\mathbf{r})$

1. When the diffusion medium has a layer structure (the planes of the layers being perpendicular to the z axis) and the scattering properties of the layers as well as the layer thicknesses vary irregularly from layer to layer, the following special correlation function is conveniently used:*

$$K_{ij}(|z-z'|) = \varepsilon_{ij} e^{-q_{ij}|z-z'|}, \tag{3.1}$$

where q_{ij}^{-1} denotes the correlation length of the fluctuations in the medium. The correlation length indicates the order of magnitude of the spatial dimensions of an inhomogeneity in a diffusion medium.

The Fourier transform of Eq. (3.1) has the form

$$k_{ij}(\mathbf{p}) = (2\pi)^2 \varepsilon_{ij} \delta(p_x) \delta(p_y) \frac{2q_{ij}}{p_z^2 + q_{ij}^2}. \tag{3.2}$$

When the correlation of the fluctuations decreases along the x axis, as well as along the z axis (the correlation does not decrease along the y axis), we can write

$$K_{ij}(\mathbf{r}) = \varepsilon_{ij} e^{-q'_{ij}|x|} e^{-q''_{ij}|z|} \tag{3.1a}$$

and

$$k_{ij}(\mathbf{p}) = 2\pi \varepsilon_{ij} \delta(p_y) \frac{2q'_{ij}}{p_x^2 + (q'_{ij})^2} \frac{2q''_{ij}}{p_z^2 + (q''_{ij})^2}. \tag{3.2a}$$

Finally, when the correlation decreases along all three axes, we have

$$K_{ij}(\mathbf{r}) = \varepsilon_{ij} e^{-q'_{ij}|x|} e^{-q''_{ij}|v|} e^{-q'''_{ij}|z|} \tag{3.1b}$$

and

$$k_{ij}(\mathbf{p}) = \varepsilon_{ij} \frac{2q'_{ij}}{p_x^2 + (q'_{ij})^2} \frac{2q''_{ij}}{p_y^2 + (q''_{ij})^2} \cdot \frac{2q'''_{ij}}{p_z^2 + (q'''_{ij})^2}. \tag{3.2b}$$

2. If the diffusion medium is not only statistically homogeneous (homogeneous on the average), but also statistically isotropic, we have

---

*Strictly speaking, the exponential function of Eq. (3.1) is a random function of a special type [15].

$$K_{ij}(\mathbf{r} - \mathbf{r}') = K_{ij}(|\mathbf{r} - \mathbf{r}'|) \tag{3.3}$$

and it is convenient to use the following correlation function:

$$K_{ij}(r) = \varepsilon_{ij} e^{-q_{ij}r}. \tag{3.4}$$

Correspondingly, we obtain

$$k_{ij}(\mathbf{p}) \equiv k_{ij}(p) = \frac{8\pi\varepsilon_{ij}q_{ij}}{(q_{ij}^2 + p^2)^2}. \tag{3.5}$$

3. If the diffusion medium contains randomly distributed localized inhomogeneities in the form of admixtures (for the sake of simplicity we assume that the admixtures are of the same type), we obtain

$$\mu_k(\mathbf{r}) = \sum_{\alpha=1}^{N} \Gamma_k(\mathbf{r} - \mathbf{R}_\alpha) - \left\langle \sum_{\alpha=1}^{N} \Gamma_k(\mathbf{r} - \mathbf{R}_\alpha) \right\rangle. \tag{3.6}$$

The summation is extended over all N admixtures. The function $\Gamma_k(\mathbf{r} - \mathbf{R}_\alpha)$ characterizes the effect of an admixture situated at the point $\mathbf{R}_\alpha$ upon the point $\mathbf{r}$. Equation (3.6) is conveniently represented in the form:

$$\mu_k(\mathbf{r}) = \int d\mathbf{r}' \Gamma_k(\mathbf{r} - \mathbf{r}') \left\{ \sum_{\alpha=1}^{N} \delta(\mathbf{r}' - \mathbf{R}_\alpha) - \left\langle \sum_{\alpha=1}^{N} \delta(\mathbf{r}' - \mathbf{R}_\alpha) \right\rangle \right\} =$$

$$= \frac{1}{(2\pi)^3} \int d\mathbf{q} \int d\mathbf{r}' \left\{ \sum_{\alpha=1}^{N} e^{i\mathbf{q}(\mathbf{r}' - \mathbf{R}_\alpha)} - \left\langle \sum_{\alpha=1}^{N} e^{i\mathbf{q}(\mathbf{r}' - \mathbf{R}_\alpha)} \right\rangle \right\} \Gamma_k(\mathbf{r} - \mathbf{r}') =$$

$$= \frac{1}{(2\pi)^3} \int d\mathbf{q} e^{-i\mathbf{q}\mathbf{r}} \gamma_k(\mathbf{q}) [\rho(\mathbf{q}) - \langle \rho(\mathbf{q}) \rangle], \tag{3.7}$$

where

$$\gamma_k(\mathbf{q}) = \int d\mathbf{r} \Gamma_k(\mathbf{r}) e^{i\mathbf{q}\mathbf{r}} \tag{3.8}$$

and

$$\rho(\mathbf{q}) = \sum_{\alpha=1}^{N} e^{i\mathbf{q}\mathbf{R}_\alpha}. \tag{3.9}$$

We assume that the admixture positions are not correlated.* The average value of the random function $f(\mathbf{R}_1,...,\mathbf{R}_N)$ has the form

$$\langle f \rangle = \frac{1}{V^N} \int d\mathbf{R}_1,\ldots, d\mathbf{R}_N f(\mathbf{R}_1, \ldots, \mathbf{R}_N), \tag{3.10}$$

where V denotes the volume of the moderating substance. Accordingly, we obtain

$$\langle \rho(\mathbf{k}) \rangle = \frac{(2\pi)^3 N}{V} \delta(\mathbf{k}), \tag{3.11}$$

$$\langle \rho(\mathbf{k}_1)\rho(\mathbf{k}_2) \rangle = \frac{N}{V}(2\pi)^3\delta(\mathbf{k}_1 + \mathbf{k}_2) + \frac{N(N-1)}{V^2}\iint d\mathbf{R}_1 d\mathbf{R}_2 e^{i\mathbf{k}_1\mathbf{R}_1}e^{i\mathbf{k}_2\mathbf{R}_2} \approx$$

$$\approx \frac{N}{V}(2\pi)^3\delta(\mathbf{k}_1 + \mathbf{k}_2) + \left(\frac{N}{V}\right)^2\iint d\mathbf{R}_1 d\mathbf{R}_2 e^{i\mathbf{k}_1\mathbf{R}_1}e^{i\mathbf{k}_2\mathbf{R}_2} = \frac{N}{V}(2\pi)^3\delta(\mathbf{k}_1 + \mathbf{k}_2) + \left(\frac{N}{V}\right)^2(2\pi)^6\delta(\mathbf{k}_1)\delta(\mathbf{k}_2). \tag{3.12}$$

*This assumption can be made at sufficiently low admixture densities when one can ignore the overlap of two or more admixtures.

With Eqs. (3.7)–(3.12), we obtain

$$K_{ij}(\mathbf{r}) = \frac{N}{V} \frac{1}{(2\pi)^3} \int d\mathbf{q}\, e^{-i\mathbf{q}\mathbf{r}} \gamma_i(\mathbf{q}) \gamma_j^*(\mathbf{q}) \tag{3.13}$$

and correspondingly,

$$k_{ij}(\mathbf{q}) = \frac{N}{V} \gamma_i(\mathbf{q}) \gamma_j^*(\mathbf{q}). \tag{3.14}$$

When each admixture occupies a volume $v_0 = (4\pi/3)R_k^3$ and when

$$\Gamma_k(\mathbf{r}) = \begin{cases} \Gamma_k^0 & r < R_k \\ 0 & r > R_k, \end{cases} \tag{3.15}$$

we obtain

$$\gamma_k(\mathbf{q}) = \int\limits_{v_0} \Gamma_k(\mathbf{r})\, e^{i\mathbf{q}\mathbf{r}}\, d\mathbf{r} = \frac{4\pi\Gamma_k^0}{q^3} [\sin qR_k - qR_k \cos qR_k]. \tag{3.16}$$

The subscript k at $\mathbf{R}_k$ points to the possibility that the "radius of action" of an admixture may differ in neutron scattering and neutron absorption.

4. As has been mentioned in the Introduction, as far as our formalism is concerned, a periodic lattice is a particular form of a diffusion medium with random inhomogeneities. The corresponding correlation functions can be calculated by using the expansion of the periodic functions $\mu_k(\mathbf{r})$ in a Fourier series:

$$\mu_k(\mathbf{r}) = \frac{1}{v_c} \sum_{\mathbf{g}} A_k(\mathbf{g})\, e^{-i\mathbf{g}\mathbf{r}}, \tag{3.17}$$

$$A_k(\mathbf{g}) = \int\limits_{v_c} \mu_k(\mathbf{r})\, e^{i\mathbf{g}\mathbf{r}}\, d\mathbf{r}, \tag{3.18}$$

where v_c denotes the volume of the elementary cell and $\mathbf{g}$ the inverse lattice vector. We recall that in the case of a medium with periodic structure the individual representatives of the statistical ensemble of inhomogeneous media differ by a shift in space; then we obtain with Eqs. (3.17) and (3.18):

$$\langle \mu_k(\mathbf{r}) \rangle = \frac{1}{v_c} \int\limits_{v_c} d\mathbf{r}' \mu_k(\mathbf{r} + \mathbf{r}') = \frac{1}{v_c} \sum_{\mathbf{g}} A_k(\mathbf{g}) e^{-i\mathbf{g}\mathbf{r}} \frac{1}{v_c} \int\limits_{v_c} e^{-i\mathbf{g}\mathbf{r}'}\, d\mathbf{r}' =$$

$$= \frac{1}{v_c} \sum_{\mathbf{g}} A_k(\mathbf{g})\, e^{-i\mathbf{g}\mathbf{r}} \delta_{\mathbf{g},0} = \frac{1}{v_c} A_k(0) = \frac{1}{v_c} \int\limits_{v_c} \mu_k(\mathbf{r})\, d\mathbf{r}. \tag{3.19}$$

With our definition of the random function $\mu_k(\mathbf{r})$, we obtain for the average $\langle \mu_k(\mathbf{r}) \rangle = 0$. This means that

$$A_k(0) = 0.$$

Let us determine the correlation function $K_{ij}(\mathbf{r} - \mathbf{r}')$. We obtain with Eqs. (3.17)–(3.19):

$$K_{ij}(\mathbf{r} - \mathbf{r}') = \frac{1}{v_c} \int\limits_{v_c} \mu_i(\mathbf{r} + \mathbf{r}'') \mu_j(\mathbf{r}' + \mathbf{r}'')\, d\mathbf{r}'' =$$

$$= \frac{1}{v_c^2} \sum_{\mathbf{g}} \sum_{\mathbf{g}'} A_i(\mathbf{g}) A_j(\mathbf{g}')\, e^{-i\mathbf{r}\mathbf{g}} e^{-i\mathbf{g}'\mathbf{r}'} \frac{1}{v_c} \int d\mathbf{r}'' e^{-i\mathbf{r}''(\mathbf{g}+\mathbf{g}')} =$$

$$= \frac{1}{v_c^2} \sum_{g} \sum_{g'} A_i(g) A_j(g') e^{-irg} e^{-ir'g'} \delta_{g, -g'} =$$

$$= \frac{1}{v_c^2} \sum_{g} A_i(g) A_j(-g) e^{-ig(r-r')} = \frac{1}{v_c^2} \sum_{g} A_i(g) A_j^*(g) e^{-ig(r-r')}. \tag{3.20}$$

If central symmetry exists in the lattice, all $A_i(g)$ are real and another representation of the function $K_{ij}(r - r')$ is possible:

$$K_{ij}(r - r') = \frac{2}{v_c^2} \sum_{g}^{\infty} A_i(g) A_j(g) \cos g (r - r'). \tag{3.21}$$

A similar expression holds for $i = j$ in some types of lattice (also when the various μ_j depend in the same fashion upon the coordinates):

$$K_{ii}(r - r') = \frac{2}{v_c^2} \sum_{g}^{\infty} |A_i(g)|^2 \cos g (r - r'). \tag{3.22}$$

The summation in Eqs. (3.21) and (3.22) must be extended over the lattice points of the inverse lattice in the right half-space. The symbol ∞ over the summation sign indicates that the effect of the lattice bounds near the edge must be included with a coefficient $1/2$.

In the case of a rectangular lattice with the periods $2a$, $2b$, and $2c$, we obtain

$$K_{ij}(r - r') = \frac{1}{(8abc)^2} \sum_{n=-\infty}^{\infty} \sum_{m=-\infty}^{\infty} \sum_{l=-\infty}^{\infty} A_i(n, m, l) \times$$

$$\times A_j^*(n, m, l) e^{-i\frac{\pi}{a} n(x-x')} e^{-i\frac{\pi}{b} m(y-y')} e^{-i\frac{\pi}{c} l(z-z')}. \tag{3.23}$$

The coefficients $A_i(n, m, l)$ are given by the equation

$$A_i(n, m, l) = \iiint_{v_c} dx\, dy\, dz\, \mu_i(x, y, z)\, e^{i\frac{\pi}{a} nx}\, e^{i\frac{\pi}{b} my}\, e^{i\frac{\pi}{c} lz}. \tag{3.24}$$

When the rectangular lattice is invariant with respect to mirror reflections at the axes Ox, Oy, and Oz, the function $K_{ij}(r - r')$ can be represented in the following form:

$$K_{ij}(r - r') = \frac{1}{8a^2b^2c^2} \sum_{n=0}^{\infty} \sum_{m=0}^{\infty} \sum_{l=0}^{\infty} \lambda_{mnl} A_i(n, m, l) \times$$

$$\times A_j(n, m, l) \cos \frac{\pi}{a} n (x - x') \cos \frac{\pi}{b} m (y - y') \cos \frac{\pi}{c} (z - z'), \tag{3.25}$$

where

$$\lambda_{nml} = \begin{cases} 0 & n = m = l = 0, \\ 1/4 & n = m = 0, \ l \neq 0 \ \text{etc.} \\ 1/2 & n \neq 0, \ m \neq 0, \ l = 0 \ \text{etc.} \\ 1 & m \neq 0, \ n \neq 0, \ l \neq 0. \end{cases}$$

For example, when a neutron-absorbing shell is located in each lattice cell so that*

$$\mu_k(r) = \begin{cases} \mu_{k0} & r_1 < r < r_2, \\ 0 & r < r_1, \, r > r_2 \end{cases} \tag{3.26}$$

(the coordinate origin coincides with the center of the shell), we have

$$A_k(n, m, l) = \frac{4\pi\mu_{k0}}{q_{nml}^2} \left[\frac{\sin q_{nml}r_2 - \sin q_{nml}r_1}{q_{nml}} - r_2 \left(\cos q_{nml}r_2 - \frac{r_1}{r_2} \cos q_{nml}r_1 \right) \right]. \tag{3.27}$$

We used the notation

$$q_{nml} = \pi \sqrt{\frac{n^2}{a^2} + \frac{m^2}{b^2} + \frac{l^2}{c^2}}. \tag{3.28}$$

For $r_1 = 0$, Eq. (3.27) is identical with the previously obtained Eq. (3.16)

$$A_k(n, m, l) = \frac{4\pi\mu_{k0}}{q_{nml}^3} [\sin q_{nml}r_2 - r_2 q_{nml} \cos q_{nml}r_2], \tag{3.27a}$$

except for the fact that in the case of a periodic structure q assumes discrete values, whereas in the case of random positions of the neutron-absorbing bodies q can assume any value.

Substituting Eq. (3.27a) into Eq. (3.25) at $i = j$, we obtain

$$K_{ii}(r - r') = \frac{2\pi^2\mu_{i0}^2}{a^2b^2c^2} \sum_{n=0}^{\infty} \sum_{m=0}^{\infty} \sum_{l=0}^{\infty} \frac{\lambda_{nml}}{q_{nml}^6} (\sin q_{nml}r_2 - q_{nml}r_2 \cos q_{nml}r_2)^2 \times$$
$$\times \cos \frac{\pi}{a} n(x - x') \cos \frac{\pi}{b} m(y - y') \cos \frac{\pi}{c} l(z - z'). \tag{3.29}$$

For $r = r_1$ we have

$$K_{ii}(0) = e_{ii} = \frac{r_2^6 \mu_{i0}^2}{72} \left(\frac{6abc}{\pi r_2^3} - 1 \right). \tag{3.30}$$

One can easily verify that the Fourier transform of the correlation function defined by Eq. (3.20) is

$$k_{ij}(p) = \frac{(2\pi)^3}{v_c^2} \sum_g \delta(p - g) A_i(g) A_j^*(g). \tag{3.31}$$

Let us switch to two-dimensional periodic lattices. In this case, the correlation function is independent of one of the coordinates (z):

$$K_{ij}(r - r') = \frac{1}{s_c} \sum_{g_2} A_i(g_2) A_j^*(g_2) e^{-ig_2(\rho - \rho')}, \tag{3.32}$$

$$A_k(g_2) = \int_{s_c} e^{ig_2 \cdot \rho} \mu_k(\rho) \, d\rho. \tag{3.33}$$

---

*It follows from Eq. (3.6) that $\langle \mu_k(\bar{r}) \rangle \neq 0$. However, because we chose $\lambda_{000} = 0$, the correlation function $K(r - r')$ is independent of $\langle \mu \rangle$.

The notations are interpreted as follows: ρ denotes the two-dimensional radius vector in a plane perpendicular to the Oz axis; $\mathbf{g}_2$ is the vector of the inverse two-dimensional lattice; and s_c denotes the area of the cell. It is easy to derive expressions which are the analog of Eqs. (3.21)–(3.25). For example, in the case of a rectangular lattice with the periods $2a$ and $2b$, we obtain

$$K_{ij}(\mathbf{r}-\mathbf{r}') = \frac{1}{4a^2b^2} \sum_{n=0}^{\infty} \sum_{m=0}^{\infty} \lambda_{mn} A_i(n,\,m)\, A_j(n,\,m) \cos\frac{\pi}{a} n\,(x-x') \cos\frac{\pi}{b} m\,(y-y'), \qquad (3.34)$$

where

$$\lambda_{mn} = \begin{cases} 0 & n=m=0, \\ {}^{1}\!/_{2} & n=0,\ m\neq 0;\ n\neq 0,\ m=0, \\ 1 & n\neq 0,\ m\neq 0 \end{cases}$$

and

$$A_k(n,\,m) = \iint\limits_{s_c} dx\, dy\, \mu_k(x,\,y)\, e^{i\frac{\pi}{a}nx}\, e^{i\frac{\pi}{b}my}. \qquad (3.35)$$

In a homogeneous, neutron–absorbing cylindrical shell with circular cross section (the coordinate origin in the cell coincides with the center of the cross section),* we have

$$\mu_k(\rho) = \begin{cases} \mu_{k1} & r_1 < \rho = \{x,\,y\} < r_2, \\ \mu_{k2} & \rho < r_1,\ \rho > r_2 \end{cases} \qquad (3.36)$$

and the coefficient $A_k(n,\,m)$ is equal to

$$A_k(n,\,m) = \frac{2\mu_{k0}}{\sqrt{\frac{n^2}{a^2}+\frac{m^2}{b^2}}} \left[r_2 J_1\!\left(\pi r_2 \sqrt{\frac{n^2}{a^2}+\frac{m^2}{b^2}}\right) - r_1 J_1\!\left(\pi r_1 \sqrt{\frac{n^2}{a^2}+\frac{m^2}{b^2}}\right) \right], \qquad (3.37)$$

where $J_1(x)$ denotes a Bessel function and $\mu_{k0} = \mu_{k1} - \mu_{k2}$.

For $r_1, = 0$, i.e., in the case of a homogeneous cylinder, we have

$$A_k(n,\,m) = \frac{2\mu_{k0}}{\sqrt{\frac{n^2}{a^2}+\frac{m^2}{b^2}}} \left[r_2 J_1\!\left(\pi r_2 \sqrt{\frac{n^2}{a^2}+\frac{m^2}{b^2}}\right) \right]. \qquad (3.37a)$$

Substituting this expression into Eq. (3.34) for $i = j$, we obtain

$$K_{ii}(\mathbf{r}-\mathbf{r}') = \frac{r_2^2\mu_{i0}^2}{a^2b^2} \sum_{n=0}^{\infty} \sum_{m=0}^{\infty} \frac{\lambda_{mn} J_1^2\!\left(\pi r_2 \sqrt{\frac{n^2}{a^2}+\frac{m^2}{b^2}}\right)}{\frac{n^2}{a^2}+\frac{m^2}{b^2}} \times$$

$$\times \cos\frac{\pi}{a} n\,(x-x') \cos\frac{\pi}{b} m\,(y-y'). \qquad (3.38)$$

---

*See preceding footnote.

It follows for $\mathbf{r} = \mathbf{r}'$ that[*]

$$K_{il}(0) = \varepsilon_{il} = \frac{\mu_{io}^2 r_2^2}{a^2 b^2} \sum_{n=0}^{\infty} \sum_{m=0}^{\infty} \lambda_{mn} \frac{J_1^2 \left(\pi r^2 \sqrt{\frac{n^2}{a^2} + \frac{m^2}{b^2}} \right)}{\frac{n^2}{a^2} + \frac{m^2}{b^2}} = \frac{\pi r_2^2 \mu_{io}^2}{4ab} \left(1 - \frac{\pi r_2^2}{4ab} \right). \tag{3.39}$$

The Fourier transform of the correlation function of Eq. (3.32) can be written in the following form:

$$k_{ij}(\mathbf{p}) = \frac{(2\pi)^3}{v_C^2} \sum_{\mathbf{g}} \delta(\mathbf{p}_2 - \mathbf{g}_2) \delta(p_z) A_i(\mathbf{g}_2) A_j^*(\mathbf{g}_2), \tag{3.40}$$

where $\mathbf{p}_2$ denotes the projection of the vector $\mathbf{p}$ onto the plane perpendicular to the Oz axis.

The correlation function of a plane lattice with the period $2a$ depends on only one coordinate:

$$K_{ij}(z - z') = \frac{1}{(2a)^2} \sum_{n=-\infty}^{\infty} A_i(n) A_j^*(n) e^{-i \frac{\pi}{a} n(z - z')}, \tag{3.41}$$

$$A_i(n) = \int_0^{2a} dz \mu_i(z) e^{i \frac{\pi}{a} nz}. \tag{3.42}$$

In a simple structure of the form[†]

$$\mu_k(z) = \begin{cases} \mu_{ko} & 2a - \alpha < z < 2a \\ 0 & 0 < z < 2a - \alpha \end{cases} \tag{3.43}$$

the coefficients $A_k(n)$ are

$$A_k(n) = \frac{a \mu_{ko}}{i \pi n} \left(1 - e^{-i \frac{\pi}{a} n \alpha} \right). \tag{3.44}$$

The correlation function of the lattice defined by Eq. (3.43) can be conveniently written in the following form:

$$K_{ij}(z - z') = \frac{\mu_{io} \mu_{jo}}{\pi^2} \sum_{n=1}^{\infty} \frac{1 - \cos \frac{\pi n}{a} \alpha}{n^2} \cos \frac{\pi}{a} n(z - z'). \tag{3.45}$$

For $i = j$ and $z = z'$ we obtain

$$K_{ii}(0) = \varepsilon_{ii} = \frac{\mu_{io}^2}{4a^2} (2a - \alpha) \alpha. \tag{3.46}$$

---

[*] Equation (3.39) (and its analog for $r_1 \neq 0$) leads to a simple expression for the sum of the double series containing the Bessel functions:

$$\sum_{n=0}^{\infty} \sum_{m=0}^{\infty} \frac{\lambda_{mn} J_1^2 \left(\pi r \sqrt{\frac{n^2}{a^2} + \frac{m^2}{b^2}} \right)}{\frac{n^2}{a^2} + \frac{m^2}{b^2}} = \frac{\pi}{4} \left(1 - \frac{\pi r^2}{4ab} \right).$$

[†] See footnote to Eq. (3.26) on p. 204.

If $\alpha = a$, Eqs. (3.43)–(3.46) simplify*:

$$\mu_k(z) = \begin{cases} \mu_{k0} & a < z < 2a, \\ 0 & 0 < z < a; \end{cases} \tag{3.43a}$$

$$A_k(n) = \begin{cases} \dfrac{2a\mu_{k0}}{i\pi n} & n \quad \text{odd,} \\ 0 & n \quad \text{even.} \end{cases} \tag{3.44a}$$

The expansion of the correlation function in a series of $\cos\frac{\pi}{a} n (z - z')$ leads to

$$K_{ij}(z - z') = \frac{2\mu_{i0}\mu_{j0}}{\pi^2} \sum_{k=1}^{\infty} \frac{\cos\dfrac{2k-1}{a}\pi(z-z')}{(2k-1)^2} \tag{3.45a}$$

and we obtain for $i = j$ and $z = z'$:

$$K_{ii}(0) = \varepsilon_{ii} = \mu_{i0}^2/4. \tag{3.46a}$$

The Fourier transform of the correlation function of a linear lattice has the form

$$k_{ij}(p) = (2\pi)^3 \frac{1}{(2a)^2} \sum_{n=-\infty}^{\infty} A_i(n) A_j^*(n) \delta\left(p_z - \frac{\pi}{a} n\right) \delta(p_x) \delta(p_y). \tag{3.47}$$

In a lattice of the type defined by Eq. (3.43), we have

$$k_{ij}(p) = (2\pi)^3 \frac{\mu_{i0}\mu_{j0}}{2\pi^2} \sum_{n=1}^{\infty} \frac{1 - \cos\dfrac{\pi n\alpha}{a}}{n^2} \left[\delta\left(p_z - \frac{\pi}{a} n\right) + \delta\left(p_z + \frac{\pi}{a} n\right)\right] \delta(p_x)\delta(p_y), \tag{3.48}$$

and for Eq. (3.43a) we obtain

$$k_{ij}(p) = (2\pi)^3 \frac{\mu_{i0}\mu_{j0}}{\pi^2} \sum_{k=1}^{\infty} \frac{1}{(2k-1)^2} \left[\delta\left(p_z - \frac{\pi}{a}(2k-1)\right) + \delta\left(p_z + \frac{\pi}{a}(2k-1)\right)\right] \delta(p_x)\delta(p_y). \tag{3.49}$$

5. The correlation functions which were considered in the preceding sections of this paragraph belong to two essentially different classes: one class describes correlations in a medium with randomly distributed inhomogeneities (nonordered systems), whereas the other class refers to media with purely periodic structure (long-range order). Corresponding differences are found in their Fourier representations (spectra). In the first case, the spectra are continuous functions of the wave vector, whereas in the second case they are linear combinations of δ functions.

In structures of an intermediate type, i.e., when the order in the medium prevails over only short distances from a particular point (short-range order) and no longer exists at large distances, this behavior is conveniently taken into account by replacing each oscillating harmonic function in the correlation function by a decreasing exponential function. The result is that the δ functions in the Fourier components are smeared out and assume a certain width. For example, if

$$\mu_i(z) = \mu_{i0} \sin qz, \tag{3.50}$$

one easily verifies that

$$K_{ii}(z) = \frac{\mu_{i0}^2}{2} \cos qz \tag{3.51}$$

---

*See footnote to Eq. (3.26) on page 204.

and

$$k_{ii}(\mathbf{p}) = (2\pi)^3 \,\delta(p_x)\,\delta(p_y)\,\frac{\mu_{i0}^2}{4}\,[\delta(p_z + q) + \delta(p_z - q)]. \tag{3.52}$$

In order to take into account long-range order in the system, we introduce in place of Eq. (3.51)

$$K_{ii}(z) = \frac{\mu_{i0}^2}{2}\,e^{-\beta|z|}\cos qz = \varepsilon_{ii}e^{-\beta|z|}\cos qz, \tag{3.53}$$

where β denotes a parameter which describes the attenuation of the oscillations in the correlation function. The Fourier transform of the function defined by Eq. (3.53) (the fluctuation spectrum) has the form

$$k_{ii}(\mathbf{p}) = \frac{\mu_{i0}^2}{2}\,(2\pi)^2\,\delta(p_x)\,\delta(p_y)\,\beta\left[\frac{1}{\beta^2 + (p_z - q)^2} + \frac{1}{\beta^2 + (p_z + q)^2}\right]. \tag{3.54}$$

For $\beta \to 0$, Eq. (3.54) coincides with Eq. (3.52). When β is large, we obtain for small p_z and for $\beta \gg q$

$$k_{ii}(\mathbf{p}) \approx (2\pi)^2\,\mu_{i0}^2\,\delta(p_x)\,\delta(p_y)\,\frac{\beta}{\beta^2 + q^2} \approx (2\pi)^2\,\varepsilon_{ii}\,\delta(p_x)\,\delta(p_y)\cdot\frac{2}{\beta}. \tag{3.55}$$

This expression coincides with Eq. (3.2), which describes the fluctuations in a nonordered medium.

As another example of disturbed periodicity in a diffusion medium we consider a lattice of neutron-absorbing rods which are not situated exactly at the lattice points, but displaced to some extent (for example, we assume that the openings for the rods are wider than the rods and that the rods can shift freely in the openings). We assume for the sake of simplicity that the distribution function $\mathbf{u}_l$ (describing the shift of the l-th rod from a lattice point) is Gaussian:

$$p(\mathbf{u}_l) = (\pi\theta)^{-1}\,e^{-u_l^2/\theta}. \tag{3.56}$$

We use the representation of Eq. (3.7) for the perturbation $\mu_k(\mathbf{r})$. The correlation function $K_{ij}(\mathbf{r} - \mathbf{r}')$ assumes the form

$$K_{ij}(\mathbf{r} - \mathbf{r}') = \frac{1}{(2\pi)^6}\iint d\mathbf{q}\,d\mathbf{q}'e^{-i\mathbf{q}\mathbf{r}}\,e^{-i\mathbf{q}'\mathbf{r}'}\,\gamma_i(\mathbf{q})\,\gamma_j(\mathbf{q}')\,[\langle\rho(\mathbf{q})\rho(\mathbf{q}')\rangle - \langle\rho(\mathbf{q})\rangle\langle\rho(\mathbf{q}')\rangle], \tag{3.57}$$

where $\gamma_k(\mathbf{q})$ is defined by the shape of the rod [see Eq. (3.8)] and

$$\rho(\mathbf{q}) = \sum_{\alpha=1}^{N} e^{i\mathbf{q}\mathbf{R}_\alpha}$$

(N denotes the number of rods in the medium).

We recall that $\mathbf{R}_\alpha = \mathbf{R}_\alpha^0 + \mathbf{u}_\alpha$ (where $\mathbf{R}_\alpha^0$ denotes the radius vector of the α-th lattice point of the ideal lattice) and take in Eq. (3.57) the average of the distribution law of the displacements defined by Eq. (3.56). We obtain

$$\langle\rho(\mathbf{q})\rangle = \left\langle\sum_{\alpha=1}^{N} e^{i\mathbf{q}\mathbf{R}_\alpha^0}\right\rangle_{\theta=0} e^{-q^2\theta/4}, \tag{3.58}$$

where the symbol $\langle...\rangle_{\theta=0}$ denotes averaging over a cell of the ideal lattice. Similarly, we have

$$\langle \rho(\mathbf{q})\, \rho(\mathbf{q}')\rangle = \left\langle \sum_\beta \sum_\alpha e^{i\mathbf{q}\mathbf{R}_\alpha} e^{i\mathbf{q}'\mathbf{R}_\beta}\right\rangle = \left\langle \sum_\alpha e^{i\mathbf{R}_\alpha\,(\mathbf{q}+\mathbf{q}')}\right\rangle + \left\langle \sum_\beta \sum_{\alpha} e^{i\mathbf{q}\mathbf{R}_\alpha} e^{i\mathbf{q}'\mathbf{R}_\beta}\right\rangle =$$

$$= e^{-\frac{(\mathbf{q}+\mathbf{q}')^2\,\theta}{4}} \left\langle \sum_\alpha e^{i\,(\mathbf{q}+\mathbf{q}')\,R_\alpha^0}\right\rangle_{\theta=0} + \left\langle \sum_{\beta\neq\alpha} e^{i\mathbf{q} R_\alpha^0} e^{i\mathbf{q}' R_\beta^0}\right\rangle_{\theta=0} e^{-\frac{q^2+q'^2}{4}\theta} =$$

$$= \left\langle \sum_\alpha \sum_\beta e^{i\mathbf{q} R_\alpha^0} e^{i\mathbf{q}' R_\beta^0}\right\rangle_{\theta=0} e^{-\frac{q^2+q'^2}{4}\theta} + \left\langle \sum_\alpha e^{i\,(\mathbf{q}+\mathbf{q}')\,R_\alpha^0}\right\rangle_{\theta=0}\left[e^{-\frac{(\mathbf{q}+\mathbf{q}')^2\theta}{4}} - e^{-\frac{q^2+q'^2}{4}\theta}\right]. \tag{3.59}$$

Thus,

$$\langle \rho(\mathbf{q})\,\rho(\mathbf{q}')\rangle - \langle\rho(\mathbf{q})\rangle\langle\rho(\mathbf{q}')\rangle = [\langle\rho(\mathbf{q})\rho(\mathbf{q}')\rangle_{\theta=0} - \langle\rho(\mathbf{q})\rangle_{\theta=0}\langle\rho(\mathbf{q})\rangle_{\theta=0}]\, e^{-\frac{q^2+q'^2}{4}\theta} +$$

$$\langle\rho(\mathbf{q})\,\rho(\mathbf{q}')\rangle - \langle\rho(\mathbf{q})\rangle\langle\rho(\mathbf{q}')\rangle + \langle\rho(\mathbf{q}+\mathbf{q}')\rangle_{\theta=0}\left[e^{-\frac{(\mathbf{q}+\mathbf{q}')^2}{4}\theta} - e^{-\frac{q^2+q'^2}{4}\theta}\right]. \tag{3.60}$$

The final expression for the fluctuation spectrum in this lattice has the form

$$k_{ij}(\mathbf{p}) = [k_{ij}(\mathbf{p})]_{\theta=0}\, e^{-\frac{p^2\theta}{2}} + \frac{N}{V}\,\gamma_i(\mathbf{p})\,\gamma_j^*(\mathbf{p})\left[1 - e^{-\frac{p^2\theta}{2}}\right]. \tag{3.61}$$

The first term of the right side of Eq. (3.61) defines the contribution of an ideal periodic structure to the correlation function. For $\theta \to \infty$ this term vanishes. The second term results from the spread of the rod coordinates: for $\theta = 0$ this term tends to zero. If $\theta \to \infty$, Eq. (3.61) coincides with the previously obtained Eq. (3.14) for a completely disordered system. We see that order in a system described by Eq. (3.56) implies a fluctuation spectrum which is different from that obtained in the case when no long-range order exists [system defined by Eq. (3.53)]. The δ-function-like peaks which are characteristic also for the ideal lattice are not broadened, but their height decreases. The intensity of the "background" resulting from deviations from strict periodicity increases at the same time.

§4. Stationary Diffusion of Thermal Neutrons

We begin our discussion of problems of neutron-transport theory in inhomogeneous media with an analysis of the stationary diffusion of thermal neutrons [10-12, 14]. In the diffusion approximation, the flux of thermal neutrons, $G(\mathbf{r}|\mathbf{r}_0)$, from a point source in an inhomogeneous medium satisfies the equation

$$\text{div}\left[\frac{D(\mathbf{r})}{D_0}\,\text{grad}\,G(\mathbf{r}\,|\,\mathbf{r}_0)\right] - \frac{\Sigma_a(\mathbf{r})}{D_0}\,G(\mathbf{r}\,|\,\mathbf{r}_0) = -\frac{S}{D_0}\,\delta(\mathbf{r} - \mathbf{r}_0), \tag{4.1}$$

where S denotes the source output. In what follows we assume for the sake of simplicity that $S = D_0$. The macroscopic absorption cross section $\Sigma_a(\mathbf{r})$ and the diffusion coefficient $D(\mathbf{r})$ are written as

$$\Sigma_a(\mathbf{r}) = \Sigma_{a0} + \sigma_a(\mathbf{r}), \quad \langle \Sigma_a(\mathbf{r})\rangle = \Sigma_{a0}, \quad \langle \sigma_a(\mathbf{r})\rangle = 0, \tag{4.2}$$

$$D(\mathbf{r}) = D_0 + d(\mathbf{r}), \quad \langle D(\mathbf{r})\rangle = D_0, \quad \langle d(\mathbf{r})\rangle = 0. \tag{4.3}$$

It follows from Eq. (4.3) that the average diffusion coefficient defined in this fashion, i.e.,

$$D_0 = \left\langle \frac{1}{3\Sigma(\mathbf{r})}\right\rangle = \frac{1}{3\Sigma_0}\left\langle \frac{1}{1 + \frac{\sigma(\mathbf{r})}{\Sigma_0}}\right\rangle \approx D_{\text{h}}\left(1 + \frac{\langle\sigma^2\rangle}{\Sigma_0^2}\right) \tag{4.4}$$

differs from $D_h = 1/3\Sigma_0$, i.e., the diffusion coefficient for a homogeneous medium ($\Sigma = \Sigma_s + \Sigma_a$; Σ_s denotes the macroscopic scattering cross section *).

It is easy to verify by substituting Eqs. (4.2) and (4.3) into Eq. (4.1) and comparing with Eq. (2.3) that in our case

$$\hat{B} = \nabla^2 - \varkappa_0^2, \tag{4.5}$$

$$\hat{\mu}(r) = -\eta(r)\nabla^2 - (\nabla\eta(r))\nabla + \nu(r), \tag{4.6}$$

where

$$\varkappa_0^2 = \frac{\Sigma_{a0}}{D_0}, \quad \eta(r) = \frac{d(r)}{D_0}, \quad \nu(r) = \frac{\sigma_a(r)}{D_0}.$$

We assume that our diffusion medium is homogeneous in space (i.e., it is homogeneous on the average). Then, the operator $M_1(\mathbf{r}|\mathbf{r'})$ [see Eqs. (2.20) and (2.30)] depends only upon the difference of the coordinates $\mathbf{r} - \mathbf{r'}$, and Eq. (2.21) can be reduced to the algebraic equation

$$[p^2 + \varkappa_0^2]\langle g(p)\rangle = 1 + m_1(p)\langle g(p)\rangle \tag{4.7}$$

by employing the Fourier transform with respect to $\mathbf{r} - \mathbf{r}_0$. We introduced the notation

$$f(\mathbf{p}) = \int d\mathbf{r} e^{i\mathbf{p}\mathbf{r}} F(\mathbf{r}). \tag{4.8}$$

Hence, we have

$$\langle g(p)\rangle = \frac{1}{p^2 + \varkappa_0^2 - m_1(p)} \tag{4.9}$$

and correspondingly,

$$\langle G(\mathbf{r}|\mathbf{r}_0)\rangle = \frac{1}{(2\pi)^3}\int \frac{d\mathbf{p} e^{-i\mathbf{p}(\mathbf{r}-\mathbf{r}_0)}}{p^2 + \varkappa_0^2 - m_1(p)}. \tag{4.10}$$

The entire information on the fluctuations in the medium is contained in the "operator" $m_1(\mathbf{p})$, which determines the expressions under the integral of Eq. (4.10). Poles $\langle g(\mathbf{p})\rangle$ correspond to the exponential decrease of $\langle G(\mathbf{r}|\mathbf{r}_0)\rangle$ far from the source.

In the zeroth approximation we set $m_1(\mathbf{p}) = 0$ and

$$g_0(\mathbf{p}) = \frac{1}{p^2 + \varkappa_0^2}; \tag{4.11}$$

the neutron flux decreases according to the law

$$G_0(\mathbf{r}|\mathbf{r}_0) = \frac{1}{4\pi|\mathbf{r}-\mathbf{r}_0|} e^{-\varkappa_0|\mathbf{r}-\mathbf{r}_0|}. \tag{4.12}$$

We substitute the expressions for $G_0(\mathbf{r}|\mathbf{r}_0)$ [Eq. (4.12)] and $\mu(r)$ [Eq. (4.6)] into the formula for the operator $M_1(\mathbf{r}|\mathbf{r'})$ [Eqs. (2.20) or (2.30)] and take the Fourier transform with respect to $\mathbf{r}-\mathbf{r'}$. We obtain after simple, but laborious calculations

---

* As usual, when an approximation for the effect of the anisotropy of the scattering is used, Σ_s must be replaced by Σ_{tr}, which denotes the transport scattering cross section. For weak absorption $\Sigma_{a0} \ll \Sigma_{s0}$ holds, and the corrections to the expression for the diffusion coefficient Σ_{a0}/Σ_{s0} can be disregarded.

$$m_1(\mathbf{p}) = p^2 \left\{ \varepsilon_{\eta\eta} + \frac{1}{(2\pi)^3} \int d\zeta \frac{-\varkappa_0^2 k_{\eta\eta}(\zeta) + 2k_{\eta\nu}(\zeta)}{(p-\zeta)^2 + \varkappa_0^2} + \right.$$
$$\left. + \frac{1}{(2\pi)^3} \int \frac{d\zeta}{p^2} \frac{[(p\zeta)^2 - p^2\zeta^2] k_{\eta\eta}(\zeta) - 2p\zeta k_{\eta\nu}(\zeta)}{(p-\zeta)^2 + \varkappa_0^2} \right\} + \frac{1}{(2\pi)^3} \int \frac{d\zeta k_{\nu\nu}(\zeta)}{(p-\zeta)^2 + \varkappa_0^2} . \tag{4.13}$$

We have

$$k_{\eta\eta}(\mathbf{p}) = \int d\mathbf{r} e^{i\mathbf{p}\mathbf{r}} K_{\eta\eta}(\mathbf{r}), \tag{4.14}$$

and

$$K_{\eta\eta}(\mathbf{r}) \equiv \langle \eta(\mathbf{r}) \eta(\mathbf{r}+\mathbf{r}') \rangle = \langle \eta(0) \eta(\mathbf{r}) \rangle = \langle \eta(\mathbf{r}) \eta(0) \rangle \tag{4.15}$$

is the correlation function for the fluctuations $\eta(\mathbf{r})$.

The other correlation functions are analogously determined:

$$\varepsilon_{\eta\eta} \equiv K_{\eta\eta}(0) = \frac{1}{(2\pi)^3} \int d\zeta k_{\eta\eta}(\zeta). \tag{4.16}$$

2. When only the absorption cross section fluctuates, we have $\eta(\mathbf{r}) \equiv 0$ and

$$m_1(\mathbf{p}) = \frac{1}{(2\pi)^3} \int \frac{d\zeta k_{\nu\nu}(\zeta)}{(p-\zeta)^2 + \varkappa_0^2} . \tag{4.17}$$

Let us consider this equation and the solution $\langle G(\mathbf{r}|\mathbf{r}_0) \rangle$ of the diffusion equation for various diffusion media (§ 3).

a) Layered medium with a correlation function as defined by Eq. (3.1). Neutrons diffuse across the layers. We obtain in this case

$$m_1(\mathbf{p}) = \varepsilon_{\nu\nu} \frac{\varkappa_1}{\varkappa_0} \frac{1}{p_z^2 + \varkappa_1^2},$$
$$\varkappa_1 = \varkappa_0 + q_\nu. \tag{4.18}$$

We can calculate the Fourier transform $\langle g(p_z) \rangle$ of the average neutron flux by substituting Eq. (4.18) into Eq. (4.9):

$$\langle g(p_z) \rangle = \frac{\varkappa_1^2 + p_z^2}{(p_z^2 + \varkappa_0^2)(p_z^2 + \varkappa_1^2) - \varepsilon_{\nu\nu} \frac{\varkappa_1}{\varkappa_0}} . \tag{4.19}$$

After calculating the integral in Eq. (4.10) with the expression for $\langle g(p_z) \rangle$ from (4.19), we obtain

$$\langle G(z|z_0) \rangle = \frac{1}{2(\varkappa_{tr}^2 - \varkappa_{as}^2)} \left[\frac{\varkappa_1^2 - \varkappa_{as}^2}{\varkappa_{as}} e^{-\varkappa_{as}|z-z_0|} - \frac{\varkappa_1^2 - \varkappa_{tr}^2}{\varkappa_{tr}} e^{-\varkappa_{tr}|z-z_0|} \right], \tag{4.20}$$

where

$$\varkappa_{as \atop tr} = \left\{ \frac{\varkappa_1^2 + \varkappa_0^2}{2} \mp \left[(\varkappa_1^2 - \varkappa_0^2)^2 + 4\varepsilon_{\nu\nu} \frac{\varkappa_1}{\varkappa_0} \right]^{1/2} \right\}^{1/2}. \tag{4.21}$$

It follows from Eqs. (4.20) and (4.21) for $\varkappa_1 \gg \varkappa_0$ that at large distances from the source $(|z - z_0|) \infty \varkappa_0^{-1})$ the average neutron flux has the form

$$\langle G(z|z_0) \rangle = \frac{1}{2\varkappa_0} e^{-\varkappa_0|z-z_0| \left(1 - \frac{\varepsilon_{\nu\nu}}{2\varkappa_1 \varkappa_0^2}\right)}, \tag{4.22}$$

i.e., the average neutron flux decreases exponentially with the relaxation length which is given

by $\varkappa_0^{-1}\left[1+\dfrac{\varepsilon_{vv}}{2\varkappa_1\varkappa_0^3}\right]^{*}$. In Eq. (4.22) we restricted ourselves to corrections in the exponent of the

exponential which are linear in $\varkappa_0/\varkappa_1 \approx \varkappa_0/q_{vv}$; the zeroth approximation was used for the factor before the exponential function. Thus, the fluctuations of the scattering properties of the medium increased the relaxation path length of the neutron flux relative to the case of diffusion in the medium without fluctuations. This is easy to understand when we recall that in heterogeneous systems the attenuation of the neutron flux takes place more slowly than in a homogeneous medium. Another feature of our solution stated in Eq. (4.20) is the "transition" term which vanishes at distances of the order of q_{vv}^{-1} from the source, i.e., at distances which correspond to the correlation length of the fluctuations in the moderator. This term is small ($\sim\varepsilon_{vv}/2\varkappa_0\varkappa_1^4$), and its influence on the absolute value of the asymptotic solution (at large distances from the neutron source) can be ignored. The attenuation of the "transition" solution is determined by the pole of $\langle g(p_z)\rangle$ in the complex p_z plane, with the pole close to the singularity of the operator $m_1(p_z)$. The behavior of the "transition" term depends substantially upon the form of the "mass operator." In order to obtain a clear idea of the behavior of the "transition" solution we must consider the contribution of higher approximations to the operator $M(r|r')$. Let us note in conclusion that in the M_1 approximation involving Eq. (2.21) one cannot describe the neutron diffusion along the layers of the medium with the correction function of Eq. (3.1). We will discuss this question in detail in **4** of §8.

b) The model of the medium with the correlation function defined by Eq. (3.1a) helps us to estimate the effect of the anisotropy of small-scale fluctuations upon the transfer of thermal neutrons. When a plane source is placed on the y_0 plane, the average neutron density is

$$\langle G\,(|\,y-y_0\,|)\rangle = \frac{1}{2\pi}\int\frac{dp_y e^{-ip_y\,(y-y_0)}}{p_y^2+\varkappa_0^2-m_1\,(0,\,p_y,\,0)}\,. \tag{4.23}$$

When the source is situated on the plane x_0, we obtain

$$\langle G\,(|\,x-x_0\,|)\rangle = \frac{1}{2\pi}\int\frac{dp_x e^{-ip_x\,(x-x_0)}}{p_x^2+\varkappa_0^2-m_1\,(p_x,\,0,\,0)}\,. \tag{4.24}$$

We have

$$m_1\,(\mathbf{p})\equiv m_1\,(p_x,\,p_y,\,p_z)=\frac{\varepsilon_{vv}\,q_{vv}'\,q_{vv}''}{\pi^2}\int\frac{dk_x\,dk_y}{(k_x^2+q_{vv}'^3)(k_y^2+q_{vv}''^2)}\frac{1}{(p_x-k_x)^2+(p_y-k_y)^2+p_z^2+\varkappa_0^2}\,. \tag{4.25}$$

We used the actual form of the correlation functions of Eqs. (3.1a) and (3.2a) for determining $m_1(\mathbf{p})$. When we integrate in Eq. (4.25) at $p_z=0$ and $q_{vv}',\,q_{vv}''\gg\varkappa_0$, we obtain

$$m_1\,(0,\,p_y,\,0)=\frac{2\varepsilon_{vv}}{\pi}\frac{\varkappa_0^2}{q_{vv}'\,q_{vv}''}\left[1+\left(\frac{p_y}{q_{vv}''}\right)^2\right]^{-1}\ln\frac{q_m}{\varkappa_0} \tag{4.26}$$

and

$$m_1\,(p_x,\,0,\,0)=\frac{2\varepsilon_{vv}}{\pi}\frac{\varkappa_0^2}{q_{vv}'\,q_{vv}''}\left[1+\left(\frac{p_x}{q_{vv}'}\right)^2\right]^{-1}\ln\frac{q_m}{\varkappa_0}\,, \tag{4.27}$$

where $q_m = \min\,(q_{vv}',\,q_{vv}'')$.

---

*The higher-order corrections for a perturbation ν must differ from the expansion of the quantities $\varkappa_{as}$ and $\varkappa_{tr}$ in series in ε^{vv}, because our initial equation (2.21) is an approximation.

We see from these expressions that the attenuation of the correlations in the medium along the layers $(\sim e^{-q_{vv'}|x|})$ reduces the correction to the attenuation constant of the average neutron flux perpendicular to the "layers" by $\frac{\varkappa_0}{q_{vv}'}\ln\frac{q_m}{\varkappa_0}$ (the attenuation of the correlations across the layers is $\sim e^{-q_{vv}''|y|}$). The neutron distribution in a medium with fluctuations of the form of Eq. (3.1a) is slightly anisotropic*

$$\frac{\varkappa_{as}^x}{\varkappa_{as}^y}=\frac{1-\dfrac{\varepsilon_{vv}}{\pi q_{vv}' q_{vv}''}\left(\ln\dfrac{q_m}{\varkappa_0}\right)\left[1-\left(\dfrac{\varkappa_0}{q_{vv}'}\right)^2\right]}{1-\dfrac{\varepsilon_{vv}}{\pi q_{vv}' q_{vv}''}\left(\ln\dfrac{q_m}{\varkappa_0}\right)\left[1-\left(\dfrac{\varkappa_0}{q_{vv}''}\right)^2\right]}, \tag{4.28}$$

where $\varkappa_{as}^i$ denotes the constant attenuations of the neutron flux at large distances from the plane source.

c) In the case of an isotropic medium of the type defined by Eq. (3.4), we have

$$m_1(p)=\frac{\varepsilon_{vv}}{p^2+\varkappa_1^2}, \tag{4.29}$$

$$\varkappa_1=\varkappa_0+q_{vv}$$

and

$$\langle g(p)\rangle=[p^2+\varkappa_0^2-\varepsilon_{vv}\overline{(p^2+\varkappa_1^2)}]^{-1}. \tag{4.30}$$

By taking the inverse Fourier transform we find the average neutron flux:

$$\langle G(\mathbf{r}\mid\mathbf{r}_0)\rangle\equiv\langle G(|\mathbf{r}-\mathbf{r}_0|)\rangle=\frac{1}{4\pi|\mathbf{r}-\mathbf{r}_0|}\frac{1}{\varkappa_{tr}^2-\varkappa_{as}^2}\{(\varkappa_1^2-\varkappa_{as}^2)\,e^{-\varkappa_{as}|\mathbf{r}-\mathbf{r}_0|}-(\varkappa_1^2-\varkappa_{tr}^2)\,e^{-\varkappa_{tr}|\mathbf{r}-\mathbf{r}_0|}\}, \tag{4.31}$$

$$\varkappa_{\substack{as\\tr}}=\left[\frac{\varkappa_1^2+\varkappa_0^2}{2}\mp\sqrt{(\varkappa_1^2-\varkappa_0^2)^2+4\varepsilon_{vv}}\right]^{1/2}. \tag{4.32}$$

We recognize immediately from Eq. (4.32) that $\varkappa_{as}\approx\varkappa_0$ and $\varkappa_{tr}\approx\varkappa_1$, and in the case of small-scale inhomogeneities we have $\varkappa_1=\varkappa_0+q_{vv}\approx q_{vv}\gg\varkappa_0$. Thus, $\varkappa_{as}$ determines the attenuation of the average neutron flux far from the source. The average neutron flux from a linear source in an isotropic medium can be calculated from Eq. (4.31):

$$\langle G(\rho)\rangle_l=\frac{1}{2\pi}\frac{1}{\varkappa_{tr}^2-\varkappa_{as}^2}\{K_0(\varkappa_{as}\rho)(\varkappa_1^2-\varkappa_{as}^2)-K_0(\varkappa_{tr}\rho)(\varkappa_1^2-\varkappa_{tr}^2)\}, \tag{4.33}$$

where ρ denotes the distance from the source and $K_0(x)$ is a modified Bessel function.

When a plane neutron source is placed into an isotropic medium, we obtain

$$\langle G(z)\rangle_{pl}=\frac{1}{2(\varkappa_{tr}^2-\varkappa_{as}^2)}\left\{\frac{\varkappa_1^2-\varkappa_{as}^2}{\varkappa_{as}}e^{-\varkappa_{as}|z|}-\frac{\varkappa_1^2-\varkappa_{tr}^2}{\varkappa_{tr}}e^{-\varkappa_{tr}|z|}\right\}, \tag{4.34}$$

where z denotes the distance from the plane on which the source is situated. In contrast to the neutron diffusion in a layered medium [Eq. (3.1), see section 1] for which the correction to the attenuation constant of the neutron flux is

---

* The anisotropy in the composition of the medium affects the neutron distribution only slightly, due to the small scale of the fluctuations $(q_{vv}', q_{vv}''\gg\varkappa_0)$. This statement is incorrect when the diffusion coefficient fluctuates (see § 8).

$$\Delta \varkappa \approx - \varkappa_0 \frac{\varepsilon_{vv}}{\varkappa_0^4} \frac{\varkappa_0}{2\varkappa_1}, \qquad (4.35)$$

we obtain in an isotropic medium

$$\Delta \varkappa \approx - \varkappa_0 \frac{\varepsilon_{vv}}{2\varkappa_0^4} \left(\frac{\varkappa_0}{\varkappa_1}\right)^2, \qquad (4.36)$$

i.e., a much smaller value. This result is easy to understand when we recall that in the layered medium of Eq. (3.1) the correlation along the layers does not decrease. In averaging over the angles (transition to an equivalent isotropic medium) a larger effective radius of the correlation of the fluctuations results.

d) In a diffusion medium with inhomogeneities in the form of localized inclusions [the correlation function has the form of Eqs. (3.13) and (3.14)] we obtain

$$m_1(\mathbf{p}) = \frac{N}{V} \frac{1}{(2\pi)^3} \int \frac{d\mathbf{q} |\gamma_v(\mathbf{q})|^2}{(\mathbf{p} - \mathbf{q})^2 + \varkappa_0^2}. \qquad (4.37)$$

When we insert $\gamma_v(q)$ in the form of Eq. (3.16) into Eq. (4.37), we obtain

$$m_1(p) = \frac{3}{2\pi} \rho \frac{(\Gamma_{v0})^2 R_v}{p} \int_0^\infty \frac{dx}{x^5} (\sin x - x \cos x)^2 \ln \frac{(x + pR_v)^2 + \varkappa_0^2 R_v^2}{(x - pR_v)^2 + \varkappa_0^2 R_v^2}, \qquad (4.38)$$

where

$$\rho = \frac{N}{V} \frac{4\pi}{3} R_v^3.$$

An approximate calculation of the integral of Eq. (4.38) is possible. The result is

$$m_1(p) = \frac{1}{\pi} \rho (\Gamma_{v0})^2 (\varkappa_0 R_v)^2 \qquad (4.39)$$

(we ignored terms of the order $(\varkappa_0 R_v)^3$ and smaller terms for $\varkappa_0 R_v \ll 1$). Accordingly, we have far from a pointlike source of neutrons:

$$\langle G(\mathbf{r}|\mathbf{r}_0) \rangle = \frac{1}{4\pi |\mathbf{r} - \mathbf{r}_0|} \exp[-\varkappa_{as} |\mathbf{r} - \mathbf{r}_0|], \qquad (4.40)$$

where

$$\varkappa_{as} = \varkappa_0 \left[1 - \frac{(\Gamma_{v0})^2}{\varkappa_0^4 2\pi} \rho (\varkappa_0 R_v)^2\right]. \qquad (4.41)$$

In boiling water (see, e.g., [16]) with a volume concentration of vapor $\rho = 0.2$, the correction to the relaxation constant of the neutron flux is

$$\frac{\Delta \varkappa}{\varkappa_0} = - \frac{\rho}{\pi} \frac{(\Gamma_{v0})^2}{\varkappa_0^4} (\varkappa_0 R_v)^2 \qquad (4.42)$$

i.e., of the order of magnitude (with the average radius of the vapor bubbles assumed as $R_\nu \approx 0.2$ cm)

$$\left|\frac{\Delta \varkappa}{\varkappa_0}\right| \lesssim 10^{-3}$$

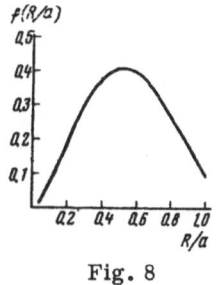

Fig. 8

(see also § 5 and the remark at the end of this section). In deriving this estimate we assumed that the neutron absorption in the vapor bubbles can be ignored. In an aqueous solution of boric acid this correction increases when the concentration of the absorber rises.

e) We take simple plane lattices as defined by Eqs. (3.43) and (3.43a) with

$$\frac{\pi}{a\varkappa_0} \gg 1$$

as an example for a medium with a periodic structure.

In this case [for a lattice as defined by Eq. (3.43)], we have

$$m_1(\mathbf{p}) \approx \frac{v_0^2 \alpha^3}{12}\left(1 - \frac{\alpha}{2a}\right)^2 = \frac{\varepsilon_{vv}}{12}\,\alpha\,(2a-\alpha) \tag{4.43}$$

and in the case of Eq. (3.43a):

$$m_1(\mathbf{p}) \approx \frac{v_0^2 a^2}{48} = \frac{\varepsilon_{vv}\,a^2}{12}. \tag{4.43a}$$

Thus, the correction to the relaxation constant of the neutron flux is in this case independent of the orientation of the plane source relative to the lattice layers. In a first approximation [with respect to $(\varkappa_0 a)^2$], the diffusion remains isotropic, and the anisotropy effect of the diffusion is of the order of magnitude of $\varepsilon_{vv} a^4 \varkappa_0^2$. The correction resulting from the inaccuracy of Eq. (2.21) amounts to $\varepsilon_{vv}^2 a^6$ (see § 8).

The expression defined by Eqs. (4.43) and (4.43a) is equal to the first correction to the relaxation constant of the neutron flux in a homogeneous medium. The same result can be obtained by applying Shevelev's work [13] to lattices of the type defined by Eqs. (3.43) and (3.43a).

f) In the case of the more complicated lattice of Eq. (3.38), we have

$$m_1(\mathbf{p}) \approx \frac{v_0^2 R^2}{4\pi^2 a^2 b^2} \sum_{n=-\infty}^{\infty} \sum_{m=-\infty}^{\infty}{}' \frac{J_1^2\left(\pi R\sqrt{\frac{n^2}{a^2} + \frac{m^2}{b^2}}\right)}{\left(\frac{n^2}{a^2} + \frac{m^2}{b^2}\right)^2}. \tag{4.44}$$

The prime at the summation sign denotes that we have to exclude the term with $n = 0$, $m = 0$. As in the preceding section, the anisotropy manifests itself by the higher order of the small terms of the parameter $\left(\frac{\varkappa_0^{-1}}{\sqrt{a^2 + b^2}}\right)^{-1}$. Figure 8 is a graphical representation of the function

$$f(R/a) = \frac{1}{4} \sum_{n=-\infty}^{\infty} \sum_{m=-\infty}^{\infty}{}' \frac{J_1^2\left(\frac{\pi R}{a}\sqrt{n^2 + m^2}\right)}{(n^2 + m^2)^2}.$$

3. When not only the absorption cross section, but also the scattering cross section fluctuates, one must take into account all terms in the expression for $m_1(\mathbf{p})$ of Eq. (4.13). Under the assumption that all correlation functions have the same dependence on the coordinates

$$K_{ij}(\mathbf{r} - \mathbf{r}') \equiv \varepsilon_{ij} B(\mathbf{r} - \mathbf{r}'), \tag{4.45}$$

the expression for $m_1(\mathbf{p})$ is conveniently rewritten in the form

$$m_1(\mathbf{p}) = p^2 \left\{ \varepsilon_{\eta\eta} + [-\varkappa_0^2 \varepsilon_{\eta\eta} + 2\varepsilon_{\eta\nu}] \frac{1}{(2\pi)^3} \int \frac{d\zeta\, b(\zeta)}{(\mathbf{p}-\zeta)^2 + \varkappa_0^2} + \right.$$

$$\left. + \frac{\varepsilon_{\eta\eta}}{p^2} \frac{1}{(2\pi)^3} \int \frac{d\zeta\, b(\zeta)}{(\mathbf{p}-\zeta)^2 + \varkappa_0^2} [(\mathbf{p}\zeta)^2 - p^2\zeta^2] - \frac{2\varepsilon_{\eta\nu}}{p^2} \frac{1}{(2\pi)^3} \int \frac{d\zeta\, b(\zeta)\,\mathbf{p}\zeta}{(\mathbf{p}-\zeta) + \varkappa_0^2} \right\} + \frac{\varepsilon_{\nu\nu}}{(2\pi)^3} \int \frac{d\zeta\, b(\zeta)}{(\mathbf{p}-\zeta)^2 + \varkappa_0^2}. \qquad (4.46)$$

Let us calculate the integrals of Eq. (4.46) for the case of a lattice defined by Eq. (3.43). At $\pi/a\varkappa_0 \gg 1$ $(p \sim \varkappa_0)$ we obtain

$$m_1(\mathbf{p}) = p^2 \left\{ \varepsilon_{\eta\eta} + \frac{\alpha(2a-\alpha)}{12} [-\varkappa_0^2 \varepsilon_{\eta\eta} + 2\varepsilon_{\eta\nu}] - \frac{p_x^2 + p_y^2}{p^2} \varepsilon_{\eta\eta} + \right.$$

$$\left. + \frac{p_x^2 + p_y^2}{12p^2} \varepsilon_{\eta\eta} \alpha(2a-\alpha)(\varkappa_0^2 + p^2 - 4p_z^2) - \frac{1}{3} \frac{p_z^2}{p^2} \varepsilon_{\eta\nu} \alpha(2a-\alpha) \right\} +$$

$$+ \frac{\varepsilon_{\nu\nu}}{12} \alpha(2a-\alpha) \approx p^2 \left\{ \varepsilon_{\eta\eta}\left(\frac{p_z^2}{p^2} - \frac{\varkappa_0^2 \alpha(2a-\alpha)}{12} \right) + \frac{\varepsilon_{\eta\nu}\alpha(2a-\alpha)}{6}\left(1 - \frac{2p_z^2}{p^2}\right) \right\} + \frac{\varepsilon_{\nu\nu}}{12}\alpha(2a-\alpha). \qquad (4.47)$$

With a lattice defined by Eq. (3.43a) we find

$$m_1(\mathbf{p}) \approx p^2 \left\{ \varepsilon_{\eta\eta}\left[\frac{p_z^2}{p^2} - \frac{\varkappa_0^2 a^2}{12} \right] + \frac{\varepsilon_{\eta\nu}}{6}\left(1 - \frac{2p_z^2}{p^2}\right) \right\} + \frac{\varepsilon_{\nu\nu}a^2}{12}. \qquad (4.47a)$$

Equations (4.47) and (4.47a) can be obtained from the results of Shevelev's work [13].

For $a\varkappa_0 \to 0$ these expressions simplify considerably:

$$m_1(\mathbf{p}) = \varepsilon_{\eta\eta}p_z^2. \qquad (4.48)$$

Thus, it follows from the equation [see Eqs. (4.9) and (4.10)]

$$p^2 + \varkappa_0^2 - m_1(\mathbf{p}) = 0, \qquad (4.49)$$

(this equation determines the attenuation constants of the neutron flux) and from Eq. (4.48) that in the limit $\varkappa_0 a \to 0$, we obtain for the diffusion along the layers $(\mathbf{p} = \{p_x, 0, 0\})$

$$p_x^2 = -\varkappa_0^2 = -\frac{\Sigma_{a0}}{D_0} = -\frac{\Sigma_{a0}}{D_h(1 + \varepsilon_{\eta\eta})} = -\frac{\varkappa_h^2}{1 + \varepsilon_{\eta\eta}}. \qquad (4.50)$$

When neutrons diffuse across the layers $(\mathbf{p} = \{0, 0, p_z\})$, we obtain for $a\varkappa_0 \to 0$

$$p_z^2 = -\frac{\varkappa_0^2}{1 - \varepsilon_{\eta\eta}} = -\frac{\Sigma_{a0}}{D_h} = -\varkappa_h^2. \qquad (4.51)$$

We used the notations $\varkappa_h^2 = \frac{\Sigma_{a0}}{D_h}$, and $D_h = \frac{1}{3\Sigma_0}$ [see Eq. (4.4)]. In the limit of thin layers, the diffusion across the layers is the same as in a homogeneous medium, whereas the correct limit value is not obtained for a constant attenuation of the neutron flux along the layers. These conclusions are identical to the results obtained in [13].

For $\varkappa_0/\beta \to 0$, Eq. (4.49) assumes the following form for an isotropic diffusion medium (isotropic on the average; β^{-1} denotes the correlation length of the fluctuations in the medium):

$$p^2\left(1 - \frac{\varepsilon_{\eta\eta}}{3}\right) + \varkappa_0^2 = 0. \qquad (4.52)$$

The "longitudinal" diffusion in a lattice consisting of cylindrical rods proceeds in the same form as in a plane lattice [see Eq. (4.50)]. We obtain for transverse diffusion

$$p_1^2 \approx - \varkappa_0^2 \left(1 + \frac{\varepsilon_{\eta\eta}}{2}\right) \approx - \varkappa_h^2 \left(1 - \frac{\varepsilon_{\eta\eta}}{2}\right), \qquad (4.53)$$

i.e., the correct limit value is not attained with the attenuation constant of the flux, as in the case of the "longitudinal" diffusion. The result is independent of the shape of the cell: $\varepsilon_{\eta\eta}$ is determined only by the volume concentration of the components of the medium.

This behavior results from the uncritical application of the diffusion equation in the limit $l/L_0 \to 0$ (l denotes the size of an inhomogeneity and L_0 the diffusion length in the inhomogeneous medium). This limit can be analyzed with the exact kinetic equation, which will be discussed in the subsequent paragraph.*

§ 5. Stationary Single-Velocity Kinetic Equation

1. The Green's function of a stationary single-velocity kinetic equation for an unlimited medium with isotropic scattering satisfies the equation

$$- \mathbf{n} \nabla G\left(\mathbf{r}, \mathbf{n} \mid \mathbf{r}_0, \mathbf{n}_0\right) - \Sigma_t(\mathbf{r}) G\left(\mathbf{r}, \mathbf{n} \mid \mathbf{r}_0, \mathbf{n}_0\right) + \frac{\Sigma_s(\mathbf{r})}{4\pi} \int d\Omega' G\left(\mathbf{r}, \mathbf{n}' \mid \mathbf{r}_0, \mathbf{n}_0\right) = - \delta(\mathbf{r} - \mathbf{r}_0) \delta(\mathbf{n} - \mathbf{n}_0) \quad (5.1)$$

($\mathbf{n}$ denotes the unit vector in the direction of the neutron motion).

We introduce an equation as defined in (2.21) for the averaged Green's function $\langle G(\mathbf{r}, \mathbf{n}, \mid \mathbf{r}_0, \mathbf{n}_0) \rangle$. The averaging is extended over the fluctuations of the total cross section $\Sigma_t(\mathbf{r})$ and the scattering cross section $\Sigma_s(\mathbf{r})$.

The operator $\hat{B}$ and the perturbation $\hat{\mu}$ are given by the following self-evident relations:

$$\hat{B} G\left(\mathbf{r}, \mathbf{n} \mid \mathbf{r}_0, \mathbf{n}_0\right) = - [\mathbf{n}\nabla + \Sigma_{t0}] G\left(\mathbf{r}, \mathbf{n} \mid \mathbf{r}_0, \mathbf{n}_0\right) + \frac{\Sigma_{s0}}{4\pi} \int d\Omega' G\left(\mathbf{r}, \mathbf{n}' \mid \mathbf{r}_0, \mathbf{n}_0\right) \qquad (5.2)$$

and

$$\hat{\mu}(\mathbf{r}) G\left(\mathbf{r}, \mathbf{n} \mid \mathbf{r}_0, \mathbf{n}_0\right) = \sigma_t(\mathbf{r}) G\left(\mathbf{r}, \mathbf{n} \mid \mathbf{r}_0, \mathbf{n}_0\right) - \frac{\sigma_s(\mathbf{r})}{4\pi} \int d\Omega' G\left(\mathbf{r}, \mathbf{n}' \mid \mathbf{r}_0, \mathbf{n}_0\right). \qquad (5.3)$$

By substituting $\hat{\mu}(\mathbf{r})$ defined by Eq. (5.3) and $G(\mathbf{r}', \mathbf{n} \mid \mathbf{r}_0, \mathbf{n}_0)$ (see, e.g., the monograph [1]) into the expression for the operator $M_1(\mathbf{r}, \mathbf{n} \mid \mathbf{r}_0, \mathbf{n}_0)$ of Eq. (2.20) we obtain from Eq. (2.21) for an isotropic source the following kinetic equation of the averaged Green's function:

$$- \mathbf{n} \nabla \langle G\left(\mathbf{r}, \mathbf{n} \mid \mathbf{r}_0\right)\rangle - \Sigma_{t0} \langle G\left(\mathbf{r}, \mathbf{n} \mid \mathbf{r}_0\right)\rangle + \frac{\Sigma_{s0}}{4\pi} \int d\Omega' \langle G\left(\mathbf{r}, \mathbf{n}' \mid \mathbf{r}_0\right)\rangle =$$

$$= - \delta(\mathbf{r} - \mathbf{r}_0) - \iint d\mathbf{r}' d\Omega' \{ K_{tt}(\mathbf{r} \mid \mathbf{r}') G_0\left(\mathbf{r}, \mathbf{n} \mid \mathbf{r}', \mathbf{n}'\right) \langle G\left(\mathbf{r}', \mathbf{n}' \mid \mathbf{r}_0\right)\rangle -$$

$$- \frac{K_{ts}(\mathbf{r} \mid \mathbf{r}')}{4\pi} \int d\Omega'' G_0\left(\mathbf{r}, \mathbf{n}'' \mid \mathbf{r}', \mathbf{n}'\right) \langle G\left(\mathbf{r}', \mathbf{n}' \mid \mathbf{r}_0\right)\rangle - \frac{K_{ts}(\mathbf{r} \mid \mathbf{r}')}{4\pi} \int d\Omega'' G_0\left(\mathbf{r}, \mathbf{n} \mid \mathbf{r}', \mathbf{n}'\right) \langle G\left(\mathbf{r}', \mathbf{n}'' \mid \mathbf{r}_0\right)\rangle +$$

---

*It follows from the results of this section that the correction (which results from fluctuations of the scattering cross section) to the constant attenuation of the neutron flux can considerably exceed the correction resulting from fluctuations of the absorption cross section. This point must be recalled when one analyzes the transition of neutrons through a boiling liquid (see 2d, § 4). In this case, the parameters are usually such that the diffusion equation cannot be employed; one must employ the kinetic equation (see 2 of § 5).

$$+ \frac{K_{ss}(\mathbf{r}\,|\,\mathbf{r}')}{4\pi} \int d\Omega'' G_0(\mathbf{r}, \mathbf{n}''\,|\,\mathbf{r}', \mathbf{n}') \int d\Omega''' \langle G(\mathbf{r}, \mathbf{n}'''\,|\,\mathbf{r}_0) \rangle \}. \tag{5.4}$$

When we take the Fourier transform with respect to $\mathbf{r} - \mathbf{r}_0$, the above equation assumes the form

$$(i\mathbf{p}\mathbf{n} - \Sigma_{t0}) \langle g(\mathbf{p}, \mathbf{n}) \rangle + \frac{\Sigma_{s0}}{4\pi} \langle g(\mathbf{p}) \rangle = -1 - \frac{1}{(2\pi)^3} \int d\zeta k_{tt}(\zeta) \int d\Omega' g_0(\mathbf{p} - \zeta, \mathbf{n}\,|\,\mathbf{n}') \langle g(\mathbf{p}, \mathbf{n}') \rangle +$$

$$+ \frac{1}{(2\pi)^3} \int d\zeta\, \frac{k_{ts}(\zeta)}{4\pi} \int d\Omega' g_0(\mathbf{p} - \zeta, \mathbf{n}') \langle g(\mathbf{p}, \mathbf{n}') \rangle +$$

$$+ \frac{1}{(2\pi)^3} \int d\zeta \frac{k_{ts}(\zeta)}{4\pi} g_0(\mathbf{p} - \zeta, \mathbf{n}) \langle g(\mathbf{p}) \rangle - \frac{1}{(2\pi)^3} \int d\zeta\, \frac{k_{ss}(\zeta)}{16\pi^2} g_0(\mathbf{p} - \zeta) \langle g(\mathbf{p}) \rangle. \tag{5.5}$$

The following notations were introduced:

$$\langle g(\mathbf{p}, \mathbf{n}) \rangle = \int d\mathbf{r} e^{i\mathbf{p}\mathbf{r}} \langle G(\mathbf{r}, \mathbf{n}) \rangle,$$

$$\langle g(\mathbf{p}) \rangle = \int d\Omega \langle g(\mathbf{p}, \mathbf{n}) \rangle.$$

In the zeroth approximation, the Fourier image of the Green's function of the kinetic equation has the form

$$g_0(\mathbf{p}, \mathbf{n}\,|\,\mathbf{n}') = -\frac{\delta(\mathbf{a} - \mathbf{n}')}{i\mathbf{p}\mathbf{n} - \Sigma_{t0}} + \frac{\Sigma_{s0}}{4\pi} \frac{1}{i\mathbf{p}\mathbf{n} - \Sigma_{t0}} \left[1 + \frac{\Sigma_{s0}}{4\pi} \int \frac{d\Omega''}{i\mathbf{p}\mathbf{n}'' - \Sigma_{t0}} \right]^{-1} \frac{1}{i\mathbf{p}\mathbf{n}' - \Sigma_{t0}}, \tag{5.6}$$

$$g_0(\mathbf{p}, \mathbf{n}) = \int d\Omega' g_0(\mathbf{p}, \mathbf{n}\,|\,\mathbf{n}'), \tag{5.6a}$$

$$g_0(\mathbf{p}) = \int d\Omega g_0(\mathbf{p}, \mathbf{n}). \tag{5.6b}$$

The correlation functions of the fluctuations of the cross sections are respectively equal to

$$k_{tt}(\mathbf{p}) = \int d\mathbf{r} e^{-i\mathbf{p}\mathbf{r}} K_{tt}(\mathbf{r}),$$

$$K_{tt}(\mathbf{r}) = \langle \sigma_t(\mathbf{r}') \sigma_t(\mathbf{r} + \mathbf{r}') \rangle$$

etc. When l (the characteristic size of the inhomogeneities of the medium) tends to zero (with the level of fluctuations ε_{ij} unchanged), the integral on the right side of Eq. (5.5) vanishes and this equation coincides with the kinetic equation for a homogeneous medium in which neutrons diffuse in all directions. This result can be obtained when we recall that $\frac{1}{(2\pi)^3} \int k_{ij}(\zeta) d\zeta = \varepsilon_{ij}$ is independent of the size of the inhomogeneities, whereas the function $g_0(\mathbf{p} - \zeta, \mathbf{n}\,|\,\mathbf{n}')$ is proportional to l for $l \to 0$. More precisely, the necessary condition is

$$\zeta \mathbf{n} \sim l_{\text{eff}}^{-1} \gg \Sigma_{t0} \gg \varkappa_0. \tag{5.7}$$

The effective path length traveled by a neutron in the homogeneous region is denoted by l_{eff}; for example, in the case of the plane lattice defined by Eq. (3.43a), we have $l_{\text{eff}} = a/\sin\theta$, where θ denotes the angle between the direction of motion of the neutron (vector $\mathbf{n}$) and the plane of the layer. Inequality (5.7) (condition of small-scale fluctuations) is not satisfied for $\zeta\mathbf{n} = 0$, i.e., for neutrons moving parallel to the lattice layers. In this particular case, a more accurate analysis must be performed (see below, 3).

2. We use spherical harmonics to find a solution of Eq. (5.5). We have, in a first

approximation

$$\langle g\,(\mathbf{p},\,\mathbf{n})\rangle = [\tfrac{1}{4\pi}\,\langle g\,(\mathbf{p})\rangle + 3\mathbf{n}\,\langle g_1\,(\mathbf{p})\rangle]. \tag{5.8}$$

With standard calculation techniques (see, e.g., [17]) we obtain the following equation, which determines the relaxation constant of the neutron flux:*

$$[\langle g\,(\mathbf{p})\rangle]^{-1} = p^2 + \varkappa_h^2 - \frac{3}{(2\pi)^3}\int \frac{d\zeta}{4\pi}\Big\{\Sigma_{t0}k_{aa}\,(\zeta)\,g_0\,(\mathbf{p}-\zeta) \,+\, 2k_{ta}\,(\zeta)\int d\Omega g_0\,(\mathbf{p}-\zeta,\,\mathbf{n})\,(i\mathbf{p}\mathbf{n}) +$$

$$+\,\frac{k_{tt}\,(\zeta)}{\Sigma_{t0}}\iint d\Omega d\Omega'\,(i\mathbf{p}\mathbf{n})\,g_0\,(\mathbf{p}-\zeta,\,\mathbf{n}\,|\,\mathbf{n}')\,(i\mathbf{p}\mathbf{n}')\Big\} = 0, \tag{5.9}$$

$$\varkappa_h^2 = 3\Sigma_{a0}\Sigma_{t0} \approx 3\Sigma_{a0}\Sigma_{s0} = 3/\lambda_{s0}\lambda_{a0}.$$

In deriving this equation, we neglected the contribution of the terms $\infty\,(\lambda_{s0}\varkappa_r)^2 \sim \lambda_{s0}/\lambda_{a0}$ to the relaxation constant, which is usually done in the diffusion approximation [1, 17]. At small inhomogeneities of the medium, Eq. (5.9), as well as Eq. (5.5), lead to the correct limit value of the relaxation constant.

Equation (5.9) accounts for the effect of the inhomogeneities with the same degree of accuracy as Eq. (5.5), i.e., in the approximation given by Eq. (2.21). The Fourier component of the neutron current $\langle g_1(\mathbf{p})\rangle$ is related to the Fourier component of the flux $\langle g(\mathbf{p})\rangle$ by the following relation:

$$\langle g_1\,(\mathbf{p})\rangle = \frac{\langle g\,(\mathbf{p})\rangle}{3\Sigma_{t0}}\Big\{i\mathbf{p} + \frac{3}{(2\pi)^3}\int\frac{d\zeta}{4\pi}\,k_{ta}\,(\zeta)\int d\Omega \mathbf{n} g_0\,(\mathbf{p}-\zeta,\,\mathbf{n}) +$$

$$+\,\frac{3}{(2\pi)^3}\int\frac{d\zeta}{4\pi}\frac{k_{tt}\,(\zeta)}{\Sigma_{t0}}\iint d\Omega d\Omega'\mathbf{n} g_0\,(\mathbf{p}-\zeta,\,\mathbf{n}\,|\,\mathbf{n}')\,(i\mathbf{p}\mathbf{n}')\Big\}, \tag{5.10}$$

i.e., we have

$$\langle g_1\,(\mathbf{p})\rangle = \hat{R}\,(\mathbf{p})\,\langle g\,(\mathbf{p})\rangle. \tag{5.10a}$$

Operator $\hat{R}$ is directly defined by Eq. (5.10). In general, in the coordinate representation, the relation between the average neutron current $\langle \mathbf{j}\rangle$ and the flux averaged over the fluctuations in the medium is an integral relation (though we use the P_1 approximation for $\langle g(\mathbf{p},\,\mathbf{n})\rangle$,

$$\langle \mathbf{j}(\mathbf{r})\rangle = \int d\mathbf{r}'\hat{R}\,(\mathbf{r}-\mathbf{r}')\,\langle G\,(\mathbf{r}')\rangle, \tag{5.10b}$$

and at the point $\mathbf{r}$ the quantity $\langle \mathbf{j}(\mathbf{r})\rangle$ depends not only on the flux at this point, but also upon the value of $\langle G\rangle$ in the region in which the operator $\hat{R}$ is nonlocal. The size of this region is defined by the scale of the fluctuations in the medium.†

We consider Eq. (5.9) in the following limits:

a) $\lambda_{t0} \approx \lambda_{s0} \ll l \ll \varkappa_h^{-1} \equiv I_h$.

In this limit, the denominator in the expression for $g_0(\mathbf{p}-\zeta,\,\mathbf{n}\,|\,\mathbf{n}')$ of Eq. (5.6) can be expanded in a series of powers of the small ratio λ_{s0}/l at $|\mathbf{p}-\zeta|,\,\mathbf{p} \ll \Sigma_{t0}$. When we consider only the large terms of this expansion, we obtain an expression for the operator of Eq. (4.13)

---

*In deriving Eq. (5.9) we omitted terms $\infty\,\epsilon_{ij}^2$.

† The operator $\hat{R}$ in Eq. (5.10b) depends upon the angle between the direction of diffusion and the orientation of the inhomogeneities in the medium. The consequence is that the extrapolated dimensions of the volume of a moderator can depend upon the positions of the inhomogeneities in the substance.

which was also obtained in § 4 by averaging over the diffusion equation. Thus, the applicability range of the approximation of Eq. (4.13) is determined and a possibility for refining the applicability range by using the ensuing terms in the series in powers of λ_{s0}/l is created.*

When only the scattering cross section fluctuates, Eq. (5.9) can be approximated in the form†

$$p^2 + \varkappa_h^2 - \frac{1}{(2\pi)^3}\int d\zeta\, \frac{k_{ss}(\zeta)}{\Sigma_{t0}^2}\left\{p^2\left[1 - \frac{2}{5}\frac{(\zeta p)^2}{\Sigma_{t0}^2 p^2} - \frac{1}{5}\frac{\zeta^2}{5\Sigma_{t0}^2}\right] + \right.$$
$$\left. + \frac{[p(p-\zeta)]^2}{\Sigma_{t0}\zeta^3}\left[1 + \frac{2p\zeta}{\zeta^2} + \frac{4(p\zeta^2)}{\zeta^4} - \frac{3}{5}\frac{\zeta^2}{\Sigma_{t0}^2}\right]\right\} = 0.$$

We omitted terms proportional to $(\zeta/\Sigma_{t0})^4$ and higher-order terms. In the case of plane lattices defined by Eqs. (3.43) and (3.43a), the convergence region of the expansion of the "mass operator" in Eq. (5.9) (expansion in powers of ζ/ζ_{t0}) shrinks to a point, and the diffusion approximation cannot be defined in a more accurate form.

b) $l \ll \lambda_{s0} \approx \lambda_{t0} \ll \varkappa_h^{-1} = L_h.$

When only the absorption cross section fluctuates, the correction to the relaxation constant has an absolute value of the order $\varkappa_h\frac{\varepsilon_{aa}l}{\Sigma_{a0}} < \varkappa_h l/\lambda_{a0}$ and for $l \ll \lambda_{s0}$ the correction is much smaller than the nondiffusion correction, whose order of magnitude is $\varkappa_h\frac{\lambda_{s0}}{\lambda_{a0}}$. Therefore, we must return to Eq. (5.5):

$$(ipn - \Sigma_{t0})\langle g(p, n)\rangle + \frac{\Sigma_{s0}}{4\pi}\langle g(p)\rangle = -1 - \frac{1}{(2\pi)^3}\int d\zeta k_{aa}(\zeta)\int d\Omega' g_0(p-\zeta;\ n\,|\,n')\langle g(p, n')\rangle. \quad (5.5a)$$

In order to obtain a first correction for the inhomogeneity ($\sim l$) it suffices to assume

$$g_0(p-\zeta,\ n\,|\,n') \approx -\frac{\delta(n-n')}{i(p-\zeta)n - \Sigma_{t0}} \approx \frac{\delta(n-n')}{i\zeta n + \Sigma_{t0}}. \quad (5.11)$$

By substituting Eq. (5.11) into Eq. (5.5a) we obtain

$$(ipn - \Sigma_{t1})\langle g(p, n)\rangle + \frac{\Sigma_{s0}}{4\pi}\langle g(p)\rangle = -1, \quad (5.5b)$$

where

$$\Sigma_{t1} = \Sigma_{t0} - \frac{1}{(2\pi)^3}\int \frac{d\zeta k_{aa}(\zeta)}{i\zeta n + \Sigma_{t0}}. \quad (5.12)$$

The form of Eq. (5.5b) agrees with the equation for the zeroth approximation. In the case of an isotropic medium and the correlation function stated in Eq. (3.4) we have

$$\Sigma_{t1} = \Sigma_{t0} - \frac{\varepsilon_{aa}}{q_{aa}}. \quad (5.13)$$

The correction to the relaxation constant amounts to several percent ($\varepsilon_{aa} \approx 0.5/\lambda_{a0}^2$) at $\lambda_{a0} \approx 5$ cm (passage of neutrons through iron) when $l \approx q_{aa}^{-1} \approx 0.5$ cm.

---

*Another way of improving the diffusion approximation is to write the Green's function of the zeroth approximation as a sum of a diffusion kernel and a single-collision kernel (see the monograph of Weinberg and Wigner [2], p. 205).

† The function g_0^{diff} was expanded in powers of p/ζ. This expansion can be made when the correlation function $k_{ss}(\zeta)$ is characterized by sharp maxima.

In the case of an anisotropic medium, Σ_{t1} of Eq. (5.12) depends upon the direction of motion of the neutron. For example, in the case of a lattice defined by Eq. (3.43a), we have

$$\Sigma_{t1} = \Sigma_{s0} + \Sigma_{a0} \left[1 - \frac{\varepsilon_{aa}}{\Sigma_{a0}^2} \left(\frac{a \varkappa_0}{6 n_z} \right)^2 \right].$$

(5.14)

We use the approximation

$$\frac{1}{4\pi} \langle g(\mathbf{p}) \rangle \approx \frac{1}{3} \sum_{i=x,\, y,\, z} \langle g(\mathbf{p}, n_i) \rangle.$$

(5.15)

This means that we classify all neutrons into three groups and that the direction of the velocity vector of each group is parallel to the corresponding coordinate axis. In this approximation, Eq. (5.5b) reduces to a system of algebraic equations. We note that in the case of lattices which are more complicated than the lattice defined by Eq. (3.43a), additional errors are made whenever the neutrons move parallel to the layers.*

When only the scattering cross section fluctuates, we obtain $k_{aa} = k_{ta} \equiv 0$ and Eq. (5.9) simplifies:

$$p^2 + \varkappa_h^2 - \frac{3 e_{ss}}{\Sigma_{t0}} \frac{1}{(2\pi)^3} \int \frac{d\zeta b(\zeta)}{4\pi} \int\int d\Omega d\Omega' \, (i\mathbf{p}\mathbf{n}) \, g_0 (\mathbf{p} - \zeta, \, \mathbf{n} \,|\, \mathbf{n}') \, (i\mathbf{p}\mathbf{n}') = 0.$$

(5.16)

$$k_{ss}(\zeta) = \varepsilon_{ss} b(\zeta).$$

In the approximation which includes the square of the size of the inhomogeneity $[\sim (l/\lambda_{s0})^2]$ this equation assumes the form†

$$p^2 + \varkappa_h^2 + \frac{3}{\Sigma_{t0}} \frac{1}{(2\pi)^3} \int \frac{d\zeta b(\zeta)}{4\pi} \left\{ \frac{\pi^2}{\zeta} \left[p^2 - \frac{(\mathbf{p}\zeta)^2}{\zeta^2} \right] - \frac{4\pi \Sigma_{t0}}{\zeta^2} \left[p^2 - \frac{2(\mathbf{p}\zeta)^2}{\zeta^2} \right] \right\} = 0,$$

(5.17)

When the approximation goes to the linear term of the dimension of the inhomogeneity, we obtain for the isotropic medium of Eq. (3.4)

$$p^2 = -\varkappa_h^2 \frac{1}{1 + \dfrac{\varepsilon_{ss}}{q \Sigma_{t0}}}.$$

(5.18)

Let us use this expression for estimating the order of magnitude of the correction to the relaxation constant of the neutron flux in a boiling liquid. We neglect the scattering by the vapor bubble and assume $q^{-1} \approx R_n$, where R_n denotes the radius of the vapor bubble. We obtain from Eq. (5.18):

$$p^2 = -\varkappa_h^2 \frac{1}{1 + \dfrac{\rho R_n}{\lambda_{s0}}},$$

(5.18a)

where ρ denotes the vapor content per unit volume. Thus, in cases of practical interest, the correction to $\varkappa_h$ can amount to several percent.

For a lattice of the type defined by Eq. (3.43a) we obtain for diffusion along the layers

---

*When the motion of neutrons along the layers of the lattice is considered, the function g_0 defined by the first equation of (5.11) must be used in the expression for Σ_{t1}.

†See footnote † on page 220.

$$p_{\parallel}^2 = -\varkappa_h^2 \left[1 + \frac{6}{\pi^2} \frac{\varepsilon_{ss} a}{\Sigma_{t0}} \sum_{n=1}^{\infty} \frac{1}{(2n-1)^3} \right]^{-1}. \tag{5.19}$$

Whereas for diffusion across the layers this approximation leads to

$$p_{\perp}^2 = -\varkappa_h^2. \tag{5.20}$$

For this system, the anisotropy of the diffusion is given by the formula

$$\frac{L_{\parallel}^2}{L_{\perp}^2} = 1 + \frac{6}{\pi^2} \frac{\varepsilon_{ss} a}{\Sigma_{t0}} \sum_{n=1}^{\infty} \frac{1}{(2n-1)^3} \approx 1 + 0.64 \frac{\varepsilon_{ss} a}{\Sigma_{t0}}. \tag{5.21}$$

3. In concluding this paragraph, let us discuss the solution to the problem of neutron transition through an inhomogeneous absorbing medium ($\Sigma_s = 0$). In this case, the first equation in (5.11) is rigorous (at $\Sigma_{t0} = \Sigma_{a0}$), and Eq. (5.5a) has the solution

$$\langle g(\mathbf{p}, \mathbf{n}) \rangle = -\left[i\mathbf{p}\mathbf{n} - \Sigma_{a0} - \frac{1}{(2\pi)^3} \int \frac{d\zeta k_{aa}(\zeta)}{i\mathbf{n}(\mathbf{p}-\zeta) - \Sigma_{a0}} \right]^{-1}. \tag{5.22}$$

In the case of small-scale fluctuations, the inverse Fourier transform can be given in analytical form and an expression for the averaged Green's function results ($\mathbf{n}\zeta \neq 0$):

$$\langle G(\mathbf{r}, \mathbf{n}) \rangle = \frac{\delta\left(\frac{\mathbf{r}}{r} - \mathbf{n}\right)}{r^2} \exp\left\{ -r\left[\Sigma_{a0} - \frac{1}{2\pi} \int \frac{d\zeta k_{aa}(\zeta)}{i\mathbf{n}\zeta + \Sigma_{a0}} \right] \right\}. \tag{5.23}$$

The same expression can be obtained by averaging the accurate solution over the normal distribution of the random function $\sigma_a(\mathbf{r})$ [18].

Special considerations are necessary for neutrons moving parallel to the layers, because in this case, Eq. (5.23) cannot be used. Equation (5.22) correctly describes the motion of neutrons moving parallel to the layers of a lattice of the type defined by Eq. (3.43a), but Eq. (5.22) is incorrect in other cases: when the lattice consists of alternating layers of three types, or when the layered medium is characterized by Gaussian fluctuations, etc., the Fourier transform of the average flux of neutrons moving parallel to the layers is given by the expression

$$\langle g(\mathbf{p}, n_x) \rangle = -\left\langle \frac{1}{i p_x n_x - \Sigma_{n0} - \sigma_i} \right\rangle, \tag{5.24}$$

where $\Sigma_{ai} = \Sigma_{a0} + \sigma_{ai}$ denotes the neutron-absorption cross section of the i-th layer.

With the self-evident equation

$$\frac{1}{a} = \int_0^{\infty} dt e^{-at}$$

we can rewrite Eq. (5.24) in the following form:

$$\langle g(\mathbf{p}, n_x) \rangle = i \int_0^{\infty} dt e^{-t(p_x n_x + i\Sigma_{a0})} F_W(-t). \tag{5.25}$$

We used the notation

$$F_W(t) = \int d\sigma_a e^{i\sigma_a t} W(\sigma_a), \tag{5.26}$$

which denotes the characteristic function of the distribution $W(\sigma_a)$. In the case of a normal distribution, we have

$$W(\sigma)\,d\sigma = (\sqrt{\pi\Delta^2})^{-1}e^{-\sigma^2/\Delta^2}d\sigma. \tag{5.27}$$

The characteristic function is

$$F_W(t) = e^{-t^2\Delta^2/4}, \tag{5.28}$$

and the Fourier transform of the average neutron flux is

$$\langle g\,(\mathbf{p},\,n_x)\rangle = e^{u^2}i\,\sqrt{\frac{\pi}{\Delta^2}}\,[1 - \Phi(u)], \tag{5.29}$$

where

$$u = \frac{p_x n_x + i\Sigma_{a0}}{\Delta} \tag{5.30}$$

and

$$\Phi(x) = \frac{2}{\sqrt{\pi}}\int_0^x e^{-t^2}dt. \tag{5.31}$$

We obtain from Eq. (5.29) for $\Delta \to 0$

$$\langle g\,(\mathbf{p},\,n_x)\rangle = e^{u^2}i\,\sqrt{\frac{\pi}{\Delta^2}}\,\frac{e^{-u^2}}{\sqrt{\pi}u} = i\,\frac{1}{\Delta u} = -\frac{1}{i p_x n_x - \Sigma_{a0}}, \tag{5.29a}$$

i.e., the Fourier transform of the neutron flux in a homogeneous medium.*

§6. Nonstationary Diffusion of Thermal Neutrons

1. In our discussion of the transition of thermal neutrons delivered by a nonstationary source in an inhomogeneous medium, we restrict ourselves to the diffusion approximation. In this approximation, the neutron flux satisfies the equation

$$-\frac{1}{vD_0}\frac{\partial}{\partial t}\,G\,(\mathbf{r},\,t\,|\,\mathbf{r}_0,\,t_0) + \mathrm{div}\left[\frac{D(\mathbf{r})}{D_0}\,\mathrm{grad}\,G\,(\mathbf{r},\,t\,|\,\mathbf{r}_0,\,t_0)\right] -$$

$$-\frac{\Sigma_a(\mathbf{r})}{D_0}\,G\,(\mathbf{r},\,t\,|\,\mathbf{r}_0,\,t_0) = -(S/D_0)\delta\,(\mathbf{r} - \mathbf{r}_0)\,\delta\,(t - t_0), \tag{6.1}$$

where S denotes the source output (we assume that $S = D_0$, for the sake of simplicity). The operator $\hat{B}$ has the form

$$\hat{B} = -\frac{1}{vD_0}\cdot\frac{\partial}{\partial t} + \nabla^2 - \varkappa_0^2, \tag{6.2}$$

where v denotes the neutron velocity.

The perturbation resulting from the inhomogeneity of the medium is described by Eq. (4.6) and is time-independent. Thus, by using the Laplace transform with respect to $t - t_0$

$$G\,(\mathbf{r}\ \mathbf{r}_0;\,\lambda) = \int_0^\infty d\,(t - t_0)\,e^{-\lambda(t-t_0)}G\,(\mathbf{r},\,t\,|\,\mathbf{r}_0,\,t_0), \tag{6.3}$$

---

*The limit $\Delta \to \infty$ has no meaning in terms of physics in the case of a purely absorbing medium.

Eq. (6.1) can be rewritten as an equation which defines the stationary diffusion of Eq. (4.1). Hence, the averaged Green's function $\langle G(\mathbf{r}|\mathbf{r}_0; \lambda)\rangle$ satisfies the equation

$$\hat{B}_\lambda \langle G(\mathbf{r}|\mathbf{r}_0; \lambda)\rangle = -\delta(\mathbf{r}-\mathbf{r}_0) - \int d\mathbf{r}' M_1^\lambda(\mathbf{r}|\mathbf{r}') \langle G(\mathbf{r}'|\mathbf{r}_0; \lambda)\rangle. \tag{6.4}$$

We have the notation

$$\hat{B}_\lambda = \nabla^2 - \varkappa_\lambda^2, \tag{6.5}$$

$$\varkappa_\lambda^2 = \varkappa_0^2 + \frac{\lambda}{vD_0}, \tag{6.6}$$

$$M_1^\lambda(\mathbf{r}|\mathbf{r}') = \langle \hat{\mu}(\mathbf{r}) G_0^\lambda(\mathbf{r}|\mathbf{r}') \hat{\mu}(\mathbf{r}')\rangle. \tag{6.7}$$

As usual, the Green's function of the zeroth approximation is the solution of the equation

$$\hat{B}_\lambda G_0^\lambda(\mathbf{r}|\mathbf{r}_0) = -\delta(\mathbf{r}-\mathbf{r}_0). \tag{6.8}$$

2. In the case of an infinite medium which is uniform in space, it is suitable to take the Fourier transform of Eq. (6.4) with respect to $\mathbf{r}-\mathbf{r}_0$ (see § 4):

$$\langle g(\mathbf{p}, \lambda)\rangle = \int e^{i\mathbf{p}(\mathbf{r}-\mathbf{r}')} \langle G(\mathbf{r}|\mathbf{r}'; \lambda)\rangle \, d(\mathbf{r}-\mathbf{r}'). \tag{6.9}$$

The result is [see Eq. (4.9)]

$$\langle g(\mathbf{p}, \lambda)\rangle = [p^2 + \varkappa_\lambda^2 - m_1(\mathbf{p}, \lambda)]^{-1}, \tag{6.10}$$

where $m_1(\mathbf{p}, \lambda)$ is defined by Eqs. (4.13) or (4.46) with the corresponding replacement of $\varkappa_0^2$ by $\varkappa_\lambda^2$. Obviously, we can use the result of § 4. When only the absorption cross section fluctuates and the fluctuations are isotropic and can be described by the correlation function of Eq. (3.4), we have

$$\langle g(\mathbf{p}, \lambda)\rangle = \left[p^2 + \varkappa_\lambda^2 - \frac{\varepsilon_{vv}}{p^2+(q+\varkappa_\lambda)^2}\right]^{-1}. \tag{6.11}$$

For small-scale fluctuations with $q \gg \varkappa_\lambda$, p, we have*

$$\langle g(\mathbf{p}, \lambda)\rangle \approx \left[p^2 + \varkappa_\lambda^2 - \frac{\varepsilon_{vv}}{q^2}\right]^{-1} = \left[p^2 + \varkappa_0^2\left(1-\frac{\varepsilon_{vv}}{q^2\varkappa_0^2}\right) + \frac{\lambda}{vD_0}\right]^{-1}. \tag{6.12}$$

We take the inverse Fourier transform and obtain

$$\langle G(\mathbf{r}|\mathbf{r}_0; \lambda)\rangle = \frac{1}{(2\pi)^3}\int \langle g(\mathbf{p}, \lambda)\rangle e^{-i\mathbf{p}(\mathbf{r}-\mathbf{r}_0)}d\mathbf{p} = [4\pi|\mathbf{r}-\mathbf{r}_0|]^{-1}\exp(-\varkappa_\lambda|\mathbf{r}-\mathbf{r}_0|), \tag{6.13}$$

where

$$\varkappa_\lambda^2 = \varkappa_0^2\left(1-\frac{\varepsilon_{vv}}{q^2\varkappa_0^2}\right) + \lambda/vD_0. \tag{6.14}$$

The integral

$$\langle G(\mathbf{r}-\mathbf{r}_0; t-t_0)\rangle = \frac{1}{2\pi i}\int_{\sigma-i\infty}^{\sigma+i\infty} d\lambda e^{\lambda(t-t_0)}\langle G(\mathbf{r}-\mathbf{r}_0; \lambda)\rangle \tag{6.15}$$

_____
*This condition is violated at small times $t-t_0$.

must be calculated in order to obtain a final expression for the average Green function $\langle G(\mathbf{r} - \mathbf{r}_0; t - t_0)\rangle$. After substituting Eq. (6.13) into Eq. (6.15) and replacing the integration variable

$$y = \frac{\lambda\,(\mathbf{r} - \mathbf{r}_0)^2}{vD_0} + \varkappa_0^2\,(\mathbf{r} - \mathbf{r}_0)^2\left(1 - \frac{\varepsilon_{vv}}{\varkappa_0^2 q^2}\right),$$

we obtain

$$\langle G(\mathbf{r} - \mathbf{r}_0;\ t - t_0)\rangle = \frac{vD_0}{4\pi|\mathbf{r} - \mathbf{r}_0|^2}\,e^{-vD_0(t-t_0)\varkappa_0^2\left(1 - \frac{\varepsilon_{vv}}{\varkappa_0^2 q^2}\right)} \times$$

$$\times \frac{1}{2\pi i}\int_{\sigma-i\infty}^{\sigma+i\infty} dy\, e^{\frac{vD_0 t}{|\mathbf{r}-\mathbf{r}_0|^2}\,y - \sqrt{y}} = e^{\varepsilon_{vv}\frac{vD_0}{q^2}(t-t_0)}\,G_0(\mathbf{r} - \mathbf{r}_0;\ t - t_0). \qquad (6.16)$$

In this approximation, the spatial dependence of the neutron flux remains constant on the average, but the attenuation rate of the flux decreases in time relative to the case of a homogeneous medium.

In the case of the periodic lattice defined by Eq. (3.43a), we have

$$m_1(\mathbf{p}, \lambda) = p^2\left[\varepsilon_{\eta\eta}\left(\frac{p_z^2}{p^2} - \frac{\varkappa_\lambda^2 a^2}{12}\right) + \frac{\varepsilon_{v\eta} a^2}{6}\left(1 - \frac{p_z^2}{p^2}\right)\right] + \frac{\varepsilon_{vv} a^2}{12} \qquad (6.17)$$

[see Eq. (4.47a)]. When the lattice is built from very thin layers ($a\varkappa_0 \ll 1$, but $a/\lambda_s \gg 1$ so that diffusion theory can be employed), we have

$$m_1(\mathbf{p}, \lambda) = \varepsilon_{\eta\eta} p_z^2, \qquad (6.18)$$

and in the case of diffusion along the layers, we find

$$\langle G(x - x_0;\ t - t_0)\rangle = vD_0\frac{e^{-\frac{(x-x_0)^2}{4vD_0\,(t-t_0)}}}{[4\pi vD_0\,(t - t_0)]^{1/2}}\,e^{-vD_0\varkappa_0^2\,(t-t_0)}. \qquad (6.19)$$

In the case of neutron diffusion across the layers, we obtain

$$\langle G(z - z_0;\ t - t_0)\rangle = vD_0\frac{e^{-\frac{(z-z_0)^2}{4vD_h\,(t-t_0)}}}{[4\pi vD_h\,(t - t_0)]^{1/2}}\,e^{-vD_0\varkappa_0^2(t-t_0)}. \qquad (6.20)$$

Thus, in the case of thin layers the time-dependent neutron-flux reduction rate is the same for longitudinal and transverse diffusion. The form of the neutron distribution depends upon the direction of diffusion in this approximation. The width of the neutron distribution depends upon the orientation of the plane source relative to the lattice planes.

3. Let us consider the neutron diffusion from a nonstationary source in a medium of finite volume. We search for solutions to Eq. (6.4) in the form of an expansion over the total system of functions $\varphi_n(\mathbf{r})$, which satisfies the boundary condition, i.e., we look for solutions

$$\langle G(\mathbf{r}\,|\,\mathbf{r}_0;\ \lambda)\rangle = \sum_n g_n(\mathbf{r}_0, \lambda)\,\varphi_n(\mathbf{r}), \qquad (6.21)$$

$$(\nabla^2 + \alpha_n^2)\,\varphi_n(\mathbf{r}) = 0. \qquad (6.22)$$

The boundary conditions are the same as for a homogeneous medium.

The expansion coefficients $g_n(r_0, \lambda)$ satisfy the following system of equations:

$$\sum_n g_n(r_0, \lambda)\left[(\varkappa_\lambda^2 + \alpha_n^2)\,\delta_{kn} - m_{kn}^\lambda\right] = \varphi_k(r_0),\tag{6.23}$$

where

$$m_{kn}^\lambda = \iint d\mathbf{r}\,d\mathbf{r}'\varphi_k(\mathbf{r})\,M_1^\lambda(\mathbf{r}\,|\,\mathbf{r}')\,\varphi_n(\mathbf{r}').\tag{6.24}$$

In a first approximation, the coefficient of the principal harmonic has the form

$$g_1(\mathbf{r}_0,\ \lambda) = \frac{\varphi_1(r_0)}{\varkappa_\lambda^2 + \alpha_1^2 - m_{11}^\lambda}.\tag{6.25}$$

Correspondingly, the attenuation constant of the principal harmonic decreases by m_{11}^λ. In the case of a plane source in a layered medium, this addition depends upon the source orientation relative to the layers, and the anisotropy of the neutron diffusion is taken into account in this fashion. When we set in the equations of this section $\lambda = 0$, we obtain expressions corresponding to stationary neutron diffusion in a finite volume of the medium.

We note in conclusion in this paragraph that in a preceding publication of the author [11] a nonstationary plane source of thermal neutrons in a layer with random inhomogeneities was discussed. It was assumed that the correlation length of the fluctuations is very small relative to the layer thickness [correction function proportional to $\sim\delta(z - z_0)$]. It was shown that the principal spatial harmonic is characterized by two attenuation constants. This results from the special form of the correlation function which was used. In the case of a periodic structure, the principal eigenvalue does not split into two close values. The attenuation constant decreases as in the stationary case.

§ 7. Energy Dependence of the Neutron
Distribution Function

1. In the aging approximation [1, 2], the neutron transfer in energy regions above the thermal range obeys a diffusion equation of the type

$$\xi_0 \frac{\partial F(\mathbf{r},\ u)}{\partial u} = \operatorname{div} D(\mathbf{r})\operatorname{grad}\left\langle\!\!\left\langle \frac{1}{\Sigma_t(\mathbf{r})} \right\rangle\!\!\right\rangle F(\mathbf{r},\ u) - \left\langle\!\!\left\langle \frac{\Sigma_a}{\Sigma_t} \right\rangle\!\!\right\rangle F(\mathbf{r},\ u) = -S\delta(\mathbf{r} - \mathbf{r}_0)\delta(u - u_0),\tag{7.1}$$

where $F(\mathbf{r},\ u) = E\Sigma_t(\mathbf{r},\ E)\,\Phi(\mathbf{r},\ E)$ denotes the collision density, u the lethargy, E the energy of the neutrons, $\xi_0 = 1 - \frac{(A-1)^2}{2A}\ln\frac{A+1}{A-1}$, and A is the mass number of the moderator atoms. The symbol $\langle\!\langle ...\rangle\!\rangle$ denotes averaging over the lethargy interval. This equation is the analog of Eq. (6.1) considered above, and the results of the preceding paragraph can be used for analyzing the behavior of the averaged Green's function of Eq. (7.1).

2. In order to take into account the energy dependence of the distribution function of diffusing neutrons, one must consider a system of coupled equations either in the so-called multigroup theory (see, e.g., [1, 2]) or in the multivelocity diffusion theory [19]. Then, both the Green's function and the operators of the kinetic equation are matrices. The equation of the zeroth approximation assumes the form

$$\sum_j \hat{B}_{lj}(x)\,G_{jk}^0(x\,|\,y) = -\delta(x - y)\,\delta_{ik}.\tag{7.2}$$

Correspondingly, the solution to this equation is

$$G_{lk}^0(x\,|\,y) = (B^{-1})_{lk}\delta(x-y). \tag{7.3}$$

Equation (2.7) must be replaced by the equation

$$G_{nk}(x\,|\,y) = G_{nk}^0(x\,|\,y) - \sum_{l,\,m}\int dx' G_{nl}^0(x\,|\,x')\,\hat{\mu}_{lm}(x')\,G_{mk}(x'\,|\,y), \tag{7.4}$$

and instead of the basic equation (2.21) we obtain

$$\sum_{n}\hat{B}_{pn}(x)\,\langle G_{nk}(x\,|\,y)\rangle = -\delta(x-y)\,\delta_{nk} - \sum_{s,\,l,\,m}\int dx'\,\langle\hat{\mu}_{ps}(x)\,G_{sl}^0(x\,|\,x')\,\hat{\mu}_{lm}(x')\rangle\,\langle G_{mk}(x'\,|\,y)\rangle. \tag{7.5}$$

Consequently, the operator $M_1(x\,|\,x')$ [see Eq. (2.20)] now has the following form:

$$M_{pm}^{(1)}(x\,|\,x') = \sum_{l,\,s}\langle\mu_{ps}(x)\,G_{sl}^0(x\,|\,x')\,\hat{\mu}_{lm}(x')\rangle. \tag{7.6}$$

In the representation in which G^0 has diagonal form:

$$G_{sl}^0(x\,|\,x') = \delta_{sl}G_s^0(x\,|\,x'),$$

$$M_{pm}^{(1)}(x\,|\,x') = \sum_{s}\langle\hat{\mu}_{ps}(x)\,G_s^0(x\,|\,x')\,\hat{\mu}_{sm}(x')\rangle \tag{7.7}$$

and

$$\hat{B}_p(x)\,\langle G_{pk}(x\,|\,y)\rangle = -\delta(x-y)\,\delta_{pk} - \sum_{s,\,m}\int dx'\,\langle\hat{\mu}_{ps}(x)\,G_s^0(x\,|\,x')\,\mu_{sm}(x')\rangle\,\langle G_{mk}(x'\,|\,y)\rangle. \tag{7.8}$$

In the two-group approximation (s, m, p, k = 1, 2), this equation was derived some time ago [10]. When the energy dependence of the average neutron flux in an inhomogeneous medium is taken into account in the m_1 approximation, the calculations become more difficult than in the single-velocity case, but pose no difficulties as far as the principles of the calculations are concerned.

3. In the method suggested in the preceding section, the energy dependence of both the average and the zeroth Green's function was approximately taken into account. It is not possible to state exact solutions to the infinite system of coupled equations defined in (7.5) and (7.8) and one must solve corresponding "shortened" equation systems. We consider in this section the energy dependence of the neutron flux in exact form (assuming that the function G_0 is known).

Let us assume for the sake of simplicity that only the absorption cross section fluctuates. Then we rewrite the kinetic equation of the diffusion approximation for the neutron flux emitted from an elementary source:

$$\nabla^2 G(\mathbf{r},\,\mathbf{r}_0;\,E,\,E_0) - \varkappa_0^2(E)G(\mathbf{r},\,\mathbf{r}_0;\,E,\,E_0) - \nu(\mathbf{r},\,E)G(\mathbf{r},\,\mathbf{r}_0;\,E,\,E_0) +$$

$$+\int G(\mathbf{r},\,\mathbf{r}_0;\,E',\,E_0)\frac{\Sigma_{s0}(E,\,E')\,dE'}{D_0(E)} = -\frac{S}{D_0(E)}\delta(E-\dot{E}_0)\delta(\mathbf{r}-\mathbf{r}_0). \tag{7.9}$$

In this equation, $\Sigma_{s0}(E,\,E')$ denotes the scattering kernel, i.e., the differential (relative to the energy) macroscopic scattering cross section of neutrons with an energy E' in a unit energy interval near E; $\varkappa_0^2(E) = \frac{\Sigma_{t0}(E)}{D_0(E)}$, $\nu(\mathbf{r},\,E) = \frac{\sigma_a(\mathbf{r},\,E)}{D_0(E)}$, where $\langle\nu(\mathbf{r},\,E)\rangle = \frac{1}{D_0(E)}\langle\sigma_a(\mathbf{r},\,E)\rangle = 0$ (see §4). In what follows we assume the source output S such that $S/D_0(E_0) = 1$.

In the M_1 approximation, the Fourier transform of the average neutron flux (the Fourier transform is taken with respect to $r - r_0$) for an infinite spatially homogeneous medium satisfies the equation

$$-[p^2 + \varkappa_0^2(E)] \langle g(p, E; E_0) \rangle + \int dE' \widetilde{\Sigma}(E, E'; p) \langle g(p; E', E_0) \rangle = -\delta(E - E_0), \qquad (7.10)$$

$$\langle g(p, E; E_0) \rangle = \int d\mathbf{r} e^{i\mathbf{p}(\mathbf{r}-\mathbf{r}_0)} \langle G(\mathbf{r}, \mathbf{r}_0; E, E_0) \rangle,$$

$$\widetilde{\Sigma}(E, E'; p) = \frac{1}{D_0(E)} \Sigma_{s0}(E, E') + \frac{1}{(2\pi)^3} \int d\zeta k_{vv}(\zeta, E, E') g_0(p - \zeta; E, E'). \qquad (7.11)$$

Thus, the fluctuations of the absorption cross section change the scattering kernel in the approximation under consideration. In view of the symmetry properties of the Green's function of the zeroth approximation, we have

$$g_0(p, E_0, E) = g_0(p, E, E_0) e^{\frac{E-E_0}{k_0 T}} \frac{E_0}{E} \frac{D_0(E)}{D_0(E_0)} \qquad (7.12)$$

(k_0 denotes the Boltzmann constant), and with the symmetry of the correlation function

$$k_{vv}(\zeta; E_0; E) = k_{vv}(\zeta; E; E_0), \qquad (7.13)$$

so that we obtain from Eq. (7.11)

$$\widetilde{\Sigma}(E', E, p) = \frac{1}{D_0(E')} \Sigma_{s0}(E', E) + \frac{1}{(2\pi)^3} \int d\zeta k_{vv}(\zeta, E', E) g_0(p - \zeta, E', E) =$$

$$= \frac{D_0(E)}{D_0(E')} \frac{E'}{E} e^{\frac{E-E'}{k_0 T}} \left\{ \frac{\Sigma_{s0}(E', E)}{D_0(E)} e^{\frac{E'-E}{k_0 T}} \frac{E}{E'} + \frac{1}{(2\pi)^3} \int d\zeta k_{vv}(\zeta, E, E') g_0(p - \zeta, E, E') \right\}. \qquad (7.14)$$

The scattering kernel $\Sigma_{s0}(E, E')$ satisfies the small-scale equilibrium principle:

$$\Sigma_{s0}(E', E) E e^{-\frac{E}{k_0 T}} = \Sigma_{s0}(E, E') E' e^{-\frac{E'}{k_0 T}}. \qquad (7.15)$$

We rewrite Eq. (7.14) with Eqs. (7.15) and (7.11) in the following form:

$$\widetilde{\Sigma}(E', E, p) = \frac{D_0(E)}{D_0(E')} \frac{E'}{E} e^{\frac{E-E'}{k_0 T}} \left[\frac{\Sigma_{s0}(E, E')}{D_0(E)} + \right.$$

$$\left. + \frac{1}{(2\pi)^3} \int d\zeta k_{vv}(\zeta, E, E') g_0(p - \zeta, E, E') \right] = \frac{D_0(E)}{D_0(E')} \frac{E'}{E} e^{\frac{E-E'}{k_0 T}} \widetilde{\Sigma}(E, E', p). \qquad (7.16)$$

This expression for $\widetilde{\Sigma}(E, E', p)$ is equivalent to the small-scale equilibrium principle of Eq. (7.15) for the scattering kernel $\Sigma_{s0}(E, E')$ and, as shown by Eq. (7.16), is not violated in the M_1 approximation.

A solution to Eq. (7.10) can be found by an expansion in Laguerre polynomials:

$$\langle g(p, E, E_0) \rangle = \sum_{i=0}^{\infty} \frac{E}{(k_0 T)^2} e^{-\frac{E}{k_0 T}} \frac{L_i^{(1)}(E/k_0 T)}{\sqrt{i!(i+1)!}} g_i(p, E_0). \qquad (7.17)$$

We can find (in approximate form) the energy dependence of the average Green's function $\langle g(p, E, E_0) \rangle$. The energy dependence of the zeroth Green's function g_0 is known and will be used in exact form. In the preceding section we used an approximation for the function g_0, and this introduced an additional error in the kernel of the equation for $\widetilde{\Sigma}(E, E', p)$.

We obtain the following equation system for the coefficients $g_i(\mathbf{p}, E_0)$:

$$\sum_{i=0}^{\infty} \{- [\delta_{ki}p^2 + (\varkappa_0^2(E))_{ki}] + \widetilde{\gamma}_{ki}(\mathbf{p})\} g_i(\mathbf{p},E_0)\} = - \frac{L_k^{(1)}(E_0/k_0 T)}{\sqrt{k!\,(k+1)!}}, \tag{7.18}$$

$$(\varkappa_0^2(E))_{ki} = \int dE \frac{E}{(k_0 T)^2} e^{-E/k_0 T} L_k^{(1)}(E/k_0 T) \frac{\varkappa_0^2(E) L_i^{(1)}(E/k_0 T)}{\sqrt{i!\,(i+1)!\,k!\,(k+1)}}, \tag{7.19}$$

$$\widetilde{\gamma}_{ki}(\mathbf{p}) = \int dE \int dE' \frac{E'}{(k_0 T)^2} e^{-E'/k_0 T} L_i^{(1)}(E'/k_0 T) \widetilde{\Sigma}(E,E',\mathbf{p}) \frac{L_k^{(1)}(E/k_0 T)}{\sqrt{i!\,(i+1)!\,k!\,(k+1)!}}. \tag{7.20}$$

In the single-group approximation, we have

$$g_1(\mathbf{p}, E_0) = - \frac{L_0^{(1)}(E/k_0 T)}{\widetilde{\gamma}_{00}(\mathbf{p}) - [p^2 + (\varkappa_0^2(E))_{00}]}. \tag{7.21}$$

When no neutron sources are present in the moderator region under consideration, the eigenvalue problem can be formulated:

$$-[p^2 + \varkappa_0^2(E)] \langle \Phi(\mathbf{p}, E)\rangle + \int dE' \widetilde{\Sigma}(E,E',\mathbf{p}) \langle \Phi(\mathbf{p},E')\rangle = 0. \tag{7.22}$$

When the energy dependence of the neutron flux $\langle \Phi(\mathbf{p}, E)\rangle$ is approximated by the Maxwell distribution $E/(k_0 T)^2 e^{-E/k_0 T}$, we obtain from Eq. (7.22) [see Eq. (7.21)]:

$$p^2 = \widetilde{\gamma}_{00}(\mathbf{p}) - (\varkappa_0^2(E))_{00} = -\left[\left(\frac{\Sigma_{a0}(E)}{D_0(E)}\right)_{00} + \frac{1}{(2\pi)^3} \iint dE\,dE' \frac{E'}{(k_0 T)^2} e^{-E/k_0 T} \int d\zeta\, k_{vv}(\zeta; E,E') g_0(\mathbf{p}-\zeta,E,E') \right]. \tag{7.23}$$

In other words, as in the single-velocity case, fluctuations of the absorption cross section reduce the attenuation constant of the neutron flux relative to the case of a homogeneous medium.

We could extend our studies of the behavior of the average neutron flux if we had an analytical expression for the Green's function of the zeroth approximation. This expression is known for a separable (or degenerate) scattering kernel

$$\Sigma_s(E,E') = \dot{A}(E') B(E). \tag{7.24}$$

It is easy to verify that in this case we obtain

$$g_0(\mathbf{p}, E; E_0) = \frac{\delta(E-E_0)}{p^2 + \varkappa_0^2(E)} + \frac{B(E)}{D_0(E)[p^2+\varkappa_0^2(E)]} \frac{A(E)}{p^2+\varkappa_0^2(E_0)} \left[1 - \int \frac{A(E')B(E')}{p^2+\varkappa_0^2(E')} \frac{dE'}{D_0(E')}\right]^{-1}. \tag{7.25}$$

By substituting Eq. (7.25) into Eq. (7.11), we can rewrite Eq. (7.10) in the following form:

$$-\left[p^2 + \varkappa_0^2(E) - \frac{1}{(2\pi)^3} \int \frac{d\zeta\, k_{vv}(\zeta,E,E')}{(\mathbf{p}-\zeta)^2+\varkappa_0^2(E)} \right] \langle g(\mathbf{p},E,E_0) +$$

$$+ \int dE' \langle g(\mathbf{p},E',E_0)\rangle \left\{ \frac{A(E')B(E)}{D_0(E)} + \frac{1}{(2\pi)^3} \int d\zeta\, k_{vv}(\zeta,E,E') \frac{A(E')B(E)}{[(\mathbf{p}-\zeta)^2+\varkappa_0^2(E)]D_0(E)} \times \right.$$

$$\left. \times \frac{1}{(\mathbf{p}-\zeta)^2+\varkappa_0^2(E)} \left[1 - \int \frac{A(E'')B(E'')}{[(\mathbf{p}-\zeta)+\varkappa_0^2(E'')]D_0(E'')} \frac{dE''}{D_0(E'')}\right]^{-1}\right\} = -\delta(E-E_0). \tag{7.26}$$

Equation (7.26) belongs to the class of integral equations with degenerate kernels (the dependence of the function $k_{vv}(\zeta, E, E')$ upon E and E' can be factorized), and the solution to this integral equation can be satisfied in analytical form. Rather cumbersome expressions are obtained, which will not be stated here.

When the characteristic size ζ_{eff}^{-1} of an inhomogeneity is small compared to λ_{s0}, the diffusion approximation cannot be employed, and one must use the exact kinetic equation which, in the M_1 approximation for isotropic scattering, has the following form (see §5):

$$[ipn - \Sigma_{t0}(E)] \langle g(p, n, E, E_0) \rangle + \frac{1}{4\pi} \iint d\Omega' dE' \Sigma_{s0}(E, E') \langle g(p, n', E' E_0) \rangle =$$

$$= -\delta(E - E_0) - \frac{1}{(2\pi)^3} \int d\zeta \int dE' k_{aa}(\zeta, E, E') \int d\Omega' g_0(p - \zeta, n, n' | E, E') \ \langle g(p, n'; \ E' E_0) \rangle. \quad (7.27)$$

The neutron source in Eq. (7.27) is isotropic; n denotes the unit vector in the direction of motion of the neutrons. In a first approximation with respect to ζ_{eff}^{-1}, the zeroth Green's function can be written in the form*

$$g_0(p - \zeta, n; \ n' | E, E') \approx \frac{\delta(E - E')\delta(n - n')}{i(p - \zeta)n - \Sigma_{t0}(E)}. \quad (7.28)$$

One cannot employ this approximation in the case of neutrons moving parallel to the planes of plane lattice or along the rods in a two-dimensional lattice (see §5). We obtain after substitution of Eq. (7.28) into Eq. (7.27)

$$[ipn - \Sigma_{t1}(E)] \langle g(p, n; \ E, E_0) \rangle + \frac{1}{4\pi} \iint d\Omega' dE' \Sigma_{s0}(E, E') \langle g(p, n'; \ E', E_0) \rangle = -\delta(E - E_0), \quad (7.29)$$

where

$$\Sigma_{t1}(E) = \Sigma_{t0}(E) - \frac{1}{(2\pi)^3} \int d\zeta \ \frac{k_{aa}(\zeta; \ E, E')}{i(p - \zeta)n - \Sigma_{t0}(E)} \approx \Sigma_{t0}(E) + \frac{1}{(2\pi)^3} \int d\zeta \ \frac{k_{aa}(\zeta; \ E, E)}{i\zeta n + \Sigma_{t0}(E)}. \quad (7.30)$$

In the approximation which is linear with respect to ζ_{eff}^{-1} (the characteristic size of the inhomogeneity), the correction to the attenuation constant of the neutron flux is the same as in the single-velocity case (see §5).

§8. Applicability Range of the M_1 Approximation

All previous calculations were made in the M_1 approximation [Eqs. (2.20) and (2.21)]. In the expansion of the "mass operator" M in moments of the random function $\hat{\mu}$, this approximation includes only the second moments, i.e., paired correlation functions. We discuss in this paragraph the applicability range of the M_1 approximation and estimate the corrections resulting from higher approximations.

1. For the sake of simplicity, let us consider the stationary diffusion equation for an infinite medium

$$[\nabla^2 + \eta(r)\nabla^2 + \nabla\eta(r)\nabla] G(r | r_0) - \varkappa_0^2 \left(1 + \frac{\nu(r)}{\varkappa_0^2}\right) G(r | r_0) = -\delta(r - r_0). \quad (8.1)$$

---

* This statement can be checked by solving the equation for the function g_0 with the successive approximation technique.

We apply the Fourier transform with respect to $\mathbf{r} - \mathbf{r_0}$:

$$g(\mathbf{p}, \mathbf{r}) = \int d(\mathbf{r} - \mathbf{r_0}) e^{i\mathbf{p}(\mathbf{r}-\mathbf{r_0})} G(\mathbf{r} \mid \mathbf{r_0}),$$

$$G(\mathbf{r} \mid \mathbf{r_0}) = \frac{1}{(2\pi)^3} \int d\mathbf{p} e^{-i\mathbf{p}(\mathbf{r}-\mathbf{r_0})} g(\mathbf{p}, \mathbf{r}). \tag{8.2}$$

The result is

$$\left\{ [1 + \eta(\mathbf{r})] [\nabla^2 - 2i\nabla\mathbf{p} - p^2] + \nabla\eta(\mathbf{r})(\nabla - i\mathbf{p}) - \varkappa_0^2\left[1 + \frac{\nu(\mathbf{r})}{\varkappa_0^2}\right] \right\} g(\mathbf{p}, \mathbf{r}) = -1. \tag{8.3}$$

The $\mathbf{r}$ dependence remains because we consider a nonaveraged Green's function $G(\mathbf{r}\mid\mathbf{r_0})$ in an inhomogeneous medium. In order to eliminate the $\mathbf{r}$ dependence, we apply once more a Fourier transform

$$\tilde{g}(\mathbf{p}, \mathbf{k}) = \int d\mathbf{r} e^{i\mathbf{k}\mathbf{r}} g(\mathbf{p}, \mathbf{r}),$$

$$g(\mathbf{p}, \mathbf{r}) = \frac{1}{(2\pi)^3} \int d\mathbf{k} e^{-i\mathbf{k}\mathbf{r}} \tilde{g}(\mathbf{p}, \mathbf{k}), \tag{8.4}$$

The transformed equation (8.3) can be rewritten in the form

$$[(\mathbf{k} + \mathbf{p})^2 + \varkappa_0^2]\, \tilde{g}(\mathbf{p}, \mathbf{k}) + \frac{1}{(2\pi)^3} \int d\mathbf{k'} \tilde{g}(\mathbf{p}, \mathbf{k} - \mathbf{k'}) \cdot \{ [(\mathbf{k} - \mathbf{k'} + \mathbf{p})^2 +$$

$$+ \mathbf{k'}(\mathbf{k} + \mathbf{p} - \mathbf{k'})]\, \eta_1(\mathbf{k'}) + \nu_1(\mathbf{k'}) \} = (2\pi)^3\, \delta(\mathbf{k}). \tag{8.5}$$

The following notation was introduced:

$$\eta_1(\mathbf{k}) = \int \eta(\mathbf{r})\, e^{i\mathbf{k}\mathbf{r}} d\mathbf{r},$$

$$\eta(\mathbf{r}) = \frac{1}{(2\pi)^3} \int \eta_1(\mathbf{k})\, e^{-i\mathbf{k}\mathbf{r}} d\mathbf{k}, \tag{8.6}$$

etc.

With the successive-approximation method, the solution of Eq. (8.5) assumes the form

$$\tilde{g}(\mathbf{p}, \mathbf{k}) = \frac{(2\pi)^3\, \delta(\mathbf{k})}{p^2 + \varkappa_0^2} - \frac{1}{p^2 + \varkappa_0^2} \frac{\mathbf{p}(\mathbf{p} + \mathbf{k})\, \eta_1(\mathbf{k}) + \nu_1(\mathbf{k})}{(\mathbf{p} + \mathbf{k})^2 + \varkappa_0^2} +$$

$$+ \frac{1}{p^2 + \varkappa_0^2} \frac{1}{(2\pi)^3} \frac{1}{(\mathbf{p} + \mathbf{k})^2 + \varkappa_0^2} \int \frac{d\mathbf{k'}}{(\mathbf{p} + \mathbf{k} - \mathbf{k'})^2 + \varkappa_0^2} [(\mathbf{k} + \mathbf{p})(\mathbf{k} + \mathbf{p} - \mathbf{k'})\, \eta_1(\mathbf{k'}) +$$

$$+ \nu(\mathbf{k'})] [\mathbf{p}(\mathbf{p} + \mathbf{k} - \mathbf{k'})\, \eta_1(\mathbf{k} - \mathbf{k'}) + \nu_1(\mathbf{k} - \mathbf{k'})] - \dots \tag{8.7}$$

With the formula for the reversal of the Eq. (8.4) Fourier transform, we obtain from Eq. (8.7):

$$g(\mathbf{p}, \mathbf{r}) = \sum_{n=0}^{\infty} \frac{(-1)^n}{p^2 + \varkappa_0^2} \frac{1}{(8\pi^3)^n} \int d\mathbf{k_1} \dots \int d\mathbf{k_n} e^{-i\mathbf{r}\sum_{i=1}^{n}\mathbf{k_i}} \frac{\mathbf{p}(\mathbf{p} + \mathbf{k_1})\, \eta_1(\mathbf{k_1}) + \nu_1(\mathbf{k_1})}{(\mathbf{p} + \mathbf{k_1})^2 + \varkappa_0^2} \times$$

$$\times \frac{(\mathbf{p} + \mathbf{k_1})(\mathbf{p} + \mathbf{k_1} + \mathbf{k_2})\, \eta_1(\mathbf{k_2}) + \nu_1(\mathbf{k_2})}{(\mathbf{p} + \mathbf{k_1} + \mathbf{k_2})^2 + \varkappa_0^2} \dots \frac{(\mathbf{p} + \sum\limits_{i=1}^{n-1}\mathbf{k_i})(\mathbf{p} + \sum\limits_{i}^{n}\mathbf{k_i})\, \eta_1(\mathbf{k_n}) + \nu_1(\mathbf{k_n})}{(\mathbf{p} + \sum\limits_{i=1}^{n}\mathbf{k_i})^2 + \varkappa_0^2}. \tag{8.8}$$

Figure 2 is a graphical representation of Eq. (8.8). The values of the various portions of the curve are easily obtained from a comparison between the figure and Eq. (8.8). When the functions η and ν have the same coordinate dependence, i.e., if

$$\eta(\mathbf{r}) = \eta_0 c(\mathbf{r}), \quad \eta_1(\mathbf{k}) = \eta_0 c_1(\mathbf{k}),$$

$$\nu(\mathbf{r}) = \nu_0 c(\mathbf{r}), \quad \nu_1(\mathbf{k}) = \nu_0 c_1(\mathbf{k}),$$

we have

$$g(\mathbf{p}, \mathbf{r}) = \sum_{n=0}^{\infty} \left(-\frac{1}{8\pi^3}\right)^n \frac{1}{p^2 + \varkappa_0^2} \int d\mathbf{k}_1 \ldots \int d\mathbf{k}_n e^{-i\mathbf{r}\sum_{i=1}^{n}\mathbf{k}_i} \frac{c_1(\mathbf{k}_1)\, c_1(\mathbf{k}_2)\, \ldots\, c_1(\mathbf{k}_n)}{(p + k_1)^2 + \varkappa_0^2} \times$$

$$\times \frac{[\mathbf{p}(\mathbf{p}+\mathbf{k}_1)\eta_0 + \nu_0] \ldots \left[\left(\mathbf{p}+\sum_{i=1}^{n-1}\mathbf{k}_i\right)\left(\mathbf{p}+\sum_{i=1}^{n}\mathbf{k}_i\right)\eta_0 + \nu_0\right]}{[(p + k_1 + k_2)^2 + \varkappa_0^2] \ldots \left[\left(\mathbf{p}+\sum_{i=1}^{n}\mathbf{k}_i\right)^2 + \varkappa_0^2\right]}. \tag{8.9}$$

In order to determine the averaged Green function $\langle g(\mathbf{p}, \mathbf{r})\rangle = \langle g(\mathbf{p})\rangle$, we must calculate the average $\langle c(\mathbf{k}_1),\ldots,c(\mathbf{k}_n)\rangle$. In a medium with spatial homogeneity, this correlation function can be written in the form

$$\langle c_1(\mathbf{k}_1)\,\ldots\,c_1(\mathbf{k}_n)\rangle = (2\pi)^3\, \delta\left(\sum_{i=1}^{n}\mathbf{k}_i\right) \int d\mathbf{r}_1 \ldots \int d\mathbf{r}_{n-1} e^{i\sum_{i=1}^{n-1}\mathbf{r}_i\mathbf{k}_i} K(\mathbf{r}_1 \ldots \mathbf{r}_{n-1},) \tag{8.10}$$

$$K(\mathbf{r}_1 \ldots \mathbf{r}_{n-1}) = \langle c(\rho)\, c(\mathbf{r}_1 + \rho)\, \ldots\, c(\mathbf{r}_{n-1} + \rho)\rangle. \tag{8.11}$$

The δ function in Eq. (8.10) makes $\langle g\rangle$ independent of $\mathbf{r}$.

We use Eq. (8.10) to expand the average Green's function in a perturbation-theory series:

$$\langle g(\mathbf{p})\rangle = \frac{1}{p^2 + \varkappa_0^2} + \sum_{n=1}^{\infty} (-)^n \frac{1}{(8\pi)^{n-1}} \int d\mathbf{k}_1 \ldots \int d\mathbf{k}_{n-1} \int d\mathbf{r}_1 \ldots \int d\mathbf{r}_{n-1} e^{i\sum_{i=1}^{n-1}\mathbf{k}_i\mathbf{r}_i} K(\mathbf{r}_1\ldots\mathbf{r}_{n-1}) \times$$

$$\times \frac{[\mathbf{p}(\mathbf{p}+\mathbf{k}_1)\eta_0 + \nu_0]\,[(\mathbf{p}+\mathbf{k}_1)(\mathbf{p}+\mathbf{k}_1+\mathbf{k}_2)\eta_0 + \nu_0] \ldots [(\mathbf{p}-\mathbf{k}_n)\mathbf{p}\eta_0 + \nu_0]}{[(p+k_1)^2 + \varkappa_0^2]\,[(p+k_1+k_2)^2 + \varkappa_0^2] \ldots [p^2+\varkappa_0^2]^3}. \tag{8.12}$$

In the M_1 approximation Eq. (8.12) reduces to a geometric progression:

$$\langle g(\mathbf{p})\rangle = \frac{1}{p^2 + \varkappa_0^2}\left\{1 + m_1(\mathbf{p})\frac{1}{p^2 + \varkappa_0^2} + \left[m_1(\mathbf{p})\frac{1}{p^2 + \varkappa_0^2}\right]^2 + \ldots\right\} = [p^2 + \varkappa_0^2 - m_1(\mathbf{p})]^{-1}, \tag{8.13}$$

where

$$m_1(\mathbf{p}) = \frac{1}{(2\pi)^3} \int \frac{k_1(\mathbf{s})\, d\mathbf{s}}{(p+s)^2 + \varkappa_0^2} [\mathbf{p}(\mathbf{p}+\mathbf{s})\eta_0 + \nu_0]^2, \tag{8.14}$$

with

$$k(\mathbf{s}) = k(-\mathbf{s}) = \int d\mathbf{r} K(\mathbf{r})\, e^{i\mathbf{s}\mathbf{r}}.$$

It is easy to verify that Eq. (8.14) coincides with the previously used equation (4.46). The relations $\eta_0^2 k(\mathbf{s}) = \varepsilon_{\eta\eta}' b(\mathbf{s})$; $\eta_0 \nu_0\, k(\mathbf{s}) = \varepsilon_{\eta\nu} b(\mathbf{s})$ and $\nu_0^2 k(\mathbf{s}) = \varepsilon_{\nu\nu} b(\mathbf{s})$ hold.

2. Let us estimate the order of magnitude of the first terms of the "mass operator" expansion which were neglected in the M_1 approximation. We assume for the sake of simplicity that only the absorption cross section fluctuates. In the case of stationary diffusion of thermal neutrons emitted from a plane source into an infinite layered medium with Gaussian fluctuations $\Sigma_a(z)$, we obtain from Eq. (8.12)

$$m_2(\mathbf{p}) = \frac{\varepsilon_{vv}^2}{(2\pi)^2} \iint \frac{ds_1 ds_2}{(p+s_1)^2 + \varkappa_0^2} \frac{b(s_1)\,b(s_2)}{(p+s_1+s_2)^2 + \varkappa_0^2} \left[\frac{1}{(p+s_1)^2+\varkappa_0^2} + \frac{1}{(p+s_2)^2+\varkappa_0^2} \right].$$

(8.15)

When the correlation function K_{vv} has the form of Eq. (3.1) we obtain at $q \gg \varkappa_0$

$$m_2(p) \sim \frac{\varepsilon_{vv}^2}{q^2 \varkappa_0^4},$$

(8.16)

whereas

$$m_1(p) \sim \frac{\varepsilon_{vv}}{\varkappa_0 q}.$$

(8.17)

Thus, at $\varepsilon_{vv}/\varkappa_0^3 q \ll 1$ (this occurs either for weak fluctuations $\varepsilon_{vv}/\varkappa_0^4 \ll 1$, or for strong, but small-scale, fluctuations $\varkappa_0/q \ll 1$) we find

$$m_2/m_1 = \frac{\varepsilon_{vv}}{\varkappa_0^3 q} \ll 1.$$

(8.18)

If $m_1 \lesssim \varkappa_0^2$, the smallness of the neglected terms (smallness relative to the terms which were taken into account) is guaranteed by the condition $\varkappa_{as}^2 \gg \dfrac{\varepsilon_{vv}^2}{\varkappa_0^4 \varkappa_1^2}$, $\varkappa^1 = \varkappa_0 + q$, where $\varkappa_{as}$ is defined by Eq. (4.21), or, for $\varkappa_0/\varkappa_1 \ll 1$, by the condition $1 - \dfrac{\varepsilon_{vv}}{\varkappa_0^3 q} \gg \dfrac{\varepsilon_{vv}^2}{\varkappa_0^6 q^2}$.

Since the correction $\sim (\varepsilon_{vv}/\varkappa_0^3 \varkappa_1)^2$ appears in the exponent with the factor $|z - z'|$, the M_1 approximation is applicable for $\varkappa_0 |z - z'| \ll \varepsilon_{vv}^{-2} \left(\dfrac{1}{\varkappa_0^3 q} \right)^{-2} \dfrac{\varkappa_{as}}{\varkappa_0}$. If $\varkappa_0/\varkappa_1 \ll 1$, this condition can be written in the form

$$\varkappa_0 |z - z'| \ll \varepsilon_{vv}^{-2} \left(\frac{1}{\varkappa_0^3 q} \right)^{-2} \sqrt{1 - \frac{\varepsilon_{vv}}{\varkappa_0^3 \varkappa_1}}.$$

In a medium with periodic structure (only the absorption cross section fluctuates), e.g., Eq. (3.43a), we have $\varkappa_0 a \ll 1$ for $m_1(p) \sim \varepsilon_{vv} a^2$, whereas $m_2 \sim \varepsilon_{vv} a^6$, i.e., $m_2(p)/m_1(p) \sim \varepsilon_{vv} a^4$. For $\varepsilon_{vv} a^2/\varkappa_0^2 \ll 1$ and $\varkappa_0 a \ll 1$ this ratio is much smaller than unity. The restrictions for the absolute value of the fluctuations are less stringent in the case of a periodic perturbation than in the previously considered case of a nonordered medium.

When not only the absorption cross section, but also the neutron-scattering cross section fluctuate, the convergence of the perturbation-theory series for the "mass operator" deteriorates. It follows from Eq. (8.12)* that for $l \ll L_0$ (l denotes the characteristic size of the inhomogeneities) in each term of the perturbation-theory series there appears a component which does not contain a small l/L_0 ratio. This limits the applicability range of the M_1 approximation to weak

---

*See footnote † on page 220.

fluctuations. These components can be accurately calculated for several types of diffusion media. Let us consider the diffusion in a layered medium in which a plane neutron source is situated perpendicular to the layers of the medium. In this case, we have $(\mathbf{p}, \mathbf{k}_i) = 0$ and the "dangerous" components do not appear. In the limit of very thin layers one must use the average diffusion coefficient D_0 in order to describe the longitudinal diffusion. In the particular case of a plane periodic lattice, this result was previously obtained by Shevelev [13]. In transverse diffusion through thin layers $(k_i/p_z \gg 1)$ we expand the function $\langle g(\mathbf{p}) \rangle$ in the perturbation-theory series of Eq. (8.12) and obtain

$$\langle g(\mathbf{p}) \rangle^{-1} = p^2 + \varkappa_0{}^2 - p^2 \{\langle \eta^2 \rangle - \langle \eta^3 \rangle + [\langle \eta^4 \rangle - \langle \eta^2 \rangle^2] - [\langle \eta^5 \rangle - 2\langle \eta^2 \rangle \langle \eta^3 \rangle] -$$
$$- [\langle \eta^6 \rangle - \langle \eta^3 \rangle^2 - 2\langle \eta^2 \rangle \langle \eta^4 \rangle + \langle \eta^2 \rangle^3] - \ldots \} = p^2 \{1 - \langle \eta^2 \rangle + \langle \eta^3 \rangle -$$
$$- [\langle \eta^4 \rangle - \langle \eta^2 \rangle] + \ldots \} + \varkappa_0{}^2 = \{1 - \langle \eta^2 \rangle + \langle \eta^3 \rangle - [\langle \eta^4 \rangle - \langle \eta^2 \rangle^2] + \ldots$$
$$\ldots \} \{p^2 + \varkappa_0{}^2 [1 - \langle \eta^2 \rangle + \langle \eta^3 \rangle - \ldots]^{-1}\}. \qquad (8.19)$$

We recall that

$$\varkappa_h^2 = 3 \langle \Sigma_a \rangle \langle \Sigma_t \rangle = \varkappa_0{}^2 [1 + \langle \eta^2 \rangle - \langle \eta^3 \rangle + \langle \eta^4 \rangle - \langle \eta^5 \rangle + \langle \eta^6 \rangle - \ldots],$$

and obtain from Eq. (8.19) the following equation for the determining the attenuation constant of the average neutron flux:

$$\langle g(\mathbf{p}) \rangle^{-1} = p^2 + \varkappa_h^2 = 0.$$

We see that for describing the transverse diffusion of neutrons we must use the diffusion coefficient for the homogeneous medium:

$$D_h = \frac{1}{3\langle \Sigma_t \rangle}.$$

In the case of a plane periodic lattice this conclusion agrees with the results of [13]. The corrections to this result include small l/L_0 ratios. The contribution of these corrections can be taken into account in the M_1 approximation even when strong fluctuations occur.

A similar result can be obtained for the diffusion of neutrons in an isotropic medium. The attenuation constant of the average flux is given by the equation*

$$p^2 + \frac{\varkappa_0^3}{3} \frac{\left\langle 1 \Big/ \left(1 + \frac{\eta}{3}\right) \right\rangle}{1 - \frac{2}{3} \left\langle \dfrac{1}{1 + \frac{\eta}{3}} \right\rangle} = 0. \qquad (8.19a)$$

In a system with axial symmetry, the diffusion along the symmetry axis takes place in the same form as the longitudinal diffusion in a plane lattice. The diffusion in the direction perpendicular to the axis is given by the equation

$$p^2 + \frac{\varkappa_0^2}{2} \frac{\left\langle \dfrac{1}{1 + \frac{\eta}{2}} \right\rangle}{1 - \frac{1}{2} \left\langle \dfrac{1}{1 + \frac{\eta}{2}} \right\rangle} = 0. \qquad (8.19b)$$

---

*In deriving this formula we replace the fraction in Eq. (8.12) by its average over the variability range of the angular variables Ω_{k_i}.

The diffusion-anisotropy factor has the form

$$L_{\parallel}^2 / L_{\perp}^2 = \frac{\left\langle \dfrac{1}{1+\dfrac{\eta}{2}} \right\rangle}{2 - \left\langle \dfrac{1}{1+\dfrac{\eta}{2}} \right\rangle} . \qquad (8.19c)$$

In the particular case of a two-component medium with free path lengths λ_1 and λ_2, we have

$$L_{\parallel}^2 / L_{\perp}^2 = \frac{\langle\lambda\rangle(\lambda_1 + \lambda_2)}{\langle\lambda\rangle^2 + \lambda_1\lambda_2} . \qquad (8.19d)$$

The results stated in Eqs. (8.19)-(8.19d) are independent of the particular structure of the cell and depend only on the volume concentration of the components of the medium (see § 4). Corrections can be found within the M_1 approximation when the Green's function of the zeroth approximation is properly chosen.

3. In the discussion above, it was everywhere assumed that the characteristic dimensions l of the inhomogeneities are small compared with the neutron diffusion length L_0 in the homogeneous medium. In this section we will consider the opposite limit, i.e., the case in which the dimensions of the inhomogeneities are large compared with L_0. This case needs special treatment because in the solution to the neutron diffusion problem in a medium with smoothly varying (varying according to a statistical law) properties, the perturbation-theory series for the "mass operator" is characterized by poor convergence. We will consider the diffusion of neutrons across layers in a medium with Gaussian fluctuations of the absorption cross section. For $l \gg L_0$ it is easy to verify that the contributions of all diagrams of the same order in the expansion of the average Green's function are approximately equal [see Eq. (2.24)], and we obtain the expression (see also [9])

$$G(z \mid z_0) = \sum_{n=0}^{\infty} (2n - 1)!! \langle G_n(z \mid z_0)\rangle_0,$$

for $G(z \mid z_0)$. The n-th term in the perturbation-theory series $\langle G(z \mid z_0)\rangle$ in the M_1 approximation at $q = 0$ is denoted by $\langle G_n(z \mid z_0)\rangle_0$ [see Eq. (4.20)]; $(2n-1)!! = 2^n \Gamma\left(n+\dfrac{1}{2}\right) \dfrac{1}{\sqrt{\pi}}$ denotes the number of diagrams of order n. We use the integral representation of the Γ function

$$\Gamma(s) = \int_0^{\infty} e^{-y} y^{s-1} \, dy$$

and change the order of summation and integration. After calculating the sum of the series, we obtain

$$\langle G(z \mid z_0)\rangle = \frac{1}{\sqrt{2\pi s}} \int_{-\infty}^{\infty} dt \, e^{-\frac{t^2}{2s}} \frac{e^{-|z-z_0|\sqrt{\varkappa_0^2 + t}}}{2\sqrt{\varkappa_0^2 + t}} , \qquad (8.20)$$

and this result agrees with the average value of $G(z \mid z_0)$ taken over the ensemble of homogeneous media with a normal distribution of the statistical quantity $\varkappa^2$ – a distribution which is sensible when the spread $\Delta\varkappa^2$ is small.

Let us refine our result. We start from the expression for the Green function of the diffusion equation in the form of the perturbation theory series of Eq. (8.9) for $\eta = 0$. In the limit of large-scale fluctuations $|\mathbf{p}| \gg |\mathbf{k}_i|$ and

$$\left(\mathbf{p} + \sum_{i=1}^{n} \mathbf{k}_i\right)^2 \approx p^2 + 2\mathbf{p} \sum_{i=1}^{n} \mathbf{k}_i. \tag{8.21}$$

In this (quasi-classical) approximation the series of Eq. (8.9) can be summed (for $\eta = 0$):

$$g(\mathbf{p}, \mathbf{r}) = i \int_0^\infty dt \exp\left[-it(p^2 + \varkappa_0^2 - i\delta)\right] \times \exp\left[-iv \int_0^t dt' c(\mathbf{r} + 2\mathbf{p}t)\right]. \tag{8.22}$$

We used the relation

$$[p^2 + \varkappa_0^2]^{-1} = i \int_0^\infty dt\, e^{-it(p^2 + \varkappa_0^2 - i\delta)}, \tag{8.23}$$

where δ denotes a small positive quantity. By averaging Eq. (8.22) over an ensemble with a normal distribution of the probabilities we obtain the following expression for the Fourier image of the average neutron flux:

$$\langle g(\mathbf{p}, \mathbf{r})\rangle \equiv \langle g(\mathbf{p})\rangle = i \int_0^\infty dt\, e^{-it(p^2 + \varkappa_0^2 - i\delta)} \times \exp\left[-\frac{v^2}{2} \int_0^t dt' \int_0^{t'} dt'' K(2\mathbf{p}(t - t'))\right]. \tag{8.24}$$

where

$$K(2\mathbf{p}(t - t')) = \langle c(\mathbf{r} + 2\mathbf{p}t) c(\mathbf{r} + 2\mathbf{p}t')\rangle = \langle c(2\mathbf{p}(t - t')) c(0)\rangle \tag{8.25}$$

denotes the correlation function of the fluctuations. We insert the Fourier image of the correlation function of Eq. (8.25)

$$k(\mathbf{s}) = \int d\mathbf{r}\, e^{i\mathbf{r}\mathbf{s}} K(\mathbf{r}) \tag{8.26}$$

into the exponent in Eq. (8.24) and integrate over t and t'. The result is

$$\langle g(\mathbf{p})\rangle = i \int_0^\infty dt\, e^{-it(p^2 + \varkappa_0^2 - i\delta)}\, e^{-\Phi(t)}, \tag{8.27}$$

with

$$\Phi(t) = \frac{v^2}{(2\pi)^3} \int \frac{d\mathbf{s}}{4(\mathbf{p}\mathbf{s})^2} k(\mathbf{s})(1 - \cos 2\mathbf{p}\mathbf{s}t). \tag{8.28}$$

For $t \to 0$ we obtain

$$\Phi(t) \approx \frac{v^2}{(2\pi)^3} \frac{t^2}{2} \int d\mathbf{s}\, k(\mathbf{s}) = \frac{v^2 t^2 K(0)}{2} = \frac{t^2 \varepsilon_{vv}}{2}. \tag{8.29}$$

In the case of an isotropic (isotropic on the average) medium we have $k(\mathbf{s}) \equiv k(s)$ and

$$\Phi(t) = \frac{v^2}{(2\pi)^2} \frac{1}{2p^2} \int_0^\infty ds\, k(s) \left[\cos 2pst - 1 + 2pst \int_0^{2pst} \frac{\sin z}{z}\right]. \tag{8.30}$$

This equation helps us to obtain an asymptotic expression for $\Phi(t)$ for large t:

$$\Phi(t) \underset{t\to\infty}{\approx} t\, \frac{v^2}{8\pi p} \int\limits_0^\infty s\, ds\, k(s). \tag{8.31}$$

When the correlations of the fluctuations in an isotropic medium can be described with the help of the function Eq. (3.4), we obtain

$$\Phi(t) \underset{t\to\infty}{\approx} \frac{\varepsilon_{vv} t}{2pq}. \tag{8.32}$$

In order to analyze Eq. (8.27) we split the region of the t integration into two portions $(0, t_c)$ and (t_c, ∞), where $t_c \approx 1/2pq$. We replace $\Phi(t)$ in the first portion of the interval by its limit for small t, and in the second portion by its asymptotic value for $t \to \infty$. The result is

$$\langle g(\mathbf{p}) \rangle \approx i \int\limits_0^{t_c} dt \exp\left[-it(p^2 + \varkappa_0^2 - i\delta) - \frac{\varepsilon_{vv}}{2} t^2 \right] + \frac{\exp\left[i\left(p^2 + \varkappa_0^2 - i\frac{\varepsilon_{vv}}{2pq} \right) \frac{1}{2pq} \right]}{p^2 + \varkappa_0^2 - i\frac{\varepsilon_{vv}}{2pq}}. \tag{8.33}$$

The first term of Eq. (8.33) is equal to

$$i \int\limits_0^{t_c} dt\, e^{-it\left(p^2 + \varkappa_0^2 \right) - \frac{\varepsilon_{vv}}{2} t^2} \approx i \int\limits_0^\infty dt\, e^{-it\left(p^2 + \varkappa_0^2 \right) - \frac{\varepsilon_{vv}}{2} t^2}, \tag{8.34}$$

and this expression coincides with the average of the solution for a homogeneous medium (average taken over an ensemble with a normal distribution). At $p^2 + \varkappa_0^2 \approx 0$ the second term has a pole. For $\varepsilon_{vv}/\varkappa_0^3 q \ll 1$ (at $\varkappa_0/q \gg 1$ this condition is a strong restriction for ε_{vv}), the values p_n of the pole are given by the expression

$$p_n = i\varkappa_0 \left(1 - \frac{\varepsilon_{vv}}{4\varkappa_0^3 q} \right). \tag{8.35}$$

When large-scale fluctuations occur in an isotropic medium, the average neutron flux can be represented by two terms, one of which decreases exponentially with increasing distance from the source, whereas the dependence of the source distance is more complicated for the other term. It follows from the above discussion that the presence of exponentially attenuated neutrons results from the asymptotic form of the function $\Phi(t)$ at large t values. When the diffusion takes place in the layered medium of Eq. (3.1) and neutrons diffuse across the layers, we have

$$\Phi(t) = \frac{\varepsilon_{vv}}{(2\pi)^3} q \int\limits_{-\infty}^{\infty} \frac{dk_z}{4\,(p_z k_z)^2} \frac{1}{q^2 + k_z^2} (1 - \cos 2\, p_z k_z t), \tag{8.36}$$

and in the region of large t values this expression tends to

$$\Phi_{as}(t) = \pm \frac{\varepsilon_{vv}}{2 p_z q} t \begin{cases} \operatorname{Re} p_z > 0, \\ \operatorname{Re} p_z < 0. \end{cases} \tag{8.37}$$

The Fourier transform of the average neutron flux contains the term

$$\langle g\left(p_z\right)\rangle = \frac{1}{p_z^2 + \varkappa_0^2 \left(1 - \dfrac{A\varepsilon_{vv}}{\varkappa_0^2 |p_z| q}\right)} \tag{8.38}$$

(A is a numerical factor) which corresponds to the exponentially attenuated part of the average flux. The exponentially attenuated part predominates at large distances from the source. Moreover, the Fourier image of the average flux contains a term corresponding to the nonexponential decrease with the distance from the source [see Eq. (8.20)].

When diffusion takes place along the layers of the medium of Eq. (3.1) we have

$$\Phi\left(t\right) = \frac{\varepsilon_{vv} t^2}{2}, \tag{8.39}$$

and the expression for the Fourier image of the average neutron flux does not contain terms with poles. This means that the attenuation of the average flux is nonexponential (the total flux, rather than some flux component, follows a nonexponential law).

4. As we found above, when we consider various problems of neutron transfer theory, the behavior of the average neutron flux in media with fluctuating parameters is essentially determined by the correlation function of the fluctuations in the medium (or by its Fourier image, i.e., the spectrum. In the M_1 approximation this is true for all types of diffusion media (in the case of large-scale fluctuations, this is true only for a normal distribution). The "mass operator" in the first approximation of the perturbation theory of Eq. (2.20) is a convolution of the Fourier image of the Green's function in the zeroth approximation and the fluctuation spectrum. When the region of large p-wave vectors gives the principal contribution to the convolution (Fourier variables which are conjugated to the radius vector), the corrections to the attenuation constant of the neutron flux are small, the perturbation-theory series for the "mass operator" converges rapidly, and the principal correction is taken into account by the M_1 approximation. When the fluctuation spectrum and the zeroth Green's function overlap strongly in the region of small wave vectors, the applicability of the M_1 approximation is limited. For example, when thermal neutrons diffuse in the layered medium of Eq. (3.1) (fluctuation of the absorption cross section) one cannot describe the "longitudinal" diffusion. In the case of the "transverse" diffusion in this medium, the perturbation-theory series for the "mass operator" converges rapidly for $l \ll L_0$, as we proved in 1 above. Finally, when the fluctuation spectrum contains a δ-like component δ (p), the M_1 approximation cannot be employed. A δ-like component in the spectrum indicates that an averaging over an ensemble of homogeneous media is made with a certain weight. As has been shown in the previous section, a nonexponential attenuation of the average neutron flux is the consequence. At the same time, the M_1 approximation leads to a strong exponential dependence upon the distance between the point of observation and the source. The same is true when the neutron motion is parallel to the layers in a plane lattice (kinetic conditions, see § 5). In this case, though no δ(p) occur in the fluctuation spectrum, the relation **pn** = 0 (n denotes the unit vector in the direction of neutron motion) implies that the inhomogeneity of the medium has no influence upon the motion of the neutrons, and the averaging must be performed over an ensemble of homogeneous media with parameters which fluctuate from medium to medium. This result is self-evident from the viewpoint of physics. The neutrons which we consider move in the material until they are absorbed and the average flux contains the neutron contribution which results from an averaging over the distribution of the fluctuating cross section.

The contribution of these neutrons can be approximately calculated with the M_1 approximation. We write the neutron flux in the form of two terms $G(x|y) = G_1(x|y) + G_2(x|y)$, where $G_1(x|y)$ denotes the flux of neutrons passing from layer to layer, and $G_2(x|y)$ the flux of neutrons moving all the time in one layer of the same material. Then G_2 satisfies the equation

$$\hat{B}_{\Sigma_i} G_2(x \,|\, y) = -\,\delta(x - y). \tag{8.40}$$

In principle, it is not difficult to find a solution to this equation (i denotes the number of the layer).

We average the equation which is satisfied by the function G_1:

$$\hat{B}_{\langle \Sigma \rangle} G_1(x \,|\, y) = -\,\delta(x - y) + \hat{\mu}(x) G_1(x \,|\, y) + \hat{\mu}(x) G_2(x \,|\, y) - \overset{\circ}{\hat{B}}_{\langle \Sigma \rangle} G_2(x \,|\, y). \tag{8.41}$$

The result is

$$\hat{B}_{\langle \Sigma \rangle} \langle G_1(x \,|\, y) \rangle = F(x \,|\, y) + \langle \hat{\mu}(x) G_1(x \,|\, y) \rangle. \tag{8.42}$$

The term F, which describes the influence of the effective sources

$$F(x \,|\, y) = -\,\delta(x - y) + \hat{B}_{\langle \Sigma \rangle}(x) \langle G_2(x \,|\, y) \rangle - \langle \hat{\mu}(x) G_2(x \,|\, y) \rangle, \tag{8.43}$$

can be found with the solution of Eq. (8.40).

Thus, we obtain

$$\langle G_1(x \,|\, y) \rangle = \int \langle G_{10}(x \,|\, x') \rangle F(x' \,|\, y)\, dx', \tag{8.44}$$

where the function $\langle G_{10}(x|x') \rangle$ satisfies the kinetic equation in the M_1 approximation:

$$\hat{B}_{\langle \Sigma \rangle}(x) \langle G_{10}(x \,|\, y) \rangle = -\,\delta(x - y) - \langle \hat{\mu}(x)\, G_{10}(x \,|\, y) \rangle \approx -\,\delta(x - y) - \int dx' M_1(x \,|\, x') \langle G_{10}(x' \,|\, y) \rangle. \tag{8.45}$$

Conclusion

In this article we obtained an approximating equation which describes the transition of neutrons through a medium containing fluctuations. The entire information which is required regarding the structure of the medium is included in the correlation function of the fluctuations or, more accurately, one must know the Fourier image of the correlation function (the spectrum). The fluctuation spectrum can be established for a large class of diffusion media ranging from a medium with random inhomogeneities [Eq. (3.1)] to periodic structures of the type defined by Eq. (3.43). However, it is very difficult to determine the spectrum for lattices with imperfections, e.g., lattices with disturbances of arbitrary amplitude or frequency.

We considered the fluctuations of the cross sections as functions of the spatial coordinates. One can expect that a similar statistical analysis of the neutron–distribution function will simplify numerical solutions to a kinetic equation whose coefficients are rapidly changing functions of energy (e.g., transport length of the free path in a crystalline moderator, or absorption cross section in the resonance region). This method may also be useful for solving certain problems of reactor kinetics.

Finally, we wish to mention that one must solve the wave equation with fluctuating coefficients in order to describe the transition of cold neutrons through an inhomogeneous medium [7-9, 20, 21].

In conclusion, the author expresses his sincere gratitude to M. V. Kazarnovskii for valuable discussions.

APPENDIX A

In this appendix we derive an equation which is satisfied by the averaged Green's function $\langle G(x \,|\, y) \rangle$. We assume that an irregular perturbation $\hat{\mu}(x)$ can be represented in the form

$$\hat{\mu}(x) = \sum_{k=1}^{N} \mu_k(x)\, \hat{a}_k(x), \tag{A.1}$$

where $\mu_k(x)$ denotes random amplitudes ($\langle \mu_k(x) \rangle = 0$) and $\hat{\alpha}_k(x)$ are operators operating on a set of variables x. In all problems considered in this article, the perturbation $\hat{\mu}(x)$ appears in the form stated in Eq. (A.1).

After inserting Eq. (A.1) into Eq. (2.1), we obtain

$$\hat{B} G(x \mid y) = -\delta(x-y) + \sum_{k=1}^{N} \mu_k(x) \hat{\alpha}_k(x) G(x \mid y). \tag{A.2}$$

We multiply this equation by

$$S = \exp\left[i \sum_{j=1}^{N} \int dz I_j(z) \mu_j(z) \right], \tag{A.3}$$

where $I_j(z)$ denotes some arbitrary functions. When we take the average of

$$\hat{B} S G(x \mid y) = -S\delta(x-y) + \sum_{k=1}^{N} \mu_k(x) S \hat{\alpha}_k(x) G(x \mid y) \tag{A.4}$$

over the random functions $\mu_k(x)$, we obtain

$$\hat{B} \langle G^I(x \mid y) \rangle = -\delta(x-y) + \sum_{k=1}^{N} \langle R_k^I(x \mid y) \rangle. \tag{A.5}$$

The following notations were introduced:

$$\langle G^I(x \mid y) \rangle = \frac{1}{\langle S \rangle} \langle S G(x \mid y) \rangle, \tag{A.6}$$

$$\langle R_k^I(x \mid y) \rangle = \frac{1}{\langle S \rangle} \langle \mu_k(x) S \hat{\alpha}_k(x) G(x \mid y) \rangle. \tag{A.7}$$

The quantity

$$\langle S \rangle = \left\langle \exp\left[i \sum_{k=1}^{N} \int dz I_k(z) \mu_k(z) \right] \right\rangle \tag{A.8}$$

is determined by the functions $I_k(z)$, i.e., $\langle S \rangle$ is a functional. In the theory of random processes, the functional $\langle S \rangle$ is termed "characteristic functional of the random functions $\mu_k(z)$." When we let $I_k(z)$ tend to zero, we obtain from Eq. (A.5) an equation for the function $\langle G \rangle$. First, let us express $\langle R_k^I(x \mid y) \rangle$ by the function $\langle G^I(x|y) \rangle$. Obviously,*

$$\begin{aligned}
\langle R_k^I(x \mid y) \rangle &= -\frac{i}{\langle S \rangle} \frac{\delta}{\delta I_k(x)} \left\{ \langle S \rangle \frac{1}{\langle S \rangle} \langle S \hat{\alpha}_k G(x \mid y) \rangle \right\} = \\
&= -\frac{i}{\langle S \rangle} \frac{\delta}{\delta I_k(x)} \{ \langle S \rangle \langle G_k^I(x \mid y) \rangle \} = \\
&= -\frac{i\delta}{\delta I_k(x)} \langle G_k^I(x \mid y) \rangle - \frac{i}{\langle S \rangle} \frac{\delta \langle S \rangle}{\delta I_k(x)} \langle G_k^I(x \mid y) \rangle,
\end{aligned} \tag{A.9}$$

where

$$\langle G_k^I(x \mid y) \rangle = \frac{1}{\langle S \rangle} \langle S \hat{\alpha}_k G(x \mid y) \rangle = \hat{\alpha}_k \langle G^I(x \mid y) \rangle. \tag{A.10}$$

---

*A simple outline of the rules of functional differentiation has been given in [22] and in the monograph [23].

After inserting Eq. (A.9) into Eq. (A.5) and taking the limit $I_k \rightarrow 0$, we obtain the following equation:

$$\hat{B}\langle G(x|y)\rangle = -\delta(x-y) - i\sum_{k=1}^{N}\frac{\delta}{\delta I_k(x)}\langle G_k^I(x|y)\rangle\Big|_{I_k=0} - i\sum_{k=1}^{N}\frac{1}{\langle S\rangle}\frac{\delta\langle S\rangle}{\delta I_k(x)}\langle G_k^I(x|y)\rangle\Big|_{I_k=0}. \tag{A.11}$$

The last term on the right side of Eq. (A.11) vanishes because

$$-i\frac{1}{\langle S\rangle}\frac{\delta\langle S\rangle}{\delta I_k(x)}\Big|_{I_k(x)=0} = -i\frac{1}{\langle S\rangle}i\langle\mu_k(x)\,S\rangle\Big|_{I_k=0} = \langle\mu_k(x)\rangle = 0. \tag{A.12}$$

Thus, the function $\langle G(x|y)\rangle$ is the solution of the following equation:

$$\hat{B}\langle G(x|y)\rangle = -\delta(x-y) - i\sum_{k=1}^{N}\frac{\delta}{\delta I_k(x)}\langle G_k^I(x|y)\rangle\Big|_{I_k=0}. \tag{A.13}$$

We simplify this equation by removing the functional derivatives by the introduction of several new quantities.* We use the rule for the variation of a complex functional [22] and replace the argument of the functional:

$$\frac{\delta\langle G_k^I(x|y)\rangle}{\delta I_k(z)} = \sum_{k'=1}^{N}\int dz'\,\frac{\delta\langle G_k^I(x|y)\rangle}{\delta\Phi_{k'}(z')}\frac{\delta\Phi_{k'}(z')}{\delta I_k(z)} \tag{A.14}$$

where

$$\Phi_k(z) = -i\frac{\delta\ln\langle S\rangle}{\delta I_k(z)} = -i\frac{1}{\langle S\rangle}\frac{\delta\langle S\rangle}{\delta I_k(z)}. \tag{A.15}$$

We express $\delta\langle G^I\rangle/\delta\Phi_k$ by the variational derivatives of the inverse function $\langle G^I\rangle^{-1}$. To do this, we perform the variation of

$$\int\langle G^I(x|x'')\rangle\langle G^I(x''|x')\rangle^{-1}dx'' = \delta(x-x'), \tag{A.16}$$

with respect to $\Phi_k(z)$. The last equation defines the inverse function $\langle G\rangle^{-1}$. We obtain

$$\int dx'\,\frac{\delta\langle G^I(x|x')\rangle}{\delta\Phi_k(z)}\langle G^I(x'|v)\rangle^{-1} = -\int dx'\,\langle G^I(x|x')\rangle\frac{\delta}{\delta\Phi_k(z)}\langle G^I(x'|v)\rangle^{-1}. \tag{A.17}$$

We multiply both sides of Eq. (A.17) from the right by $\langle G^I(v|y)\rangle$ and integrate over v. With Eq. (A.16), we obtain the desired expression for the functional derivatives

$$\frac{\delta\langle G^I(x|y)\rangle}{\delta\Phi_k(z)} = -\int dx'\int dx''\,\langle G^I(x|x')\rangle\frac{\delta}{\delta\Phi_k(z)}\langle G^I(x'|x'')\rangle^{-1}\langle G^I(x''|y)\rangle. \tag{A.18}$$

Similarly, one easily verifies that

$$\frac{\delta\langle G_k^I(x|y)\rangle}{\delta\Phi_k(z)} = -\iint dx'dx''\,\hat{\alpha}_k(x)\langle G^I(x|x')\rangle\frac{\delta}{\delta\Phi_k(z)}\langle G^I(x'|x'')\rangle^{-1}\langle G^I(x''|y)\rangle. \tag{A.19}$$

---

*A similar method is used in quantum field theory [3, 4] and in statistical physics [5].

Thus, it follows from Eqs. (A.14) and (A.19) that

$$\frac{\delta \langle G_k^I (x\,|\,y)\rangle}{\delta I_k(z)} = -\sum_{k=1}^N \iiint dz' dx' dx'' \hat{\alpha}_k(x) \langle G^I(x\,|\,x')\rangle \frac{\delta}{\delta \Phi_{k'}(z')} \langle G^I(x'\,|\,x'')\rangle^{-1} \langle G^I(x''\,|\,y)\rangle \frac{\delta \Phi_{k'}(z')}{\delta I_k(z)}. \qquad \text{(A.20)}$$

We transform the variational derivative $\delta \Phi_{k'}(z')/\delta I_k(z)$:

$$\frac{\delta \Phi_{k'}(z')}{\delta I_k(z)} = -i\frac{\delta^2 \ln \langle S\rangle}{\delta I_k(z)\delta I_{k'}(z')} = -i\frac{\delta^2}{\delta I_k(z)\delta I_{k'}(z')} \left\{\ln \left\langle \exp\left[i\sum_{l=1}^N \int I_l(v)\mu_l(v)\,dv\right]\right\rangle\right\} =$$

$$= -\frac{i}{\langle S\rangle^2}\left\langle \mu_{k'}(z')\exp\left[i\sum_{l=1}^N \int I_l(v)\mu_l(v)\,dv\right]\right\rangle\left\langle \mu_k(z)\exp\left[i\sum_{l=1}^N \int I_l(v)\mu_l(v)\,dv\right]\right\rangle +$$

$$+ i\frac{1}{\langle S\rangle}\left\langle \mu_{k'}(z')\mu_k(z)\exp\left[i\sum_{l=1}^N \int I_l(v)\mu_l(v)\,dv\right]\right\rangle. \qquad \text{(A.21)}$$

With $\langle \mu_k\rangle = 0$, we obtain for $I_k \to 0$:

$$\left.\frac{\delta \Phi_{k'}(z')}{\delta I_k(z)}\right|_{I_k=0} = i\langle \mu_{k'}(z')\mu_k(z)\rangle. \qquad \text{(A.22)}$$

After substituting Eq. (A.20) for $I_k = 0$ into Eq. (A.13), we obtain the final equation for the averaged Green's function $\langle G\rangle$:

$$\hat{B}\langle G(x\,|\,y)\rangle = -\delta(x-y) - \int dx' M(x\,|\,x')\langle G(x'\,|\,y)\rangle, \qquad \text{(A.23)}$$

where

$$M(x\,|\,x') = -\sum_{k=1}^N \sum_{k'=1}^N \int dz' \int dz \langle \mu_k(z)\mu_{k'}(z')\rangle \hat{\alpha}_k(x)\langle G(x\,|\,z)\rangle \Gamma_{k'}(z, x'\,|\,z) \qquad \text{(A.24)}$$

and

$$\Gamma_{k'}(z, x'\,|\,z') = -\left.\frac{\delta}{\delta \Phi_{k'}(z')}\langle G^I(z\,|\,x')\rangle^{-1}\right|_{I_k=0} \qquad \text{(A.25)}$$

are, respectively, the analogs of the mass operator and the apex function (these quantities are used in quantum field theory [3-6]). On the first nonvanishing approximation of perturbation theory [5] we have

$$\langle G(x\,|\,y)\rangle \approx G_0(x\,|\,y) \qquad \text{(A.26)}$$

and

$$\Gamma_k^0(x, y\,|\,z) \approx -\delta(x-z)\hat{\alpha}_k(x)\delta(x-y). \qquad \text{(A.27)}$$

In this approximation, the "mass operator" is equal to

$$M_1(x\,|\,x') = \sum_{k=1}^N \sum_{k'=1}^N \int dz \int dz' \langle \mu_k(x)\mu_{k'}(z')\rangle \hat{\alpha}_k(x) G_0(x\,|\,z)\delta(z-z')\hat{\alpha}_{k'}(z)\delta(z-x') =$$

$$= \sum_{k=1}^N \sum_{k'=1}^N \int dz \langle \mu_k(x)\mu_{k'}(z)\rangle \hat{\alpha}_k(x) G_0(x\,|\,z)\hat{\alpha}_{k'}(z)\delta(z-x'), \qquad \text{(A.28)}$$

and, accordingly,

$$\int dx' M_1(x\,|\,x')\,\langle G\,(x'\,|\,y)\rangle = \sum_{k=1}^{N}\sum_{k'=1}^{N}\int dx'\,\langle\mu_k(x)\,\mu_{k'}(x')\rangle\,\hat{\alpha}_k(x)\,G_0(x\,|\,x')\,\hat{\alpha}_{k'}(x')\,\langle G(x'\,|\,y)\rangle. \tag{A.29}$$

In the first nonvanishing approximation of perturbation theory the function $\langle G(x|\,y)\rangle$ satisfies the equation

$$\hat{B}\,\langle G\,(x\,|\,y)\rangle = -\,\delta\,(x - y) - \int dx'\sum_{k=1}^{N}\sum_{k'=1}^{N}\langle\mu_k(x)\,\mu_{k'}(x')\rangle\,\hat{\alpha}_k(x)\,G_0(x\,|\,x')\,\hat{\alpha}_{k'}(x')\,\langle G(x'\,|\,y)\rangle, \tag{A.30}$$

which we previously obtained by another method.

It is easy to verify that there exists a correspondence between the problem in the M_1 approximation and the exact formulation of the problem with the characteristic functional

$$\langle S\rangle = 1 - \frac{1}{2}\sum_{k,\,j}\iint dz_1 dz_2 I_k(z_1)\,I_j(z_2)\,\langle\hat{\mu}_k(z)\,\hat{\mu}_j(z_2)\rangle +$$

$$+\frac{1}{24}\sum_{k,\,j,\,l,\,m}\int\cdots\int dz_1\ldots dz_4 I_k(z_1)\ldots I_m(z_4)\,\langle\hat{\mu}_k(z_1)\,\hat{\mu}_j(z_2)\rangle\,\langle\hat{\mu}_l(z_3)\,\hat{\mu}_m(z_4)\rangle - \ldots \tag{A.31}$$

This expansion can be represented in the following symbolic form:

$$\langle S\rangle = \cos\sqrt{\sum_{k,\,j}\iint dz_1 dz_2 I_k(z_1)\,I_j(z_2)\,\langle\mu_k(z_1)\,\mu_j(z_2)\rangle}\,. \tag{A.32}$$

In the case of an ensemble of homogeneous media, the last expression becomes:

$$\langle S\rangle = \cos I\Delta, \tag{A.33}$$

which corresponds to the distribution law of Eq. (1.3).

APPENDIX B

We will derive an equation for the correlation function

$$\langle D(x_2,\,x_1\,|\,x_2',\,x_1')\rangle = \langle G(x_2\,|\,x_1)\,G(x_2'\,|\,x_1')\rangle. \tag{B.1}$$

After multiplying Eq. (2.10) by a similar equation and averaging the equation for the correlator $\langle D\rangle$, we obtain

$$\langle D(x_2,\,x_1\,|\,x_2',\,x_1')\rangle = G_0(x_2\,|\,x_1)\,G_0(x_2'\,|\,x_1') +$$

$$+\iint dy_1 dy_1'\,G_0(x_2\,|\,y_1)\,G_0(x_2'\,|\,y_1')\,\langle\hat{\mu}(y_1)\,\hat{\mu}(y_1')\rangle\,G_0(y_1\,|\,x_1)\,G_0(y_1'\,|\,x_1') +$$

$$+\iint dy_1 dy_2\,G_0(x_2\,|\,y_1)\,\langle\hat{\mu}(y_1)\,G_0(y_1\,|\,y_2)\,\hat{\mu}(y_2)\rangle\,G_0(y_2\,|\,x_1)\,G_0(x_2'\,|\,x_1') +$$

$$+\,G_0(x_2\,|\,x_1)\iint dy_1' dy_2'\,G_0(x_2'\,|\,y_1')\,\langle\hat{\mu}(y_1')\,G_0(y_1'\,|\,y_2')\,\hat{\mu}(y_1')\rangle\,G_0(y_2'\,|\,x_1') + \ldots \tag{B.2}$$

We denote $\langle D(x_2,\,x_1|,\,x_2' x_2'|)\rangle$ by the symbol

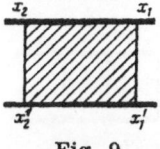

Fig. 9

and obtain the following graphical representation of Eq. (B.2):

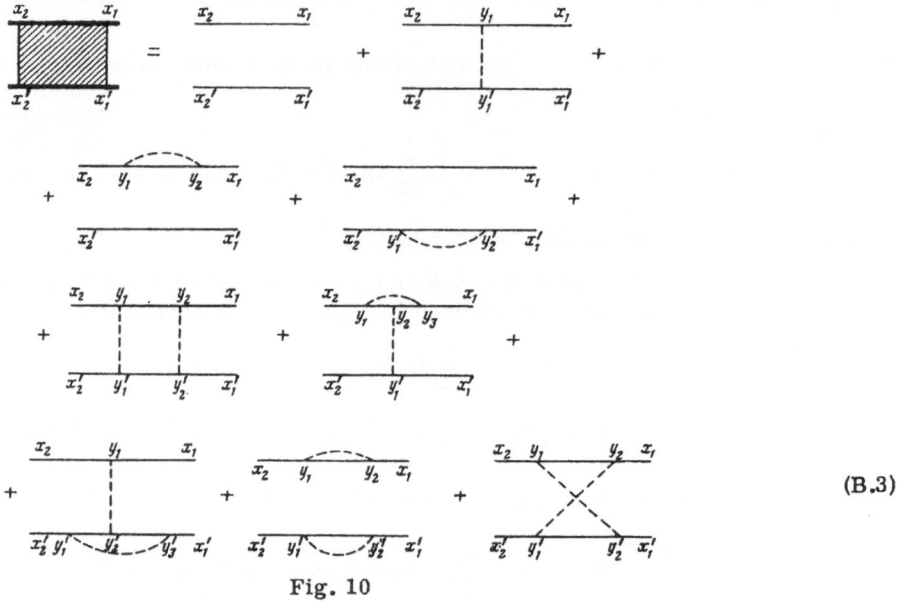

$$(B.3)$$

Fig. 10

The notation is that introduced in § 2.

The graphs with dashed lines drawn to one of the solid lines can be summed by making use of the results of § 2. We obtain in the approximation of Eq. (2.20):

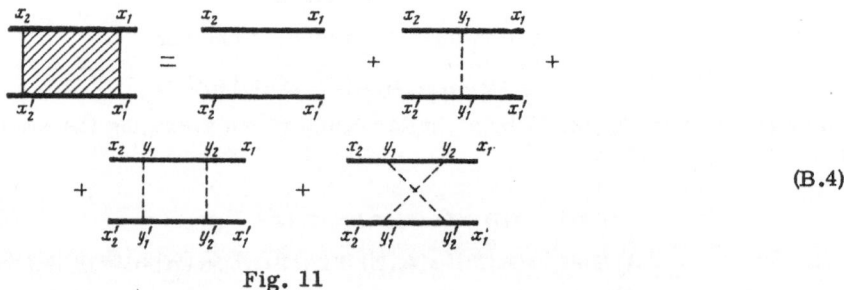

$$(B.4)$$

Fig. 11

The first term in Eq. (B.4) is equal to $\langle G(x_2, x_1) \rangle \langle G(x_2'|x_1') \rangle$ and does not contain correlations between $G(x_2|x_1)$ and $G(x_2'|x_1')$. These correlations can be taken into account in approximate form by summing diagrams of the type

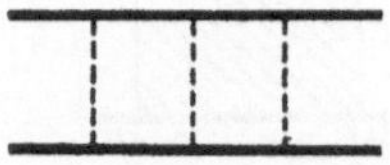

$$(B.5)$$

Fig. 12

In the so-called staircase approximation, the correlation function satisfies the equation

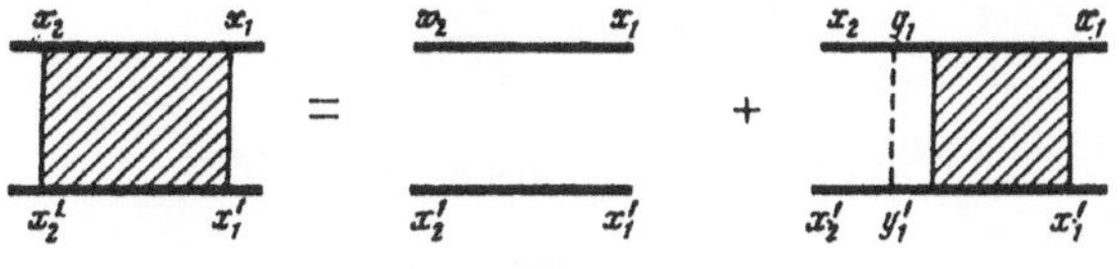

$$\tag{B.6}$$

Fig. 13

or

$$\langle D(x_2, x_1 \mid x_2', x_1') \rangle = \langle G(x_2 \mid x_1) \rangle \langle G(x_2' \mid x_1') \rangle +$$

$$+ \iint dy_1 dy_1' \langle G(x_2 \mid y_1) \rangle \langle G(x_1' \mid y_1') \rangle \langle \hat{\mu}(y_1) \, \hat{\mu}(y_1') \rangle \langle D(y_1, x_1 \mid y_1', x_1') \rangle. \tag{B.7}$$

The mean square neutron flux [for a source $\sim \delta(x - y)$] is equal to the correlation function $\langle D \rangle$ at pairwise coinciding arguments $\langle D(x, y \mid x, y) \rangle$.

APPENDIX C

In this appendix we derive an expression for the neutron distribution over the energy ψ, and the distribution function is averaged over the fluctuations of the absorption cross section at a constant scattering cross section. The approximated differential equation describing neutron moderation in an inhomogeneous moderator has the form [1, 2]

$$\frac{d}{dE} (E \Sigma \xi \psi) - \Sigma_a \psi + \frac{\Sigma_s}{\Sigma} S = 0, \tag{C.1}$$

where $\xi = 2/A$; A denotes the mass number of the moderator atoms; and S is the neutron output of the source.

In the case of weak absorption we have $\langle \Sigma_a \rangle \ll \Sigma_s$ and $\Sigma_s / \Sigma \approx 1$, and Eq. (C.1) assumes the form

$$E \frac{d\psi}{dE} + \left(1 - \frac{\Sigma_a}{\xi \Sigma_s}\right) \psi = - \frac{1}{\xi \Sigma_s} S. \tag{C.2}$$

In the case of a monoenergetic source, this equation is

$$E \frac{d\psi}{dE} - [a - 1 + b(E)] \psi = - q \delta(E - E_0), \tag{C.3}$$

$$a = \frac{1}{\xi \Sigma_s} \langle \Sigma_a \rangle, \quad b(E) = \frac{\Sigma_a(E)}{\xi \Sigma_s} - a, \quad \langle b(E) \rangle = 0 \tag{C.4}$$

and has the solution

$$\psi(E) = q \frac{E_0^{-a}}{E^{1-a}} e^{- \int\limits_{E}^{E_0} \frac{b(E') \, dE'}{E'}} \tag{C.5}$$

By averaging Eq. (C.5) over the fluctuations of the absorption cross section when the energy changes, we obtain

$$\langle \psi(E \mid E_0) \rangle = q \frac{E_0^{-a}}{E^{1-a}} \left\langle e^{\int\limits_{E}^{E_0} \frac{b(E')}{E'} dE'} \right\rangle. \tag{C.6}$$

In the zeroth approximation, we have $b(E) = 0$ and therefore,

$$\psi_0(E \mid E_0) = q \, \frac{E_0^{-a}}{E^{1-a}}. \tag{C.7}$$

In the case of a normal distribution of the fluctuations, we obtain

$$\langle \psi(E \mid E_0) \rangle = q \, \frac{E_0^{-a}}{E^{1-a}} \exp \left\{ \frac{1}{2} \int\limits_{E}^{E_0} \int\limits_{E}^{E_0} \frac{dE' dE''}{E' E''} \langle b(E') b(E'') \rangle \right\} =$$

$$= \psi_0(E \mid E_0) \exp \left\{ \frac{1}{2} \int\limits_{E}^{E_0} \int\limits_{E}^{E_0} \frac{dE' dE''}{E' E''} \langle b(E') b(E'') \rangle \right\}. \tag{C.8}$$

When the correlation length is small relative to the characteristic energy interval in which the neutron distribution in a moderator with constant cross sections changes substantially, then we use the approximation

$$\langle b(E) b(E') \rangle \approx \alpha \delta(E - E'), \quad \alpha = \text{const}, \tag{C.9}$$

and obtain the following expression from Eq. (C.8):

$$\langle \psi(E \mid E_0) \rangle = \psi_0(E \mid E_0) e^{\frac{\alpha}{2} \int\limits_{E}^{E_0} \frac{dE'}{(E')^2}} = \psi_0(E \mid E_0) e^{\frac{\alpha}{2} \left(\frac{1}{E} - \frac{1}{E_0} \right)}. \tag{C.10}$$

When the correlation function changes slowly with the energy (large-scale fluctuations) we have

$$\langle b(E) b(E') \rangle \approx A = \text{const} \tag{C.11}$$

and

$$\langle \psi(E \mid E_0) \rangle = \psi_0(E \mid E_0) e^{\frac{A}{2} (\ln E_0 - \ln E)^2}. \tag{C.12}$$

The corresponding equation in the M_1 approximation with the correlation Eq. (C.9) has the form form

$$\left(E \, \frac{d}{dE} + 1 - a \right) \langle \psi(E \mid E_0) \rangle = -q \delta(E - E_0) - \frac{\alpha}{2E} \langle \psi(E \mid E_0) \rangle. \tag{C.13}$$

The solution of this equation is

$$\langle \psi(E \mid E_0) \rangle = \psi_0(E \mid E_0) e^{\frac{\alpha}{2} \left(\frac{1}{E} - \frac{1}{E_0} \right)}$$

and coincides with Eq. (C.10), which was previously obtained by another method.

References

1. A. D. Galanin, Theory of Nuclear Reactors Operating with Thermal Neutrons, Atomizdat, Moscow (1959).
2. A Weinberg and E. Wigner, Physical Theory of Neutron Chain Reactors. (II). U. of Chicago Press, Chicago (1958).
3. A. I. Akhiezer and A. B. Berestetskii, Quantum Electrodynamics, Fizmatgiz, Moscow (1959).
4. E. S. Fradkin, Trudy FIAN, 29:7 (1965).
5. V. I. Bonch-Bruevich and S. V. Tyablikov, The Green's Function in Statistical Mechanics, Fizmatgiz, Moscow (1961).

6. D. A. Kirzhnits, Multiparticle Field-Theory Methods, Atomizdat, Moscow (1963).

7. R. C. Bourrett, Nuovo Cim., 26:1 (1962).

8. V. Finkel'berg, Zh. Eksp. Teor. Fiz., 46:725 (1964).

9. V. I. Tatarskii, Zh. Eksp. Teor. Fiz., 46:1399 (1964).

10. A. V. Stepanov, Pulsed Neutron Research, IAEA Vienna, Vol. 1 (1965), p. 339.

11. A. V. Stepanov, Atomnaya Energiya, 20:265 (1966).

12. A. V. Stepanov, Atomnaya Energiya, 21:292 (1966).

13. Ya. V. Shevelev, Atomnaya Energiya, 2:224 (1957).

14. A. V. Stepanov, Atomnaya Energiya, 22:271 (1967).

15. B. R. Levin, Theory of Random Processes and Its Application in Radio Engineering, Sovet-skoe Radio, Moscow (1960).

16. V. Zavoiskii, Atomnaya Energiya, 10:272, 521 (1961).

17. A. Akhizer and I. Pomeranchuk, Some Problems of Nuclear Theory, Gostekhizdat, Moscow (1950).

18. R. Meghreblain and D. Holmes, Reactor Analysis, McGraw-Hill, New York (1960).

19. M. V. Kazarnovskii, A. V. Stepanov, and F. L. Shapiro, Reports of the Second International Conference on the Peaceful Use of Atomic Energy (Geneva 1958), Reports of Soviet Scientists, Vol. I, Atomizdat, Moscow (1959), p. 469.

20. I. V. Andreev, Zh. Eksp. Teor. Fiz., 48:1437 (1965).

21. A. V. Stepanov, Pulsed Neutron Research, IAEA Vienna, Vol. 1 (1965), p. 355.

22. A. I. Alekseev, Usp. Fiz. Nauk, 23:41 (1961).

23. A. S. Monin and A. M. Yaglom, Statistical Hydromechanics, Vol. 1, Nauka, Moscow (1965); V. I. Tatarskii, Wave Propagation in a Turbulent Atmosphere, Nauka, Moscow (1967).

MODEL OF A HEAVY MODERATOR AND AN ANALYTICAL SOLUTION OF THE NEUTRON THERMALIZATION PROBLEM

M. V. Kazarnovskii

Introduction

In recent years, solutions of the neutron-transfer equation have been obtained with computers in comprehensive form, so that it has become possible to solve almost any concrete problem in the theory of neutron thermalization, provided that the differential characteristics of the interaction between the neutrons and matter are known with sufficient accuracy, i.e., when the scattering probability and the absorption cross section are known. However, even in the case of moderators which were most extensively researched, the neutron-scattering probabilities are still incompletely and inexactly known. Therefore, calculations are of any value only when the possible influence of errors and of the incompleteness of the data can be assessed. Apart from this, it is frequently necessary to study neutron distributions for various geometries and parameters of the moderator, for example, when the optimum parameters of a moderator are to be determined. Numerical solutions to problems of this type require very laborious calculations and the numerical results are sometimes rather sketchy.

However, on the other hand, in the majority of cases of practical interest, the neutron distributions are weakly dependent upon the scattering probabilities, but are mainly determined by a relatively small number of neutron-dependent parameters of the medium, for example, the total scattering cross section, the transfer cross section, the first one or two transferred energy (or velocity) moments of scattering, the absorption cross section, etc. This condition is particularly applicable to stationary neutron spectra which are usually smooth curves of a very simple form. In these cases one can obtain (by calculations) a satisfactory agreement between the solution to the neutron-transfer equation and models of scattering operators, with models rather far from reality, yet including the proper parameters mentioned above and satisfying conditions which result for the scattering operator from the general scattering laws for neutrons in matter. The results can at least serve as a zeroth approximation.

Of particular interest are models of a scattering operator which facilitates the solution of the neutron-transfer equation in various cases of practical importance and which at the same time satisfies the general requirements. These models include some arbitrary functions or parameters, and by varying these parameters one can approximate the neutron scattering in a real moderator and examine the influence of various scattering parameters upon the neutron distribution.

248

Some time ago [1], a model for a neutron-scattering kernel in a heavy moderator was proposed. In this model the scattering operator has the desired properties when the diffusion approximation is employed. However, this model did not take into consideration the anisotropy of neutron scattering. The present article suggests and discusses a more general model of a neutron-scattering operator in a heavy moderator. Our model makes it possible to take the neutron-scattering anisotropy into account.

In §1 we consider the general properties of a neutron-scattering operator in thermalization theory. The following conditions must be fulfilled by the operator: a) at large velocities, the neutron-scattering law in the matter transforms into the scattering law which is valid for free atoms at rest; b) the Maxwell distribution of neutrons is preserved in the scattering process; c) the scattering probability cannot be negative; and d) the total number of neutrons remains conserved in the scattering process. It could be shown that these conditions to the scattering operator are less stringent than the conditions which must be used for the scattering probability.

A model of the scattering operator is suggested in §2. This model satisfies all the conditions mentioned above and includes three arbitrary functions of the velocity and the scattering angles of the neutrons. The functions can be chosen so that the best approximation to the neutron-scattering law in a real moderator is obtained.

Possible approximations are discussed in §3.

In §4, we determine for this model the distribution function of the neutron velocities in integral form. This is discussed for the case of an arbitrary (nonstationary) source. We give a short discussion of the eigenvalues and eigenfunctions of the neutron-transfer equation.

§1. Neutron-Transfer Equation and General Properties of the Scattering Operator

The neutron distribution $N(\mathbf{r}, \mathbf{v}, t)$ in a moderator, i.e., the number of neutrons in a unit of volume near the point defined by $\mathbf{r}$ and in a unit velocity interval near $\mathbf{v}$ at the time t, must satisfy the transfer equation

$$\frac{\partial N}{\partial t} + \mathbf{v}\nabla N = -\frac{1}{\tau_a}N + \hat{H}N + S \tag{1.1}$$

in the general case. Here $\tau_a(v)$ denotes the lifetime of neutrons with the velocity v (limited by absorption); $S(\mathbf{r}, \mathbf{v}, t)$ is the neutron density in the source; and $\hat{H}$ denotes the scattering operator defined by

$$\hat{H}\psi(\mathbf{v}) = \int \psi(\mathbf{v}')w(\mathbf{v}' \to \mathbf{v})d\mathbf{v}' - \psi(\mathbf{v})\int w(\mathbf{v} \to \mathbf{v}')d\mathbf{v}', \tag{1.2}$$

where $w(\mathbf{v}' \to \mathbf{v})$ denotes the scattering pattern, i.e., the probability that a neutron with velocity $\mathbf{v}'$ passes after one second into a unit interval of velocities around $\mathbf{v}$. The boundary conditions for the solution to Eq. (1.1) result from the neutron-flow balance on the surface of the moderator.*

Thus, the scattering properties of a moderator affect the neutron distribution only through the scattering operator. The scattering operator itself must satisfy the following conditions, which result from the overall properties of the scattering process:

---

* For example, if all neutron sources are situated within a moderator, on the moderator surface $N(\mathbf{r}, \mathbf{v}, t)$ must tend to zero for all vectors $\mathbf{v}$ directed inside the moderator (for all t).

1. For $v' \to \infty$, the quantity $w(v' \to v)$ transforms into the probability for scattering at free atoms at rest, i.e., in the general case of a moderator consisting of a mixture of various isotopes, we have

$$\lim_{v' \to \infty} w(v' \to v) = \sum_j \frac{(A_j + 1)^2}{2\pi A_j^2} \Sigma_{jf} \, \delta\left[v'^2 - v^2 - \frac{(\mathbf{v} - \mathbf{v}')^2}{A_j}\right], \tag{1.3}$$

where A_j denotes the mass number of atoms of type j, and Σ_{jf} the particle macroscopical scattering cross section for these atoms (the atoms are considered as free and at rest).

Accordingly, in the limit $v \to \infty$, the scattering operator of Eq. (1.2) assumes the form

$$\lim_{v \to \infty} \hat{H}\psi(\mathbf{v}) = \hat{H}_{as}\psi(\mathbf{v}) = \sum_j \left\{ \int d\mathbf{v}'\psi(\mathbf{v}') \frac{(A_j + 1)^2}{2\pi A_j^2} \Sigma_{jf} \, \delta\left[v'^2 - v^2 - \frac{(\mathbf{v} - \mathbf{v}')^2}{A_j}\right] - \psi(\mathbf{v}) \, v \, \Sigma_{jf} \right\}. \tag{1.4}$$

In the case of a heavy moderator, i.e., for $A_j \to \infty$, we obtain by an expansion of the δ function into a power series of $1/A_j$ and by some simple calculations

$$\hat{H}_{as}\psi(\mathbf{v}) = \int \frac{d\Omega'}{4\pi} \psi(v\Omega') v\Sigma_f - \psi(v\Omega) v\Sigma_f + \int \frac{d\Omega'}{4\pi} \psi(v\Omega') 4 \frac{v\Sigma_f}{A}\left(1 - \frac{\Omega\Omega'}{2}\right) + \int \frac{d\Omega}{4\pi} \frac{\partial\psi(v\Omega')}{\partial v} \frac{v^2\Sigma_f}{A}(1 - \Omega\Omega'), \tag{1.5}$$

where we introduced the notations

$$\Sigma_f = \sum_j \Sigma_{jf}, \quad \frac{1}{A} = \sum_j \frac{1}{A_j} \frac{\Sigma_{jf}}{\Sigma_f}, \tag{1.6}$$

$$\mathbf{v} = v\Omega. \tag{1.7}$$

The terms which we omitted in Eq. (1.5) are of the order of $1/A^2$. However, the sum of the first two terms of the right side of Eq. (1.5) is often very small (when the angular distribution of the neutrons is close to an isotropic distribution). Moreover, this sum describes only the spatial diffusion of the neutrons, but not their moderation. It is therefore of principal importance to include the third and the fourth terms. We have to consider in detail the role of the terms which were neglected in Eq. (1.5). We find that the neglected terms may be important at high velocities. They provide a contribution of the order $\frac{v}{A}\left|\frac{\partial \ln \psi}{\partial v}\right|$ at $v \gg 1$. Then Eq. (1.5) is valid if

$$v\left|\frac{\partial \ln \psi}{\partial v}\right| \ll A. \tag{1.8}$$

In heavy moderators ($A \gg 1$), this condition is usually satisfied in the entire velocity region of practical interest.* Obviously, the "asymptotic" expressions of the various quantities must be defined as the values of the quantities at large v values; yet the v values must be such that they satisfy condition (1.8).

2. The scattering probability must satisfy the small-scale equilibrium condition[†]

---

* For example, in the case of a Maxwell distribution, condition (1.8) is violated for $v \sim \sqrt{A}$, i.e., in a region in which the number of neutrons is exponentially small.

[†] It is assumed that the velocity is measured in units of $\sqrt{2kT/m}$, where T denotes the temperature of the moderator, k the Boltzmann constant, and m the neutron mass.

$$e^{-v'^2} w(\mathbf{v}' \to \mathbf{v}) = e^{-v^2} w(\mathbf{v} \to \mathbf{v}').$$

(1.9)

This means that the scattering operator must make the function e^{-v^2} vanish:

$$\hat{H} e^{-v^2} \equiv \int e^{-v'^2} w(\mathbf{v}' \to \mathbf{v}) d\mathbf{v}' - e^{-v^2} \int w(\mathbf{v} \to \mathbf{v}') d\mathbf{v}' = 0.$$

(1.10)

We note that condition (1.10) is less stringent than condition (1.9). The condition was discussed in greater detail in [1].

3. The function $w(\mathbf{v}' \to \mathbf{v})$ must be positive for all $\mathbf{v}$ and $\mathbf{v}'$. When this condition is violated, the neutron distribution can become negative and this is obviously an absurdity. As a matter of fact, when we assume that $w(\mathbf{v}' \to \mathbf{v})$ is negative for $\mathbf{v}' = \mathbf{v}_0$, we have $\mathbf{v} = \mathbf{v}_1$ (it was assumed that $\mathbf{v}_0 \neq \mathbf{v}_1$, since $w(\mathbf{v}' \to \mathbf{v})$ for $\mathbf{v}' = \mathbf{v}$ means that no scattering occurs and this case is meaningless). Furthermore, it is assumed that at the time $t = 0$ sources uniformly distributed in space emit neutrons with a velocity $\mathbf{v}_0$ in pulses, i.e., for $t \to 0$ ($t > 0$) the neutron distribution has the form

$$N(\mathbf{v}) = N_0 \delta(\mathbf{v} - \mathbf{v}_0), \quad N_0 = \text{const.}$$

(1.11)

Then Eq. (1.1) assumes the following form for $t \to 0$:

$$\frac{\partial N(\mathbf{v}, t)}{\partial t} = N_0 w(\mathbf{v}_0 \to \mathbf{v}') - \left[\frac{1}{\tau_a} + \int w(\mathbf{v} \to \mathbf{v}') d\mathbf{v}' \right] N_0 \delta(\mathbf{v} - \mathbf{v}_0).$$

(1.12)

This means that $\dfrac{\partial N(\mathbf{v}_1, t)}{\partial t}$ is negative at that time. But since $N(\mathbf{v}_1, t)$ is zero at that time [according to Eq. (1.11)], $N(\mathbf{v}_1, t)$ becomes negative in the following moment.

Thus, in the general case the condition

$$w(\mathbf{v}' \to \mathbf{v}) > 0 \quad \text{for all} \quad \mathbf{v}' \neq \mathbf{v}$$

(1.13)

is one of the most important properties of the scattering operator. However, when we restrict our considerations to the case of greatest practical interest, namely to continuous and differentiable neutron distributions [the approximation Eq. (1.5) is sensible only for this type of distributions], we can use a condition which is less rigorous than that of (1.13). Then, the condition that the neutron distribution be nonnegative (for nonnegative sources) means that the condition

$$\left. \frac{\partial N(\mathbf{r}_0, \mathbf{v}_0, t)}{\partial t} \right|_{t=t_0} > 0$$

(1.14)

must hold if at some moment t_0 the neutron distribution N tends to zero at some point $\mathbf{r}_0$ in space at some velocity $\mathbf{v}_0$ (in the preceding moments, N was a positive function of all its variables). Moreover, since the function $N(\mathbf{r}, \mathbf{v}, t_0)$ has a minimum at $\mathbf{r} = \mathbf{r}_0$, $\mathbf{v} = \mathbf{v}_0$, we have

$$\nabla N(\mathbf{r}, \mathbf{v}_0, t_0)|_{\mathbf{r}=\mathbf{r}_0} = \left. \frac{\partial N(\mathbf{r}_0, \mathbf{v}, t_0)}{\partial \mathbf{v}} \right|_{\mathbf{v}=\mathbf{v}_0} = 0$$

(1.15)

$\left(\dfrac{\partial}{\partial \mathbf{v}} = \mathbf{i} \dfrac{\partial}{\partial v_x} + \mathbf{j} \dfrac{\partial}{\partial v_y} + \mathbf{k} \dfrac{\partial}{\partial v_z} \right.$, where $\mathbf{i}, \mathbf{j}$, and $\mathbf{k}$ denote the unit vectors in x, y, and z direction, respectively). When we insert Eqs. (1.14) and (1.15) into Eq. (1.1) for the most "dangerous" case

$$S(\mathbf{r}_0, \mathbf{v}_0, t_0) = 0,$$

(1.16)

we obtain

$$\hat{H} N(\mathbf{r}_0, \mathbf{v}, t_0)|_{\mathbf{v}=\mathbf{v}_0} \geqslant 0.$$

(1.17)

In other words, we must postulate

$$\hat{H}\psi(\mathbf{v})\big|_{\mathbf{v}=\mathbf{v}_0} \geqslant 0, \tag{1.18}$$

if

$$\psi(\mathbf{v}_0) = 0, \quad \frac{\partial\psi(\mathbf{v})}{\partial\mathbf{v}}\Big|_{\mathbf{v}=\mathbf{v}_0} = 0. \tag{1.19}$$

Strictly speaking, Eqs. (1.14) and (1.17) are necessary but not sufficient conditions for the neutron distribution to be nonnegative because the case

$$\frac{\partial N(\mathbf{r}_0, \ \mathbf{v}_0, \ t)}{\partial t}\Big|_{t=t_0} = 0 \tag{1.20}$$

requires special considerations. It is easy to recognize that in the cases of practical interest* Eq. (1.18) will suffice when we postulate that in Eq. (1.18) the equality holds only in the trivial case $\psi \equiv 0$ for all $\mathbf{v}$, or in the limit $v_0 \to \infty$.

The operator $\hat{H}$ in the form of Eq. (1.5) is of particular interest. Let us consider this operator in the case of an isotropic neutron distribution $\psi(\mathbf{v})$ (for example, in the case of a homogeneous infinite medium with uniformly distributed isotropic sources). When the conditions (1.19) are satisfied at some velocity $v = v_0$, we can easily verify that

$$\hat{H}_{as}\psi(v)\big|_{v=v_0} = 0. \tag{1.21}$$

This means that for real heavy moderators the operator of Eq. (1.5) adequately describes the scattering only in the limit $v_0 \to \infty$, and at finite neutron velocities one must add to this operator a term $\Delta\hat{H}$ such that

$$\hat{H}\psi(v)\big|_{v=v_0} = (\hat{H}_{as} + \Delta\hat{H})\psi(v)\big|_{v=v_0} > 0. \tag{1.22}$$

However, the replacement of $\hat{H}$ by $\hat{H}_{as}$ (progressive approximation) does not lead to negative neutron distributions.

When the condition (1.19) is fulfilled for an arbitrary anisotropic neutron distribution, the first and third terms of the right side of Eq. (1.5) give an essentially positive contribution, whereas the second term vanishes and only the fourth term can become negative. But, compared to the first term the fourth term is of the order of $1/A$ and can be important in practical cases only if ψ depends strongly upon the velocity at large v. For example, if

$$\psi(\mathbf{v}) \infty e^{-v^2}[1 + \Omega\mathbf{a}(v)], \tag{1.23}$$

where $\mathbf{a}(v)$ denotes some vector which depends slightly on the velocity (the average value of this vector is smaller than unity everywhere, except at the point $v = v_0$†), the quantity $\hat{H}_{as}\psi(\mathbf{v})_{v=v_0}$ becomes negative for $v \sim \sqrt{A}$, i.e., when Eq. (1.5) cannot be employed (see footnote *, page 250).

4. Finally, when the scattering probability is completely arbitrary, it follows from the definition of the scattering parameter given in Eq. (1.2) that

$$\int \hat{H}\psi(\mathbf{v})d\mathbf{v} = 0. \tag{1.24}$$

Obviously, this is the conservation law for the total number of neutrons in the scattering process.

Thus, in the general case, the scattering operator for any heavy moderator must satisfy the four basic conditions (1.5), (1.10), (1.18), and (1.24).

---

* This is the case in all real moderators at temperatures $T > 0$.
† In other words, $\psi(\mathbf{v})$ tends to zero for $v = v_0$, $\Omega \equiv \Omega_0 = -\mathbf{a}(v_0)$.

§2. Model of a Scattering Operator

for a Heavy Moderator

In several cases of practical importance (for example, in the diffusion approximation), the spatial diffusion of neutrons can be stated in analytical form in the single-velocity approximation. In the general case, one cannot give an analytical solution to the multivelocity neutron-transfer problem. Therefore, models of a scattering operator are of great importance if they can describe the properties of real moderators and, at the same time, help to simplify the transfer equations.

One can attempt to design a model of the scattering operator from the general properties of the scattering probability, i.e., on the basis of conditions (1.3), (1.9), and (1.13). When this is done (see [1]) in the case of a heavy moderator, a heavy monoatomic gas (or its generalization) is the simplest model, since the scattering operator reduces to a second-order differential operator. For example, if the absorption takes place according to a 1/v law, an analytical solution to the neutron-transfer equation cannot be obtained, even when no diffusion in space occurs.

The problem of obtaining an analytical solution to the transfer equation simplifies considerably, if one discards one of the conditions (1.3) or (1.9). For example, when one discards the correct asymptotic behavior for v → ∞, one can replace the scattering probability by a product of two functions, one of which depends only on v, and the other one only on v' (separable scattering pattern of isotropic scattering) [2]. For example, in the diffusion approximation the solution to the neutron-transfer equation is easily found. However, the solution is valid only for v, v' ≲ 1, and then the problem is to match the solution with the neutron distribution in the subthermal region. The situation is only slightly improved if the scattering probability is approximated by the sum of the probability for elastic scattering (in the laboratory system), which is proportional to δ(v − v'), and for the separable probability for isotropic scattering (this was done in a recent publication by Corngold and Durgan [3]*). The reason is that elastic scattering affects mainly the diffusion in space, but does not directly influence the moderation and thermalization processes.

Another possibility is to give up the small-scale equilibrium condition (1.9) for the scattering probability and to replace it by the integral condition (1.10). This condition is obtained when one adds to the scattering probability of the correct asymptotic scattering operator an adequately selected, separable part. This approach was taken in our preceding work [1]. The result was a scattering operator of the form[†]

$$\hat{H}'\widetilde{\psi}(v) = \frac{d\,[\mu(v)\,\widetilde{\psi}(v)]}{dv} - a(v)\,\widetilde{\psi}(v) + I\left\{a(v)\,M(v) - \frac{d\,[\mu(v)\,M(v)]}{dv}\right\}, \qquad (2.1)$$

where

$$I = \frac{\displaystyle\int_0^\infty a(v)\,\widetilde{\psi}(v)\,dv}{\displaystyle\int_0^\infty a(v)\,M(v)\,dv}, \qquad (2.2)$$

---

*In the model of Corngold and Durgan, the asymptotic form of the scattering operator for v → ∞ results when one neglects terms of the order 1/A in the approximation (1.5).

[†] We have used the symbol $\widetilde{\psi}(v)$ for a distribution function over v such that $\widetilde{\psi}(v)dv$ denotes the number of neutrons with velocities in the interval (v, v + dv), since this function was used in the article cited. The function $\widetilde{\psi}(v)$ is related to the previously introduced function by

$$\widetilde{\psi}(v) = 4\pi v^2 \psi(v).$$

$$M(v) = 4\pi^{-1/2}v^2 e^{-v^2}; \tag{2.3}$$

$a(v)$ and $\mu(v)$ are largely arbitrary functions of the velocity v.* With this model one finds in the diffusion approximation analytical expressions of the neutron distribution functions in the general case of a stationary or nonstationary source, as well as the eigenfunctions and the eigenvalues of the neutron-transfer equation.

However, this model does not make it possible to take into consideration the anisotropy of the neutron scattering. This shortcoming can be removed if instead of the asymptotic scattering operator used in [1] in the form

$$\hat{H}'_{as}\widetilde{\psi}(v) = \frac{d}{dv}\left[\frac{v^2\Sigma_f}{A}\widetilde{\psi}(v)\right], \tag{2.4}$$

one employs an asymptotic operator in the form of Eq. (1.5), because the latter operator describes the anisotropy of the scattering process at large distances in correct form. The logical generalization of this operator to the small-velocity region is

$$\hat{H}_0\psi(v) = \int \psi(v\Omega')\omega(v, \ \Omega\Omega')d\Omega' - \omega(v)\psi(v\Omega) + \frac{1}{v^2}\frac{\partial}{\partial v}\int v^2\psi(v\Omega')\mu(v, \ \Omega\Omega')d\Omega', \tag{2.5}$$

$$\omega(v) = \int \omega(v, \ \Omega\Omega')d\Omega', \tag{2.6}$$

provided that the functions $\omega(v,\Omega\Omega')$ and $\mu(v,\Omega\Omega')$ fulfill the conditions

$$\lim_{v\to\infty}\omega(v, \ \Omega\Omega') = \frac{1}{4\pi}v\Sigma_f\left(1 + \frac{2}{A}\Omega\Omega'\right), \tag{2.7a}$$

$$\lim_{v\to\infty}\mu(v, \ \Omega\Omega') = \frac{1}{4\pi A}v^2\Sigma_f(1 - \Omega\Omega'). \tag{2.7b}$$

Obviously, the operator of Eq. (2.5) does not satisfy the condition (1.10). The simplest model of a scattering operator which satisfies this condition can be obtained † if one adds to Eq. (2.5) an operator $\Delta\hat{H}$ which corresponds to a separable scattering probability

$$\Delta\hat{H}\psi(v) = \frac{b(v)}{v^2}\int a(v', \ \Omega\Omega')\psi(v')dv' - \psi(v)a(v)\int_0^\infty b(v')dv', \tag{2.8}$$

$$a(v) = \int a(v, \ \Omega\Omega')d\Omega', \tag{2.9}$$

where $a(v,\Omega\Omega')$ denotes some arbitrary function of the neutron velocity and the scattering angle b(v) is an unknown function of the velocity v alone (below, this function will be uniquely determined). Without loss of generality, we can postulate

$$\int_0^\infty b(v)dv = 1. \tag{2.10}$$

The result is a scattering operator of the form

$$\hat{H}\psi(v) = \int \psi(v\Omega')\omega(v, \ \Omega\Omega')d\Omega' + \frac{b(v)}{v^2}\int a(v', \ \Omega\Omega')\psi(v')dv' - $$
$$-[\omega(v) + a(v)]\psi(v) + \frac{1}{v^2}\frac{\partial}{\partial v}\int v^2\psi(v\Omega')\mu(v, \ \Omega\Omega')d\Omega'. \tag{2.11}$$

---

* The properties of these functions were discussed in detail in [1].

† In complete analogy with [1]; the notation has been so chosen that the functions $\omega(v)$, $\mu(b)$, $a(v)$, and b(v) here and there coincide.

As has been outlined in § 1, the operator $\hat{H}$ must satisfy the conditions (1.5), (1.10), (1.18), and (1.24) in the general case.

Condition (1.5) is satisfied, if

$$\lim_{v \to \infty} \frac{a(v)}{v} = 0 \tag{2.12}$$

holds [as can be inferred from Eq. (2.13), the function b(v) decreases rather rapidly with increasing v].

It is obviated by direct substitution that condition (1.10) is satisfied, provided that

$$b(v) = \frac{a(v)\, v^2 e^{-v^2} - \dfrac{d}{dv}[\mu(v)\, v^2 e^{-v^2}]}{\displaystyle\int_0^\infty a(v')\, v'^2 e^{-v'^2}\, dv'}, \tag{2.13}$$

where we introduced the notation (see preceding footnote)

$$\mu(v) = \int \mu(v,\ \Omega\Omega')\, d\Omega'. \tag{2.14}$$

We note that Eq. (2.13) satisfies Eq. (2.10) only if

$$\lim_{v \to 0} v^2 \mu(v) = 0. \tag{2.15}$$

When Eq. (2.15) holds, the operator defined by Eq. (2.11) automatically satisfies condition (1.24) for all functions $\omega(v, \Omega\Omega')$, $\mu(v, \Omega\Omega')$, and $a(v, \Omega\Omega')$.

It is recommended to consider condition (1.18) separately for the cases of isotropic and anisotropic neutron distributions. In the first case, i.e., for $\psi(\mathbf{v}) \equiv \psi(v)$, the operator defined by Eq. (2.11) assumes the form

$$\hat{H}\psi(v) = \frac{b(v)}{v^2}\int_0^\infty a(v')\psi(v')\,v'^2 dv' - a(v)\psi(v) + \frac{1}{v^2}\frac{d}{dv}[v^2\psi(v)\mu(v)]. \tag{2.16}$$

If at some velocity v_0 the relation (1.19) is fulfilled, i.e., if

$$\psi(v_0) = \frac{d\psi(v)}{dv}\bigg|_{v=v_0} = 0, \tag{2.17}$$

we obtain

$$\hat{H}\psi(v)\big|_{v=v_0} = \frac{b(v_0)}{v_0^2}\int_0^\infty a(v')\psi(v')\,v'^2 dv'. \tag{2.18}$$

Thus, condition (1.18) will be satisfied if*

$$b(v) > 0,\ a(v) > 0 \tag{2.19}$$

for all v. This means, according to Eq. (2.13), that for all v the inequality

$$a(v) + 2\mu(v)\left(v - \frac{1}{v}\right) - \frac{d\mu(v)}{dv} > 0 \tag{2.20}$$

must be satisfied.

---

*The case b(v) < 0, $a(v)$ < 0 is inconsistent with Eq. (2.10).

In order to satisfy Eq. (1.18) for arbitrary neutron distributions the self-evident relation

$$\omega(v, \ \Omega\Omega') > 0, \tag{2.21}$$

must be satisfied in addition to the conditions (2.19), i.e., the probability of elastic scattering must always be positive. But even this is not sufficient. A simple and accurate quantitative criterion for the applicability of condition (1.18) cannot be stated for the general case. Therefore, we will separately consider the regions $v \gg 1$, $v \sim 1$, and $v \ll 1$. In the high-velocity range, the operator $\hat{H}$ transforms into $\hat{H}_{as}$ and the results of pp. 151-152 are applicable. Since, in our case of a heavy moderator, the probability of elastic scattering at $v \sim 1$ is much greater than the probability of inelastic scattering, i.e., since

$$\omega(v) \gg \mu(v), \tag{2.22}$$

the results can be generalized to the region of intermediate velocities.* Thus, for the condition (1.18) to hold at $v \gtrsim 1$, the neutron distributions must not be strongly dependent upon the velocity near the points at which the distributions tend to zero. However, in the cases of practical interest, such events are rare (strong resonances in the cross sections are the main reason for a sharp velocity dependence of the neutron distributions).

The region of low velocities deserves particular attention. In that region, no elastic scattering may occur (for example, in crystals beyond the Bragg cutoff), and the above arguments are not applicable. Furthermore, the neutron distributions can have velocity-dependent singularities in that region. Thus, one can expect that the condition of a nonnegative neutron distribution is violated and special considerations are required in each individual case.

We see that the operator of Eq. (2.11) satisfies all basic requirements which the scattering operator must fulfill at relatively weak restrictions for the functions[†] $\omega(v, \Omega\Omega')$, $\mu(v, \Omega\Omega')$, and $a(v, \Omega\Omega')$ and for the neutron distribution. Below, we will assume in addition that the function $\mu(v)$ does not tend to zero for any finite velocity, because the neutron-transfer equation of isotropic neutron distributions has a singularity for that value, and this singularity can cause a singularity in the neutron distribution, which is inconsistent in terms of physics. According to Eq. (2.7b), this means that for all v

$$\mu(v) > 0 \tag{2.23}$$

holds. With Eqs. (2.15) and (2.20) we obtain that if $a(0) \neq 0$,

$$\lim_{v \to 0} \frac{\mu(v)}{v} < \frac{a(0)}{3} \tag{2.24}$$

must hold; if a vanishes for $v \to 0$, the function $\mu(v)$ must drop off even more rapidly with decreasing v [more rapidly than $va(v)$].

No further restrictions need be imposed on the functions $\omega(v, \Omega\Omega')$, $\mu(v, \Omega\Omega')$, and $a(v, \Omega\Omega')$ and they can be chosen so that the operator of Eq. (2.11) results in the best approximation for the scattering operator of a real moderator.

§3. Selection of the Functions ω, μ, and a

In this chapter, we will discuss several methods for approximating the real scattering operator by our operator.

---

*When, in the operator defined in Eq. (2.11), a term corresponding to a separable probability is missing, the calculations are easier to perform.

[†] It follows from general physical considerations that the functions ω, μ, and a must be bound for all finite velocity values; apart from this, the function $\mu(v, \Omega\Omega'$ must have a bound derivative with respect to v.

Moments of Transferred Velocity. The following is one of the most logical ways of selecting the functions $\omega(v, \Omega\Omega')$, $\mu(v, \Omega\Omega')$, and $a(v, \Omega\Omega')$. We search for the effective scattering probability for which the scattering operator has the form stated in Eq. (2.11). It is easy to verify that the probability can be represented by the δ function of the argument $v' - v$ and the derivative of this δ function:

$$w_{ef}(\mathbf{v}' \to \mathbf{v}) = \frac{1}{v^2}\omega(v', \Omega\Omega')\delta(v' - v) + a(v', \Omega\Omega')\frac{b(v)}{v^2} - \frac{1}{v^3}\mu(v', \Omega\Omega')\delta'(v' - v). \tag{3.1}$$

After that, in accordance with [1], one can attempt to relate any parameter of the true scattering probability $w_R(\mathbf{v}' \to \mathbf{v})$ with the functions ω, μ, and a. For example, one can postulate that the zeroth, first, and second velocity-transfer moments of w_{ef} and w_R must be identical for a particular scattering angle, i.e., that*

$$m_n(v, \Omega\Omega') = \int_0^\infty w(\mathbf{v} \to \mathbf{v}')(v' - v)^n v'^2 dv' \tag{3.2}$$

for n = 0, 1, 2. [We note that $m_0(v, \Omega\Omega')$ has the meaning of the total probability for scattering at a particular angle.] When we recall Eq. (2.13) and insert Eq. (3.1) into Eq. (3.2), we obtain that in our model the moments have the form

$$m_0(v, \Omega\Omega') = \omega(v, \Omega\Omega') + a(v, \Omega\Omega'), \tag{3.3}$$

$$m_1(v, \Omega\Omega') = -\mu(v, \Omega\Omega') + a(v, \Omega\Omega')\frac{\langle a(v)v\rangle - v\langle a(v)\rangle + \langle\mu(v)\rangle}{\langle a(v)\rangle}, \tag{3.4}$$

and for n ≥ 2

$$m_n(v, \Omega\Omega') = \frac{a(v, \Omega\Omega')}{\langle a(v)\rangle}\sum_{k=0}^n \frac{n!(-v)^{n-k}}{k!(n-k)!}\{\langle a(v)v^k\rangle + k\langle\mu(v)v^{k-1}\rangle\}. \tag{3.5}$$

In particular, for n = 2 we obtain

$$m_2(v, \Omega\Omega') = \frac{a(v, \Omega\Omega')}{\langle a(v)\rangle}\{\langle v^2 a(v)\rangle + 2\langle v\mu(v)\rangle - 2v[\langle va(v)\rangle + \langle\mu(v)\rangle] + v^2\langle a(v)\rangle\}. \tag{3.6}$$

Here, and in the following discussion, we denote by the symbol $\langle\ldots\rangle$ averaging over a Maxwell distribution, i.e., for some function $f(v)$

$$\langle f(v)\rangle = \frac{4}{\sqrt{\pi}}\int_0^\infty f(v)v^2 e^{-v^2}dv \tag{3.7}$$

holds. After solving Eqs. (3.3), (3.4), and (3.6) for ω, μ, and a, we obtain

$$\omega(v, \Omega\Omega') = m_0(v, \Omega\Omega') - \frac{m_2(v, \Omega\Omega')}{v^2 - 2vq + p}, \tag{3.8}$$

$$\mu(v, \Omega\Omega') = \frac{(q - v)m_2(v, \Omega\Omega')}{v^2 - 2vq + p} - m_1(v, \Omega\Omega'), \tag{3.9}$$

---

*For n ≠ 0, the n-th moment of transferred velocity is defined as the ratio $\frac{m_n(v_1, \Omega\Omega')}{m_0(v, \Omega\Omega')}$. However, this normalization leads to very cumbersome formulas in our case.

$$a(v, \Omega\Omega') = \frac{m^2(v, \Omega\Omega')}{v^2 - 2vq + p}. \tag{3.10}$$

The parameters q and p are defined by the expressions

$$q = \frac{\langle va(v)\rangle + \langle \mu(v)\rangle}{\langle a(v)\rangle}, \tag{3.11a}$$

$$p = \frac{\langle v^2 a(v)\rangle + 2\langle v\mu(v)\rangle}{\langle a(v)\rangle}. \tag{3.11b}$$

However, it is easily verified that if the functions $m_1(v, \Omega\Omega')$ and $m_2(v, \Omega\Omega')$ satisfy the relations*

$$M_1 \equiv \frac{4}{\sqrt{\pi}} \int m_1(v, \Omega\Omega') e^{-v^2} d\mathbf{v} = 0, \tag{3.12}$$

$$M_2 \equiv \frac{4}{\sqrt{\pi}} \int m_2(v, \Omega\Omega') e^{-v^2} d\mathbf{v} = -\frac{8}{\sqrt{\pi}} \int v m_1(v, \Omega\Omega') e^{-v^2} d\mathbf{v}, \tag{3.13}$$

substitution of Eqs. (3.9) and (3.10) into Eqs. (3.11a) and (3.11b) transforms these expressions into identities for all values of the parameters q and p. We note that for even n the quantities

$$M_n = \frac{4}{\sqrt{\pi}} \int_0^{\infty} v^2 e^{-v^2} dv \int w(\mathbf{v} \to \mathbf{v}')(v' - v)^n dv' = \frac{4}{\sqrt{\pi}} \int m_n(v, \Omega\Omega') e^{-v^2} d\mathbf{v}, \tag{3.14}$$

i.e., the transferred velocity moments averaged over a Maxwell distribution, are important neutron parameters of a moderator.† Thus, the quantity M_0 is the average scattering probability per second of neutrons obeying a Maxwell distribution, while M_2 denotes the well-known Nelkin parameter (more accurately, its analog in a velocity representation). In our model, we have with Eqs. (3.3) and (3.5)

$$M_0 = \langle \omega(v)\rangle + \langle a(v)\rangle, \tag{3.15}$$

$$M_n = \sum_{k=0}^{n} \frac{n!\,(-1)^{n-k}}{k!\,(n-k)!} \frac{\langle a(v)\,v^{n-k}\rangle}{\langle a(v)\rangle} \{\langle a(v)\,v^k\rangle + k\langle \mu(v)\,v^{k-1}\rangle\}, \quad n > 2; \tag{3.16}$$

and specifically,

$$M_2 = 2\langle a(v)\,v^2\rangle + 2\langle \mu(v)\,v\rangle - \frac{2\langle a(v)\,v\rangle}{\langle a(v)\rangle} \{\langle a(v)\,v\rangle + \langle \mu(v)\rangle\}. \tag{3.17}$$

When Eqs. (3.8)–(3.10) are used for the calculation of the functions ω, μ, and a from given velocity-transfer moments, the quantities q and p can be considered as additional dimensionless, positive, free parameters.‡ However, these parameters are not independent. Substituting Eq. (3.9) into Eq. (2.24), we obtain

---

*These relations result from the small-scale equilibrium principle; it suffices that condition (1.10) is satisfied for these equations to be valid. Therefore, the equations must hold for any real scattering probability, hence also for a scattering probability as defined by Eq. (3.1).

†These moments vanish for all odd n. This follows from the small-scale equilibrium principle and from Eq. (1.10).

‡According to Eqs. (2.19), (2.23), (3.11a), and (3.11b), the parameters are integrals of positive quantities.

$$\frac{q}{p} = \lim_{v \to 0} \frac{m_1(v)}{m_2(v)}, \tag{3.18}$$

where

$$m_{1,2}(v) = \int m_{1,2}(v, \Omega\Omega')\, d\Omega'. \tag{3.19}$$

The other requirements listed in § 2 impose certain restrictions on the parameters. For example, the condition that $a(v, \Omega\Omega')$ is bounded and positive leads to

$$q^2 < p. \tag{3.20}$$

Particularly restrictive is the inequality

$$\frac{2(p - q^2) m_2(v)}{(v^2 - 2qv + p)^2} + \frac{v - q}{v^2 - 2qv + p} \left[\frac{dm_2(v)}{dv} + 2\left(\frac{1}{v} - v\right) m_2(v) \right] + \frac{dm_1(v)}{dv} + 2\left(\frac{1}{v} - v\right) m_1(v) > 0, \tag{3.21}$$

which is obtained when Eqs. (3.9) and (3.10) are inserted into Eq. (2.20). This inequality imposes restriction not only on the parameters p and q, but also on the transferred velocity moments for which a similar calculation of the functions ω, μ, and a can be made in principle. For example, it is easy to verify that at certain velocities this inequality is not satisfied (for any q and p values) by the transferred velocity moments calculated with the heavy gas model for which

$$m_1(v) = \frac{\Sigma_f}{A}\left(\frac{3}{2} - v^2\right), \tag{3.22}$$

$$m_2(v) = \frac{\Sigma_f}{A} v \tag{3.23}$$

(strictly speaking, these expressions are valid only in the velocity range $v \gg A^{-\frac{1}{2}}$; however, for $A \gg 1$ the restrictions referring to the region $v \lesssim A^{-\frac{1}{2}}$ do not affect the result).

We wish to emphasize that this behavior does by no means indicate a shortcoming of our model, but rather of the heavy-gas moderator model, since according to the heavy gas model, the energy exchange between neutrons and moderator takes place in infinitely small portions. For this reason, Eq. (3.23) tends to zero for $v \to 0$, whereas in the case of real moderators $m_2(0)$ is a finite quantity comparable to $m_1(0)$. Since in our model the scattering probability is separable, the scattering must involve a jump-like change of the neutron energy at all velocities, and the quantity $m_2(v)$ does not vanish for all v. It is a shortcoming of our model that the transferred energy portions can be too large, at least in the range of high velocities. In this region, the heavy gas model is superior to our model. Naturally, the heavy gas model is one of the least "successful" approximations to our model.

Let us note in conclusion that a restriction for the quantity p results from Eqs. (3.18) and (3.20):

$$p < \frac{m_2^2(0)}{m_1^2(0)}. \tag{3.24}$$

On the other hand, when we expand the inequality (3.21) in a power series of v and restrict ourselves to the first two terms of the expansion, we obtain with Eq. (3.18), after simple calculations, another restriction for p:

$$p < \frac{4}{3} \frac{m_2^2(0)}{m_1^2(0)} \left\{ 2 + \lim_{v \to 0} \frac{d}{dv}\left[\frac{m_2(v)}{m_1(v)}\right] \right\}^{-1}; \tag{3.25}$$

in certain cases, this restriction can be more rigorous than Eq. (3.24). But we emphasize that independent of these two inequalities, one must always check the condition (3.21) in the entire velocity range.

Conjugated Moments of Transferred Velocity. In the preceding article [1], the assumption was made that the functions $\omega(v, \Omega\Omega')$, $\mu(v, \Omega\Omega')$, and $a(v, \Omega\Omega')$ (more precisely, their analogs which are independent of the scattering angle) can be selected from the condition that the conjugated moments of transferred velocity at some scattering angle agree for the scattering probability defined by Eq. (3.1) and the true scattering probability, i.e., that

$$\widetilde{m}_n(v, \Omega\Omega') = \int_0^\infty w(\mathbf{v}' \to \mathbf{v})(v' - v)^n v'^2 \, dv' \tag{3.26}$$

for n = 0, 1, and 2. However, as we will see below, this method is suitable only for conjugated moments of transferred velocity under restrictions which are not feasible in terms of physics. Yet it is worthwhile to consider this problem because this helps to illustrate the degree of inadequacy of our model.

For our scattering probabilities of Eqs. (3.1), some simple calculations lead to

$$\widetilde{m}_0(v, \Omega\Omega') = \omega(v, \Omega\Omega') + \frac{1}{v^2} \frac{\partial}{\partial v}[v^2 \mu(v, \Omega\Omega')] + \frac{b(v)}{v^2} a_2(\Omega\Omega'), \tag{3.27}$$

$$\widetilde{m}_1(v, \Omega\Omega') = \mu(v, \Omega\Omega') + \frac{b(v)}{v^2}[a_3(\Omega\Omega') - va_2(\Omega\Omega')], \tag{3.28}$$

and for $n \geq 2$

$$\widetilde{m}_n(v, \Omega\Omega') = \frac{b(v)}{v^2} \sum_{k=0}^{n} \frac{n!\,(-1)^k}{k!\,(n-k)!} v^k a_{2+n-k}(\Omega\Omega'). \tag{3.29}$$

In particular, we obtain

$$\widetilde{m}_2(v, \Omega\Omega') = \frac{b(v)}{v^2}[a_4(\Omega\Omega') - 2va_3(\Omega\Omega') + v^2 a_2(\Omega\Omega')], \tag{3.30}$$

where

$$a_n(\Omega\Omega') = \int_0^\infty v^n a(v, \Omega\Omega') \, dv. \tag{3.31}$$

We note that according to Eq. (3.31) the n-th conjugated moment exists in our model only if

$$\lim_{v \to \infty} \frac{a(v, \Omega\Omega')}{v^{3+n}} = 0. \tag{3.32}$$

For any real moderator (except for a hydrogen-containing moderator), conjugated moments of transferred velocity exist for all n (this means that $a(v, \Omega\Omega')$ must decrease with increasing v at least exponentially). The fact that the conjugated moments of transferred velocity tend to infinity in the case of a heavy moderator is characteristic for the crudeness of our model.

When we require that Eqs. (3.27), (3.28), and (3.30) agree with the true conjugated moments of transferred velocity and when we solve these equations for $\omega(v, \Omega\Omega')$, $\mu(v, \Omega\Omega')$, and b(v), we obtain

$$\omega(v, \Omega\Omega') = \widetilde{m}_0(v, \Omega\Omega') - \frac{a_2(\Omega\Omega')\,\widetilde{m}_2(v, \Omega\Omega')}{a_4(\Omega\Omega') - 2va_3(\Omega\Omega') + v^2a_2(\Omega\Omega')} -$$

$$- \frac{1}{v^2}\frac{\partial}{\partial v}\left\{v^2\widetilde{m}_1(v, \Omega\Omega') - v^2\frac{[a_3(\Omega\Omega') - va_2(\Omega\Omega')]\,\widetilde{m}_2(v, \Omega\Omega')}{a_4(\Omega\Omega') - 2va_3(\Omega\Omega') + v^2a_2(\Omega\Omega')}\right\}. \tag{3.33}$$

$$\mu(v, \Omega\Omega') = \widetilde{m}_1(v, \Omega\Omega') - \frac{[a_3(\Omega\Omega') - va_2(\Omega\Omega')]\,\widetilde{m}_2(v, \Omega\Omega')}{a_4(\Omega\Omega') - 2va_3(\Omega\Omega') + v^2a_2(\Omega\Omega')}, \tag{3.34}$$

$$b(v) = \frac{v^2\widetilde{m}_2(v, \Omega\Omega')}{a_4(\Omega\Omega') - 2va_3(\Omega\Omega') + v^2a_2(\Omega\Omega')}. \tag{3.35}$$

It follows from Eq. (3.35) according to Eq. (2.10), that the relation

$$\int_0^\infty \frac{v^2\widetilde{m}_2(v, \Omega\Omega')\,dv}{a_4(\Omega\Omega') - 2va_3(\Omega\Omega') + v^2a_2(\Omega\Omega')} = 1, \tag{3.36}$$

must hold, i.e.,

$$\lim_{v\to\infty} v\widetilde{m}_2(v, \Omega\Omega') = 0. \tag{3.37}$$

It is easy to show that in the case of real moderators this is equivalent to

$$\lim_{v\to\infty} \widetilde{m}_2(v, \Omega\Omega') = \frac{1}{6\pi}\frac{v\Sigma_f}{A}K_{Av}(1 - \Omega\Omega'), \tag{3.38}$$

where K_{Av} denotes the average kinetic energy of the scattered atoms (in units of kT). For cases of practical interest, this procedure is actually not applicable.

From the viewpoint of physics, these results are obtained because at large neutron velocities our model increases the probability for large energy transfers (this fact was noted before — see page 256). Therefore, in cases in which the neutron distributions have sharp maxima at high energies (for example, in the initial stages of the moderation process of neutrons from a pulsed source), the model leads to incorrect results. On the other hand, in the case of rather smooth and rapidly moderated neutron distributions (stationary spectra), we will see that our model is adequate. This is due to the fact that the principal term of the asymptotic expansion (for $v \to \infty$) of our scattering operator renders the correct dependence.*

Simplest Version of the Model. In order to analyze the basic properties of our model, it is convenient to consider the case in which the functions $\omega(v, \Omega\Omega')$, $\mu(v, \Omega\Omega')$, and $a(v, \Omega\Omega')$ are defined by very simple analytical expressions. Therefore, we use Eqs. (2.7a) and (2.7b) and set

$$\omega(v, \Omega\Omega') = \frac{1}{4\pi}v\Sigma_f\left(1 + \frac{2}{A}\Omega\Omega'\right), \tag{3.39}$$

$$\mu(v, \Omega\Omega') = \frac{1}{4\pi}\frac{v^2\Sigma_f}{A}(1 - \Omega\Omega'), \tag{3.40}$$

and assume that the function $a(v, \Omega\Omega')$ is constant:

$$a(v, \Omega\Omega') = a_0/4\pi. \tag{3.41}$$

Then, the conditions (2.12) and (2.21)-(2.24) are automatically fulfilled. For the inequality (2.20) to hold, for all v

---

*Difficulties in the calculation of conjugated moments of transferred velocity result from the fact that the calculations require higher terms of the asymptotic expansion.

$$\frac{Aa_0}{2\Sigma_f} > 2v - v^3, \tag{3.42}$$

i.e.,

$$\frac{Aa_0}{\Sigma_f} > \frac{8}{3}\sqrt{\frac{2}{3}} \tag{3.43}$$

must hold. An arbitrary constant a_0 can be chosen.

It is easy to verify that in our case the expressions for the transferred velocity moment $m_n(v, \Omega\Omega')$ of Eq. (3.2) have the form

$$m_0(v, \Omega\Omega') = \frac{1}{4\pi} v\Sigma_f\left(1 + \frac{2}{A}\Omega\Omega'\right) + \frac{1}{4\pi}a_0, \tag{3.44}$$

$$m_1(v, \Omega\Omega') = \frac{1}{4\pi}\left[\frac{2}{\sqrt{\pi}}a_0 + \frac{3}{2}\frac{\Sigma_f}{A} - va_0 - \frac{v^2\Sigma_f}{A}(1 - \Omega\Omega')\right], \tag{3.45}$$

and for $n \geq 2$

$$m_n(v, \Omega\Omega') = \frac{1}{2\pi^{3/2}}\sum_{k=0}^{n}\frac{n!\,(-v)^{n-k}}{k!\,(n-k)!}\left[a_0\Gamma\left(\frac{k+3}{2}\right) + k\frac{\Sigma_f}{A}\Gamma\left(\frac{k+4}{2}\right)\right]; \tag{3.46}$$

in particular, we have

$$m_2(v, \Omega\Omega') = \frac{2}{\pi^{3/2}}\frac{\Sigma_f}{A} + \frac{3}{8\pi}a_0 - v\left(\frac{3\Sigma_f}{4\pi A} + \frac{a_0}{\pi^{3/2}}\right) + v^3\frac{a_0}{4\pi}. \tag{3.47}$$

Accordingly, the parameters M_n of Eq. (3.14) are defined by the expressions

$$M_0 = \frac{2}{\sqrt{\pi}}\Sigma_f + a_0, \tag{3.48}$$

and for $n \geq 2$ we have

$$M_n = \frac{4}{\pi}\sum_{k=0}^{n}\frac{n!\,(-1)^{n-k}}{k!\,(n-k)!}\,\Gamma\left(\frac{n-k+3}{2}\right)\left[a_0\Gamma\left(\frac{k+3}{2}\right) + k\frac{\Sigma_f}{A}\Gamma\left(\frac{k+4}{2}\right)\right]; \tag{3.49}$$

a particular case is

$$M_2 = \frac{2}{\sqrt{\pi}}\frac{\Sigma_f}{A} + \left(3 - \frac{8}{\pi}\right)a_0. \tag{3.50}$$

We recall that this version of the model is too crude to permit us to expect a quantitative agreement with real moderators. Specifically, the model leads to a too high value of the usual Nelkin parameter $M_2^{(N)}$ [4] which, in our notations assumes the form

$$M_2^{(N)} = \sqrt{\frac{\pi m}{8kT}}\left\langle\int w_s(\mathbf{v} \to \mathbf{v}')(v'^2 - v^2)^2\,d\mathbf{v}'\right\rangle \tag{3.51}$$

(the factor $\sqrt{\pi m/8kT}$ is equal to the inverse average velocity of thermal neutrons and results from the transition of the differential scattering cross section used by Nelkin to the scattering probability per second). In our case we obtain for the dimensionless quantity $M_2^{(N)}/\Sigma_f$:

$$M_2^{(N)}/\Sigma_f = \frac{12}{A}\left(1 + \frac{\sqrt{\pi}}{8}\frac{Aa_0}{\Sigma_f}\right).$$ (3.52)

This means that the quantity $AM_2^{(N)}/12\Sigma_f$ must be greater than unity (the constant a_0 is positive by definition). In reality, this quantity is much smaller than unity, as, for example, in the cases of beryllium and graphite. This is due to the fact that our version of the model implies an excessive fraction of large velocity transfers, because we assume the function a_0 to be constant, whereas according to Eq. (3.10), one must anticipate that the function a_0 decreases like $1/v$ with increasing v. Thus, it is convenient to use for quantitative estimates the smallest possible a_0 values, i.e., a_0 values for which relation (3.43) becomes an equality.

Other Methods for Selecting the Model Parameters. Since our immediate goal is the approximation of the operator rather than the determination of the scattering probability, the functions $\omega(v, \Omega\Omega')$, $\mu(v, \Omega\Omega')$, and $a(v, \Omega\Omega')$ can be chosen from the condition that the application of our operator to somehow determined functions $\psi_n(v)$ give the same result as the application of the true scattering operator; the functions $\psi_n(v)$ are conveniently taken close to typical neutron distribution functions. The relation between the model parameters and the scattering angles can be determined only if the functions $\psi_n(v)$ depend upon the direction of the vector v, and the method is the more sensitive to the above condition, the stronger the dependence of $\psi_n(v)$ upon the direction of v. Therefore, let us assume $\psi_n(v)$ in the form

$$\psi_n(v) = \delta(\Omega - \Omega_0)\varphi_n(v),$$ (3.53)

where Ω_0 denotes some constant unit vector which does not affect the result. Our scattering operator applied to these functions leads to

$$\hat{H}\psi_n(v) = \varphi_n(v)\omega(v, \Omega\Omega_0) + \frac{b(v)}{v^2}\int_0^\infty a(v', \Omega\Omega_0)\varphi_n(v')v'^2 dv' -$$

$$- \delta(\Omega - \Omega_0)\varphi_n(v)[\omega(v) + a(v)] + \frac{1}{v^2}\frac{\partial}{\partial v}[v^2\varphi_n(v)\mu(v, \Omega\Omega_0)],$$ (3.54)

whereas application of the scattering operator $\hat{H}_R$ corresponding to the real scattering probability $W_R(v \to v')$ results in

$$\hat{H}_R\psi_n(v) = \int_0^\infty \varphi_n(v')w_R(\Omega_0v' \to \Omega v)v'^2 dv' - \delta(\Omega - \Omega_0)\varphi_n(v)\int w_R(\Omega_0 v \to v')dv'.$$ (3.55)

We see from Eqs. (3.54) and (3.55) that the application of the operators $\hat{H}$ and $\hat{H}_R$ to the functions $\psi_n(v)$ leads to the same result if the equalities

$$\omega(v) + a(v) = \int w_R(\Omega_0 v \to v')\,dv',$$ (3.56)

$$\varphi_n(v)\omega(v, \Omega\Omega_0) + \frac{b(v)}{v^2}\int_0^\infty a(v', \Omega\Omega_0)\varphi_n(v')v'^2 dv' +$$

$$+ \frac{1}{v^2}\frac{\partial}{\partial v}[v^2\varphi_n(v)\mu(v, \Omega\Omega_0)] = \int_0^\infty \varphi_n(v')w_R(\Omega_0 v' \to \Omega v)v'^2 dv'$$ (3.57)

are satisfied. When we insert into Eq. (3.57) two functions $\varphi_n(v)$, we obtain three equations for the determination of $\omega(v, \Omega\Omega_0)$, $\mu(v, \Omega\Omega_0)$, and $a(v, \Omega\Omega_0)$ and the solution of these equations does not pose major difficulties.* In the general case, one of these functions is undetermined. The function can be determined from some other additional conditions, for example when we replace Eq. (3.56) by the condition

$$\omega(v, \Omega\Omega_0) + a(v, \Omega\Omega_0) = \int_0^\infty w_R(\Omega_0 v \to \Omega v') v'^2 dv',\tag{3.58}$$

or, in order to simplify things, when we assume that the function a is completely independent of the scattering angle.

A wide variety of considerations can be employed in selecting functions $\varphi_n(v)$, and only the result decides the advantage of one method over another. One of the simple possibilities for selecting $\varphi_n(v)$ is

$$\varphi_n(v) = v^n e^{-v^2}, \quad n = 0, 1.\tag{3.59}$$

Finally, in the case in which the angular dependence of the functions μ and a is immaterial,[†] the method mentioned in [1] can prove useful for the solution of stationary problems. This method requires that the first three terms of the expansion of the neutron-velocity distribution function in a series of inverse powers of the neutron lifetime be the same for the true scattering operator and our scattering operator in the case of an infinite, homogeneous moderator.

In other words, many possibilities exist for approximating with our model the most important properties of the true scattering operator. On the other hand, the neutron spectra encountered in the neutron-thermalization theory are usually weakly dependent upon the small details of the scattering probability. There is reason to hope that our model will help to obtain an approximation to a scattering operator which is fully adequate for the solution of numerous problems in neutron-thermalization theory.

§4. Solution of the Neutron-Transfer Equation in the Case of an Infinite, Homogeneous Moderator

In the case of an infinite, homogeneous moderating medium with a uniform distribution of isotropic sources, the transfer equation (1.1) assumes the form

$$\frac{\partial \widetilde{N}(v, t)}{\partial t} - \frac{\partial}{\partial v}[\mu(v)\widetilde{N}(v, t)] + \left[a(v) + \frac{1}{\tau_a}\right]\widetilde{N}(v, t) =$$

$$= \frac{4}{\sqrt{\pi}}\left\{a(v)v^2 e^{-v^2} - \frac{d}{dv}[\mu(v)v^2 e^{-v^2}]\right\}I(t) + 4\pi v^2 S(v, t),\tag{4.1}$$

$$I(t) = \frac{1}{\langle a(v)\rangle}\int_0^\infty a(v')\widetilde{N}(v', t) dv'\tag{4.2}$$

when a scattering operator in the form of Eqs. (2.11) and (2.13) is used. We have the following notation: $\widetilde{N}(v, t)dv = 4\pi v^2 N(v, t)dv$ denotes the number of neutrons in the velocity interval

*The integral

$$\int_0^\infty a(v', \Omega\Omega_0)\varphi_n(v') v'^2 dv'$$

must be considered as a parameter and b must be calculated with Eq. (2.13). The result is a linear first-order differential equation for μ and an algebraic equation for a.

† This is the case when the diffusion approximation can be employed.

(v , v + dv) per unit volume of the moderator at the time t. The neutron sources are normalized in similar fashion. The solution to this equation is easily obtained in the general case and has the form:*

$$\widetilde{N}(v, t) = \frac{1}{2\pi i} \int_{\alpha-i\infty}^{\alpha+i\infty} n(v, x) e^{xt} dx, \quad \alpha > 0,$$ (4.3)

$$n(v, x) = \frac{1}{\mu(v)} \int_{v}^{\infty} s(v', x) F(x, v', v) dv' + i(x) \left\{ M(v) - \frac{1}{\mu(v)} \int_{0}^{\infty} \left[x + \frac{1}{\tau_a(v')} \right] M(v') F(x, v', v) dv' \right\},$$ (4.4)

$$i(x) = \frac{\int_{0}^{\infty} \frac{a(v)}{\mu(v)} dv \int_{v}^{\infty} s(v', x) F(x, v', v) dv'}{\int_{0}^{\infty} \frac{a(v)}{\mu(v)} dv \int_{x}^{\infty} \left[x + \frac{1}{\tau_a(v')} \right] M(v') F(x, v', v) dv'},$$ (4.5)

$$F(x, v', v) = \exp \left\{ \int_{v'}^{v} \left[x + \frac{1}{\tau_a(v'')} + a(v'') \right] \frac{dv''}{\mu(v'')} \right\},$$ (4.6)

$$s(v, x) = 4\pi v^2 \int_{0}^{\infty} S(v, t) e^{-tx} dt.$$ (4.7)

The time at which the source was switched on (time t) was taken as the origin of the time count, i.e., S(v, t) = 0 for t < 0.

Let us consider this solution in the most important special cases of a monochromatic source, namely in the case of a stationary and a pulsed monochromatic source.

Stationary Monochromatic Source. This case is obtained when we set in Eq. (4.7):

$$S(v, t) = \frac{\sigma}{4\pi v_0^2} \delta(v - v_0)$$ (4.8)

(σ denotes the source intensity and v_0 the velocity of the neutrons emitted), and when we try to determine N in the limit t → ∞. The result is

$$\widetilde{N}(v) = \sigma \frac{\theta(v_0 - v)}{\mu(v)} \exp \left\{ -\int_{v}^{v_0} \left[a(v') + \frac{1}{\tau_a(v')} \right] \frac{dv'}{\mu(v')} \right\} +$$

$$+ \sigma \left\{ M(v) - \frac{1}{\mu(v)} \int_{v}^{\infty} \frac{M(v'')}{\tau_a(v'')} \exp \left\{ -\int_{v}^{v''} \left[a(v') + \frac{1}{\tau_a(v')} \right] \frac{dv'}{\mu(v')} \right\} dv'' \right\} \times$$

$$\times \frac{\int_{0}^{v_0} \frac{a(v)}{\mu(v)} \exp \left\{ -\int_{v}^{v_0} \left[a(v') + \frac{1}{\tau_a(v')} \right] \frac{dv'}{\tau_a(v')} \right\} dv}{\int_{0}^{\infty} \frac{a(v)}{\mu(v)} dv \int_{v}^{\infty} \frac{M(v'')}{\tau_a(v'')} \exp \left\{ -\int_{v}^{v''} \left[a(v') + \frac{1}{\tau_a(v')} \right] \frac{dv'}{\mu(v')} \right\} dv''},$$ (4.9)

where

$$\theta(x) = \begin{cases} 0, & x < 0, \\ 1, & x > 0. \end{cases}$$ (4.10)

---

*The solution can be obtained by using the Laplace transform (in t) in Eq. (4.1). The result is an ordinary first-order linear differential equation (for details, see [1]).

The first term in Eq. (4.9) describes the distribution of the neutrons which were never scattered and relates to the separable part of the scattering probability, i.e., a scattering involving a large energy transfer. The distribution of the neutrons which were scattered in some fashion is described by the second term. The second term accounts for the Maxwell distribution (the importance of the second term increases with the lifetime τ_a of the neutrons), as well as for the corrections to the Fermi distribution at high velocities. For $\tau_a \gg A/\Sigma_f$, the neutron spectrum of Eq. (4.9) is characterized by a smooth transition from a Fermi distribution (for $v \gg 1$) to a Maxwell distribution (for $v \sim 1$).

Equation (4.9) assumes its simplest form when we assume that the absorption obeys a $1/v$ law (τ_a independent of v), and when we use Eqs. (3.40) and (3.41) for the functions $\mu(v)$ and $a(v)$ [Eqs. (3.40) and (3.41) refer to the simplest version of the model]. We obtain

$$\widetilde{N}(v) = \frac{\sigma A}{\Sigma_f} \frac{\theta(v_0 - v)}{v^2} \exp\left\{ -\left(a_0 + \frac{1}{\tau_a}\right) \frac{A}{\Sigma_f}\left(\frac{1}{v} - \frac{1}{v_0}\right) \right\} +$$

$$+ \sigma\tau_a M(v) - \frac{\sigma A}{\Sigma_f} \frac{1}{v^2} \int_v^\infty M(v') \exp\left\{ -\left(a_0 + \frac{1}{\tau_a}\right) \frac{A}{\Sigma_f}\left(\frac{1}{v} - \frac{1}{v'}\right) \right\} dv'. \tag{4.11}$$

From the viewpoint of physics, the meaning of each term is self-evident.

Finally, let us compare the asymptotic (for $v \to \infty$) expansions of the neutron distributions obtained from an accurate calculation for a heavy moderator and from a calculation according to our model. According to [4], the accurate calculation leads to

$$\widetilde{N}_{\text{acc}}(v) = \frac{\sigma A}{\Sigma_f} \frac{1}{v^2}\left[1 - \frac{A}{\Sigma_f} \frac{1}{\tau_a} \frac{1}{v} + \frac{A^2}{\Sigma_f^2}\left(\frac{1}{2\tau_a^2} + \frac{2\Sigma_f^2 K_{A\tau}}{3A^2}\right)\frac{1}{v^2} + \cdots \right]. \tag{4.12}$$

We obtain from Eq. (4.11) with our model:

$$\widetilde{N}(v) = \frac{\sigma A}{\Sigma_f} \frac{1}{v^2}\left[1 - \frac{A}{\Sigma_f}\left(\frac{1}{\tau_a} + a_0\right)\frac{1}{v} + \frac{A^2}{\Sigma_f^2}\left(\frac{1}{2\tau_a^2} + \frac{a_0}{\tau_a} + \frac{a_0^2}{2}\right)\frac{1}{v^2} + \cdots \right]. \tag{4.13}$$

Thus, it is not possible to select the constants a_0 so that Eqs. (4.12) and (4.13) coincide with an accuracy extending to terms of the order $1/v$.*

Pulsed Monoenergetic Source. In this case, we have

$$S(v, t) = \frac{\sigma}{4\pi v_0^2} \delta(v - v_0)\delta(t), \tag{4.14}$$

i.e., according to Eq. (4.7)

$$s(v, x) = \sigma\delta(v - v_0). \tag{4.15}$$

We obtain from Eqs. (4.3)–(4.7)

$$\widetilde{N}(v, t) = \frac{\sigma}{\mu(v)} \delta\left[t - \int_v^{v_0} \frac{dv'}{\mu(v')} \right] \exp\left\{ -\int_v^{v_0} \left[\frac{1}{\tau_a(v')} + a(v')\right] \frac{dv'}{\mu(v')} \right\} +$$

---

*When we do not restrict ourselves to the simplest version of the model, but assume that $\mu(v)$ and $a(v)$ are arbitrary functions of the velocity, then, as has been outlined on page 255 and in [1], we can obtain coincidence $N(v)$ and $N_{\text{acc}}(v)$ for all v with an accuracy reaching to terms of the order $1/\tau_d^2$, provided that τ_a is small.

$$+ \frac{\sigma}{2\pi i} \int\limits_{a-i\infty}^{a+i\infty} dx e^{xt} \int\limits_0^{v_0} \frac{a(v'')}{\mu(v'')} F(x, v_0, v'') dv'' \cdot \frac{M(v) - \frac{1}{\mu(v)} \int\limits_v^\infty \left[x + \frac{1}{\tau_a(v')}\right] M(v') F(x, v', v) dv'}{\int\limits_0^\infty \frac{a(V)}{\mu(V)} dV \int\limits_V^\infty \left[x + \frac{1}{\tau_a(V)}\right] M(V) F(x, v', V) dv'}. \qquad (4.16)$$

To illustrate our case, we write this expression for the simplest version of our model, i.e., when the relations (3.40) and (3.41) hold and when τ_a = const. We obtain

$$\widetilde{N}(v, t) = \frac{\sigma_0}{v^2} \delta\left(\frac{\Sigma_f}{A} t + \frac{1}{v_0} - \frac{1}{v}\right) e^{-\left(a_0 + \frac{1}{\tau_a}\right)t} +$$

$$+ \sigma e^{-t/\tau_a} \left\{ M(v) - \theta\left(1 - \frac{\Sigma_f tv}{A}\right) \frac{1}{\left(1 - \frac{\Sigma_f tv}{A}\right)^2} M\left(\frac{v}{1 - \frac{\Sigma_f tv}{A}}\right) e^{-a_0 t} \right\}. \qquad (4.17)$$

The first terms on the right sides of Eqs. (4.16) and (4.17) describe the distribution of the neutrons which did not experience a single collision, i.e., these terms result from the separate part of the scattering probability [as well as the first term of Eq. (4.9)]. In other words, the neutron distribution is proportional to

$$\delta\left[t - \int\limits_v^{v_0} \frac{dv'}{\mu(v')}\right] \qquad (4.18)$$

(for the simplest version of the model, this factor has the form $\delta\left[t - \frac{A}{\Sigma_f}\left(\frac{1}{v} - \frac{1}{v_0}\right)\right]$). We

see that the velocity of neutrons of this type is uniquely related to the moderation time, i.e., to the neutron "age." This is easy to understand when one recalls that the scattering operator of Eq. (2.5) is equivalent to the scattering operator in the approximation involving neutron aging and some effective mass of the scattering atoms. The relation between the effective mass and the velocity is given by the function $\mu(v)$. The contribution of this term is very important at small times, but decreases exponentially with time (in the simplest version of the model, like e^{-t/a_0}). Naturally, the true neutron distribution must have a continuous velocity dependence and cannot be a δ function of the type defined by Eq. (4.18). However, at sufficiently small times, in a heavy moderator, the neutron-distribution function (over the velocities) differs from zero at any moment only in a very narrow velocity interval (with a width of the order $v/\sqrt{A}$) near some average velocity. The average velocity is given by the δ function of Eq. (4.18). In the limit $A \to \infty$, the true neutron distribution tends to an expression which is proportional to Eq. (4.18).

The second terms in Eqs. (4.16) and (4.17) describe the neutron distributions resulting from collisions (separable part of the scattering probability). These probabilities tend to their asymptotic expressions, i.e., to Maxwell distributions, with increasing time. For large t values, the second terms in Eqs. (4.16) and (4.17) are practically the only terms because the contributions of the first terms become negligibly small.

Our model is capable of adequately describing the intrinsic neutron distribution for rather small as well as for rather large times. However, in intermediate times (t ~ $1/a_0$), our model does not adequately describe the thermalization process. When the true neutron spectrum transforms from a narrow initial line to a broad Maxwell distribution, basically by gradually increasing the width of the maximum, in our model the effective width increase in the neutron spectrum with time results from an increase in the broad component of the spectrum. Therefore, the calculation of nonstationary neutron spectra with our model can lead to gross errors.

But one can expect that our model will prove adequate and convenient for the calculation of various integral characteristics of nonstationary neutron distributions, of average velocities $\overline{v(t)}$, of the average square of the velocity $\overline{v^2(t)}$, etc. Let us state the expression for the n-th velocity moment for the neutron distribution of Eq. (4.18), i.e., for the simplest version of our model:

$$\overline{v^n(t)} \equiv \frac{\int\limits_0^\infty \widetilde{N}(v,t)\,v^n\,dv}{\int\limits_0^\infty \widetilde{N}(v,t)\,dv} = \langle v^n \rangle - \left\langle \frac{v^n}{(1 + \Sigma_f t/A)^n} \right\rangle e^{-t/\tau_a} + \left(\frac{1}{v_0} + \frac{\Sigma_f t}{A} \right)^{-n} e^{-t/\tau_a}. \tag{4.19}$$

Eigenvalues and Eigenfunctions of the Neutron-Transfer Equation.   It is easy to verify that the poles of the expression under the integral of Eq. (4.3) are the (discrete) eigenvalues of Eq. (4.1) (as functions of x). The corresponding eigenfunctions (except for a normalizing factor) are the residues of the integral (4.3) calculated for those poles. Since by definition of the function $\mu(v)$ does not vanish at any finite v value (see page 257), only the poles of the function i(x) of Eq. (4.5) can be the poles x_j of the expression under the integral of Eq. (4.3), i.e., the x_j are the roots of the transcendental equation

$$\int\limits_0^\infty \frac{a(v)}{\mu(v)}\,dv \int\limits_v^\infty \left[x_j + \frac{1}{\tau_a(v)} \right] M(v')\,F(x_j, v', v)\,dv' = 0. \tag{4.20}$$

Since the expression under the integral is positive for all v values equation (4.20) has a single trivial root. For constant τ_a the root has the form

$$x_0 = -\frac{1}{\tau_a}, \tag{4.21}$$

and the Maxwell distribution is the corresponding eigenfunction.

The continuous part of the eigenvalue spectrum of Eq. (4.1) is situated in the region

$$x \leqslant -\frac{1}{\tau_a}. \tag{4.22}$$

Conclusion

It was shown above that it is possible to design a scattering-operator model which helps to describe the neutron-transfer equation in relative simple form, and which at the same time satisfies all basic conditions which must be imposed on the scattering operator: correct asymptotic behavior at large neutron velocities, conservation of a Maxwell distribution, exclusion of negative neutron distributions, and conservation of the total number of neutrons.

The model comprises three arbitrary functions of the neutron velocity and the scattering angles. These functions can be determined, for example, from the condition that the true scattering probability and the scattering probability of our model lead to a coincidence of the zeroth, first, and second moments of transferred velocity for a given scattering angle. The functions can also be determined from other integral characteristics defining neutron distributions. More particularly, the basic properties of the model can be obtained by considering its simplest version, in which the moderator is characterized by a single parameter in addition to the mass number and the scattering cross section of the neutrons for free atoms at rest.

Our model helps to determine analytical expressions for neutron spectra generated by arbitrary (nonstationary) sources in an infinite homogeneous medium with an arbitrary velocity dependence of the absorption cross section (one can obtain from these expressions the solution of the neutron transfer equation in the diffusion approximation by employing standard methods). Inspection of the results reveals that even the simplest version of the model makes it possible to describe in satisfactory fashion neutron distributions in cases in which the velocity dependence is continuous. However, for neutron distributions resulting from pulsed monoenergetic sources, our model leads to qualitatively incorrect results. In real moderators, the neutron-spectrum transformation from an initial narrow line to a broad asymptotic ($t \to \infty$) Maxwell distribution occurs basically by a gradual increase in line width. In our model, this process implies an increase in the fraction of the asymptotic distribution which was previously generated. However, the integral characteristics of the neutron spectrum (average velocity or average square of the velocity, etc., as functions of time) are described by our model in this case, too. The divergence of the integrals

$$\int w(\mathbf{v}' \to \mathbf{v})(v' - v)^n \, dv',$$

is another proof of the crudeness of our model. In other words, the conjugated moments of the transferred velocity are divergent, whereas in the case of a heavy moderator these integrals converge for all n. The principle of our model implies that not even the first three of these integrals ($n = 0, 1,$ and 2) converge and behave in a sensible fashion. All this means that our model can lead to erroneous results in cases in which the neutron distributions are strongly velocity-dependent, especially in the case of large velocities. From the viewpoint of physics, this results from the fact that the model leads to too large probabilities for great energy changes resulting from the scattering of neutrons with high velocities.

The overall conclusion is that our model is quite a satisfactory zeroth approximation to the solution of a very broad class of neutron-thermalization problems of practical importance. Rather reliable quantitative results can be obtained with our model in a number of cases (when the neutron distributions are characterized by a rather smooth velocity dependence). Naturally, additional studies are required in order to determine more accurately the problems which can be satisfactorily solved with our model.

References

1. M. V. Kazarnovskii, Atomnaya Énergiya, 22:100 (1967).
2. N. Corngold, P. Michael, and W. Wollman, Proc. BNL Conf. on Neutron Thermalization, Vol. IV, p. 1103 (1962).
3. N. Corngold and K. Durgan, Nucl. Sci. Eng., 25:450 (1966).
4. G. Placzek, Phys. Rev., 86:377 (1952).